An Introduction to Discrete Mathematics

Second Edition

Steven Roman

Emeritus Professor
California State University, Fullerton

White

Innovative Textbooks
www.innovativetextbooks.com

ISBN 1-878015-29-X

To order phone 949-854-5667

Innovative Textbooks
www.innovativetextbooks.com

To Donna

That is a good book which is opened with expectation and closed with profit. (Amos Bronson Alcott, *Table Talk*)

Seeing there is nothing (right well beloved Students in the Mathematickes) that is so troublesome to Mathematicall practise, nor that doth more molest and hinder Calculators, then the Multiplications, Divisions, square and cubical Extractions of great numbers, which besides the tedious expence of time are for the most part subject to many slippery errors, I began therefore to consider in my minde, by what certaine and ready Art I might remove those hindrances. (John Napier, 1614)

And Lucy, dear child, mind your arithmetic. . . . What would life be without arithmetic, but a scene of horrors? (Sidney Smith, *Letters: To miss —*)

Preface

There has been a considerable increase in interest in Discrete Mathematics as a result of the recent upsurge in the number of computer science majors. I hope this book will fill the needs not only of these students, but of mathematics students as well. The book is intended to be an elementary introduction to certain topics in Discrete Mathematics, designed primarily for freshmen and sophomores, with no college-level mathematics prerequisites.

In writing this book, I have tried to keep the following two main goals in mind:

1. Acquaint the students with a variety of mathematical concepts that will be needed in the study of computer science and in the further study of mathematics.
2. Introduce the students to the "mathematical" way of thinking, that is, the ideas of a definition, a theorem, and a proof.

How many times have we, as instructors of mathematics or computer science, had to stop in the middle of a class to explain what the term "converse" means, let alone to explain that the truth of a statement does not imply the truth of its converse? How many of our students know what it means to prove a theorem by mathematical induction or by contradiction? How many know what an equivalence relation is, or a partial order, or a permutation, or a combination, and so on? There is little doubt in my mind that students of computer science *and* students of mathematics can benefit immensely from a course that is designed to introduce them to these and other important concepts *in an elementary way*.

The second goal of this book is every bit as important as the first, and this is where I think the present book differs from many of the recent books intended for the same audience. Rather than racing through as much material as is humanly possible for an instructor to cover in one or two semesters, I feel it is more important to spend a little extra time helping the students gain some understanding of how we learn mathematics. I have tried to reach this goal by including more explanation and motivation than is typically included in many textbooks. Above all, I have tried to make the students feel "involved" in the discussions, rather than simply being outside spectators.

For those who have used the first edition of this book, there have been three major changes in the second edition, each requested by a number of users. First, Section 1.5 on mathematical induction has been enlarged to include the strong form of induction. Second, a new section (4.11) has been added on second order linear nonhomogeneous recurrence relations. Finally, a new section (5.8) on elementary probability theory has been included.

Let me briefly discuss the contents of the book. Chapter 1 covers the basic prerequisites for the course. As indicated in the table of contents, the section on countable and uncountable sets is optional.

Chapter 2 is devoted to elementary logic. There are two main goals here. One is to discuss the concept of logical equivalence, so that the relationships between a statement, its converse, inverse, and contrapositive can be made clear. The other goal is to discuss the relationship between logic and elementary circuit design. This includes a discussion of Boolean functions. Also, Karnaugh maps are discussed in the last section of the chapter.

In Chapter 3 we discuss relations on sets, including equivalence relations and partially ordered sets. Topological sorting is included as an application. The last section of the chapter introduces the student to the concept of a morphism (of order).

Chapters 4 and 5 contain a fairly complete introduction to elementary combinatorics. Four sections of Chapter 4 are devoted to the important topic of recurrence relations. A thorough discussion of the principle of inclusion-exclusion is contained in Chapter 5, along with an introduction to probability.

Finally, Chapter 6 is devoted to graph theory, where we discuss both directed and undirected graphs. Some of the more applied topics included are the depth first search, binary search trees, Huffman codes, the minimal spanning tree problem, and the shortest path problem. Several algorithms are discussed, but they are given in English rather than in pseudocode; thus, no knowledge of pseudocode is necessary. The final section is devoted to finite state machines.

The chapters of the book are independent of one another, with the following exceptions:

1. Sections 1.1, 1.2, 1.4, and 1.5 are assumed throughout the book.
2. Sections 2.1, 2.2, and 2.3 (which contain information on the converse, inverse, and so on of a statement) are recommended for the rest of the book.
3. Chapter 5 requires a knowledge of the first eight sections of Chapter 4.

The book can be used in courses of rather different emphasis. For example, if one wishes to teach a course on combinatorics and graph theory, there is ample material for this. One possibility is the following:

Sections 1.1, 1.2, 1.4, 1.5	4 lectures
Sections 2.1–2.4	4 lectures
Chapter 4	14 lectures
Sections 5.1–5.6	8 lectures
Sections 6.1–6.9	12 lectures
	42 lectures

On the other hand, a course could be designed with more emphasis on logic and algebra. One possibility is the following:

Sections 1.1, 1.2, 1.4, 1.5	4 lectures
Sections 2.1–2.6	8 lectures
Sections 3.1–3.5	7 lectures
Sections 4.1–4.10	12 lectures
Sections 6.1–6.9	12 lectures
	43 lectures

I would like to thank the following individuals, who reviewed all or part of the manuscript.

V. K. Balakrishnan, University of Maine
Ken Bogart, Dartmouth College
Curtis Cook, Oregon State University
Margaret B. Cozzens, Northeastern University
Richard Grassl, University of New Mexico
Kent Harris, Western Illinois University
Seth Hochwald, Saddleback Community College
Nick Krier, Colorado State University
Timothy Lance, State University of New York at Albany
Anthony Ralston, State University of New York at Buffalo
E. G. Whitehead, Jr., University of Pittsburgh

Users of the first edition who provided valuable information during preparation of this revision are:

James P. Conlan, Menlo College
Lawrence Davis, Southeastern Louisiana University
Jean Droch, College of Mount St. Joseph
Ira Gessel, Brandeis University
Marylou Gibson, Virginia Commonwealth University
Richard Griego, University of New Mexico
Jo Anne Growney, Bloomsburg University
Robert Hecht, Philadelphia High School of Engineering and Science
David John, Missouri Western State College
Robert Johnson, Indiana State University
John Leonard, University of Arizona
Truett Mathis, William Jewell College
William McIntosh, Central Methodist College
John Michaels, State University of New York at Brockport
Gilbert Orr, University of Southern Colorado
Paul Pederson, University of Denver
Glen Powers, Western Kentucky University
Richard Pulskamp, Xavier University
Sally Sestini, California State University at Fullerton
Patrick Smiley, Glassboro State College
Charles Suffel, Stevens Institute of Technology
Kenneth Wiggins, Walla Walla College

Steven Roman

Contents

For a "mixt company" implies, that, save
Yourself and friends, and half a hundred
more,
Whom you may bow to without looking
grave,
The rest are but a vulgar set.
—Byron, *Beppo*

What is now proved was once only
imagin'd.
—William Blake, *Proverbs of Hell*

We must never assume that which is
incapable of proof.
—G. H. Lewes, *Physiology of Common Life*

Be sure of it; give me the ocular proof.
—Shakespeare, *Othello*

Chapter ONE

Sets, Functions, and Proof Techniques

1.1 The Language of Sets

Virtually all of mathematics is concerned, in one way or another, with the concept of a set, and so it is a good idea for us to begin our study of discrete mathematics with a brief discussion of sets. You may already be familiar with much of this material, but a short discussion will serve to make sure that we are all speaking the same language.

The concept of a set is so fundamental that we will not attempt to give it a precise definition. There have been many attempts to find such a definition, but problems always seem to arise.

For example, if you look up the word *set* in the dictionary, you may find something like this: "a collection of mathematical elements. . . ." But, what is a collection? If you look up this word in the dictionary, you might find: "an accumulation of objects. . . ." But, what is an accumulation?

By now you can see the problem. It is not possible to give a simple definition of the term *set* without using other terms that, in turn, need to be defined. In this way, we end up in a vicious circle of definitions.

For our purposes, we will consider the concept of a set as being so basic that it does not have a definition. We will not run into any trouble in this book if we simply rely on our intuitive understanding of what a set is.

Let us consider some examples of sets.

Example 1.1.1

The following are sets.

$A = \{1, 5, a, w, >\}$, $B = \{x \mid x = 2 \text{ or } x = 5\}$

$C = \{\varnothing, \{1, 2\}, 5\}$, $D = \{B, A, S, I, C\}$, $E = \{BASIC\}$

We should make a few comments about the notation used to describe sets. The bracket notation $\{. . .\}$ is read "the set containing." Thus, for example, D is *the set containing* the letters B, A, S, I, and C, and the set E is *the set containing* the word BASIC. (Thus, D contains five elements, but E contains only one element.) When brackets are combined with the vertical bar $\{. . . \mid . . .\}$ we read this as "the set of all . . . for which. . . ." As an example, we would read the set B as "the set of all x for which x is equal to 2 or x is equal to 5."

The set C has the interesting feature that some of its elements are themselves sets. In particular, the elements of C are the empty set $\varnothing$, the set $\{1, 2\}$, and the number 5. As you can see from this example, it is perfectly acceptable for the elements of a set to be other sets.

Example 1.1.2

The set that has no elements is called the empty set. It is denoted by the symbol $\varnothing$. (This symbol is a small circle with a line through it, and *not* the Greek letter phi.) One should be careful not to confuse the empty set $\varnothing$ with the set $\{0\}$, which contains the number 0.

Example 1.1.3

There are several sets of numbers that we should discuss, since we will be using them throughout the book.

The number 0 and the counting numbers 1, 2, 3, 4, . . ., are called **natural numbers**. The set of all natural numbers is denoted by **N**,

$$\mathbf{N} = \{0, 1, 2, 3, . . .\}$$

The series of dots is meant to express the fact that there are other elements in this set besides 0, 1, 2, and 3 and that, from the pattern in the first few elements, it is easy to tell what they are.

The set consisting of the natural numbers and the negative counting numbers $-1, -2, -3, -4, . . .$ is the set of **integers**. We denote this set by **Z**,

$$\mathbf{Z} = \{. . ., -3, -2, -1, 0, 1, 2, 3, . . .\}$$

Any number that can be written in the form p/q, where p and q are integers and q is not equal to zero, is called a **rational number**. For example, the numbers

$$\tfrac{1}{2},\ -\tfrac{5}{3},\ \tfrac{6}{2},\ \tfrac{3}{1}$$

are rational numbers. Notice that any integer is a rational number. For example, the integer 5 can be written in the form $5/1$, which is the quotient of two integers. We denote the set of rational numbers by **Q**,

$$\mathbf{Q} = \{p/q \mid p \text{ and } q \text{ are integers and } q \neq 0\}$$

Any rational number can be expressed as a repeating decimal, and any repeating decimal is a rational number. For example, we have

$$\tfrac{4}{3} = 1.3333 \cdots$$

$$\tfrac{1}{7} = .142857142857 \cdots$$

$$2 = 2.0000 \cdots$$

It can be shown that the numbers π and $\sqrt{2}$ cannot be expressed in the form p/q, where p and q are integers. Hence, these numbers are not rational. They can, however, be expressed as *nonrepeating* decimals.

Any number that can be expressed as a nonrepeating decimal is called an **irrational number**. The set of irrational numbers is denoted by **I**,

$$\mathbf{I} = \{x \mid x \text{ is a nonrepeating decimal}\}$$

The rational and irrational numbers together form the set **R** of **real numbers**,

$$\mathbf{R} = \{x \mid x \text{ is a rational number or } x \text{ is an irrational number}\}$$

Example 1.1.4

We will use the following example throughout the book. Let $\Sigma = \{a, b, c, d\}$ be the set containing the first four letters of the alphabet. (Σ is the upper case Greek letter sigma.) Then we can form a new set, denoted by Σ^*, consisting of all "words" that can be made from the letters in Σ. In this case, a "word" can be any finite arrangement of letters taken from Σ—it does not have to be a meaningful word in any language. For example, the expressions

$$a,\ aa,\ aaa,\ cababc,\ bbbaaabbbcdcd$$

are words. In this context, the set Σ is usually called the **alphabet**, and Σ^* is referred to as the set of all words *over* the alphabet Σ. It is also convenient to include in Σ^* an **empty word**, denoted by θ (the Greek letter *theta*), and defined to be the word with no letters.

Of course, we can use other sets as alphabets. When we use the alphabet $\Sigma = \{0, 1\}$, then the set Σ^* of all words over Σ is the set of all possible **binary words**. For example,

$$0,\ 01,\ 1101,\ 0011000111$$

are binary words. Binary words are extremely important since, as you probably know, computers store data in the form of binary words.

When we use the alphabet $\Sigma = \{0, 1, 2\}$, then Σ^* is the set of all **ternary words**. For example,

2, 210, 00000, 21012021

are ternary words.

The **length** of a word is simply the number of "letters" in the word, counting repetitions of course. (We will use the term *letter* for the elements of any alphabet Σ, even if Σ does not contain actual letters of the Roman alphabet.) For example, the length of the word *aabab* is 5 and the length of the binary word 11001010 is 8. Binary words of length 8 are especially important in computer science, since they are frequently the smallest unit that can be easily manipulated within the computer. A binary word of length 8 is called a *byte*. □

Example 1.1.5

Consider the set

$$S = \{n \mid n \text{ is an integer, } n > 2, \text{ and there exist positive integers } x, y, \text{ and } z \text{ for which } x^n + y^n = z^n\}$$

The set S is a truly fascinating set. As you probably know, there are many positive integers x, y, and z that satisfy the equation

$$x^2 + y^2 = z^2$$

For example, $3^2 + 4^2 = 5^2$ and $12^2 + 16^2 = 20^2$. (Just think about the Pythagorean Theorem.)

This led some people to wonder whether there were any positive integers x, y, and z that satisfied the equation

$$x^3 + y^3 = z^3$$

or

$$x^4 + y^4 = z^4$$

or, for that matter,

$$x^n + y^n = z^n \tag{1.1.1}$$

for any integer n larger than 2.

A brilliant mathematician named Pierre de Fermat (1601?–1665) was the first to give this question serious thought. In fact, he claimed to be able to show that there were no positive integers x, y, and z that would satisfy Equation 1.1.1, for *any* integer n greater than 2. In other words, he claimed to be able to show that the set S is the empty set.

Fermat wrote in the margin of one of his books, "To divide a cube into two cubes, a fourth power, or in general any power whatever into two powers of the same denomination above the second is impossible, and I have assuredly found an admirable proof of this, but the margin is too narrow to contain it." (By "divide," Fermat means to write as a sum of.)

It is unfortunate that Fermat did not take the time to write down his proof in another book, because to this very day, no one has ever been able to prove that Fermat was correct! Despite the fact that many other mathematicians and scientists have worked on this problem, no one knows whether there is an integer n, greater than 2, and positive integers x, y, and z that satisfy Equation 1.1.1.

Actually, with the help of modern-day computers, we do know that no positive integers x, y, and z exist that satisfy Equation 1.1.1, for any exponent n that is less than 100,000, but of course this does not completely solve the problem. Thus, while we do know that no integer n between 3 and 100,000 is in the set S, we still have no idea whether or not the set S is the empty set!

By the way, Fermat's statement that S is indeed the empty set has since become known as **Fermat's Last Theorem** and is perhaps the most famous unsolved problem in all of mathematics. (The term *theorem* is usually reserved for statements that have already been proved, but this is an exception.) Also, as Howard Eves puts it in his book *An Introduction to the History of Mathematics* (Saunders College Publishing, Philadelphia), "Fermat's last 'Theorem' has the peculiar distinction of being the mathematical problem for which the greatest number of incorrect proofs have been published." □

Let us now give some elementary definitions associated with sets. To begin with, we will denote the fact that x is an element of a set A by $x \in A$. If x is not in A, we denote this by $x \notin A$.

A set A is said to be a **subset** of a set B if every element of A is also an element of B. In this case, we say that A is *contained in* B, and write $A \subset B$. If $A \subset B$, we also say that B is a **superset** of A, and write $B \supset A$.

Example 1.1.6

a) $\{1, 4, a\} \subset \{0, 1, 3, 4, a, b\}$

b) The sets **N**, **Z**, **Q**, and **R** in Example 1.1.3 form a hierarchy of subsets, in the sense that

$$\mathbf{N} \subset \mathbf{Z} \subset \mathbf{Q} \subset \mathbf{R}$$

Example 1.1.7

a) Let Σ be an alphabet. For each $n = 0, 1, 2, 3, \ldots$, let Σ_n be the set of all n-letter words over Σ. Then Σ_n is a subset of Σ^*. (Σ_0 consists of just the empty word θ.)

b) Let Σ be an alphabet. For each $n = 0, 1, 2, 3, \ldots$ let Γ_n be the set of all words over Σ that have *at most* n letters. Then Γ_n is also a subset of Σ^*.

c) Any subset of Σ^* is called a **language** over Σ. For example, the set of all English words over the set $\Sigma = \{a, b, \ldots, z\}$ is a language over Σ (The English Language). Also, the set of all bytes is a language over $\Sigma = \{0, 1\}$. □

It is important to notice that, according to the definition of subset, the empty set $\varnothing$ is a subset of any set, and a set is always a subset of itself.

A subset B of a set A is said to be a **proper subset** of A if it is *not* equal to A itself. Thus, all subsets of A are proper subsets *except* the set A itself, which is referred to as the **improper subset** of A.

Of course, two sets A and B are equal if they contain exactly the same elements. This can be expressed in terms of subsets by saying that two sets A and B are equal if and only if each one is a subset of the other, in symbols, $A = B$ if and only if $A \subset B$ and $B \subset A$.

Given a set A, we can always form a new set whose elements are all the possible *subsets* of A. This set is called the **power set** of A and is denoted by $\mathcal{P}(A)$. Let us have some examples of power sets.

Example 1.1.8

a) The empty set $\emptyset$ has exactly one subset, namely the empty set itself. Hence, the power set of the empty set has exactly one element, namely, the empty set,

$$\mathcal{P}(\emptyset) = \{\emptyset\}$$

(Notice that there is a difference between the set $\{\emptyset\}$, whose only element is the empty set, and the empty set $\emptyset$, which has no elements.)

b) The set $A = \{1\}$ has exactly 2 subsets, namely, the empty set $\emptyset$ and the set A itself. Hence,

$$\mathcal{P}(A) = \{\, \emptyset, \{1\} \,\}$$

c) The set $B = \{1, 2\}$ has 4 subsets

$$\emptyset, \{1\}, \{2\}, \{1, 2\}$$

and so

$$\mathcal{P}(B) = \{\, \emptyset, \{1\}, \{2\}, \{1, 2\} \,\}$$

d) The set $C = \{1, 2, 3\}$ has 8 subsets

$$\emptyset, \{1\}, \{2\}, \{3\}$$

$$\{1, 2\}, \{1, 3\}, \{2, 3\}, \{1, 2, 3\}$$

and so

$$\mathcal{P}(C) = \{\, \emptyset, \{1\}, \{2\}, \{3\}, \{1, 2\}, \{1, 3\}, \{2, 3\}, \{1, 2, 3\} \,\}$$ □

The number of elements in a finite set A will be denoted by $|A|$. Thus, for example, if $A = \{1, 3, a, b\}$, then $|A| = 4$. Of course, $|\emptyset| = 0$. The number $|A|$ is called the **size** of A, or the **cardinality** of $|A|$.

The **union** of two sets A and B is defined to be the set of all elements that are in either of the sets A or B. (This includes elements that are in both sets.) The union of A and B is denoted by $A \cup B$. In symbols, we have

$$A \cup B = \{x \mid x \in A \text{ or } x \in B \text{ or both}\}$$

The **intersection** of the sets A and B is defined to be the set of all elements that are in *both* of the sets A and B. The intersection is denoted by $A \cap B$. In symbols,

$$A \cap B = \{x \mid x \in A \text{ and } x \in B\}$$

The **difference** A − B of two sets A and B is defined to be the set of all elements that are in A but that are *not* in B. The expression A − B is usually read "A minus B" and is called the **relative complement** of B in A. (Notice the order of the sets A and B.) In symbols,

$$A - B = \{x \mid x \in A \text{ and } x \notin B\}$$

Let us have some examples of unions, intersections, and relative complements.

Example 1.1.9

a) $\{1, 2, 3\} \cup \{2, 4, 6\} = \{1, 2, 3, 4, 6\}$ and $\{1, 2, 3\} \cap \{2, 4, 6\} = \{2\}$
b) $\{1, 2, 3\} - \{2, 4, 6\} = \{1, 3\}$ and $\{2, 4, 6\} - \{1, 2, 3\} = \{4, 6\}$
c) $\{a, c\} - \{a, c\} = \varnothing$
d) $\{1, 2, 3\} - \{4, 5, 6\} = \{1, 2, 3\}$
e) $\Sigma_0 \cup \Sigma_1 \cup \Sigma_2 \cup \cdots \cup \Sigma_n = \Gamma_n$ for all $n \geq 0$ (See Example 1.1.7)
f) $\Gamma_n - \Sigma_n = \Gamma_{n-1}$ for all $n \geq 1$ (See Example 1.1.7) □

If two sets A and B have no elements in common, then we say that they are **disjoint**. In symbols, A and B are disjoint if and only if $A \cap B = \varnothing$. For example, the sets Σ_n and Σ_m defined in Example 1.1.7 are disjoint if $n \neq m$. (Are the sets Γ_n and Γ_m defined in Example 1.1.7 disjoint?)

If $A_1, A_2, \ldots, A_n$ are sets, and if no *two* of these sets have any elements in common, then we say that these sets are **mutually disjoint**. (The term **pairwise disjoint** is also used.) In symbols, the sets $A_1, A_2, \ldots, A_n$ are mutually disjoint if

$$A_i \cap A_j = \varnothing$$

for *all* choices of $i, j = 1, 2, \ldots, n$ with $i \neq j$.

Example 1.1.10

As an example of mutual disjointness, consider the set Σ^* of all words over an alphabet Σ and the sets Σ_n defined in Example 1.1.7. Then since no word can be in more than one of these sets, we see that the sets $\Sigma_0, \Sigma_1, \Sigma_2, \ldots$ are mutually disjoint. □

Given a set A, we would like to be able to form the set of all elements that are not in A. Before we can do this, however, we must first specify a set U of elements that contains the "candidates," so to speak, for not being in A. Of course, A must be a subset of U. Then we can form the set of all elements of U that are not in A; that is, we can form the set U − A. This set U − A is called the **complement** of A and is also denoted by A^c.

The set U is generally called a **universal set**. This is a somewhat unfortunate choice of terms, because it implies that there is only one such set. But the fact is that any set can be a universal set. All that matters is that, before we take the complement of any set, we agree on what the universal set will be.

As an example, let $A = \{1, 2, 3\}$. If we take the universal set U to be the set of all *positive* natural numbers, then we would have

$$A^c = U - A = \{4, 5, 6, \ldots\}$$

On the other hand, if we take the universal set U to be the set of all integers, then we would have

$$A^c = U - A = \{\ldots, -3, -2, -1, 0, 4, 5, 6, \ldots\}$$

The moral of this is simply that *the complement of a set depends on what the universal set is.*

The relative complement can be expressed in terms of the complement by the formula

$$A - B = A \cap B^c$$

where A and B are subsets of U.

The concepts of union, intersection, relative complement, and complement can be described by using certain pictures, called **Venn diagrams**. [These are named after the English mathematician John Venn (1834–1923).]

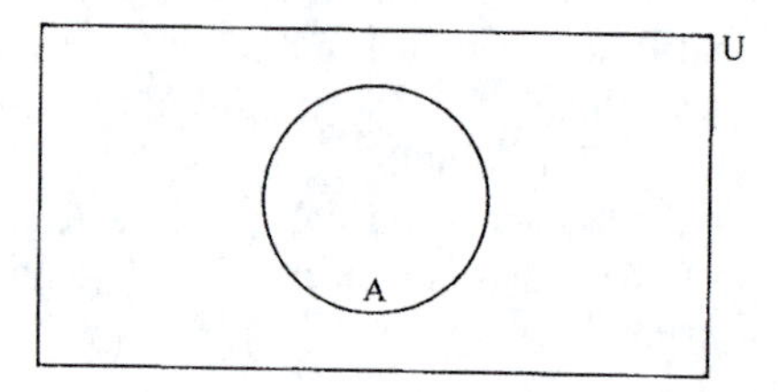

Figure 1.1.1
The Venn diagram for one set A.

The Venn diagram for one set A is pictured in Figure 1.1.1. Notice that the set is contained in a universal set U. We can describe the complement of A simply by shading in the appropriate region in the Venn diagram. This is done in Figure 1.1.2.

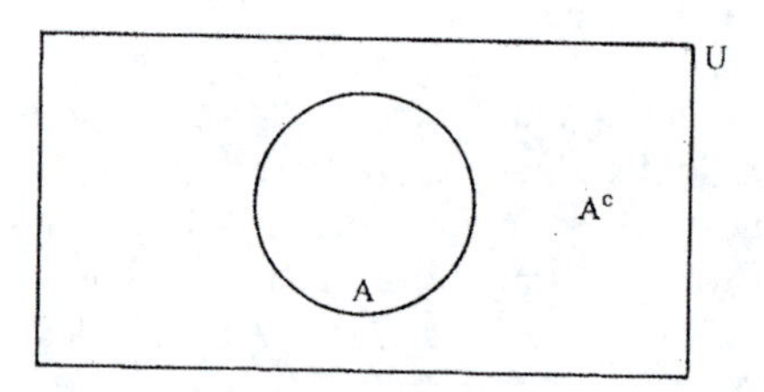

Figure 1.1.2
The shaded region is A^c.

The Venn diagram for two sets A and B is pictured in Figure 1.1.3. Again, both of the sets are contained in a universal set U, and we can shade in the regions that represent the union $A \cup B$, the intersection $A \cap B$, the relative complement $A - B$, and the complement A. This is done in Figures 1.1.4 through 1.1.7.

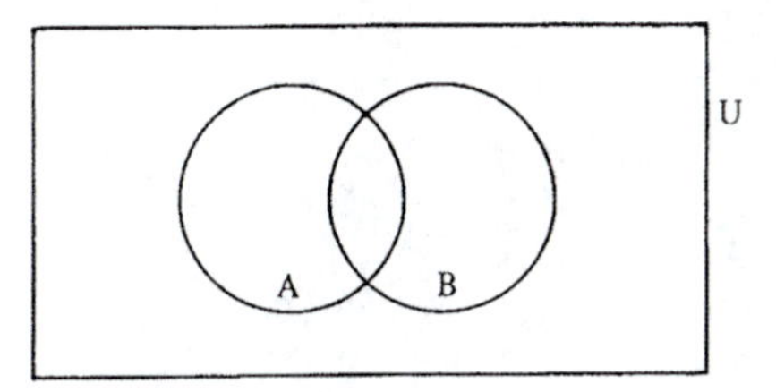

Figure 1.1.3
The Venn diagram for two sets A and B.

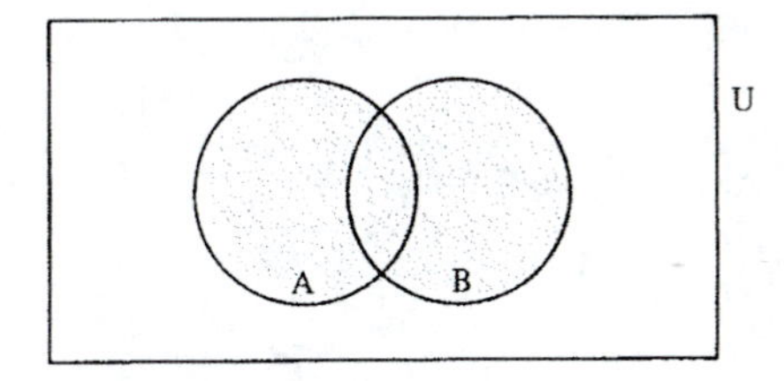

Figure 1.1.4
The shaded region is $A \cup B$.

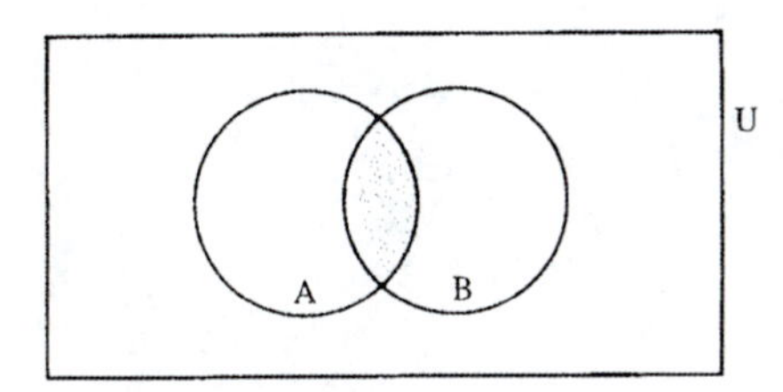

Figure 1.1.5
The shaded region is $A \cap B$.

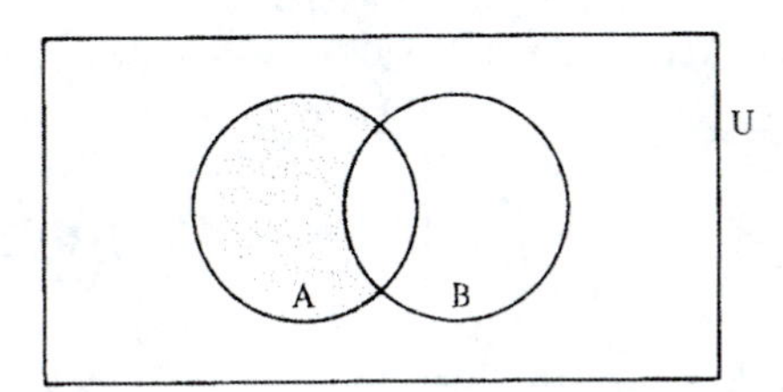

Figure 1.1.6
The shaded region is $A - B$.

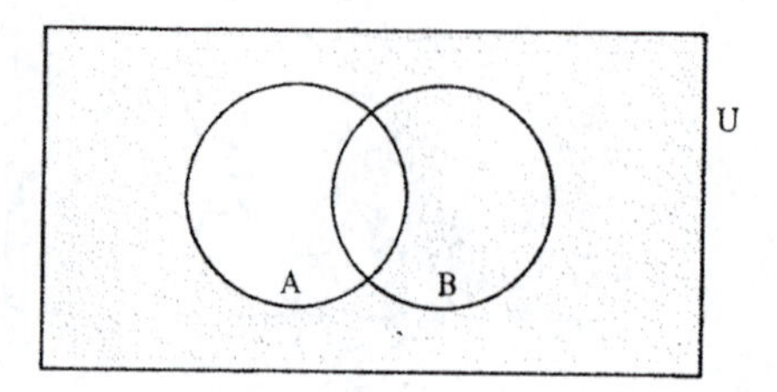

Figure 1.1.7
The shaded region is A^c.

The Venn diagram for three sets A, B, and C is pictured in Figure 1.1.8. In Figure 1.1.9 we shade in the region that represents $(A \cap B) \cup C$.

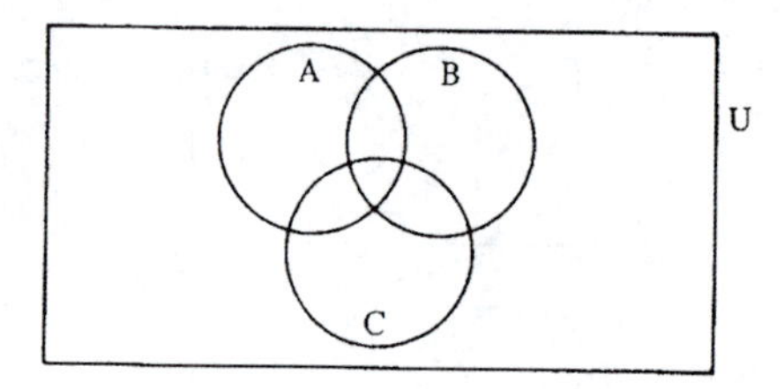

Figure 1.1.8
The Venn diagram for three sets A, B, and C.

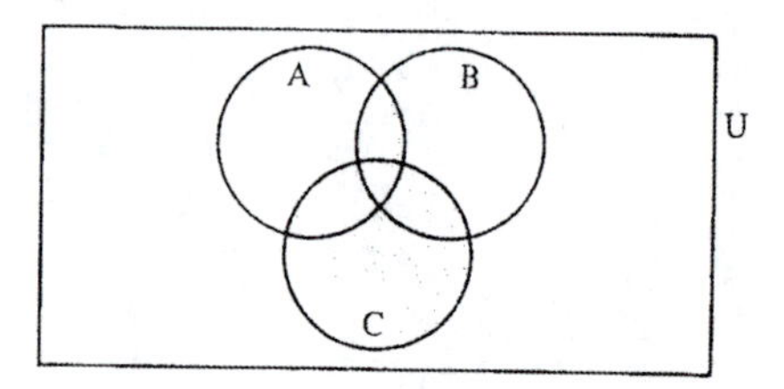

Figure 1.1.9
The shaded region is $(A \cap B) \cup C$.

Given two sets A and B, we can form the set consisting of all *ordered pairs* of the form (a, b), where a is any element of A and b is any element of B. This set is called the **cartesian product** of A and B and is denoted by $A \times B$,

$$A \times B = \{(a, b) \mid a \in A \text{ and } b \in B\}$$

Of course, there is no reason why A and B cannot be the same set. In fact, the set $A \times A$ is sometimes denoted by A^2.

Example 1.1.11

Let $A = \{1, 2\}$ and $B = \{1, 4, 7\}$. Then

$$A \times B = \{(1, 1), (1, 4), (1, 7), (2, 1), (2, 4), (2, 7)\}$$

$$B \times A = \{(1, 1), (1, 2), (4, 1), (4, 2), (7, 1), (7, 2)\}$$

and

$$A^2 = \{ (1, 1), (1, 2), (2, 1), (2, 2) \}$$

It is important to keep in mind that when A and B are different sets, the cartesian product A × B is *not* equal to the cartesian product B × A.

The cartesian product can be extended to more than two sets. For example, if A, B, and C are sets, then the cartesian product A × B × C is the set of all *ordered triples* (a, b, c), where a is any element of A, b is any element of B, and c is any element of C. In symbols,

$$A \times B \times C = \{(a, b, c) \mid a \in A, b \in B, \text{ and } c \in C\}$$

Let A be a nonempty set (that is, $A \neq \emptyset$), and suppose that $B_1, B_2, \ldots, B_n$ are subsets of A, with the property that

1) None of the sets $B_1, B_2, \ldots, B_n$ is empty;

2) The sets $B_1, B_2, \ldots, B_n$ are mutually disjoint. That is, no two of these sets have any elements in common;

3) The union of all of the sets $B_1, B_2, \ldots, B_n$ is equal to A. In symbols, $B_1 \cup B_2 \cup \ldots \cup B_n = A$.

Then we say that the sets $B_1, B_2, \ldots, B_n$ form a **partition** of the set A. The subsets $B_1, B_2, \ldots, B_n$ are called the **blocks** of the partition.

Loosely speaking, a partition of A is a "breaking up" of the set A into nonempty, nonoverlapping subsets.

Let us consider some examples of partitions.

Example 1.1.12

Let A = {1, 2, 3, 4, 5, 6, 7, 8}. Then the sets

$$B_1 = \{1, 2, 3\} \quad , \quad B_2 = \{4, 5\} \quad , \quad B_3 = \{6, 7, 8\}$$

form a partition of A. Also, the sets

$$C_1 = \{1, 5, 8\} \quad , \quad C_2 = \{4\} \quad , \quad C_3 = \{6, 7\} \quad , \quad C_4 = \{2, 3\}$$

form a partition of A. The partition

$$D_1 = \{4\} \quad , \quad D_2 = \{6, 7\} \quad , \quad D_3 = \{2, 3\} \quad , \quad D_4 = \{1, 5, 8\}$$

is the same as the previous partition, since it has exactly the same blocks. In other words, the order in which the blocks of the partition are written makes no difference.

Example 1.1.13

Let Σ, Σ_n, and Γ_n be the sets defined in Example 1.1.7. Then the sets $\Sigma_0, \Sigma_1, \Sigma_2, \ldots, \Sigma_n$ form a partition of the set Γ_n. (Do the sets $\Gamma_0, \Gamma_1, \Gamma_2, \ldots, \Gamma_n$ form a partition of Γ_n?)

Example 1.1.14

In FORTRAN, all data must be one of the following six basic types: integer, single precision real, double precision real, complex, logical, or character. All variables in a given program must be assigned one of these six types (either explicitly or by virtue of the first letter in the variable name). In other words, the set of all variables in a given program is *partitioned* into *at most* six blocks, according to the six data types. Of course, there may be fewer than six blocks if a program does not contain variables of all of the data types. Thus, we can see that the concept of a data type in FORTRAN, as well as in BASIC, Pascal, and other high-level languages, is related to the concept of a partition. □

Now let us discuss some of the properties of the union, intersection, and complement. Actually, there are many similarities between the union and intersection of sets and the addition and multiplication of real numbers, especially when it comes to the properties that these operations satisfy. (However, there is no counterpart of the complement within the arithmetic of real numbers.)

This similarity can be seen by relating union to addition, intersection to multiplication, the empty set $\emptyset$ to the number 0, and the universal set U to the number 1. For example, the property that

$$A \cup B = B \cup A$$

for all sets A and B is analogous to the property that

$$a + b = b + a$$

for all real numbers a and b.

Let us now make a list of the most important properties of the operations of union, intersection, and complement. Some of these properties have names, which we will include in the list. As you read this list, you might want to see if you can write down the analogous property of addition and multiplication of real numbers.

Properties of Union, Intersection, and Complement

Let U be a universal set. Then the following properties hold for *all* subsets A, B, and C of U.

Property 1 (Properties of $\emptyset$ and U)

$$A \cup \emptyset = A \quad , \quad A \cap U = A$$

$$A \cup U = U \quad , \quad A \cap \emptyset = \emptyset$$

Property 2 (The idempotent properties)

$$A \cup A = A \quad , \quad A \cap A = A$$

Property 3 (The commutative properties)

$$A \cup B = B \cup A \quad , \quad A \cap B = B \cap A$$

Property 4 (The associative properties)

$$A \cup (B \cup C) = (A \cup B) \cup C$$

$$A \cap (B \cap C) = (A \cap B) \cap C$$

Property 5 (The distributive properties)

$$A \cup (B \cap C) = (A \cup B) \cap (A \cup C)$$

$$A \cap (B \cup C) = (A \cap B) \cup (A \cap C)$$

Property 6 (Properties of the complement)

$$\varnothing^c = U \quad , \quad U^c = \varnothing$$

$$A \cup A^c = U \quad , \quad A \cap A^c = \varnothing$$

$$(A^c)^c = A$$

Property 7 (De Morgan's laws)

$$(A \cup B)^c = A^c \cap B^c \quad , \quad (A \cap B)^c = A^c \cup B^c$$

Properties 1 through 7 can be very helpful when it comes to simplifying expressions involving unions, intersections, and complements.

Example 1.1.15

a) $(A^c \cap B^c)^c = (A^c)^c \cup (B^c)^c = A \cup B$

↑ *De Morgan's law* ↑ *Property 6*

b) $[(A \cup B^c) \cup (A \cup B)]^c$

De Morgan's law → $= (A \cup B^c)^c \cap (A \cup B)^c$
De Morgan's law → $= (A^c \cap B) \cap (A^c \cap B^c)$
Associativity → $= A^c \cap B \cap A^c \cap B^c$
Commutativity → $= A^c \cap A^c \cap B \cap B^c$
Associativity → $= (A^c \cap A^c) \cap (B \cap B^c)$
Properties 2 and 6 → $= A^c \cap \varnothing$
Property 1 → $= \varnothing$

Now let us consider the question of how to verify the properties of union, intersection, and complement. In general, the easiest way to show that two sets are equal is to show that each one of the sets is a subset of the other. Let us demonstrate this method by proving one of the distributive properties, which we put into a theorem.

Theorem 1.1.1
For any three sets A, B, and C, we have

$$A \cup (B \cap C) = (A \cup B) \cap (A \cup C) \tag{1.1.2}$$

PROOF As we said, in general the easiest way to show that two sets are equal is to show that each one is a subset of the other. In this case, then, we want to show that

$$A \cup (B \cap C) \subset (A \cup B) \cap (A \cup C) \tag{1.1.3}$$

and

$$(A \cup B) \cap (A \cup C) \subset A \cup (B \cap C) \tag{1.1.4}$$

In order to prove 1.1.3, we must show that any element of $A \cup (B \cap C)$ is also in $(A \cup B) \cap (A \cup C)$. So, let x be any element of $A \cup (B \cap C)$, that is,

$$x \in A \cup (B \cap C)$$

Then, according to the definitions of union and intersection,

$$x \in A \quad \text{or} \quad x \in B \cap C$$

and so

$$x \in A \quad \text{or} \quad (x \in B \text{ and } x \in C)$$

In either case,

$$x \in (A \cup B) \quad \text{and} \quad x \in (A \cup C)$$

and so

$$x \in (A \cup B) \cap (A \cup C)$$

which is what we wanted to show. Hence, 1.1.3 holds.

In order to prove 1.1.4, we must show that any element of $(A \cup B) \cap (A \cup C)$ is also an element of $A \cup (B \cap C)$. So, let x be any element of $(A \cup B) \cap (A \cup C)$,

$$x \in (A \cup B) \cap (A \cup C)$$

Then, according to the definitions of union and intersection,

$$x \in A \cup B \quad \text{and} \quad x \in A \cup C \tag{1.1.5}$$

Now, if $x \in A$ then

$$x \in A \cup (B \cap C)$$

which is what we wanted to show. On the other hand, if $x \notin A$, then, according to the two equations in 1.1.5, we must have

$$x \in B \quad \text{and} \quad x \in C$$

Hence,

$$x \in B \cap C$$

and so

$$x \in A \cup (B \cap C)$$

which is again what we wanted to show. This proves 1.1.4 and completes the proof of the theorem. ∎

As you can see, this proof is not really very difficult, for it simply amounts to applying the definitions of union and intersection several times.

Exercises

1. Let A = {1, 3, 5, 7}, B = {3, 4, 5, 6, 7}, and C = {0, 1, 4}. Find
a) $A \cup B$ b) $A \cap B$ c) $A - C$
d) $B - (A \cup C)$ e) $A - (B - C)$ f) $(A \cap B) - (A \cup C)$

2. Let A = $\{B, A, S, I, C\}$, B = $\{P, A, S, C, A, L\}$, and C = $\{F, O, R, T, R, A, N\}$. Find
a) $A \cup B$ b) $A \times C$ c) $A - (B \cap C)$
d) Show that $A \cup (B \cap C) = (A \cup B) \cap (A \cup C)$.

3. Let S = $\{a, b, c\}$, T = $\{1, a\}$, and V = $\{1, 2, 3, c\}$. Find $(S - T) - V$ and $S - (T - V)$. Are they equal?

4. Let A = {1, 2, 3}.
a) If U = {0, 1, 2, 3, 4, 5}, find A^c.
b) If U = {1, 2, 3}, find A^c.
c) If U = **R**, find A^c.
d) If U = {2, 3, 4}, does A^c exist? If so, find it and if not, explain why not.

5. If A = {1, 2, 3} and B = $\{x, y, z\}$, find $A \times B$ and $B \times A$. Is $A \times B$ the same as $B \times A$?

6. Find A^2, where A = {1, 2, 3}.

7. Let A = {1, 2}, B = {1, 4}, and C = {2, 4}. Find $A \times B \times C$.

For Exercises 8 through 11, let Σ_n and Γ_n be as in Example 1.1.7.

8. Let Σ = {1, 2, 3}. Find
a) Σ_1 b) Σ_2 c) Σ_3
d) Γ_1 e) Γ_2 f) Γ_3

9. Let $\Sigma = \{a, b, 1, 2\}$. Find
a) Σ_3 b) Γ_3

10. Simplify the expression $(\Gamma_n - \Sigma_n) - \Sigma_{n-1}$.

11. Simplify the expression $\Gamma_n - \Gamma_{n-1}$.

12. If $A \subset B$, what is $A \cup B$? What is $A \cap B$? What is $A - B$? What is $B - A$?

13. List the elements in each of the following sets.
a) $\{x^2 + 1 \mid x \text{ is an integer and } -3 < x \leq 2\}$
b) $\{x \mid x \text{ is an integer and } x = 2y, y^2 \leq 17\}$
c) $\{x \mid x \text{ is a rational number and } 2x = \sqrt{2}\}$

14. State whether or not the sets A and B are equal. Give a reason for your answer.
a) $A = \{3x + 1 \mid x \text{ is an integer}\}$
$B = \{4y - 12 \mid y \text{ is an integer}\}$
b) $A = \{2n + 1 \mid n \text{ is an even integer}\}$
$B = \{4n + 1 \mid n \text{ is an even integer}\}$

15. List all of the subsets of the set S = $\{1, a, z\}$.

16. Find the power set of {1, 2, 3, 4}.

17. Find $\mathcal{P}(\{\emptyset, \{1\}\})$.

18. Find $\mathcal{P}(\{\{\emptyset\}, \{\{\emptyset\}\}\})$.

19. Let A = {1, 2} and B = $\{a, b\}$.

a) Find $\mathcal{P}(A \times B)$. b) Find $\mathcal{P}(A) \times \mathcal{P}(B)$.
Are these two sets the same?

20. Draw a Venn diagram for three sets A, B, and C.
a) Shade in the region that represents $(A \cup B) \cap C$.
b) Shade in the region that represents $A \cup (B \cap C)$.
Are these two sets the same? Give a reason for your answer.

21. Does the expression $A \cup B \cap C$ make sense? Justify your answer.

22. Do the chapters form a partition of the set of distinct words in a given book? Explain.

23. Does the set of even integers and the set of odd integers form a partition of the set of all integers? Explain.

24. Does the set of positive real numbers and the set of negative real numbers form a partition of the set of all real numbers? Explain.

25. Let $\Sigma = \{a, b, \ldots, z\}$. Let Σ_a be the set of all words of length n that contain the letter a, and similarly for $\Sigma_b, \Sigma_c, \ldots, \Sigma_z$. Do the sets $\Sigma_a, \Sigma_b, \ldots, \Sigma_z$ form a partition of the set Σ_n? Explain.

26. Using Properties 1 through 7, show that

$$(A \cup B) \cap C = (A \cap C) \cup (B \cap C)$$

Specify which properties you actually used.

27. Using Properties 1 through 7, simplify each of the following expressions.
a) $(A \cup B^c)^c$ b) $((A \cup B^c)^c \cup A^c)^c$
c) $((A^c \cap (A^c \cup C)^c)^c \cap D^c)^c$

28. Find analogous equations, within the arithmetic of real numbers, if they exist, of the equations in Property 1.

29. Find analogous equations, within the arithmetic of real numbers, if they exist, of the equations in Property 2.

30. Find analogous equations, within the arithmetic of real numbers, if they exist, of the equations in Property 3.

31. Find analogous equations, within the arithmetic of real numbers, if they exist, of the equations in Property 4.

32. Find analogous equations, within the arithmetic of real numbers, if they exist, of the equations in Property 5.

33. Is it true that $A - (B - C) = (A - B) - C$? If so, prove it, and if not, find an example of three sets for which it does not hold. How would you describe this in words?

34. As we have already said, the sets A, B, and C are mutually disjoint if no two of them have any elements in common; that is, if $A \cap B = \varnothing$, $A \cap C = \varnothing$, and $B \cap C = \varnothing$.
a) Show that if A, B, and C are mutually disjoint, then $A \cap B \cap C = \varnothing$.
b) On the other hand, find three sets A, B, and C for which $A \cap B \cap C = \varnothing$, but that are *not* mutually disjoint.

35. Find a sequence of sets $A_1, A_2, A_3, \ldots$ for which
1) $A_i \subset Z$ for all $i = 1, 2, 3, \ldots$ and
2) $A_1 \supset A_2 \supset A_3 \supset \cdots$

36. Prove that if A and B are subsets of a universal set U, then $A - B = A \cap B^c$.

37. Verify the equations in Property 1.

38. Verify the idempotent properties.

39. Verify the commutative properties.

40. Verify the associative properties.

41. Verify the second distributive property.

42. Verify the equations in Property 6.

43. Verify De Morgan's laws.

44. Generalize De Morgan's Laws to the case of three sets. That is, find formulas for $(A \cup B \cup C)^c$ and $(A \cap B \cap C)^c$, and prove that your formulas are correct. (You may use De Morgan's Laws for two sets in your proof.)

If A and B are sets, we define the **symmetric difference** *of A and B by*

$$A \Delta B = (A - B) \cup (B - A)$$

Exercises 45 through 48 concern the symmetric difference.

45. Let $A = \{1, 2, 3, 4, 5, 6\}$, $B = \{4, 5, 6, 7, 8\}$, and $C = \{0, 1, 2, 8\}$. Find

a) $A \Delta B$ b) $B \Delta C$ c) $A \Delta A$
d) $A \Delta \emptyset$ e) $(A \Delta B) \Delta C$ f) $(A \Delta A) \Delta A$

46. What is $\Gamma_n \Delta \Gamma_{n-1}$?

47. Prove that

$$A \Delta B = B \Delta A$$

How would you put this property in words?

48. Is it true that

$$A \Delta (B \Delta C) = (A \Delta B) \Delta C$$

If so, prove it, and if not, find an example of three sets A, B, and C for which it does not hold. How would you describe this in words?

1.2
One-to-one Correspondences

Now that we have discussed some of the basic concepts of set theory, we want to consider the following questions.

1) What does it mean to say that a set is *finite*? What does it mean to say that a set is *infinite*?
2) What does it mean to say that two sets have the same size?

You probably already have an intuitive idea of what it means to say that a set is finite, or to say that two sets have the same size. However, in order to work with these concepts in the setting of mathematics or science, we must formulate very precise, careful definitions.

For example, to say that a set is finite if it has a finite number of elements is not at all satisfactory, for it uses the word finite to define the concept of a finite set. This type of definition is said to be *circular* and is useless.

As another example, to say that two sets have the same size if they have the same number of elements is also not satisfactory, since we do not know what it means to talk about the number of elements in an infinite set.

So, at this point, it is best to set aside any preconceived ideas as to what these concepts might mean, so that we can "start fresh" and decide how to define them in a precise way.

Let us consider the second question first. What does it mean to say that two sets have the same size?

As an example, consider the following sets:

$$A = \{a, c, g, h, j, k, m, n, q\}$$
$$B = \{2, 3, 6, 7, 8, 9, 0, +, -\} \tag{1.2.1}$$

The most direct way to tell whether or not these two sets have the same size is *not* to count the number of elements in each set, but rather to compare the two sets element for element, or in other words, to "pair off" the elements of one set with the elements of the other set. If one set runs out of elements before the other one does, then the sets do *not* have the same size. (In fact, the set that runs out of elements first is the smaller set, but we don't care about that now.) On the other hand, if the two sets run out of elements at exactly the same time, then the sets do have the same size.

Such a pairing of the elements of A and B is illustrated below.

$$a \leftrightarrow 2$$
$$c \leftrightarrow 3$$
$$g \leftrightarrow 6$$
$$h \leftrightarrow 7$$
$$j \leftrightarrow 8$$
$$k \leftrightarrow 9$$
$$m \leftrightarrow 0$$
$$n \leftrightarrow +$$
$$q \leftrightarrow -$$

Figure 1.2.1

This pairing shows that A and B do have the same size.

As another example, consider the sets

$$C = \{a, b, c, g, 7, +\} \quad \text{and} \quad D = \{4, 8, k, e\} \tag{1.2.2}$$

In this case, no matter how we try to pair off the elements of the two sets (never using an element of either set more than once, of course), we will run out of the elements in D before we run out of the elements in C, and so the two sets do not have the same size.

Now we are ready for a definition.

Definition

Let A and B be sets. Then a pairing of the elements of A with the elements of B is called a **one-to-one correspondence** between the elements of A and the elements of B if it satisfies the following two conditions.

1) Every element of A appears once, and *only* once, on the left side of the pairing.

2) Every element of B appears once, and *only* once, on the right side of the pairing.

Loosely speaking, a pairing of the elements of A with the elements of B is a one-to-one correspondence if every element of the two sets is used once, and only once. However, we had to be a bit more careful when it came to the wording in the actual definition because we had to allow for the possibility that an element of A may also be in B.

For example, consider the sets A = $\{a, b, c\}$ and B = $\{c, d, e\}$. Then the pairing

$$a \leftrightarrow c$$

$$b \leftrightarrow d$$

$$c \leftrightarrow e$$

does fit the definition of a one-to-one correspondence as we have written it, even though the element c appears more than once in the pairing. The point is that, as an element of A, it appears only once (on the left side), and as an element of B it appears only once (on the right side).

Thus, Figure 1.2.1 describes a *one-to-one correspondence* between the elements of the sets A and B given in 1.2.1. On the other hand, there is no one-to-one correspondence between the elements of the sets C and D in 1.2.2.

Now we are in a position to give a precise definition of what it means to say that two sets have the same size.

Definition

Two sets A and B are said to have the **same size** if there is a one-to-one correspondence between the elements of A and the elements of B.

It is very important to keep in mind that this is the *definition* of the phrase "the same size" as it applies to sets. From now on, whenever we say that two sets have the same size, this is what we mean. If this definition seems a bit more complicated than necessary (it really isn't), then at least you can take comfort in the fact that it does make perfectly good sense.

When two sets have the same size, we also say that they have the same cardinality (recall that cardinality means the same thing as size). Also, two sets that are the same size are said to be **equipollent**. The terms **equipotent, equivalent**, and **equinumerous** are also used. Thus, we can state that *two sets are equipollent if and only if there is a one-to-one correspondence between their elements.*

Example 1.2.1

Let us have some more examples of one-to-one correspondences.

Consider the sets A = $\{1, 3, 5, 7\}$ and B = $\{2, 4, 6, 8\}$. Then the pairing described by

$$1 \leftrightarrow 2$$

$$3 \leftrightarrow 4$$

$$5 \leftrightarrow 6$$

$$7 \leftrightarrow 8$$

is a one-to-one correspondence between the elements of A and the elements of B.

Therefore, according to the definition, these two sets have the same size, or are equipollent. We can describe this correspondence more efficiently by the formula

$$n \leftrightarrow n + 1 \qquad \text{for} \qquad n = 1, 3, 5, 7 \tag{1.2.3}$$

Another one-to-one correspondence between the elements of A and the elements of B is given by

$$1 \leftrightarrow 8$$
$$3 \leftrightarrow 4$$
$$5 \leftrightarrow 2$$
$$7 \leftrightarrow 6$$

Unfortunately, this correspondence does not have a simple description such as 1.2.3.

On the other hand, the pairing

$$1 \leftrightarrow 4$$
$$3 \leftrightarrow 8$$
$$5 \leftrightarrow 2$$
$$7 \leftrightarrow 8$$

is *not* a one-to-one correspondence between the elements of A and the elements of B, since it violates the second condition of the definition. □

Example 1.2.2

Consider the set A of all students at your school, and the set B of all student identification numbers. Then there is a one-to-one correspondence between these two sets, given by pairing off each student with his or her identification number.

On the other hand, if we pair off each student with his or her birthdate, we do not get a one-to-one correspondence, since there are many students with the same birthdate. (Provided your school has at least 367 students, that is.) □

Example 1.2.3

One-to-one correspondences are used by a computer to store information. As you probably know, all information is stored in the computer in the form of binary numbers. Therefore, whenever you enter a character, such as a letter, a digit, or a punctuation mark into a computer, it is converted into a binary word. This is done by using a **character code**, which is nothing more than a one-to-one correspondence between a certain set of characters and a certain set of binary numbers.

One of the most commonly used character codes is the ASCII code, shown in the table below. (ASCII stands for American Standard Code for Information Interchange.) We have given the character code for each character as a decimal number, rather than a binary number. (Humans prefer decimal numbers; computers prefer binary numbers.)

ASCII Table

ASCII Code (Decimal)	*Character**	*ASCII Code (Decimal)*	*Character*	*ASCII Code (Decimal)*	*Character*	*ASCII Code (Decimal)*	*Character*
0	NUL	32	SPACE	64	@	96	‘
1	SOH	33	!	65	A	97	a
2	STX	34	"	66	B	98	b
3	ETX	35	#	67	C	99	c
4	EOT	36	$	68	D	100	d
5	ENQ	37	%	69	E	101	e
6	ACK	38	&	70	F	102	f
7	BEL	39	'	71	G	103	g
8	BS	40	(	72	H	104	h
9	HT	41	)	73	I	105	i
10	LF	42	*	74	J	106	j
11	VT	43	+	75	K	107	k
12	FF	44	,	76	L	108	l
13	CR	45	-	77	M	109	m
14	SO	46	.	78	N	110	n
15	SI	47	/	79	O	111	o
16	DLE	48	0	80	P	112	p
17	DC1	49	1	81	Q	113	q
18	DC2	50	2	82	R	114	r
19	DC3	51	3	83	S	115	s
20	DC4	52	4	84	T	116	t
21	NAK	53	5	85	U	117	u
22	SYN	54	6	86	V	118	v
23	ETB	55	7	87	W	119	w
24	CAN	56	8	88	X	120	x
25	EM	57	9	89	Y	121	y
26	SUB	58	:	90	Z	122	z
27	ESC	59	;	91	[	123	{
28	FS	60	<	92	\	124	|
29	GS	61	=	93	]	125	}
30	RS	62	>	94	^	126	~
31	US	63	?	95	_	127	DEL

*Codes 0 to 31 are nonprinting control codes for printers and other devices.

As you can see, the ASCII code is nothing but a one-to-one correspondence between the set of characters and the set of integers between 0 and 127. This code can be used to encode any expression for storage in the computer. For example, the ASCII code for the word NEXT is obtained as follows

$$N \to 78$$

$$E \to 69$$

$$X \to 88$$

$$T \to 84$$

and the ASCII code for the expression $(x + y)$ is

$$(\to 40$$

$$x \to 120$$

$$\text{(space)} \to 32$$

$$+ \to 43$$

$$\text{(space)} \to 32$$

$$y \to 121$$

$$) \to 41$$

There are other character codes in common use today. For example, IBM uses a character code of its own invention, whose acronym is EBCDIC (Extended Binary Coded Decimal Interchange Code). However, the essential feature of all these character codes is that they are one-to-one correspondences. This implies that each expression has a unique code, so that the computer can translate back and forth from expressions to their codes. □

Example 1.2.4

Let S be a set of size n. Then there is a very important one-to-one correspondence between the subsets of S and the binary words of length n. Let us illustrate this with an example. Let $\Sigma = \{0, 1\}$ and suppose that S is any set with 5 elements, say

$$S = \{a_1, a_2, a_3, a_4, a_5\}$$

Then we can define a one-to-one correspondence between the elements of $\mathcal{P}(S)$ (the subsets of S) and the elements of Σ_5 (the binary words of length 5) as follows.

If T is a subset of S, then we pair T with the binary word of length 5 that has a 1 in the i-th position if and only if a_i is in the subset T. Some examples of pairings will help make this correspondence clear.

$$\{a_1\} \leftrightarrow 10000$$

$$\{a_2, a_3\} \leftrightarrow 01100$$

$$\{a_1, a_3, a_5\} \leftrightarrow 10101$$

$$\{a_1, a_2, a_3, a_4, a_5\} \leftrightarrow 11111$$

It is not hard to see that this pairing is indeed a one-to-one correspondence between the elements of $\mathcal{P}(S)$ and the elements of Σ_5. Thus, according to the definition, these two sets have the same size, or are equipollent.

More generally, if S is any set with n elements, then we can use the same approach to define a one-to-one correspondence between the elements of the power set $\mathcal{P}(S)$ and the elements of the set Σ_n of all binary words of length n. (Incidentally, this is one way to store the subsets of a finite set in a computer.) □

Example 1.2.5

If Σ and Ω (omega) are sets of the same size, then the sets Σ^* and Ω^* also have the same size. In fact, any one-to-one correspondence between Σ and Ω can be "extended" to a one-to-one correspondence between Σ^* and Ω^*. Let us demonstrate this with an example.

Suppose that $\Sigma = \{1, 2, 3, 4\}$ and $\Omega = \{a, b, c, d\}$, and consider the one-to-one correspondence

$$\begin{aligned} 1 &\leftrightarrow a \\ 2 &\leftrightarrow b \\ 3 &\leftrightarrow c \\ 4 &\leftrightarrow d \end{aligned} \tag{1.2.4}$$

This correspondence can be extended to a one-to-one correspondence between Σ^* and Ω^* by matching the letters in any given word according to the pairing in 1.2.4. For instance, we have

$$\begin{aligned} 21 &\leftrightarrow ba \\ 431 &\leftrightarrow dca \\ 22244 &\leftrightarrow bbbdd \\ 12342314 &\leftrightarrow abcdbcad \end{aligned}$$

□

Before considering any more examples of one-to-one correspondences, let us make one more observation. Consider the set $A = \{a, b, 4, +, =\}$. How do we count the number of elements in this set?

The answer is that we associate with each element in this set a natural number, as follows

$$\begin{aligned} a &\leftrightarrow 1 \\ b &\leftrightarrow 2 \\ 4 &\leftrightarrow 3 \\ + &\leftrightarrow 4 \\ = &\leftrightarrow 5 \end{aligned}$$

(This is usually done mentally, of course.) Now, this is exactly the same as finding a one-to-one correspondence between the elements of A and the elements of the set $S = \{1, 2, 3, 4, 5\}$.

As a matter of fact, counting the number of elements in any nonempty finite set A amounts to nothing more than finding a one-to-one correspondence between the elements of A and the elements of a set of the form

$$S = \{1, 2, 3, \ldots, n\}$$

for some positive integer n.

Once we have found such a one-to-one correspondence, we know that the set A contains exactly n elements. In symbols, $|A| = n$.

Thus, we see that counting actually amounts to finding a special type of one-to-one correspondence.

Our remarks about counting make it easy for us to decide how we should define the term "finite." Intuitively speaking, we would like to say that a set is finite if we can count the number of elements in it. Using the concept of a one-to-one correspondence, we can make this more precise.

Definition

A set A is said to be **finite** if it is either the empty set or if there is a one-to-one correspondence between the elements of A and the elements of a set of the form $S = \{1, 2, 3, \ldots, n\}$, for some positive integer n. If a set is not finite, then we say that it is **infinite.** □

Example 1.2.6

The set $A = \{a, 7, >, 3, ?\}$ is finite since, as you can check for yourself, there is a one-to-one correspondence between the elements of A and the elements of the set $S = \{1, 2, 3, 4, 5\}$.

The set $\mathbf{N} = \{0, 1, 2, \ldots\}$ of natural numbers is infinite, since there is no one-to-one correspondence between its elements and the elements of any set of the form $S = \{1, 2, 3, \ldots, n\}$, for any value of n. (Of course, we have not proved this, but it certainly seems reasonable.) □

In general, it is not always easy to show that a given set is finite, or infinite, but at least now we know exactly what is involved in trying to do so. The next result can be very helpful in this regard.

Theorem 1.2.1

1) If A is a finite set, and B is a *subset* of A, then B is also finite.

2) If A is an infinite set, and if B is a *superset* of A, then B is also infinite.

We will not prove this theorem, but let us at least give some examples of it.

Example 1.2.7

The set $B = \{a, 3, ?\}$ is a subset of the set A in Example 1.2.6, and so, according to Theorem 1.2.1, it too is finite. Actually, part 2 of Theorem 1.2.1 is more interesting, for it shows us that the set **Z** of all integers, the set **Q** of all rational numbers, and the set **R** of all real numbers are infinite sets. This follows from the fact that these sets are supersets of the set **N** of natural numbers. □

Up to now we have been concentrating mostly on one-to-one correspondences between finite sets (with the exception of Example 1.2.5). However, the subject of one-to-one correspondences between infinite sets is truly fascinating, and so we will discuss it in the next section.

Exercises

1. Find two different one-to-one correspondences between the elements of the set A = {1, 5, y, z} and B = {1, 5, x, y}.
2. Find all of the one-to-one correspondences between the elements of the sets A = {1, 2} and B = {a, b}. Find all of the pairings of the elements of these two sets that are not one-to-one correspondences.
3. Use the ASCII code given in Example 1.2.3 to encode the following expressions.
   a) FOR b) For c) Input
   d) Save e) BASIC f) Discrete Math
4. Use the ASCII code given in Example 1.2.3 to encode the following expressions.
   a) $x + y$ b) xy c) $x * y$
   d) $(x + y) * (x - y)$
5. Use the ASCII code given in Example 1.2.3 to decode the following expression

   77|65|84|72|32|73|83|32|70|85|78

   (The bars are used to separate the individual characters.)

In Exercises 6 through 10, find a one-to-one correspondence between the elements of the set A and the set B. Describe the correspondence by a formula as in 1.2.3.

6. A = {0, 1, . . ., 10}, B = {2, 3, . . ., 12}
7. A = {0, 1, 2, 3, 4}, B = {0, 2, 4, 6, 8}
8. A = {1, 2, 5, 7}, B = {1, 4, 25, 49}
9. A = {0, 1, 2, 3, 4, 5}, B = {7, 70, 700, 7000, 70000, 700000}
10. A = {2, 3, 5, 7}, B = {4, 8, 32, 128}
11. Find a one-to-one correspondence between the elements of the sets $\mathcal{P}(\{a, b\})$ and Σ_2, where $\Sigma = \{0, 1\}$.
12. Find a one-to-one correspondence between the elements of the sets $\mathcal{P}(\{a, b, c\})$ and Σ_3, where $\Sigma = \{0, 1\}$.
13. Let $\Sigma = \{1, 2\}$ and $\Omega = \{a, b\}$. Find a one-to-one correspondence between the elements of the sets Σ^* and Ω^*.
14. Suppose that you have n different switches and that each switch can be either on or off. Find a one-to-one correspondence between the possible switch configurations and the subsets of the set $\{1, 2, \ldots, n\}$.
15. Suppose that you have n different switches, each of which can be in one of three possible states. Find a one-to-one correspondence between the set of possible switch configurations and the set S_n of all n-letter words over the set S = {0, 1, 2}.
16. Find a one-to-one correspondence between the elements of the sets

   $$A = \mathcal{P}(\{0, 1, \ldots, n\}) \quad \text{and} \quad B = \mathcal{P}(\{1, 2, \ldots, n + 1\}).$$

17. Find a one-to-one correspondence between the elements of the sets

   $$A = \mathcal{P}(\{1, 2, \ldots, n\}) \quad \text{and} \quad B = \mathcal{P}(\{2, 4, \ldots, 2n\}).$$

18. Find a one-to-one correspondence between the first 11 natural numbers and the powers of 2 between 1 and 2^{10} (inclusive).
19. Let $0 \le k \le n$. Find a one-to-one correspondence between the subsets of size k of the set $S = \{1, 2, \ldots, n\}$ and the subsets of size $k + 1$ of the set $T = \{1, 2, \ldots, n + 1\}$ that contain the integer $n + 1$. Describe the correspondence by a formula involving unions.

20. Find a one-to-one correspondence between the subsets of the set $S = \{1, 2, \ldots, n\}$ that contain an even number of elements and the subsets of the set $T = \{1, 2, \ldots, n + 1\}$ that contain $n + 1$ and that contain an odd number of elements.
21. Let $S = \{1, 2, \ldots, 100\}$. Suppose that we define a pairing between the subsets of S that contain an odd number of elements and the subsets of S that contain an even number of elements as follows. If T is a subset of S that contains an odd number of elements, then we pair T with the subset $T \cup \{a\}$, where a is the smallest integer in S that is not in T. Since T contains an odd number of elements, the set $T \cup \{a\}$ contains an even number of elements. Is this pairing a one-to-one correspondence? Justify your answer.
22. Suppose that A, B, and C are sets. Suppose also that there is a one-to-one correspondence between the elements of A and B and that there is a one-to-one correspondence between the elements of B and C. Can you use these correspondences to find a one-to-one correspondence between the elements of A and C? Explain your answer. What does this say about the sizes of sets?

1.3 Countable and Uncountable Sets

In this section, we want to continue our discussion of one-to-one correspondences, especially those involving infinite sets. Let us begin with a very important observation.

We have already decided what it means to say that two sets have the same size. This is given in the second definition of the previous section, and it seems like a reasonable definition; that is, it seems to make perfectly good sense.

Now, this definition applies to *all* sets. It does not discriminate between finite sets and infinite sets. This is how it should be, since we would not want it to mean one thing for two *finite* sets to have the same size, and another thing for two *infinite* sets to have the same size.

The point that we want to make is this. Now that we have decided to accept this as our definition, *we must also accept whatever consequences may arise from it.*

We are about to discuss some examples of one-to-one correspondences between infinite sets, and in doing so, we will arrive at some rather peculiar-looking results concerning the sizes of infinite sets.

For example, we will soon see that the set

$$\mathbf{E} = \{0, 2, 4, 6, \ldots\}$$

of even natural numbers has the same size as the set

$$\mathbf{N} = \{0, 1, 2, 3, \ldots\}$$

of *all* natural numbers!!

This may seem like a very peculiar result. After all, how can these two sets have the same size when **N** contains all of the elements of **E**, and many other elements as well?

The answer is simple. As we will see in a moment, there is a one-to-one correspondence between the elements of **E** and the elements of **N**. Therefore, *we have no choice but to say that these two sets have the same size—that is our definition!*

In a sense, we have found the only reasonable definition of what it means for two sets to be the same size, and we must accept the consequences. We should hasten to point out, however, that once you get a feel for this definition you will probably discover that the consequences are not as strange as you may have thought at first.

Let us now consider the promised examples.

Example 1.3.1

There is a one-to-one correspondence between the elements of the set

$$\mathbf{N} = \{0, 1, 2, 3, \ldots\}$$

of all natural numbers, and the elements of the set

$$\mathbf{E} = \{0, 2, 4, 6, \ldots\}$$

of all even natural numbers.

We can write down the first part of this correspondence as follows

$$\begin{aligned} 0 &\leftrightarrow 0 \\ 1 &\leftrightarrow 2 \\ 2 &\leftrightarrow 4 \\ 3 &\leftrightarrow 6 \\ 4 &\leftrightarrow 8 \\ &\vdots \end{aligned} \tag{1.3.1}$$

Now, we cannot write down the entire one-to-one correspondence, since the sets **N** and **E** are infinite. (We will always have this problem when dealing with infinite sets.) Therefore, we must resort to a formula for describing the correspondence, similar to the one given in 1.2.3.

In this case, by looking for the pattern in 1.3.1, we see that the *entire* correspondence can be described by the formula

$$n \leftrightarrow 2n \qquad \text{for} \qquad n = 0, 1, 2, 3, \ldots \tag{1.3.2}$$

Let us emphasize that, in this case, a formula such as 1.3.2 is much more important than it is in the case of finite sets, since it is really the only way to describe the entire correspondence.

It is not hard to see that the pairing described in 1.3.2 is indeed a one-to-one correspondence between the elements of the sets **N** and **E**. Clearly, every element of **N** is used once, and only once, on the left side, and every element of **E** is used once, and only once, on the right side. (Of course, the even numbers do appear on both sides of the correspondence. For example, we have the pairings $1 \leftrightarrow 2$ and $2 \leftrightarrow 4$. But, as we pointed out in the last section, this does *not* violate the definition of one-to-one correspondence.)

As a result of this one-to-one correspondence, *according to the definition*, the sets **N** and **E** have the same size, or are equipollent. □

Example 1.3.2

There is a one-to-one correspondence between the elements of the set

$$\mathbf{N} = \{0, 1, 2, 3, \ldots\}$$

of all natural numbers, and the elements of the set

$$\mathbf{Z} = \{\ldots, -3, -2, -1, 0, 1, 2, 3, \ldots\}$$

of all integers.

We can describe the first part of this correspondence as follows.

$$\begin{array}{rcr} 0 & \leftrightarrow & 0 \\ 1 & \leftrightarrow & -1 \\ 2 & \leftrightarrow & 1 \\ 3 & \leftrightarrow & -2 \\ 4 & \leftrightarrow & 2 \\ 5 & \leftrightarrow & -3 \\ 6 & \leftrightarrow & 3 \\ & \vdots & \end{array}$$

If you look at each side of this pairing, you can see the pattern. The left side consists simply of the elements of **N**, written in their usual order, and the right side consists of the elements of **Z**, written in a pattern that alternates positive and negative integers.

It is rather clear from this pattern that the pairing does form a one-to-one correspondence. Every natural number appears once, and only once, on the left side of the correspondence, and every integer appears once, and only once, on the right side. Thus, according to the definition, the sets **N** and **Z** are equipollent.

In order to describe the entire correspondence, however, we must resort to a formula, such as the one in 1.3.2. In this case, the formula is a bit more involved. Let us see if we can find it.

The pattern in this correspondence can be described by saying that the *even* natural numbers on the left side are paired with the *nonnegative* integers on the right, and the *odd* natural numbers on the left are paired with the *negative* integers on the right. So let us split the correspondence into two parts

$$\begin{array}{rcl} 0 & \leftrightarrow & 0 \\ 2 & \leftrightarrow & 1 \\ 4 & \leftrightarrow & 2 \\ 6 & \leftrightarrow & 3 \\ & \vdots & \end{array} \qquad \begin{array}{rcr} 1 & \leftrightarrow & -1 \\ 3 & \leftrightarrow & -2 \\ 5 & \leftrightarrow & -3 \\ 7 & \leftrightarrow & -4 \\ & \vdots & \end{array}$$

Now, the left hand part of this correspondence can be described by the formula

$$n \leftrightarrow \frac{n}{2} \qquad \text{for} \qquad n = 0, 2, 4, 6, \ldots$$

and the right hand part can be described by the formula

$$n \leftrightarrow -\frac{n+1}{2} \qquad \text{for} \qquad n = 1, 3, 5, 7, \ldots$$

(Try this for the first few values of n to convince yourself.)

Finally, we can combine these two formulas to get

$$n \leftrightarrow \begin{cases} \dfrac{n}{2} & \text{for } n = 0, 2, 4, 6, \ldots \\ -\dfrac{n+1}{2} & \text{for } n = 1, 3, 5, 7, \ldots \end{cases} \tag{1.3.3}$$

Actually, this formula can be simplified a bit more by using the greatest integer function (see Exercise 7).

Example 1.3.3

There is even a one-to-one correspondence between the elements of the set **N** of natural numbers and the elements of the set **Q** of all rational numbers! Hence, these two sets have the same size; that is, they are equipollent. We will indicate how to find such a one-to-one correspondence in Exercises 9 and 10.

At this point, it may seem as though all infinite sets have the same size as the set **N** of natural numbers. However, this is not the case.

Example 1.3.4

Let us show that the set **S** of all *real* numbers between 0 and 1 (not including 0 or 1) is *not* the same size as the set **N** of all natural numbers.

In order to do this, we must show that there *cannot* be a one-to-one correspondence between the elements of these two sets. We will do this by showing that, given *any* pairing

$$\begin{aligned} 0 &\leftrightarrow r_1 \\ 1 &\leftrightarrow r_2 \\ 2 &\leftrightarrow r_3 \\ 3 &\leftrightarrow r_4 \\ &\ \vdots \end{aligned}$$

of the elements of **N** with the elements of **S**, there is always some element of **S** that does not appear on the right side. Hence, there is no one-to-one correspondence between the elements of **N** and the elements of **S**.

The easiest way to describe what we have in mind is to use an example. Consider a pairing that begins as follows:

$$\begin{aligned} 0 &\leftrightarrow r_1 = .45263 \cdots \\ 1 &\leftrightarrow r_2 = .39820 \cdots \\ 2 &\leftrightarrow r_3 = .67552 \cdots \\ 3 &\leftrightarrow r_4 = .11111 \cdots \\ 4 &\leftrightarrow r_5 = .33915 \cdots \\ &\vdots \end{aligned} \tag{1.3.4}$$

(We are writing all elements of **S** as decimals.) At this point, we must use our imagination to think of this pairing as continuing indefinitely, so that all of the natural numbers appear on the left-hand side.

Of course, this is just one possible pairing, but the point is that what we are about to do will apply equally well to *any other* pairing that we might wish to consider. In other words, it applies to *all* pairings.

We want to show that there is some element of **S** that does not appear on the right side of 1.3.4. In order to find such an element, call it x, we first place the pairing into a rectangular grid, as shown in Figure 1.3.1(a), and then we circle those digits that lie on the diagonal of the grid.

Now, let us write the number x at the top of the grid. Our plan is to take the circled digits, change each one of them, and use the resulting digits to form x. More specifically, the digits in x are obtained by the following rule. If a circled digit is *not* equal to 5, then put a 5 directly above it, at the top of the grid. On the other hand, if a circled digit is equal to 5, then put a 3 directly above it, at the top of the grid. In this way, we obtain the digits for the number x. [See Figure 1.3.1(b).]

For this particular pairing, we get

$$x = .55353 \cdots$$

Why does this number x not appear on the right side of the pairing 1.3.4?

How could it? It is certainly not the first number, r_1, because the first digit in x is not the same as the first digit in r_1. It is certainly not the second number, r_2, because the second digit in x is not the same as the second digit in r_2.

Continuing in this way, we see that x cannot be equal to any of the numbers r_1, $r_2, r_3, r_4, r_5, \ldots$. In general, x cannot be equal to the n-th number r_n in the pairing 1.3.4 because the n-th digit of x is not the same as the n-th digit of r_n. After all, this is exactly how we planned it when we formed the number x. When we chose the n-th digit for x, we looked at the n-th circled digit in the grid, which happens to be the n-th digit in r_n, and deliberately picked a different digit to use in x!

Thus, we have found an element of **S** that does not appear on the right side of the pairing 1.3.4. It is easy to see that this same reasoning will work for *any* pairing of the elements of **N** and elements of **S**, and so we can conclude that no such pairing

(a)

0 ⟷ $r_1 = .$	(4)	5	2	6	3	. . .
1 ⟷ $r_2 = .$	3	(9)	8	2	0	. . .
2 ⟷ $r_3 = .$	6	7	(5)	5	2	. . .
3 ⟷ $r_4 = .$	1	1	1	(1)	1	. . .
4 ⟷ $r_5 = .$	3	3	9	1	(5)	. . .
⋮			⋮			⋮

(b)

$x =$	5	5	3	5	3	. . .
0 ⟷ $r_1 = .$	(4)	5	2	6	3	. . .
1 ⟷ $r_2 = .$	3	(9)	8	2	0	. . .
2 ⟷ $r_3 = .$	6	7	(5)	5	2	. . .
3 ⟷ $r_4 = .$	1	1	1	(1)	1	. . .
4 ⟷ $r_5 = .$	3	3	9	1	(5)	. . .
⋮				⋮		⋮

Figure 1.3.1

can possibly be a one-to-one correspondence. In other words, the sets **N** and **S** do not have the same size. □

Of course, one of the consequences of the previous example is the remarkable statement that *not all infinite sets have the same size*!

It can be shown that the set **S** has the same size as the set **R** of all real numbers (see Exercise 16), and we can conclude from this that *the sets* **N** *and* **R** *do not have the same size*.

Up to now, we have discussed only what it means for two sets to have the same size, or not to have the same size. We have not discussed what it means for one infinite set to be larger (or smaller) than another. Let us do that now. The following definition applies to both finite and infinite sets.

Definition

Let A and B be sets. Then we say that the set B is **larger than** the set A if the following two conditions hold:

1) A does not have the same size as B; that is, there is no one-to-one correspondence between the elements of A and the elements of B.
2) There is a one-to-one correspondence between the elements of A and the elements of some *proper* subset of B.

If B is larger than A, we also say that A is **smaller than** B. □

At this point, you may be wondering why we need both conditions in this definition. Isn't it true that if condition 2 holds, then condition 1 must also hold?

Contrary to what we might think at first, the answer to this question is no. As an example, let A be the set **E** of all even natural numbers and let B be the set **N** of all natural numbers.

Then these sets do satisfy the second condition, since **E** is itself a proper subset of **N**. (There certainly is a one-to-one correspondence between the elements of **E** and the elements of **E**.) On the other hand, as we saw in Example 1.3.1, there is also a one-to-one correspondence between the elements of **E** and the elements of **N**.

Thus, it is possible for there to be a one-to-one correspondence between the elements of a set A and the elements of a *proper* subset of a set B and, at the same time, a one-to-one correspondence between the elements of A and the elements of (all of) B! This is precisely why we require that both of the conditions in the definition hold before saying that one set is larger than another. (Incidentally, for *finite* sets, condition 2 does imply condition 1.)

Using the previous definition, we can state that the set **R** of all real numbers is *larger than* the set **N** of all natural numbers, although we will not take the time to actually prove this.

Let us pause now to see where we stand. The sets **E**, **N**, **Z** and **Q** all have the same size, but the set **R** is larger. This naturally leads us to wonder whether or not there are any infinite sets that are even larger than **R**.

It so happens that there are. In fact, a mathematician named Georg Cantor (1845–1918) who, incidentally, developed the theory of infinite sets that we are now discussing, showed that the power set $\mathcal{P}(S)$ of any infinite set S is actually larger than the set S itself. (This is also true for finite sets.) This result is now known as **Cantor's Theorem**.

Theorem 1.3.1 (Cantor's Theorem)

The power set $\mathcal{P}(S)$ of any set S is larger than the set S.

We will not take the time to prove this theorem, but as a result of it, we can find a whole sequence of infinite sets, each one larger than the one before it! For example, consider the sequence

$$\mathbf{N}, \mathcal{P}(\mathbf{N}), \mathcal{P}(\mathcal{P}(\mathbf{N})), \mathcal{P}(\mathcal{P}(\mathcal{P}(\mathbf{N}))), \ldots$$

where **N** is, as always, the set of natural numbers. Each set in this sequence is simply the power set of the set immediately preceding it. Thus, according to Cantor's Theorem, each set is larger than the one preceding it.

Before Cantor's time, mathematicians thought that there was only one "infinity" and that all sets were either finite or infinite. Put another way, they knew that there were many different sizes of finite sets, in fact, an infinite number of different sizes of finite sets, but they thought that there was only one size of infinite set.

However, Cantor's Theorem shows that there are actually an infinite number of different sizes of infinite sets as well! If you find this fact difficult to accept, then you might consider asking yourself why that should be any stranger than the fact that there are an infinite number of different sizes of finite sets. Just because we cannot easily see the difference in the sizes of the sets **N** and **R**, for example, does not automatically mean that there is no difference. In fact, we have shown by a perfectly reasonable argument that there is a difference in the sizes of these two sets.

Infinite sets that have the same size as the set **N** are very special. If S is a set with the same size as **N**, then we can find a one-to-one correspondence between the elements of S and the elements of **N**,

$$s_1 \leftrightarrow 0$$
$$s_2 \leftrightarrow 1$$
$$s_3 \leftrightarrow 2$$
$$s_4 \leftrightarrow 3$$
$$\vdots$$

What we are doing when we describe this one-to-one correspondence between S and **N** is *making a list* of the elements of S. Of course, this is an infinite list, and so we can never actually complete it. However, we can say that given any element of S, it will eventually appear on the list.

This statement cannot be made for any listing of the real numbers, since the set of real numbers does not have the same size as the set **N**. As we have seen, no matter how we try to list the real numbers, some real numbers will never appear. That is precisely why the set **R** is larger than the set **N**.

Of course, if S is a finite set, then its elements can also be listed. Therefore, we can say that *the elements of a set S can be listed if and only if S either is finite or has the same size as the set* **N**. Because this fact is so important, the following definition is made.

Definition

Any set that has the same size as the set **N** of all natural numbers is said to be **countably infinite**. Any set that is either finite or countably infinite is said to be **countable**. Any set that is not countable is said to be **uncountable.** □

Sometimes the word **denumerable** is used in place of countable. (The term *listable* is not used.)

According to this definition, the sets **E**, **N**, **Z**, and **Q** are countably infinite (they are also countable). Thus, it is (theoretically) possible to make a list of the elements in these sets. However, the set **R** is uncountable.

Incidentally, the set **I** of all irrational numbers is also uncountable. In fact, it has the same size as the set **R**. It is interesting to notice, therefore, that the set **I** of irrational numbers is actually larger than the set **Q** of rational numbers.

Example 1.3.5

If S is a finite set, then, although we will not prove it, the set S* of all words over S happens to be countable. This means that it is theoretically possible to make a list of the words in S* in such a way that every word will eventually appear on the list. One way to make such a list is to first list all of the one-letter words, then list all of the two-letter words, then list all of the three-letter words, and so on. In this way, we will not miss any words. □

The subject of infinite sets is one of the most fascinating in all of mathematics and, as you can see, it contains some big surprises. Perhaps one of its biggest surprises is that the subject has many important *practical* applications, especially to computer science. Unfortunately, we have only had time to scratch the surface here. (The subject is usually covered in much more depth in a course in set theory or logic.)

Exercises

In Exercises 1 through 6, find a one-to-one correspondence between the elements of the set A and the elements of the set B. Give a formula for the correspondence, as we did in Examples 1.3.1 and 1.3.2.

1. A = {0, 1, 2, . . .} , B = {1, 2, 3, . . .}
2. A = {0, 1, 2, . . .} , B = {5, 6, 7, . . .}
3. A = {3, 4, 5, . . .} , B = {−2, −1, 0, 1, 2, . . .}
4. A = {0, 2, 4, 6, . . .} , B = {1, 3, 5, 7, . . .}
5. A = the set of all even integers , B = the set of all odd integers
6. A = the set of all even integers , B = **Z**
7. If x is any real number, then the *greatest integer in* x is defined to be the largest integer n that is less than or equal to x. The greatest integer in x is denoted by $[x]$. Thus, for example, we have $[\frac{5}{2}] = 2$, $[\frac{1}{4}] = 0$, $[-\frac{7}{3}] = -3$, and $[n] = n$ for all integers n. Show that formula 1.3.3 in Example 1.3.2 can be written

$$n \leftrightarrow (-1)^n\left[\frac{n+1}{2}\right]$$

Hint, consider separately the cases n even and n odd.

8. Show that there is a one-to-one correspondence between the elements of the set **N** of all natural numbers and the elements of the set **N** × **N** of all ordered pairs of natural numbers. *Hint*, there is no simple formula for such a correspondence. However, you can give a plausible argument for the existence of such a correspondence by using the following picture.

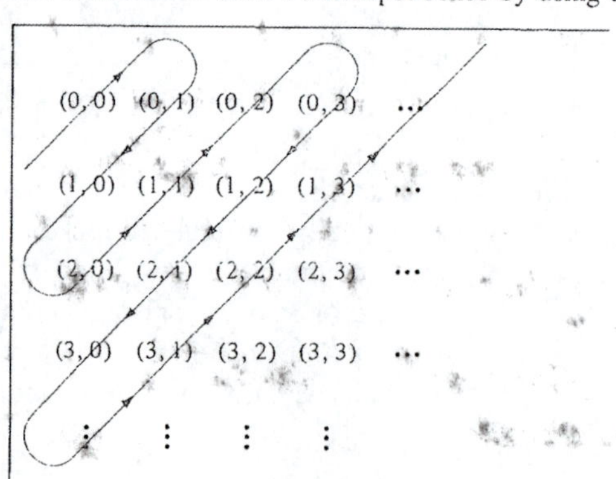

By following the indicated path, describe in words how you would construct a one-to-one correspondence between the elements of the sets **N** and **N** × **N**. This shows that **N** and **N** × **N** have the same size. In other words, the set **N** × **N** is countably infinite.

9. By considering the following picture

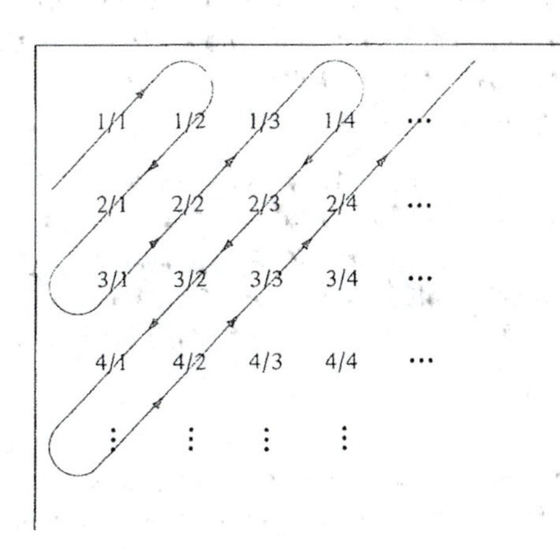

give a plausible argument to show that there is a one-to-one correspondence between the natural numbers and the positive rational numbers. *Hint,* you will have to be more careful in this exercise than in the previous one, since some rational numbers are repeated more than once in this picture.

10. a) By modifying the argument you used in the previous exercise, show that there is a one-to-one correspondence between the even natural numbers and the positive rational numbers.

b) Show also that there is a one-to-one correspondence between the odd natural numbers and the negative rational numbers.

c) Put these two correspondences together to form a one-to-one correspondence between all natural numbers and all rational numbers.

This shows that **N** and **Q** have the same size. In other words, **Q** is countably infinite.

11. Show that the set **R** of all real numbers has the same size as the set of all points on a line. *Hint,* use the real number line.

12. Show that the set **R** × **R** of all ordered pairs of real numbers has the same size as the set of all points in a plane.

13. Show that the set of all circles in the plane is equipollent to the set

$$\{(r, c) \mid r > 0 \text{ is a real number and } c = (x, y) \text{ is an ordered pair of real numbers}\}$$

14. Show that **N** is equipollent to the set of all perfect squares. (A natural number is a *perfect square* if it can be written in the form k^2, for some integer k.)

15. Show that the set A of all quadratic equations with real solutions is equipollent to the set

$$\{(a, r_1, r_2) \mid a, r_1, \text{ and } r_2 \text{ are real numbers and } a \neq 0\}$$

16. Show that the function $f(x) = \tan(\pi x + \pi/2)$ gives a one-to-one correspondence between the elements of the set

$$\mathbf{S} = \{x \mid x \text{ is a real number and } 0 < x < 1\}$$

and the elements of the set **R**.

1.4

Functions

One of the most important concepts in all of mathematics is that of a function. Although you probably already have some familiarity with functions, we do want to spend a little time discussing those aspects that we will need in this book.

Let us begin with the definition.

Definition

A function consists of three things:

1) A nonempty set A, called the **domain** of the function.
2) A nonempty set B, called the **range** of the function.
3) A rule that assigns to each element of A one, and *only* one, element of B. □

We will use letters such as f, g and h to denote functions, and we will write $f{:}A \to B$ to denote the fact that f is a function with domain A and range B. (The expression $f{:}A \to B$ is read "f is a function from A to B.") We will also denote the domain of a function f by dom(f), and the range by ran(f).

If $f{:}A \to B$, and if a is an element of A, then we will denote by $f(a)$ the element of B that is assigned to a by the function f. The element $f(a)$ is called the **image** of a under the function f.

Sometimes it is useful to visualize functions by means of a simple picture, such as the one in Figure 1.4.1. (This is similar to representing sets by Venn diagrams.) Figure 1.4.2 illustrates the property that a rule *cannot* have in order to be called a

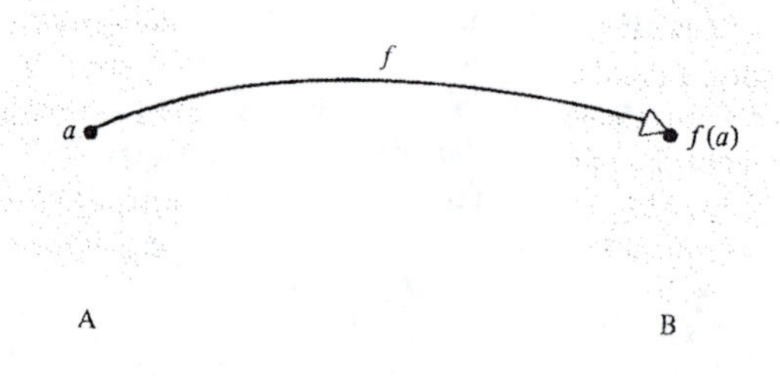

Figure 1.4.1

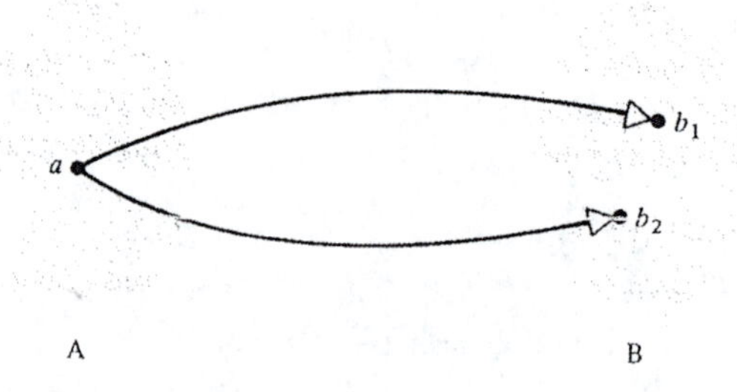

Figure 1.4.2
A rule that does *not* define a function.

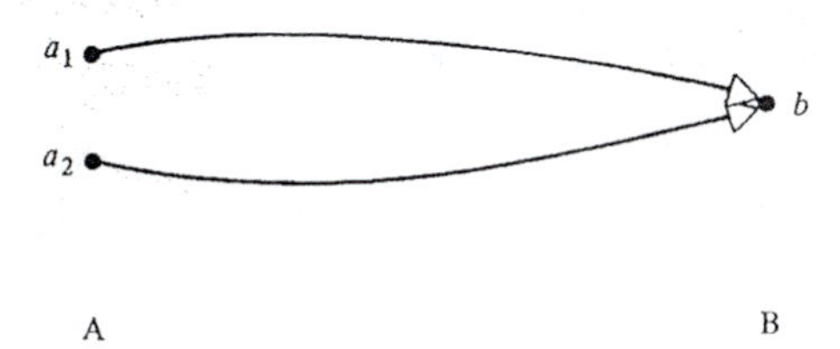

Figure 1.4.3
This is permissible for a function.

function. In this figure, a single element a in the domain A is assigned two different elements b_1 and b_2 in the range B.

On the other hand, there is nothing in the definition of a function that prohibits the behavior illustrated in Figure 1.4.3. In this figure, two different elements, a_1 and a_2, of the domain A are assigned the same element b in the range. We will discuss this behavior again later in the section.

Let us consider some examples of functions.

Example 1.4.1

Let A = {1, 3, 5} and B = $\{a, b, c, d\}$. Then the rule

$$f(1) = d \quad , \quad f(3) = a \quad , \quad f(5) = a$$

is a function from A to B. □

Example 1.4.2

Let A = **R** be the set of all real numbers, and let B = **R**′ be the set of all nonnegative real numbers. Then the formula

$$f(x) = x^2$$

defines a function $f:\mathbf{R} \to \mathbf{R}'$ from **R** to **R**′. For example, we have $f(2) = 4$, $f(-2) = 4$, $f(0) = 0$ and $f(3) = 9$. □

Example 1.4.3

Let S be any finite set, and let $\mathcal{P}(S)$ be its power set. Then we can define a function $f:\mathcal{P}(S) \to \mathbf{N}$ by

$$f(A) = |A|$$

for all subsets A of S. In words, $f(A)$ is the size of the set A. □

Example 1.4.4

Let Char be the set of all characters that have an ASCII code. (See Example 1.2.3.) Then we can define a function ASC:Char → {0, 1, . . ., 127} by defining ASC(α) to be the ASCII code for the character α (the Greek letter *alpha*). For example, we have

$$\text{ASC}(+) = 43 \quad , \quad \text{ASC}(]) = 93 \quad , \quad \text{ASC}(g) = 103 \quad , \quad \text{ASC}(4) = 52$$

Similarly, we can define a function CHR:$\{0, 1, \ldots, 127\} \to$ Char by defining CHR(n) to be that character whose ASCII code is the integer n. For example,

$$\text{CHR}(43) = + \quad , \quad \text{CHR}(93) =] \quad , \quad \text{CHR}(103) = g \quad , \quad \text{CHR}(52) = 4$$

□

Example 1.4.5

Let Σ be an alphabet. Then we can define a function $\ell:\Sigma^* \to \mathbf{N}$ by

$$\ell(w) = \text{length of } w$$

where w is any word in Σ^*. In other words, $l(w)$ is just the length of w. (Recall that the length of a word is the number of "letters" in the word. Thus, for example, we have

$$\ell(aabab) = 5 \quad , \quad \ell(\theta) = 0 \quad , \quad \ell(100001100) = 9$$ □

Example 1.4.6

Let $\Sigma = \{0, 1\}$, and consider the set Σ_n of all binary words of length n. Then we can define a function $H:\Sigma_n \times \Sigma_n \to \mathbf{N}$ as follows. If w and u are words in Σ_n, then we compare them place by place and define $H(w, u)$ to be the number of places where the words differ. For example, for the function $H:\Sigma_5 \times \Sigma_5 \to \mathbf{N}$ we have

$$H(10000, 00101) = 3$$

since the words $w = 10000$ and $u = 00101$ differ in exactly three places, namely, the first, third, and fifth places. Similarly, we have

$$H(00000, 11111) = 5$$

$$H(10101, 01010) = 5$$

$$H(11001, 01000) = 2$$

and so on. The function H is called the **Hamming distance function**, and the number $H(w, u)$ is called the **Hamming distance** between w and u (named after the computer scientist Richard W. Hamming). As the name implies, the Hamming distance is a measure of how "close" two binary words are to each other. In fact, it is easy to see that two binary words are equal if and only if their Hamming distance is equal to 0. The Hamming distance is very important in a branch of computer science known as *Coding Theory*. □

Example 1.4.7

Let A and B be nonempty sets. Then a function from A $\times$ B to A is sometimes called a **transition function**. Transition functions are denoted by the Greek letter δ (*delta*). One reason that such functions are called transition functions can be seen from the following simple example.

Let

$$A = \{\text{sad, happy, ecstatic}\}$$

and

$$B = \{\text{bad news, good news, wonderful news}\}$$

and let $\delta: A \times B \to A$ be defined by

$$\delta[(\text{sad, bad news})] = \text{sad}$$
$$\delta[(\text{sad, good news})] = \text{happy}$$
$$\delta[(\text{sad, wonderful news})] = \text{ecstatic}$$
$$\delta[(\text{happy, bad news})] = \text{sad}$$
$$\delta[(\text{happy, good news})] = \text{ecstatic}$$
$$\delta[(\text{happy, wonderful news})] = \text{ecstatic}$$
$$\delta[(\text{ecstatic, bad news})] = \text{happy}$$
$$\delta[(\text{ecstatic, good news})] = \text{ecstatic}$$
$$\delta[(\text{ecstatic, wonderful news})] = \text{ecstatic}$$

As you can see, the transition function δ describes the transition in the state of a person who receives certain news. For example, the fact that

$$\delta[(\text{sad, good news})] = \text{happy}$$

says that if the person is in the sad state and receives good news, he (or she) moves into the happy state. We will see some more serious examples of transition functions in Chapter 5. □

Example 1.4.8

Consider an electric circuit that has a battery, two light bulbs, and three switches. The state of the switches can be described by a binary word of length 3. For example, the word $w = 100$ indicates that switch 1 is on, but switches 2 and 3 are off. Similarly, the state of the light bulbs can be indicated by a binary word of length 2.

If $\Sigma = \{0, 1\}$, then we can define a function $f: \Sigma_3 \to \Sigma_2$ to indicate the various possible light bulb configurations. For example, consider the function f defined as follows

$$f(000) = 00 \quad , \quad f(001) = 11 \quad , \quad f(010) = 10 \quad , \quad f(100) = 01$$
$$f(011) = 00 \quad , \quad f(101) = 00 \quad , \quad f(110) = 00 \quad , \quad f(111) = 11$$

According to this function, when switches 1 and 3 are off but switch 2 is on, then since $f(010) = 10$, light bulb 1 is on and light bulb 2 is off. □

These examples should convince you that functions play an important role in applications of mathematics. Now let us consider a few properties that a function may or may not possess.

Example 1.4.1 illustrates the fact that not all elements in the range of a function need be assigned to some element in the domain. In this example, the element c in the range is not assigned to any element in the domain.

Those functions that have the property that every element of their range is assigned to some element of the domain are special, and so we make the following definition.

Definition

Let f:A → B be a function from A to B. Then the **image** of f is the set of all elements of B that are assigned to some element of A by the function. In symbols,

$$\text{Im}(f) = \{b \in \text{B} \mid b = f(a) \text{ for some } a \in \text{A}\}$$

Furthermore, if a function f:A → B has the property that Im(f) = Ran(f), then we say that the function is **onto** B. A function that is onto is called a **surjective** function. □

The definition of a surjective function can also be phrased as follows. A function f:A → B is **surjective** if for every element b in B, there is at least one element a in A for which $f(a) = b$. After all, this is exactly what it means to say that the image of f is all of B.

Example 1.4.9

Let A = {1, 2, 3} and B = {1, 2}. Then the function f:A → B defined by

$$f(1) = 2 \quad , \quad f(2) = 2 \quad , \quad f(3) = 1$$

is surjective, since Im(f) = {1, 2} = B. On the other hand, the function g:A → B defined by

$$g(1) = 2 \quad , \quad g(2) = 2 \quad , \quad g(3) = 2$$

is not surjective, since Im(g) = {2} ≠ B. □

We will leave it to you, in the exercises, to decide which of the functions in Examples 1.4.2 through 1.4.7 are surjective.

Example 1.4.1 also illustrates the fact, mentioned earlier, that two different elements in the domain of a function can be assigned to the same element in the range. In this case, the elements 3 and 5 are both assigned the same element a in the range. Functions that do *not* have this property are very important, and they deserve a special name.

Definition

Let f:A → B be a function. If no two *different* elements of A are assigned to the same element of B, we say that the function is **one-to-one**. In symbols, a function is one-to-one if

$$a_1 \neq a_2 \quad \text{implies that} \quad f(a_1) \neq f(a_2)$$

A function that is one-to-one is called an **injective** function. □

In other words, we can say that a function is injective if it does *not* have the property shown in Figure 1.4.3. In this case, two distinct elements a_1 and a_2 in A are assigned to the same element b in B. Another way to phrase the definition of one-to-one is to say that a function f:A → B is one-to-one if whenever $f(a_1) = f(a_2)$, then we must have $a_1 = a_2$. Sometimes this description is easier to apply than the one given in the definition.

Example 1.4.10

Let A = {1, 2, 3} and B = {1, 2, 3, 4}. Then the function f:A → B defined by

$$f(1) = 1 \quad , \quad f(2) = 3 \quad , \quad f(3) = 4$$

is injective. In this case, it is easy to see that if a_1 and a_2 are elements of A and if $a_1 \neq a_2$, then $f(a_1) \neq f(a_2)$. □

Example 1.4.11

The function f:**N** → **N** defined by

$$f(n) = n^2$$

for all n in **N** is an injective function, since if n_1 and n_2 are two different natural numbers, then $f(n_1) = n_1^2$ and $f(n_2) = n_2^2$ are also different.

On the other hand, the function g:**Z** → **N** defined by

$$g(z) = z^2$$

is *not* injective since, for example, we have $-1 \neq 1$, but $g(-1) = g(1)$. □

Again we will leave it to you in the exercises to determine which of the functions in Examples 1.4.2 through 1.4.7 are injective.

A function that is both injective and surjective is called a **bijective** function (or simply a one-to-one and onto function). We have already spent considerable time studying bijective functions in Sections 1.2 and 1.3. After all, *a bijective function f:A → B from A to B is nothing more than a one-to-one correspondence between the elements of A and the elements of B*.

Suppose that f:A → B is a function from A to B, and that g:B → C is a function from B to C, as pictured in Figure 1.4.4. Then we can define a new function, from A to C, denoted by $g \circ f$, by the formula

$$(g \circ f)(a) = g(f(a))$$

for all elements a in A. This function is called the **composition** of g with f and is also pictured in Figure 1.4.4.

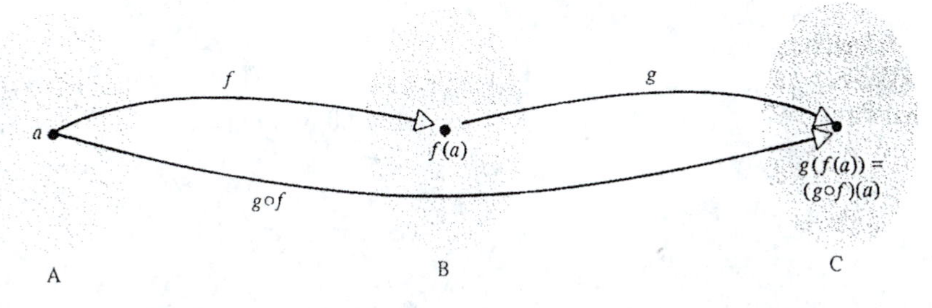

Figure 1.4.4

Applying the function $g \circ f$ to an element a in A is the same as first applying the function f to a, to get $f(a)$, and then applying the function g to that, to get $g(f(a))$. In short, it is "first f, then g."

Example 1.4.12

Let A = $\{a, b, c\}$, B = $\{1, 2, 3, 4\}$, and C = $\{a, 2, \quad 6\}$. Let f:A → B be defined by

$$f(a) = 2 \quad , \quad f(b) = 2 \quad , \quad f(c) = 1$$

and let g:B → C be defined by

$$g(1) = a \quad , \quad g(2) = -6 \quad , \quad g(3) = -6 \quad , \quad g(4) = a$$

Then $g \circ f$:A → C is the function given by

$$(g \circ f)(a) = g(f(a)) = g(2) = -6$$

$$(g \circ f)(b) = g(f(b)) = g(2) = -6$$

$$(g \circ f)(c) = g(f(c)) = g(1) = a$$

Example 1.4.13

Let S be a finite set and x be an element *not* in S. Let $f:\mathcal{P}(S) \to \mathcal{P}(S \cup \{x\})$ be the function defined by

$$f(T) = T \cup \{x\}$$

for all T in $\mathcal{P}(S)$. Let $g:\mathcal{P}(S \cup \{x\}) \to \mathbf{N}$ be the function defined by

$$g(V) = |V|$$

for all V in $\mathcal{P}(S \cup \{x\})$. Then the composition $g \circ f:\mathcal{P}(S) \to \mathbf{N}$ is given by

$$\begin{aligned}(g \circ f)(T) &= g(f(T)) \\ &= g(T \cup \{x\}) \\ &= |T \cup \{x\}| \\ &= |T| + 1\end{aligned}$$

for all T in $\mathcal{P}(S)$. □

Example 1.4.14

Let A = B = C = **R**, where **R** is the set of all real numbers. Let $f:\mathbf{R} \to \mathbf{R}$ be defined by

$$f(x) = x^2 + 1$$

and let $g:\mathbf{R} \to \mathbf{R}$ be defined by

$$g(x) = 4x - 6$$

Then $g \circ f: \mathbf{R} \rightarrow \mathbf{R}$ is defined by

$$\begin{aligned}(g \circ f)(x) &= g(f(x)) \\ &= g(x^2 + 1) \\ &= 4(x^2 + 1) - 6 \\ &= 4x^2 - 2\end{aligned}$$

It is important to keep in mind that, in general, the composition $g \circ f$ is not the same function as the composition $f \circ g$. In fact, it is usually the case that only one of these compositions can even be defined (why?). However, even if they can both be defined, they are generally not equal. (As an example, try computing the composition $f \circ g$ for the functions of Example 1.4.14.)

If f:A $\rightarrow$ B is a bijective function from A to B, then it is possible to define a function g:B $\rightarrow$ A, from B to A, with the property that $g \circ f$:A $\rightarrow$ A is the identity function on A and $f \circ g$:B $\rightarrow$ B is the identity function on B. In symbols,

$$(g \circ f)(a) = a \tag{1.4.1}$$

for all a in A, and

$$(f \circ g)(b) = b \tag{1.4.2}$$

for all b in B.

Equation 1.4.1 says that if we first apply f to any element a in A and then apply g to that, we return to the original element a. Similarly, Equation 1.4.2 says that if we first apply g to an element b in B and then apply f to that, we again return to the original element b.

In loose terms then, the functions f and g act in opposite manners to each other—g "undoes" whatever f "does," and vice versa.

The function g is called the **inverse** of f and is usually denoted by the symbol f^{-1}. (The inverse of f is pictured in Figure 1.4.5.) We should emphasize that a function f:A $\rightarrow$ B has an inverse f^{-1}:B $\rightarrow$ A if and only if it is bijective.

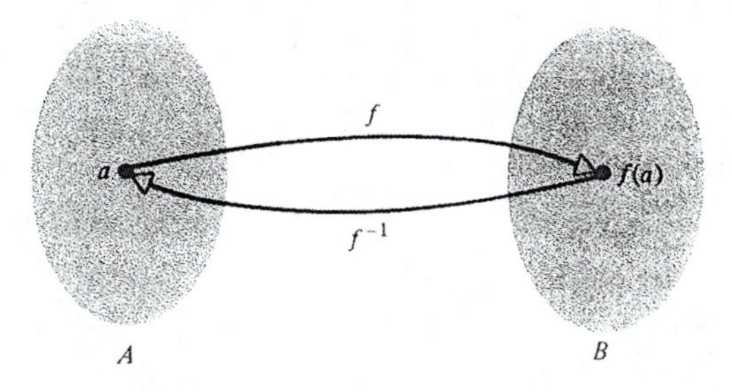

Figure 1.4.5
A picture of the inverse function f^{-1}.

Example 1.4.15

Let A = $\{a, b, c\}$ and B = $\{5, -9, 0\}$. Then the function f:A $\rightarrow$ B defined by

$$f(a) = -9 \quad , \quad f(b) = 0 \quad , \quad f(c) = 5$$

is bijective. Hence, it has an inverse $f^{-1}:\mathrm{B} \rightarrow \mathrm{A}$, which is given by

$$f^{-1}(5) = c \quad , \quad f^{-1}(-9) = a \quad , \quad f^{-1}(0) = b$$

□

■ Example 1.4.16

The functions ASC:Char $\rightarrow \{0, 1, \ldots, 127\}$ and CHR:$\{0, 1, \ldots, 127\} \rightarrow$ Char defined in Example 1.4.4 are inverses of one another. Thus, we can write

$$\mathrm{ACS}(\mathrm{CHR}(n)) = n \quad \text{and} \quad \mathrm{CHR}(\mathrm{ASC}(\alpha)) = \alpha$$

□

□ □ Exercises

In Exercises 1 through 14, decide which of the given rules are functions. For those that are functions, find the image and decide whether or not the function is surjective, injective, or bijective.

1. $f:\{1, 2, 3\} \rightarrow \{1, 2, 3, 4\}$
   $f(1) = 1 \quad , \quad f(2) = 1 \quad , \quad f(3) = 1$
2. $f:\mathbf{N} \rightarrow \mathbf{N}$
   $f(n) = -n^2$ for all $n \in \mathbf{N}$
3. $f:\mathbf{N} \rightarrow \mathbf{N}$
   $f(n) = n + 1$ for all $n \in \mathbf{N}$
4. $f:\mathbf{Z} \rightarrow \mathbf{Z}$
   $f(n) = n + 1$ for all $n \in \mathbf{Z}$
5. $\Sigma = \{0, 1\} \quad , \quad f:\Sigma^* \rightarrow \Sigma$
   $f(w) = 0$ if w contains an odd number of 1's
   $f(w) = 1$ if w contains an even number of 1's
6. $\Sigma = \{0, 1\} \quad , \quad f:\Sigma^* \rightarrow \mathbf{N}$
   $f(w) =$ the number of 1's in the word w
7. Σ is an alphabet, $a \in \Sigma$, $f:\Sigma^* \rightarrow \Sigma^*$,
   $f(w) = aw$, where aw is simply the word obtained by putting the letter a in front of the word w. (This operation is called *juxtaposing* a and w, and we will discuss it in more detail later in the book.)
8. $f:\mathbf{N} \rightarrow$ Char
   $f(n) = \mathrm{CHR}(n + 1)$
9. $f:\mathbf{Q} \rightarrow \mathbf{Q}$
   $f(p/q) = p$ for all $p/q \in \mathbf{Q}$
10. $f:\mathcal{P}(\{0, 1, \ldots, n - 1\}) \rightarrow \mathcal{P}(\{1, 2, \ldots, n\})$
    $f(\mathrm{S}) = \{x \in \{1, 2, \ldots, n\} \mid x - 1 \in \mathrm{S}\}$
11. Let S be a set and suppose that $x \notin \mathrm{S}$.
    $f:\mathcal{P}(\mathrm{S}) \rightarrow \mathcal{P}(\mathrm{S} \cup \{x\})$
    $f(\mathrm{A}) = \mathrm{A} \cup \{x\}$ for all $\mathrm{A} \in \mathcal{P}(\mathrm{S})$
12. Let S be a set and let $x \in \mathrm{S}$.
    $f:\mathcal{P}(\mathrm{S}) \rightarrow \mathcal{P}(\mathrm{S})$
    $f(\mathrm{A}) = \mathrm{A} - \{x\}$ for all $\mathrm{A} \in \mathcal{P}(\mathrm{S})$
13. $f:\mathrm{N} \rightarrow \mathbf{Q}$
    $f(n) = 1/n$ for all $n \in \mathbf{N}$
14. $f:\mathrm{R} \rightarrow \mathbf{R}$
    $$f(x) = \begin{cases} |x| & \text{if} \quad x \geq 0 \\ -|x| & \text{if} \quad x < 0 \end{cases} \quad \text{for all} \quad x \in \mathbf{R}$$
    ($|x|$ is the absolute value of x)

15. Is the function defined in Example 1.4.2 surjective, injective, bijective, or none of these? Justify your answer.

16. Is the function defined in Example 1.4.3 surjective, injective, bijective, or none of these? Justify your answer.

17. Are the functions ASC and CHR defined in Example 1.4.4 surjective, injective, bijective, or none of these? Justify your answer.

18. Is the function defined in Example 1.4.5 surjective, injective, bijective, or none of these? Justify your answer.

19. Is the function defined in Example 1.4.6 surjective, injective, bijective, or none of these? Justify your answer.

20. Is the function defined in Example 1.4.8 surjective, injective, bijective, or none of these? Justify your answer.

21. Let H be the Hamming distance function. Find H(w, u) for the given values of w and u.
a) H(11110, 00001) b) H(101, 010) c) H(101010, 010110)

22. The Hamming distance function can be defined on words over any alphabet. In each case, find H(w, u).
a) H($aabb$, $abab$) b) H(xyz, yxz) c) H($aaaaa$, $bbbbb$)

23. Let $\Sigma = \{0, 1\}$. Is the Hamming distance function $H:\Sigma_n \times \Sigma_n \rightarrow \mathbf{N}$ injective? Explain your answer.

24. Is the Hamming distance function $H:\Sigma_n \times \Sigma_n \rightarrow \mathbf{N}$ surjective? Explain your answer.

25. It is common for rooms to have one overhead light and two wall switches that control the light. If both switches are off, then so is the light. However, regardless of the state of the two switches, whenever one of the switches is changed, the status of the light is changed. Find a transition function that describes this situation.

26. Referring to Example 1.4.14, compute $f \circ g$. Does $f \circ g = g \circ f$?

27. Let $A = \{1, 2, 3\}$, $B = \{a, b, c\}$, $C = \{x, y, z\}$. Let $f:A \rightarrow B$ be defined by

$$f(1) = b \quad , \quad f(2) = f(3) = c$$

Let $g:B \rightarrow C$ be defined by

$$g(a) = g(b) = z \quad , \quad g(c) = y$$

Compute $g \circ f$. Is $f \circ g$ defined? Explain.

28. Let $A = \{1, 2, 3, 4, 5\}$, $B = \{2, 3, 4, 5, 6\}$. Let $f:A \rightarrow B$ be defined by

$$f(a) = a + 1$$

for all $a \in A$, and let $g:B \rightarrow A$ be defined by

$$g(b) = \begin{cases} b & \text{if } b < 5 \\ b - 1 & \text{if } b \geq 5 \end{cases}$$

Compute $g \circ f$ and $f \circ g$.

29. Let $f:\mathbf{N} \rightarrow \mathbf{N}$ be defined by

$$f(n) = n^2 + 1$$

for all $n \in \mathbf{N}$. Compute $f \circ f$.

30. Let $f:\mathbf{N} \rightarrow \mathbf{N}$ be defined by

$$f(n) = 3n + 4$$

Does f have an inverse? If so, find it.

31. Let $f:\mathbf{R}^+ \to \mathbf{R}^+$ be defined by

$$f(x) = x^2$$

for all $x \in \mathbf{R}^+$. ($\mathbf{R}^+$ is the set of all positive real numbers.) Does f have an inverse? If so, find it.

32. Let $f:\mathbf{R} \to \mathbf{R}$ be defined by

$$f(x) = x^2$$

for all $x \in \mathbf{R}$. Does f have an inverse? If so, find it.

33. Show that a function $f:A \to B$ is injective if and only if whenever $f(a_1) = f(a_2)$, then $a_1 = a_2$.

34. Let A and B be finite sets.

a) If there is a bijective function $f:A \to B$ from A to B, what can you say about $|A|$ and $|B|$?

b) If there is a surjective function $g:A \to B$ from A to B, what can you say about $|A|$ and $|B|$?

c) If there is an injective function $h:A \to B$ from A to B, what can you say about $|A|$ and $|B|$?

35. Let A and B be finite sets, with $|A| = |B|$.

a) Show that any surjective function $f:A \to B$ is also injective.

b) Show that any injective function $g:A \to B$ is also surjective.

36. Let $f:A \to B$ be a bijective function. Show that $f^{-1}:B \to B$ is also a bijective function.

1.5 Inductive Proofs and Inductive Definitions

It is quite common in mathematics and its applications to deal with sums of various kinds. (We will see some examples to support this fact a bit later in the section.) For this reason, a special notation has been developed, known as **summation notation**.

As a simple example of summation notation, the sum $1+2+3+\cdots+100$ of the first 100 positive integers can be written in the form

$$1 + 2 + 3 + \cdots + 100 = \sum_{i=1}^{100} i$$

The Greek letter Σ is the equivalent of the letter S and stands for sum. The letter i is simply a variable that ranges over all of the *integers* from 1 to 100. This variable is called the **index of summation**, and the numbers 1 and 100 are called the **lower limit** and **upper limit** of the summation, respectively. Of course, we could use other letters for the index of summation, and the letters i, j, k, n, and m seem to be the most common.

In order to evaluate an expression of the form

$$\sum_{i=l}^{u} p(i)$$

where $p(i)$ is an algebraic expression involving i, we simply substitute consecutive integers for i, starting with the lower limit l and ending with the upper limit u, and then we add the results. Some additional examples should make the idea clear.

Example 1.5.1

a) $\sum_{i=1}^{5} i^2 = 1^2 + 2^2 + 3^2 + 4^2 + 5^2$

b) $\sum_{k=2}^{6} \frac{1}{k(k+1)} = \frac{1}{2(2+1)} + \frac{1}{3(3+1)} + \frac{1}{4(4+1)} + \frac{1}{5(5+1)} + \frac{1}{6(6+1)}$

c) $\sum_{j=-m}^{m} j^n = (-m)^n + (-m+1)^n + \cdots + 0^n + \cdots + (m-1)^n + m^n$

d) The sum of the first five even integers can be written

$$0 + 2 + 4 + 6 + 8 = \sum_{i=0}^{4} 2i$$

e) The sum of the first n positive perfect squares can be written

$$1^2 + 2^2 + \cdots + n^2 = \sum_{i=1}^{n} i^2$$

(A **perfect square** is an integer that can be written in the form k^2 for some integer k.) ▯

Now let us turn to the main topic of this section. There is a very special type of proof technique, known as the technique of **mathematical induction**, that is very effective in proving certain types of mathematical statements, including many that involve sums. Let us first illustrate the technique with some examples.

Suppose we wish to prove that

$$n < 2^n \tag{1.5.1}$$

for all positive integers n. Of course, inequality 1.5.1 is actually an infinite number of different inequalities—one for each value of n—and so we cannot prove it simply by substituting each possible value of n.

But, suppose we can show that

1) inequality 1.5.1 holds for $n = 1$, and

2) *if* inequality 1.5.1 holds for any positive integer $n = k$, then it must also hold for the next larger positive integer $n = k + 1$

Then, by repeatedly using statement 2, we can conclude that inequality 1.5.1 must hold for *all* positive integers n. After all, it holds for $n = 1$, by statement 1. Then statement 2 tells us that 1.5.1 holds for the next larger integer, namely, $n = 2$. Using statement 2 again, we see that 1.5.1 holds for the next larger integer, namely, $n = 3$, and so on. Continuing in this way, we see that 1.5.1 holds for all positive integers n. This is the basic idea behind mathematical induction.

So, let us show that statements 1 and 2 are true. Statement 1 amounts to saying that $1 < 2$, which is certainly true. In order to show that statement 2 is true, we must show that *if* inequality 1.5.1 holds for a positive integer $n = k$, then it also holds for the next larger integer $n = k + 1$. That is, if

$$k < 2^k$$

then

$$k + 1 < 2^{k+1}$$

But this is certainly true, for if $k < 2^k$, then

$$k + 1 < 2^k + 1 < 2^k + 2 \leq 2^{k+1}$$

Hence, statement 2 is also true. Therefore, by the reasoning above, we can conclude that inequality 1.5.1 holds for all positive integers n. This completes the proof by mathematical induction.

Let us try an example involving a sum.

Example 1.5.2

We wish to prove that

$$1 + 2 + 3 + \cdots + n = \frac{n(n+1)}{2} \tag{1.5.2}$$

for all positive integers n. (The left-hand side of Equation 1.5.2 is meant to denote the sum of the positive integers between 1 and n. Thus, for example, if $n = 2$, the left side is $1 + 2$.) Using summation notation, Equation 1.5.2 can be written

$$\sum_{i=1}^{n} i = \frac{n(n+1)}{2}$$

As in the last example, we cannot prove that 1.5.2 holds for all positive integers n by checking all possible values for n. However, if we can show that

1) Equation 1.5.2 holds for $n = 1$, and
2) *if* Equation 1.5.2 holds for any positive integer $n = k$, then it must also hold for the next larger positive integer $n = k + 1$

then Equation 1.5.2 will hold for all positive integers n.

Statement 1 follows easily by substituting $n = 1$ in 1.5.2, which gives $1 = 1$. In order to prove statement 2, we assume that Equation 1.5.2 holds for the integer $n = k$ and show that it must then also hold for $n = k + 1$. In other words, we assume that

$$1 + 2 + 3 + \cdots + k = \frac{k(k+1)}{2} \tag{1.5.3}$$

and show that

$$1 + 2 + 3 + \cdots + (k+1) = \frac{(k+1)(k+2)}{2} \tag{1.5.4}$$

But, using Equation 1.5.3, we have

$$\begin{aligned} 1 + 2 + 3 + \cdots + (k + 1) &= 1 + 2 + 3 + \cdots + k + (k + 1) \\ &= \frac{k(k + 1)}{2} + (k + 1) \\ &= \frac{(k + 1)(k + 2)}{2} \end{aligned}$$

which does show that Equation 1.5.4 holds.

Thus, statement 2 is true, and so we can conclude that Equation 1.5.2 holds for all positive integers n. This completes the proof. □

As you can see from these examples, the technique of mathematical induction applies in situations where we want to prove a whole collection of "propositions" $P(n)$ that depend on the positive integers n. In the first example, $P(n)$ is the proposition that $n < 2^n$, and in the second example, $P(n)$ is the proposition that $1 + 2 + 3 + \cdots + n = n(n + 1)/2$.

Using this terminology, we can describe the **principle of mathematical induction**.

Principle of Mathematical Induction. Suppose that for every positive integer n, $P(n)$ is a proposition that involves n. If we can show that

1) the proposition $P(1)$ is true, and

2) if the proposition $P(k)$ is true for any positive integer k, then so is the proposition $P(k + 1)$

then $P(n)$ must be true for *all* positive integers n.

One of the major points of confusion concerning the principle of mathematical induction is the following. Are we not, in the process of showing that statement 2 is true, in fact *assuming* that $P(k)$ holds for all k? Surely, we cannot be allowed to assume that what we are trying to prove is already true!

The answer to this question is no, we are not assuming that $P(k)$ is true for all k. What we are doing when we show that statement 2 is true is showing that *if* $P(k)$ *happens to be true*, then $P(k + 1)$ is also true. This is quite different from assuming that $P(k)$ is true for all k.

Let us consider another example of the use of mathematical induction. Then we will consider an important application.

Example 1.5.3

Let us prove the formula

$$1 \cdot 2 + 2 \cdot 3 + 3 \cdot 4 + \cdots + n \cdot (n + 1) = \frac{n(n + 1)(n + 2)}{3}$$

In summation notation, this can be written

$$\sum_{i=1}^{n} i(i + 1) = \frac{n(n + 1)(n + 2)}{3} \tag{1.5.5}$$

In order to prove this by mathematical induction, we let P(n) be the proposition that Equation 1.5.5 holds. Then P(1) is the proposition that Equation 1.5.5 holds for $n = 1$; that is, that

$$1 \cdot 2 = \frac{1(1 + 1)(1 + 2)}{3}$$

which is certainly true. Now we must show that if P(k) is true, then so is P($k + 1$). That is, we must show that if

$$\sum_{i=1}^{k} i(i + 1) = \frac{k(k + 1)(k + 2)}{3} \tag{1.5.6}$$

then

$$\sum_{i=1}^{k+1} i(i + 1) = \frac{(k + 1)(k + 2)(k + 3)}{3} \tag{1.5.7}$$

But, using Equation 1.5.6, we have

$$\sum_{i=1}^{k+1} i(i + 1) = \sum_{i=1}^{k} i(i + 1) + (k + 1)(k + 2)$$

(We have "broken off" the last term in the sum, reducing the upper index of summation by 1. Of course, we must then add on this last term, obtained by setting $i = k + 1$.)

$$= \frac{k(k + 1)(k + 2)}{3} + (k + 1)(k + 2)$$

$$= \frac{k(k + 1)(k + 2) + 3(k + 1)(k + 2)}{3}$$

$$= \frac{(k + 1)(k + 2)(k + 3)}{3}$$

which gives Equation 1.5.7. Thus, if P(k) is true, then so is P($k + 1$), and according to the principle of mathematical induction, P(n) is true for all positive integers n. This concludes the proof by induction. ∎

At this point, you may be wondering why we are interested in sums of the form 1.5.2 and 1.5.5. The reason is that they occur in various applications. One such application comes in determining the value of certain variables in a computer program. Let us illustrate this with some examples.

Example 1.5.4

Consider the computer program segment

```
10 X := 0;
20 FOR I := 1 TO 100 DO
30          X := X + I;
40 PRINT X
```

What value will be printed for the variable X?

We can answer this question by reasoning as follows. When I = 1, instruction 30 makes X = 1. When I = 2, instruction 30 makes X := X + I = 1 + 2. When I = 3, instruction 30 makes X := X + I = 1 + 2 + 3, and so on. From this pattern we can see that when I = 100, instruction 30 will make

$$X = 1 + 2 + 3 + \cdots + 100$$

But, according to the formula 1.5.2, which we proved by induction, this sum is equal to 100(101)/2 = 5050, and so the computer will print the value X = 5050. (Try it for yourself!) □

Example 1.5.5

Consider the program segment

```
X := 0;
FOR I := 1 TO N DO
    FOR J := 1 TO I DO
        X := X + J;
PRINT X
```

What value will the computer "print" for the variable X? (Of course, the answer in this case will be expressed in terms of N.)

In this case, when I = 1, we have X = 1, when I = 2, we have X = 1 + (1 + 2), and when I = 3, we get X = 1 + (1 + 2) + (1 + 2 + 3). From this pattern, we see that when I = N,

$$X = 1 + (1 + 2) + (1 + 2 + 3) + \cdots + (1 + 2 + \cdots + N)$$

Now, we can use Equation 1.5.2 several times to write this sum in the form

$$X = \frac{1 \cdot 2}{2} + \frac{2 \cdot 3}{2} + \frac{3 \cdot 4}{2} + \cdots + \frac{N(N + 1)}{2}$$

$$= \left(\frac{1}{2}\right)[1 \cdot 2 + 2 \cdot 3 + 3 \cdot 4 + \cdots + N(N + 1)]$$

and this can be simplified further by using Equation 1.5.5, to give

$$X = \frac{N(N + 1)(N + 2)}{6}$$

□

Sometimes we want to prove that the propositions $P(n)$ are true, where n ranges over all integers greater than or equal to a specific integer m, which could be positive, zero, or negative. We can easily modify the principle of mathematical induction to accommodate this situation as follows. If we can show that

1) the proposition $P(m)$ is true, and

2) if the proposition $P(k)$ is true for any integer $k \geq m$ then so is the proposition $P(k + 1)$

then $P(n)$ must be true for all integers $n \geq m$.

Let us illustrate this for the case $m = 0$ by proving that any set that contains exactly n elements has exactly 2^n subsets.

■ Theorem 1.5.2

Any set of size n has exactly 2^n subsets.

PROOF For any $n \geq 0$, let $P(n)$ be the proposition that any set of size n has exactly n elements. Then $P(0)$ is the proposition that any set of size 0 has exactly $2^0 = 1$ subset. But the only set of size 0 is the empty set $\emptyset$, which does indeed have exactly 1 subset, namely, the empty set itself. Hence, $P(0)$ is true.

Now we must show that if $P(k)$ is true, then so is $P(k + 1)$. That is, we must show that if any set of size k has exactly 2^k subsets, then any set of size $k + 1$ has exactly 2^{k+1} subsets.

So let $S = \{x_1, x_2, \ldots, x_{k+1}\}$ be any set of size $k + 1$. Then, we may classify all of the subsets of S into two groups—those subsets that contain the element x_{k+1}, and those that do not. Now, the subsets of S that do not contain the element x_{k+1} are precisely the subsets of the set $\{x_1, x_2, \ldots, x_k\}$. Furthermore, since this set has size k, we may use the fact that $P(k)$ is true to deduce that there are 2^k such subsets. In other words, there are 2^k subsets of S that do not contain the element x_{k+1}.

On the other hand, the number of subsets of S that contain the element x_{k+1} is the same as the number of subsets of S that do not contain x_{k+1}. One way to see this is to observe that if we take all of the subsets of S that do not contain x_{k+1} and add x_{k+1} to each of these subsets, then we will get a complete list of all of the subsets of S that do contain the element x_{k+1}.

Hence, there are also 2^k subsets of S that do contain the element x_{k+1}, and so the total number of subsets of S is $2^k + 2^k = 2^{k+1}$. Thus, $P(k + 1)$ is true, and so the theorem is proved. ■

Recall from Section 1.1 that if A and B are subsets of a universal set U, then one of De Morgan's laws says that

$$(A \cup B)^c = A^c \cap B^c$$

The principle of mathematical induction is very useful in extending equations such as this from two sets A and B to an arbitrary (finite) number of sets $A_1, A_2, \ldots, A_n$. In particular, we have the following theorem.

■ Theorem 1.5.3

If $A_1, A_2, \ldots, A_n$ are subsets of the universal set U (where $n \geq 2$), then

$$(A_1 \cup A_2 \cup \cdots \cup A_n)^c = A_1^c \cap A_2^c \cap \cdots \cap A_n^c \tag{1.5.8}$$

PROOF In order to prove this by induction, we let $P(n)$ be the proposition that Equation 1.5.8 holds. Then, $P(2)$ is the proposition that

$$(A_1 \cup A_2)^c = A_1^c \cap A_2^c$$

which is just De Morgan's law. Hence, $P(2)$ is true.

Now we must show that if $P(k)$ is true, then so is $P(k + 1)$. First, we have

$$(A_1 \cup A_2 \cup \cdots \cup A_{k+1})^c = ([A_1 \cup A_2 \cup \cdots \cup A_k] \cup A_{k+1})^c$$

Using De Morgan's law on the right side of this equation (with $A = A_1 \cup A_2 \cup \cdots \cup A_k$ and $B = A_{k+1}$), we get

$$(A_1 \cup A_2 \cup \cdots \cup A_{k+1})^c = [A_1 \cup A_2 \cup \cdots \cup A_k]^c \cap A_{k+1}^c \tag{1.5.9}$$

Now, since $P(k)$ is true, we have $[A_1 \cup A_2 \cup \cdots \cup A_k]^c = A_1^c \cap A_2^c \cap \cdots \cap A_k^c$, and so Equation 1.5.9 can be written

$$(A_1 \cup A_2 \cup \cdots \cup A_{k+1})^c = [A_1^c \cap A_2^c \cap \cdots \cap A_k^c] \cap A_{k+1}^c$$

or

$$(A_1 \cup A_2 \cup \cdots \cup A_{k+1})^c = A_1^c \cap A_2^c \cap \cdots \cap A_{k+1}^c$$

which is $P(k + 1)$. Thus, if $P(k)$ is true, so is $P(k + 1)$. This completes the proof by induction. ∎

There is another form of the principle of mathematical induction, known as the *strong form,* that is sometimes more useful than the form we have discussed.

The Strong Form of the Principle of Mathematical Induction. Let m be an integer. Suppose that for every integer $n \geq m$, $P(n)$ is a proposition that involves n. If we can show that

1) the proposition $P(m)$ is true
2) if the propositions $P(m), \ldots, P(k)$ are true for any $k \geq m$, then so is the proposition $P(k + 1)$

then $P(n)$ must be true for all integers $n \geq m$.

The difference between the strong form of induction and the previous form (sometimes called the *weak form*) is that in the strong form, in order to prove that $P(k + 1)$ is true, we assume not just that the previous proposition $P(k)$ is true, but that *all* of the previous propositions $P(m), \ldots, P(k)$ are true. We should point out, however, that the strong form of induction is equivalent to the weak form. Put another way, either form of induction can be used to establish the other.

Example 1.5.6

An integer $p \geq 2$ is **prime** if it has no positive integral factors other than 1 and itself. Thus, a prime number is a positive integer that cannot be factored into the product of two integers both greater than 1. The first few prime numbers are 2, 3, 5, 7, 11, 13, 17, and 19.

One of the most fundamental properties of the integers is the fact that any integer $n \geq 2$ can be written as a product of prime numbers. (Actually, this factoring of n into prime numbers is *unique*, except for the order of the factors, but we will not worry about that here.)

In order to prove this fact about integers, we use the strong form of mathematical induction. Let $P(n)$ be the proposition that the integer n can be written as a product of prime numbers. Then P(2) is certainly true, since the number 2 is itself prime. So let us assume that P(2), . . ., P(k) are true, and show that P(k + 1) must then also be true.

The proposition P(k + 1) is the proposition that $k + 1$ can be written as a product of prime numbers. If $k + 1$ is prime, then P(k + 1) is certainly true. However, if $k + 1$ is not prime, then it can be factored

$$k + 1 = a \cdot b$$

where a and b are integers satisfying

$$2 \leq a \leq k,\ 2 \leq b \leq k$$

Hence, according to our assumption, P(a) and P(b) are both true. In other words, a and b can both be written as a product of prime numbers, say

$$a = p_1 p_2 \cdots p_i,\ b = q_1 q_2 \cdots q_j$$

But then

$$k + 1 = a \cdot b = p_1 p_2 \cdots p_i q_1 q_2 \cdots q_j$$

is a product of primes, showing that P(k + 1) is true. Thus, by the strong form of mathematical induction, we may conclude that P(n) is true for all $n \geq 2$. This completes the proof.

Notice that the weak form of induction would not have been sufficient for this example. The reason is that assuming only that P(k) is true does not tell us whether P(a) or P(b) are true, which is what we need in this case. Hence, we must use the strong form. □

Example 1.5.7

Suppose that the post office issued only 3¢ and 5¢ stamps. The strong form of induction can be used to prove that any amount of postage over 7 cents can be obtained with an appropriate combination of such stamps.

Let P(n) be the proposition that postage of n cents can be obtained using 3¢ and 5¢ stamps, where $n \geq 8$. It is easy to see that postage of 8 cents, 9 cents, or 10 cents can be obtained. Hence, P(8), P(9) and P(10) are true. (As you will see, we require all three for the coming discussion.)

Now let us suppose that P(8), . . ., P(k) are true, where we may assume that $k \geq 10$, and show that P(k + 1) must also be true. Since $k \geq 10$, we have $8 \leq k - 2 \leq k$, and so according to our assumption, P(k − 2) is true. That is, postage of $k - 2$ cents can be obtained using 3¢ and 5¢ stamps. Hence, by including an additional 3¢ stamp, we can make postage of $k - 2 + 3 = k + 1$ cents. Thus, P(k + 1) is true, and according to the strong form of induction, P(n) is true for all $n \geq 8$. This completes the proof by induction.

We should observe that, as is common with proofs by mathematical induction, the proof does not show us *how* to combine stamps to obtain any given amount of postage over 7 cents—it only tells us that it can be done. □

Even though the principle of mathematical induction applies to a very specific type of problem, it is still one of the most powerful techniques that we have. For, in a sense, it is a technique that allows us to prove an infinite number of different results at one time.

A concept related to the proof by induction is *definition by induction*. This type of definition is frequently used in mathematics and computer science to define the elements of a particular set. As a simple example, we can define the set of positive even integers inductively as follows.

Definition

The set $\mathbf{E}^+$ of positive even integers is the set defined by the following conditions.

1) $0 \in \mathbf{E}^+$

2) If n is in $\mathbf{E}^+$, then so is $n + 2$

3) Nothing is in $\mathbf{E}^+$ unless it can be obtained by a finite number of applications of statements 1 and 2. □

As you can see, this definition tells us exactly how to determine what the elements of $\mathbf{E}^+$ are. Statement 1 tells us that 0 is an element of $\mathbf{E}^+$. Statement 2 tells us that if n is an element of $\mathbf{E}^+$, then so is $n + 2$. Putting these two statements together, we can conclude that 0, 2, 4, 6, . . ., and so on, are elements of $\mathbf{E}^+$. The third statement is a *limiting* statement, in that it tells us that no other "objects" are in the set $\mathbf{E}^+$ other than those that can be obtained using statements 1 and 2 a finite number of times.

Now let us give a formal definition of an inductive definition.

Definition

An **inductive definition** consists of three parts.

1) *A basis clause*. This tells us that certain elements belong to the set in question. The basis clause "gets us started."

2) *An inductive clause*. This tells us how to use elements that are in the set to get other elements that are in the set.

3) *A limiting clause*. This tells us that the only elements in the set are those that can be obtained using statements 1 and 2. □

Let us consider some additional examples of inductive definitions.

Example 1.5.8

Let $\Sigma = \{0, 1\}$. Then we can define the set Σ^* of all binary words inductively as follows.

1) The empty word θ is in Σ^*

2) If w is in Σ^* then so are $0w$ and $1w$.

3) Nothing is in Σ^* unless it can be obtained by a finite number of applications of statements 1 and 2.

(If w is a binary word, then by $0w$, we mean the binary word obtained by putting 0 in front of w, and similarly for $1w$.) □

Example 1.5.9

Intuitively speaking, an *arithmetic expression* is an expression formed by adding, subtracting, multiplying, and dividing a finite number of real numbers. Of course, parentheses may be needed in order to convey the intended meaning of the expression. [For example, the expression $2 - (3 + 4)$ has a different meaning than the expression $(2 - 3) + 4$.] The concept of an arithmetic expression is best defined using an inductive definition as follows.

Definition

The set S of all arithmetic expressions can be defined as follows. Let **R** be the set of all real numbers.

1) If $r \in \mathbf{R}$ then $r \in S$.
2) If $x, y \in S$ then
(i) $(+x) \in S$
(ii) $(-x) \in S$
(iii) $(x + y) \in S$
(iv) $(x - y) \in S$
(v) $(x/y) \in S$, provided $y \neq 0$.
(vi) $(x \cdot y) \in S$
3) Nothing is in S unless it can be obtained by a finite number of applications of statements 1 and 2.

Exercises

In Exercises 1 through 7, write the given summation without using summation notation.

1. $\sum_{i=0}^{5} i^5$

2. $\sum_{i=1}^{4} i^i$

3. $\sum_{j=-6}^{6} (-1)^j$

4. $\sum_{k=0}^{8} (-1)^k \frac{k}{k+1}$

5. $\left(\sum_{i=0}^{3} i\right)\left(\sum_{j=0}^{3} j\right)$

6. $\sum_{i=1}^{4} \left(\sum_{j=1}^{3} i^j\right)$

7. $\sum_{i=1}^{2} \left(\sum_{j=1}^{2} \left(\sum_{k=1}^{2} (i + j + k)\right)\right)$

In Exercises 8 through 15, write the given summation using summation notation.

8. $0 + 1 + 2 + 3 + 4 + 5$
9. $0 + 2 + 4 + 6 + 8 + 10$
10. $1 + 3 + 5 + 7 + 9 + 11$
11. The sum of the first 100 even nonnegative integers.
12. The sum of the first 10 nonnegative perfect cubes. (A number is a **perfect cube** if it has the form n^3, where n is an integer.)
13. The sum of n consecutive powers of 2, starting with 1.
14. $17 + 31 + \cdots + (2n^2 - 1)$
15. $\frac{1}{4} + \frac{1}{36} + \cdots + \frac{1}{m^2(m-1)^2}$

In Exercises 16 through 26, prove the given propositions by using mathematical induction.

16. $\sum_{k=1}^{n} (2k - 1) = n^2$

17. $1^2 + 2^2 + 3^2 + \cdots + n^2 = \dfrac{n(n+1)(2n+1)}{6}$

18. $\sum_{i=1}^{n} i^3 = \dfrac{n^2(n+1)^2}{4}$

19. $\dfrac{1}{1 \cdot 2} + \dfrac{1}{2 \cdot 3} + \dfrac{1}{3 \cdot 4} + \cdots + \dfrac{1}{n(n+1)} = \dfrac{n}{n+1}$

20. $\sum_{k=1}^{n} \dfrac{1}{(2k-1)(2k+1)} = \dfrac{n}{2n+1}$

21. $\sum_{j=1}^{n} j^3 = \left(\sum_{j=1}^{n} j\right)^2$

22. $1 + 2 + 2^2 + \cdots + 2^n = 2^{n+1} - 1$

23. $1^2 \cdot 2 + 2^2 \cdot 3 + 3^2 \cdot 4 + \cdots + n^2(n+1) = \dfrac{n(n+1)(3n^2 + 7n + 2)}{12}$

24. $\sum_{i=1}^{n} (-1)^i = \dfrac{(-1)^n - 1}{2}$

25. $1 \cdot 2^0 + 2 \cdot 2^1 + 3 \cdot 2^2 + \cdots + n \cdot 2^{n-1} = (n-1) \cdot 2^n + 1$

26. $\sum_{j=0}^{n} r^j = \dfrac{r^{n+1} - 1}{r - 1} \qquad (r \neq 1)$

27. Use induction to prove that if $A_1, A_2, \ldots, A_n$ are subsets of a universal set U, then

$$(A_1 \cap A_2 \cap \cdots \cap A_n)^c = A_1^c \cup A_2^c \cup \cdots \cup A_n^c$$

28. Using the fact that $|x + y| \leq |x| + |y|$ for all real numbers, prove by induction that

$$|x_1 + x_2 + \cdots + x_n| \leq |x_1| + |x_2| + \cdots + |x_n|$$

for all real numbers $x_1, x_2, \ldots, x_n$.

*In Exercises 29 through 38, determine the value that a computer would print for the variable X. The notation A * B stands for the product of A and B. The notation ^ means exponentiation; that is, A ^ B stands for A^B. You may need to refer to Exercises 16 through 26.*

29.
```
X := 0;
FOR I := 1 TO 1000 DO
    FOR J := I to 1000 DO
        X := X + 1;
PRINT X
```

30.
```
X := 0;
FOR I := 1 TO N DO
    X := X + 2 ^ I;
PRINT X
```

31.
```
X := 0;
FOR I := 1 TO N DO
    FOR J := 1 TO M DO
        X := X + 1;
PRINT X
```

32.
```
X := 0;
FOR I := 1 TO 10 DO
    X := X + I ^ 2;
PRINT X
```

33.
```
X := 0;
FOR I := 1 TO N DO
    X := X + I + I ^ 2;
PRINT X
```

34.
```
X := 0;
FOR I := 1 TO N DO
    FOR J := I TO N DO
        X := X + J;
PRINT X
```

35.
```
X := 0;
FOR I := 1 TO N DO
    FOR J := 1 TO I DO
        X := X + I;
PRINT X
```

36.
```
X := 0;
FOR I := 1 TO N DO
    FOR J := 1 TO I DO
        X := X + I * J;
PRINT X
```

37.
```
X := 0;
FOR I := 1 TO N DO
    X := X + (-1) ^ I;
PRINT X
```

38.
```
X := 0;
FOR I := 1 TO N DO
    FOR J := 1 TO I DO
        X := X + I * (2 ^ J);
PRINT X
```

39. Show that any amount of postage over 3 cents can be obtained by combining 2¢ and 5¢ stamps.

40. Show that any number greater than 23 can be written as a sum of 5's and/or 7's.

41. Consider the sequence of numbers $a_0, a_1, a_2, \ldots$ with the property that $a_0 = 1$, $a_1 = 1$ and

$$a_n = a_{n-1} + a_{n-2} \text{ for } n \geq 2$$

In words, each number (beginning with the third number) is the sum of the two previous numbers. These numbers are called the **Fibonacci numbers**. The first few Fibonacci numbers are 1, 1, 2, 3, 5, 8 and 13. Show that $a_n < 2^n$ for all $n \geq 1$.

In Exercises 42 through 48, give an inductive definition for the indicated set.

42. The set of all positive odd integers.

43. The set of all negative even integers.

44. The set of all even integers.

45. The set of all integers.

46. The set of all words over a given alphabet Σ.

47. The set of all words over Σ that contain a particular element $a \in \Sigma$.

48. Let Σ be the set of all letters and digits, along with the decimal point ".". In Microsoft BASIC (for the IBM Personal Computer), a variable name consists of a word over Σ whose first character must be a letter. Give an inductive definition of the set of variable names in Microsoft BASIC.

49. Referring to the previous exercise, suppose that we are further limited in our choice of variable names by the fact that no name can have length greater than 40. Give an inductive definition of the set of all variable names in this case.

1.6

Proof by Contradiction

Proofs play a very important role in mathematics. In fact, it is fair to say that, if it were not for proofs, there would be very little, if any, mathematics. The reason for this is quite simple. We cannot always trust our intuition when it comes to deciding what is true and what is not true.

Section 1.3 provides a good example of this. If we had let our intuition be our only guide, we would probably have concluded that the set **E** of even natural numbers was smaller than the set **N** of all natural numbers. However, as we *proved* in Example 1.3.1, these two sets actually have the same size.

Even though there is no magic formula for finding proofs, there are some general techniques that can be very helpful. We discussed one of these techniques in the previous section. However, the technique of mathematical induction is rather limited in scope. In this section, we want to discuss another technique, called **proof by contradiction**, that can be applied to more general situations.

Perhaps the best way to describe this technique is by giving some examples of how it is used. The idea behind a proof by contradiction is to *assume* that the conclusion we want to prove is *false* and derive from this a contradiction to something that we know is true. If we can succeed in doing this, then we will have shown that the conclusion must have been true.

Consider the following statements, which we put into the form of theorems.

Theorem 1.6.1

The product of any two rational numbers is a rational number.

Theorem 1.6.2

The product of any nonzero rational number and any irrational number is an irrational number.

We will give a proof of the first theorem that does not use the method of contradiction and a proof by contradiction of the second theorem.

PROOF OF THEOREM 1.6.1 Let r_1 and r_2 be rational numbers. Then, according to the definition of rational number, r_1 and r_2 have the form

$$r_1 = \frac{p_1}{q_1} \quad \text{and} \quad r_2 = \frac{p_2}{q_2}$$

where p_1, p_2, q_1, and q_2 are integers, and q_1 and q_2 are not equal to zero. But then

$$r_1 r_2 = \frac{p_1}{q_1}\frac{p_2}{q_2} = \frac{p_1 p_2}{q_1 q_2}$$

and so the product $r_1 r_2$ is a rational number. This concludes the proof. ▮

PROOF OF THEOREM 1.6.2 Let us assume that it is *not* true that the product of any nonzero rational number and any irrational number is irrational. Put another way, let us assume that there is a nonzero rational number r, and an irrational number x, with the property that the product rx is a rational number. Using this assumption, we want to derive a contradiction.

Now, since the product rx is rational, it has the form

$$rx = \frac{p}{q}$$

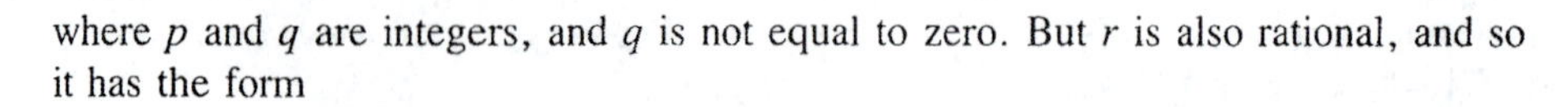

where p and q are integers, and q is not equal to zero. But r is also rational, and so it has the form

$$r = \frac{m}{n}$$

where m and n are integers, both different from zero. Substituting this into the first equation gives

$$\frac{m}{n} x = \frac{p}{q}$$

Multiplying both sides of this by n and dividing both sides by m gives

$$x = \frac{np}{mq}$$

But this says that x is a rational number, which is a contradiction to the fact that x is an irrational number! (After all, a number cannot be both rational and irrational.)

Because we have arrived at a contradiction, the fault must be in the only assumption that we have made, namely, that there is a rational number and an irrational number whose product is rational. Thus, we can conclude that the product of any rational number and any irrational number must be irrational. This concludes the proof. ■

As you can see, there is quite a bit of difference between these two proofs. In the first one, we made no assumptions, whereas in the second one, we assumed that what we wanted to show was false! This assumption led us to a contradiction, and so we concluded that what we wanted to show must have been true after all.

Let us do another example of a proof by contradiction.

■ Theorem 1.6.3

For all positive integers n, we have

$$n < 2^n \tag{1.6.1}$$

PROOF Let us assume that 1.6.1 does *not* hold for all positive integers n and try to arrive at a contradiction.

If 1.6.1 does not hold for all positive integers, then there must be a *smallest* positive integer for which it does not hold. Let us denote this integer by k. Clearly, we have $k \geq 2$. Thus, we have

$$2^k \leq k \tag{1.6.2}$$

Since k is the *smallest* such integer and since $k - 1 > 0$, we also have

$$k - 1 < 2^{k-1} \tag{1.6.3}$$

Adding 1 to both sides of the inequality 1.6.3 gives

$$k < 2^{k-1} + 1$$

and combining this with 1.6.2, we get

$$2^k < 2^{k-1} + 1$$

Subtracting 2^{k-1} from both sides of this gives

$$2^k - 2^{k-1} < 1$$

But since $2^k - 2^{k-1} = 2^{k-1}(2 - 1) = 2^{k-1}$, this becomes

$$2^{k-1} < 1$$

which is certainly not true for any positive integer k. This is the contradiction that we have been looking for, and so the assumption that the inequality 1.6.1 does not hold must be in error. In other words, 1.6.1 does hold, and the proof is complete. ∎

Exercises

In Exercises 1 through 4, prove the given theorem in two ways, one that uses the technique of proof by contradiction and one that does not.

1. *Theorem* If $0 < x < y$ then $x^2 < y^2$.
2. *Theorem* If an integer n is a multiple of 4, then it is also a multiple of 2.
3. *Theorem* If $x^2 = x$, and $x \neq 1$, then $x = 0$.
4. *Theorem* $n^2 + 1 < (n + 1)^2$ for all positive integers n.
5. Give a proof by contradiction of the following theorem.
   Theorem $n < 3^n - 1$ for all positive integers n.
6. Prove the following theorem by contradiction.
   Theorem For all real numbers x and y,

   $$|x + y| \leq \max\{ 2|x|, 2|y| \}$$

   where $\max\{a, b\}$ stands for the maximum of the numbers a and b.
   Hint, use the fact that $|x + y| \leq |x| + |y|$.
7. Prove the following theorem by contradiction.
   Theorem Given any positive real number M, and any positive real number e, there is a natural number n with the property that $ne > M$.
   You may use the fact that for any real number r, there is a natural number k for which $k > r$. Incidentally, the property stated in this theorem is known as the **Archimedean property of the real numbers.**
8. a) Give a proof that does not use contradiction of the fact that the product of two even integers is even. *Hint*, an integer n is even if and only if it can be written in the form $n = 2k$ for some integer k.
   b) Give a proof that does not use contradiction of the fact that the product of two odd integers is odd. *Hint*, an integer n is odd if and only if it can be written in the form $n = 2k + 1$ for some integer k.
9. Give a proof by contradiction of the following theorem.
   Theorem There is no rational number r for which $r^2 = 2$.
   Hints:
   1) assume that there is a rational number $r = p/q$ where p and q have no common factors, with the property that $(p/q)^2 = 2$.

2) use this to show that p^2 is even.
3) use the results of the last exercise to show that p must also be even.
4) use the fact that p is even, and that $(p/q)^2 = 2$, to show that q^2 is even.
5) deduce from this that q is even.
6) show that this is a contradiction.

He was in Logic, a great critic,
Profoundly skill'd in Analytic;
He could distinguish and divide
A hair 'twixt south and south-west side.
—Butler, *Hudibras*

Logic and Logic Circuits

2.1 Statements, Connectives, and Symbolic Language

It is certainly true that anyone who hopes to be successful in mathematics or science must be able to think logically. After all, both mathematics and science are based on our ability to create sound, logical arguments. But how do we know whether or not a given argument is logically sound?

The answer to this question lies in a branch of mathematics known as logic. It is this branch of mathematics that tells us what constitutes a sound, logical argument, and what does not. Of course, it is up to us to find the arguments in the first place, but once we have found them, the principles of mathematical logic can be used to decide whether or not they are logically sound.

Also, it may surprise you to learn that logic has some very important *practical* applications, for example, to the design of so-called *logic circuits*, which are used in computers. In this chapter, we will take a look not only at those aspects of logic that will help us recognize logical arguments but also at the application of logic to circuit design.

Let us begin by making a few simple definitions. A **statement** is any sentence that has a *truth value*, that is, any sentence that is either true or false.

Of course, not all sentences have a truth value, and so not all sentences are statements. For example, the sentences "Go home" and "Is it raining?" do not have a truth value, and so they are not statements. However, sentences such as "I have a home" and "It is raining" do have a truth value, and so they are statements.

Statements can be divided into two types. The first type is called a **simple statement**. Some examples of simple statements are

1) Today is Monday.
2) John is 4 years old.
3) It is raining.
4) $3 + 4 = 7$
5) $1 + 1 = 5$

Notice that each of these sentences has a truth value, and so each is a statement. Notice also that the truth value of a statement may change as, for example, with statement 1, which is true on Mondays and false on the other days.

The second type of statement is the **compound statement**, which is made up of one or more simple statements that are linked together by **connectives**. The connectives that we will be considering are

not

and

or (2.1.1)

if . . . then . . .

if and only if

We will refer to these connectives as the *five basic connectives*. The following are examples of compound statements.

1) It is not Monday.
2) It is raining and the sky is cloudy.
3) If today is a weekday, then I must not go to work.
4) $2x = 2$ if and only if $x = 1$
5) If $x = 2$ and $y = 3$ then $x + y = 100$
6) The sun is not shining if and only if tomorrow is Friday and there are no clouds in the sky.

(As you can see from some of these examples, a compound statement does not have to make much sense.)

We can make the intuitive idea of a compound statement more precise by giving an inductive definition.

Definition

Compound statements can be defined as follows.

1) All simple statements are compound statements.
2) If S and T are compound statements, then

(i) (not S) is a compound statement
(ii) (S and T) is a compound statement
(iii) (S or T) is a compound statement
(iv) (if S then T) is a compound statement
(v) (S if and only if T) is a compound statement

3) Nothing is a compound statement unless it can be obtained by a finite number of applications of statements 1 and 2 above. □

Now, mathematical logic can tell us nothing about whether a *simple* statement is true or false. This we must decide by other means. However, one of the main goals of logic is to tell us whether a *compound* statement is true or false, once we know the truth values of the simple statements that make it up.

For example, consider the compound statement

Today is Tuesday and it is raining. (2.1.2)

Mathematical logic will not tell us whether the simple statements "Today is Tuesday" and "It is raining" are true or false. That is up to us to determine. However, once we have decided whether these simple statements are true or false, logic will tell us whether Compound Statement 2.1.2 is true or false.

In this particular case, as we will see, the principles of mathematical logic tell us that Compound Statement 2.1.2 is true when, and only when, *both* of the simple statements that make it up are true. Of course, this is just what we would expect from our intuitive understanding of the word "and," but mathematical logic gives this connective a precise meaning.

One of our goals then is to give precise meanings to each of the five basic connectives, so that we will be able to tell whether a given compound statement is true or false, based on the truth values of the simple statements that make it up.

Before doing this, however, we need to find a more efficient way to express compound statements. Consider the statement

I am a mathematician and $2 + 2 = 5$. (2.1.3)

Logic tells us the same thing about this statement as it does about Statement 2.1.2, namely, that it is true when, and only when, *both* of the simple statements that make it up are true.

The reason for this is that Statements 2.1.2 and 2.1.3 have the same *logical structure*. This structure can be exhibited by writing these statements in the *symbolic form*

p and q (2.1.4)

where p and q are simple statements.

Now, from the point of view of mathematical logic, it is the logical structure of a compound statement that is important, and not the individual simple statements that make it up. Put another way, if two compound statements have the same symbolic form, as is the case with Statements 2.1.2 and 2.1.3, then from the point of view of mathematical logic, they may as well be the same statement.

Furthermore, once a compound statement has been written in symbolic form, it becomes very easy to determine its truth value. This is usually done in a table, called the **truth table** for the compound statement. For example, the truth table for the Compound Statement 2.1.4 is

p	q	p *and* q
T	T	T
T	F	F
F	T	F
F	F	F

Notice that we have used the letter T to stand for true and the letter F to stand for false. This table gives us the truth value of the Compound Statement 2.1.4 for each of the possibilities of the truth values of the simple statements p and q. As an example, if we happen to know that p is true and q is false, then the second row of this table tells us that the statement "p and q" is false. In a sense, this truth table is actually giving us the logical meaning of the connective "and."

We will devote the rest of this section to the question of how to translate compound statements from English into symbolic form, and vice versa. Then, in the next section, we will learn how to find the truth tables of compound statements.

It is customary to represent simple statements by letters, such as p, q, r, s, and so on. It is also customary to use a special notation for each of the connectives. For easy reference, let us put this notation in a table. Each of the connectives also has a name, which we include in the table.

Connective	*Notation*	*Name of Connective*
not	$\sim$	negation
and	$\wedge$	conjunction
or	$\vee$	disjunction
if . . . then . . .	$\rightarrow$	conditional
if and only if	$\leftrightarrow$	biconditional

Before turning to examples, we should point out that the statements

p implies q

and

p only if q

have the same logical meaning as the statement

if p then q

Thus, for example, the following statements

If it is sunny then I will go to school

It is sunny implies that I will go to school

It is sunny only if I will go to school

all have the same meaning.

Example 2.1.1

Let us begin with some examples of statements and their negations.

Statement	*Negation*
John is sick	John is not sick
Some dogs have fleas	No dogs have fleas
$2 + 5 = 7$	$2 + 5 \neq 7$
There exists an integer that is not divisible by 5	All integers are divisible by 5

□

Incidentally, the statements

All men are animals,

Every man is an animal,

Each man is an animal,

Any man is an animal,

all have the same meaning, and the negation of any one of these statements can be written

Not all men are animals,

or

Some men are not animals,

or

There exists a man who is not an animal.

Example 2.1.2

The conjunction of the two statements "John is sick" and "Some dogs have fleas" is the statement

John is sick and some dogs have fleas;

and the disjunction of these two statements is

John is sick or some dogs have fleas. □

Example 2.1.3

In order to translate the statement

John is at the library or he is not at the bookstore (2.1.3)

into symbolic form, we let p be the simple statement "John is at the library," and we let q be the simple statement "He is at the bookstore." Then, the symbolic form of statement 2.1.3 is

$$p \vee \sim q$$

□

Example 2.1.4

In order to translate the statement

If today is Monday or it is not raining,
then we can go to the store (2.1.4)

into symbolic form, we let p be the simple statement "Today is Monday," we let q be the simple statement "It is raining," and we let r be the simple statement "We can go to the store." Then the symbolic form of Statement 2.1.4 is

$$(p \vee \sim q) \rightarrow r$$

Notice that parentheses are needed in this case in order to distinguish it from $p \vee (\sim q \rightarrow r)$, which is the symbolic form of the statement "Today is Monday, or if it is not raining, then we can go to the store." □

Example 2.1.5

Consider the statement

It is not true that today is Friday or we are unhappy.

Letting p be the statement "Today is Friday" and q be the statement "We are happy," the symbolic form of this statement is

$$\sim(p \vee \sim q)$$

Strictly speaking, if we were to translate this back into English, it would read "It is not true that today is Friday or we are *not* happy." However, there is essentially no difference between being "not happy" and being "unhappy." This example shows that it is acceptable to make modifications in the wording of a statement in order to put it into symbolic form, as long as we do not change the meaning of the statement. □

Example 2.1.6

Let p be the simple statement "I study," and let q be the simple statement "I will pass this course." Then

$$(p \rightarrow q) \wedge (\sim p \rightarrow \sim q)$$

is the symbolic form for the statement

If I study, then I will pass this course, and if I do not

study, then I will not pass this course.

When a compound statement is written in symbolic form, the letters that represent the simple statements are generally called the **statement variables** of the compound statement. Thus, for example, the statement $p \wedge (q \rightarrow \sim p)$ has statement variables p and q, and the statement $(p \vee r) \wedge (q \vee \sim s)$ has statement variables p, q, r, and s.

Exercises

1. Which of the following sentences are statements? For those that are statements, decide whether they are simple or compound.
   a) All men are created equal
   b) Mary
   c) Four score and seven years ago
   d) If you do not work the exercises, you will have a hard time passing this course
   e) Did you vote?
   f) No smoking
   g) John is short, but he is over 21 years of age
   h) $x^3 = 6$
   i) The President implies that all is well
   j) The fact that the weather is nice implies that John is sailing
   k) Stop talking and listen to me
   l) If today is Friday, then I will go, otherwise I will not go

2. Find the negation of the following statements.
   a) All fish can swim.
   b) All even integers n have the form $n = 2k$ for some integer k.
   c) Today is not Friday.
   d) It is not true that today is not Friday.
   e) Some dogs eat fish.
   f) It is not true that today is not a holiday.

3. Let p be the statement "John loves Mary," and let q be the statement "Mary loves John." Find the symbolic form for the following compound statements.
   a) John loves Mary and Mary loves John.
   b) John and Mary love each other.
   c) John does not love Mary, but Mary does love John.
   d) It is not true that John loves Mary and Mary does not love John.
   e) John loves Mary if and only if Mary does not love John.
   f) If Mary loves John, then John does not love Mary.

4. Let p be the statement "I am a mathematician," and let q be the statement $2 + 2 = 5$. Find the symbolic form for the following statements.
   a) I am not a mathematician.
   b) I am not a mathematician nor is $2 + 2 = 5$.
   c) Neither of the statements "I am a mathematician" nor "$2 + 2 = 5$" is true.
   d) If I am not a mathematician, then $2 + 2 \neq 5$.
   e) I am a mathematician if and only if I am not a mathematician.

5. Let p be the statement "it is raining," and let q be the statement "the sky is falling." Translate the following symbolic statements into grammatically correct English sentences.
   a) $p \wedge \sim q$ b) $\sim(\sim p)$ c) $\sim(p \vee q)$
   d) $\sim p \vee (p \wedge \sim q)$ e) $\sim q \rightarrow (p \vee \sim q)$
   f) $\sim q \leftrightarrow \sim p$ g) $(\sim p \rightarrow q) \leftrightarrow (\sim q \rightarrow p)$
6. Let p be the statement "all rational numbers are real," let q be the statement "some real numbers are rational," and let r be the statement "some rational numbers are real." Translate the following symbolic statements into grammatically correct English sentences.
   a) $\sim q$ b) $\sim p \rightarrow \sim r$
   c) $p \rightarrow (q \wedge r)$ d) $p \wedge q \wedge r$
   e) $p \leftrightarrow \sim(\sim p)$ f) $\sim(\sim\sim p \wedge \sim q) \vee \sim r$
7. Translate the following statements into symbolic form. Be sure to indicate what the statement variables stand for.
   a) John is not in love with Mary.
   b) Today is Sunday, but I must still go to school.
   c) Frank is interested in Judy, and vice versa.
   d) Either Frank loves Judy, or Judy loves Frank, or both.
   e) If it is raining next Sunday, then we cannot go to the park.
   f) The grass will get watered if and only if it rains.
8. Translate the following statements into symbolic form. Be sure to indicate what the statement variables stand for.
   a) If it is raining, then I will not go to the store, and if it is not raining, then I will go to the store.
   b) If today is Friday, then I will go to school, otherwise I will not go. (*Hint*, use the biconditional.)
   c) Today is not a good day to jog, so I will swim instead.
   d) Some cats are pets, and all cats are neither dogs nor rabbits.
   e) If the sky is cloudy, then it rains, and if it rains, then the sky is cloudy.
   f) Either Frank loves Mary, or Mary loves Frank, but not both.
9. Consider the following statement:

   If $x = 0$, then $x^2 = 0$ and if $x^2 = 0$, then $x = 0$

   a) Express this statement in symbolic form using only the conjunction and the conditional.
   b) Express this statement in symbolic form using only the biconditional.

2.2 Truth Tables, Tautologies, and Contradictions

Now that we know how to translate compound statements into symbolic form, we can turn our attention to finding the truth tables of statements that are in symbolic form. As we have seen, the truth table for a given statement will tell us when that statement is true and when it is false, depending on the truth values of the simple statements that make it up.

We should begin our discussion of truth tables with the simplest possible cases, namely, those compound statements that involve only one connective. In a sense, these truth tables give us the logical meaning of each of the connectives.

□ □ Negation

The truth table that describes the logical meaning of the connective "not" is

p	$\sim p$
T	F
F	T

This truth table certainly seems reasonable, since our intuitive feeling about the word "not" tells us that if the statement p is true, then the statement "not p" should be false, and vice versa.

□ □ Conjunction

The truth table that describes the logical meaning of the connective "and" is

p	q	$p \wedge q$
T	T	T
T	F	F
F	T	F
F	F	F

This truth table also seems reasonable, since our intuition tells us that the statement "p and q" should be true when, and *only* when, both p and q are true.

□ □ Disjunction

The truth table that describes the logical meaning of the connective "or" is

p	q	$p \vee q$
T	T	T
T	F	T
F	T	T
F	F	F

Actually, there are two common meanings associated with the word "or" in the English language. In ordinary usage, the phrase "p or q" can mean either "p or q or both," or it can mean "p or q but not both." When the first meaning is intended for the word "or," it is called the **inclusive or**, and when the second meaning is intended, it is called the **exclusive or**.

Of course, we cannot allow this ambiguity in mathematical logic, and so we must choose which "or" to use. Fortunately, the choice has been made for us. *In mathematical logic, the connective "or" always refers to the inclusive or.* Thus, the statement "p or q" always means "p or q or both." This coincides with the meaning given in the previous truth table. (Can you write down the truth table for the exclusive "or"?)

□ □ Conditional

The truth table that describes the logical meaning of the connective "if . . . then . . ." or "implies" is

p	q	$p \rightarrow q$
T	T	T
T	F	F
F	T	T
F	F	T

The first two rows of this table seem quite reasonable. If p is true and q is true, then the statement "if p then q" should certainly be true. Similarly, if p is true and q is false, then the statement "if p then q" should certainly be false.

On the other hand, the last two rows of this table may seem a bit surprising. Why should it be that if p is false, then the statement "if p then q" is true?

There are a variety of arguments given to justify the last two rows of this table, and it seems as though every book on mathematical logic gives a slightly different one. These arguments range from "there is no other choice since . . ." (see Exercise 29) to "it doesn't really matter since no one would bother to consider the statement 'if p then q' when they knew that p was false."

In any case, we can be comforted by the fact that it certainly does no harm to give it this meaning. Also, it does make it easy to remember the truth table. All we need to remember is that the statement "if p then q" is false *only* when p is true and q is false.

□ □ Biconditional

The truth table that describes the logical meaning of the connective "if and only if" is

p	q	$p \leftrightarrow q$
T	T	T
T	F	F
F	T	F
F	F	T

This table expresses the fact that the statement "p if and only if q" is true precisely when the statement variables p and q have *the same* truth value, regardless of whether that truth value is true or false.

Now that we have considered the truth tables that describe the five basic connectives, we can use them to find the truth tables of more complex compound statements. Some examples will make the method clear.

Example 2.2.1

The truth table for the compound statement $p \wedge (\sim q)$ can be constructed as follows.

p	q	$\sim q$	$p \wedge (\sim q)$
T	T	F	F
T	F	T	T
F	T	F	F
F	F	T	F

Notice that the first two columns of this table list the four possibilities for the truth values of the statement variables p and q. The third column is an "intermediate step" in the computation. After all, before we can compute the truth value of $p \wedge (\sim q)$, we need the truth value of $\sim q$. Then we can use the first and third columns to obtain the last column.

Strictly speaking, the truth table for the statement $p \wedge (\sim q)$ is thus

p	q	$p \wedge (\sim q)$
T	T	F
T	F	T
F	T	F
F	F	F

However, there is really no point in going to the trouble of removing the intermediate steps, and so from now on we will simply leave them in. □

Example 2.2.2

The truth table for the compound statement $p \to (p \vee q)$ can be constructed as follows.

p	q	$p \vee q$	$p \to (p \vee q)$
T	T	T	T
T	F	T	T
F	T	T	T
F	F	F	T

□

Example 2.2.3

The truth table for the compound statement $\sim[(p \wedge \sim p) \rightarrow q]$ is

p	q	$\sim p$	$p \wedge \sim p$	$(p \wedge \sim p) \rightarrow q$	$\sim[(p \wedge \sim p) \rightarrow q]$
T	T	F	F	T	F
T	F	F	F	T	F
F	T	T	F	T	F
F	F	T	F	T	F

Example 2.2.4

The truth table for the compound statement $(p \wedge \sim q) \leftrightarrow (\sim p \rightarrow q)$ is

p	q	$\sim p$	$\sim q$	$p \wedge \sim q$	$\sim p \rightarrow q$	$(p \wedge \sim q) \leftrightarrow (\sim p \rightarrow q)$
T	T	F	F	F	T	F
T	F	F	T	T	T	T
F	T	T	F	F	T	F
F	F	T	T	F	F	T

Example 2.2.5

The truth table for the compound statement $(p \wedge q) \rightarrow (p \wedge r)$ is

p	q	r	$p \wedge q$	$p \wedge r$	$(p \wedge q) \rightarrow (p \wedge r)$
T	T	T	T	T	T
T	T	F	T	F	F
T	F	T	F	T	T
T	F	F	F	F	T
F	T	T	F	F	T
F	T	F	F	F	T
F	F	T	F	F	T
F	F	F	F	F	T

Notice that, in this case, the compound statement contains three different statement variables p, q, and r. Therefore, the truth table must have eight rows in order to include all of the possibilities for the truth values of these variables.

It would be helpful to know exactly how many rows are required for the truth table of a compound statement that contains exactly *n different* statement variables.

As we can see from our examples, if $n = 1$, that is, if the compound statement contains only one statement variable, as is the case for negation, then the truth table must have two rows. On the other hand, if $n = 2$, then the truth table must have $4 = 2^2$ rows, and if $n = 3$, then the truth table must have $8 = 2^3$ rows.

It seems reasonable to guess that if a compound statement contains exactly n different statement variables, then the truth table for that statement must have 2^n rows. We will leave it for an exercise to show, by using mathematical induction, that this is indeed the case (see Exercise 28).

The compound statements in Examples 2.2.2 and 2.2.3 are rather special, and we should make a few comments about them. According to its truth table, the statement in Example 2.2.2

$$p \rightarrow (p \vee q)$$

is always true, regardless of what truth values we assign to the statement variables p and q! (Actually, it does seem reasonable that this statement, which reads "if p then p or q" should always be true.)

In a similar way, we see from the truth table that the statement in Example 2.2.3

$$\sim[(p \wedge \sim p) \rightarrow q]$$

is always false, regardless of what truth values we assign to the statement variables p and q. (This may not seem so obvious. Nevertheless, the truth table shows that this is the case. In fact, this is precisely the point of having truth tables!)

These two examples prompt us to make the following definition.

■ Definition

A compound statement that is always true, regardless of what truth values are assigned to its statement variables, is called a **tautology**. We also say that such a statement is **logically true**.

A compound statement that is always false, regardless of what truth values are assigned to its statement variables, is called a **contradiction**. We also say that such a statement is **logically false**. □

It is important to keep in mind that a compound statement does not have to be either a tautology or a contradiction. For example, the statements in Examples 2.2.1, 2.2.4, and 2.2.5 are neither tautologies nor contradictions, since they can be true at times and false at other times, depending on the truth values of their statement variables. Also, just because a statement is *not* a tautology does not mean that it must be a contradiction.

The term *logically true* is really very appropriate. For, in a sense, a compound statement that is logically true is true *because of its logical structure and not because of the individual simple statements that make it up.*

Perhaps this is easiest to see in the example $p \rightarrow p$, which is logically true. This statement simply cannot help but be true, because of the way it is structured. Any statement of the form "if p then p" has simply got to be true, regardless of what p is. On the other hand, a statement such as $p \rightarrow q$ may be true, or it may be false, depending on the particular truth values of p and q.

Of course, a similar discussion applies to logically false statements. That is, they are false because of their logical structure and not because of the simple statements that make them up. (Consider the statement $p \wedge \sim p$.)

Sometimes it is possible to show that a compound statement is a tautology without having to construct its truth table. Before we can make this precise, we need to observe that if A and B are compound statements, then so are $\sim$A, A $\wedge$ B, A $\vee$ B, A $\rightarrow$ B, and A $\leftrightarrow$ B.

For example, if A is the statement $p \wedge (q \rightarrow \sim p)$ and B is the statement $p \leftrightarrow \sim q$, then $\sim$A is the statement

$$\sim(p \wedge (q \rightarrow \sim p))$$

A $\vee$ B is the statement

$$(p \wedge (q \rightarrow \sim p)) \vee (p \leftrightarrow \sim q)$$

and A $\rightarrow$ B is the statement

$$(p \wedge (q \rightarrow \sim p)) \rightarrow (p \leftrightarrow \sim q)$$

Now we can give two important theorems about tautologies. The first theorem is more interesting from a theoretical point of view than from a practical one. But the second theorem is very useful from a practical point of view, as we will see in the next section.

Theorem 2.2.1

Let A and B be compound statements. If the compound statements A and A $\rightarrow$ B are tautologies, then so is the compound statement B.

PROOF Let us prove this statement by the method of proof by contradiction. Suppose that A and A $\rightarrow$ B are tautologies, but that B is *not* a tautology. Then we want to arrive at a contradiction.

Since B is not a tautology, it must be possible to assign truth values to the statement variables in B that make it false. Suppose we assign these truth values to the statement variables in A and A $\rightarrow$ B. (If A has some statement variables that do not appear in B, then we may assign any truth value to those variables.)

Now, since A $\rightarrow$ B is a tautology, it must be true for this (or in fact any) assignment of truth values to its statement variables. However, since B is false, according to the truth table for the conditional, the only way that A $\rightarrow$ B can be true is if A is false. This is a contradiction to the fact that A is a tautology, and so our assumption that B is not a tautology is in error. Thus, B must be a tautology, and the proof is complete. ∎

Before giving the second theorem, let us consider a simple example. As you can easily check, the statement

$$p \rightarrow (q \rightarrow p) \tag{2.2.1}$$

is a tautology. Now suppose that B_1 and B_2 are any compound statements. Then, if we replace p by B_1 and q by B_2, the resulting statement

$$B_1 \rightarrow (B_2 \rightarrow B_1) \tag{2.2.2}$$

is also a tautology. In order to see this, we must show that any assignment of truth values to the statement variables in Statement 2.2.2 makes it true.

But, assigning a truth value to the statement variables in Statement 2.2.2 simply amounts to giving each of the statements B_1 and B_2 a truth value. Therefore, the truth value of Statement 2.2.2 is exactly the same as the truth value of Statement 2.2.1, when p is given the same truth value as B_1 and q is given the same truth value as B_2. But 2.2.1 is a tautology, and so it is always true. Hence, 2.2.2 is always true, and so it is a tautology.

There is nothing special about the tautology 2.2.1. The fact is that we can replace the statement variables in any tautology and get another tautology. Of course, we must be careful to replace *all* the occurrences of a given statement variable by the *same* compound statement. Let us put this into a theorem, which we call the **substitution theorem for tautologies**.

Theorem 2.2.2

Let A be a tautology, and suppose that A contains the distinct statement variables $p_1, p_2, \ldots, p_n$ (and perhaps others as well). Suppose also that $B_1, B_2, \ldots, B_n$ are compound statements. Then, if in the tautology A, we replace p_1 by B_1, p_2 by B_2, and so on, the resulting statement is also a tautology.

The proof is quite similar to the argument that we have just given in the special case of Statement 2.2.1, and so we will leave it as an exercise.

Example 2.2.6

It is easy to see that the statement $p \lor \sim p$ is a tautology. If we replace p by $p \to \sim q$, then we get the statement

$$(p \to \sim q) \lor \sim(p \to \sim q)$$

which, according to Theorem 2.2.2, must also be a tautology. □

Example 2.2.7

The statement $p \to (q \to p)$ is a tautology, as you can easily check by computing its truth table. If we replace p by $(q \to (p \land \sim r))$, we get the statement

$$(q \to (p \land \sim r)) \to (q \to (q \to (p \land \sim r)))$$

which, according to Theorem 2.2.2, is also a tautology. □

Let us conclude this section with one final definition. If S is a compound statement, then we can define a function, f_S called the **truth function** of S, in the following way.

1) If S contains the n distinct statement variables $p_1, p_2, \ldots, p_n$, and no others, then the truth function f_S is a function of n variables $x_1, x_2, \ldots, x_n$, each of which can take on the values T or F.
2) The value of $f_S(x_1, x_2, \ldots, x_n)$ is defined to be the truth value of the statement S, when the simple statements $p_1, p_2, \ldots, p_n$ are given the truth values $x_1, x_2, \ldots, x_n$, respectively.

The truth function of a statement is easy to determine from the truth table for that statement.

Example 2.2.8

The truth function f_S of the statement S in Example 2.2.4 is the function of two variables given by

$$f_S(T, T) = F$$

$$f_S(T, F) = T$$

$$f_S(F, T) = F$$

$$f_S(F, F) = T$$

□

Example 2.2.9

The truth function f_S of the statement S in Example 2.2.5 is the function of three variables given by

$$f_S(T, T, T) = T$$

$$f_S(T, T, F) = F$$

$$f_S(T, F, T) = T$$

$$f_S(T, F, F) = T$$

$$f_S(F, T, T) = T$$

$$f_S(F, T, F) = T$$

$$f_S(F, F, T) = T$$

$$f_S(F, F, F) = T$$

□

As you can see from these examples, there is really very little difference between the truth table and the truth function of a compound statement. Mostly, it is a matter of notation. Nevertheless, we wanted to define truth functions here because we will have use for them later in the chapter.

Exercises

In Exercises 1 through 16, construct the truth table and the truth function for the given statement. Also, determine whether the statement is a tautology, a contradiction, or neither.

1. $\sim(\sim p)$
2. $p \wedge \sim p$
3. $p \vee \sim p$
4. $\sim p \to \sim q$
5. $\sim p \leftrightarrow (p \wedge q)$
6. $(p \wedge q) \to q$
7. $\sim(\sim p) \to \sim(\sim p \to p)$
8. $(p \wedge q) \to r$
9. $[(p \to q) \wedge (q \to p)] \leftrightarrow (p \leftrightarrow q)$
10. $\sim[p \wedge (q \vee p) \leftrightarrow p]$

11. $[(p \to \sim q) \wedge (q \leftrightarrow \sim p)] \to \sim(\sim p \vee q)$
12. $(p \wedge r) \to (\sim r \vee \sim q)$
13. $(p \to q) \to (r \to q)$
14. $[(p \to q) \wedge (q \to r)] \to (p \to r)$
15. $(p \wedge q) \leftrightarrow (r \vee s)$
16. $[p \to (q \to r)] \to [(p \to q) \to (p \to r)]$
17. Prove that the statement

$$p \to (q \to p)$$

is a tautology.
18. Use Theorem 2.2.2 to prove that the statement

$$(p \wedge \sim r) \to ((p \to q) \to (p \wedge \sim r))$$

is a tautology. (Hint, see Exercise 17.)
19. Use Theorem 2.2.2 to prove that the statement

$$(p \wedge q \wedge r \wedge s \wedge t) \to (p \wedge q \wedge r \wedge s \wedge t)$$

is a tautology. If you tried to do this by constructing the truth table, how many rows would the table have?
20. a) Prove that the statement

$$(p \to q) \to (\sim p \vee q)$$

is a tautology.

b) Prove that the statement

$$((p \to q) \to q) \to (\sim(p \to q) \vee q)$$

is also a tautology.
21. Prove that the statement

$$(p \wedge q) \wedge \sim(p \vee q)$$

is a contradiction.
22. Which of the following are tautologies?

a) $\sim(q \to p) \to \sim p$ b) $(\sim p \to q) \to (p \to \sim q)$

c) $(q \vee p) \to (\sim p \to q)$ d) $(p \to (q \to r)) \to ((p \wedge q) \vee r)$
23. Show that the statements $p \to q$ and $(\sim p) \vee q$ have the same truth table. (Hence, they also have the same truth function.)
24. Show that the statements $p \leftrightarrow q$ and $(p \to q) \wedge (q \to p)$ have the same truth function.
25. Find the truth table for the exclusive "or."
26. Let A and B be compound statements. If A is a contradiction, what can you say about the statement A $\to$ B?
27. Suppose that A, B, and C are compound statements. Prove that if A $\to$ B and B $\to$ C are tautologies, then so is A $\to$ C.
28. Prove, by using mathematical induction, that if a compound statement contains exactly *n different* statement variables, then its truth table must have 2^n rows in order to cover all of the possibilities. *Hint*, you might want to show first that finding the number of rows in such a truth table amounts to finding the number of ways to fill n slots

____ ____ ____ ____ . . . ____

with T's and F's in such a way that each slot gets exactly one T or one F.

29. Let us discuss one reason why the conditional has the truth table that it does. Given that we want the first two rows to be as they are, the possible choices for ways to fill in the rest of the table are given below.

		Possible Choices for $p \to q$			
p	q	(1)	(2)	(3)	(4)
T	T	T	T	T	T
T	F	F	F	F	F
F	T	T	T	F	F
F	F	T	F	T	F

Now, our common sense tells us that the statement $p \to (p \vee q)$ should be a tautology; that is, it should always be true. Show that only under choice (1) is $p \to (p \vee q)$ a tautology.

2.3 Logical Equivalence

Let us construct the truth table for the statement $\sim p \vee q$.

p	q	$\sim p$	$\sim p \vee q$
T	T	F	T
T	F	F	F
F	T	T	T
F	F	T	T

As you can see, this is exactly the same as the truth table for the statement $p \to q$.

It turns out that many different compound statements have the same truth table. In fact, for every compound statement, there are an infinite number of other compound statements that have the same truth table (see Exercise 19).

Of course, two compound statements have the same truth table if and only if they *always have the same truth value*, regardless of what truth values are assigned to their statement variables.

This leads us to make the following definition.

Definition

Two compound statements are said to be **logically equivalent** if they always have the same truth value; that is, if whenever one of the statements is true, then so is the other. Whenever two statements A and B are logically equivalent, we denote this by writing A ≡ B. □

Thus, we can write

$$\sim p \lor q \equiv p \to q$$

Of course, one way to tell whether or not two compound statements are logically equivalent is simply to compare their truth tables. We must be very careful, however, in computing the truth tables to be sure that we take the statement variables in the same order in each table. Otherwise, we cannot compare the truth tables.

The concepts of logical equivalence and tautology are closely related, as the next theorem shows.

Theorem 2.3.1

Two compound statements A and B are logically equivalent if and only if the compound statement A $\leftrightarrow$ B is a tautology.

PROOF The proof of this is very easy. If the statements A and B are logically equivalent, then any assignment of truth values to the statement variables gives the same truth value to both A and B, and so, according to the logical meaning of the biconditional, it makes the statement A $\leftrightarrow$ B true. Hence, A $\leftrightarrow$ B is a tautology.

On the other hand, if A $\leftrightarrow$ B is a tautology, then any assignment of truth values to the statement variables makes the statement A $\leftrightarrow$ B true; that is, it gives the same truth value to both A and B. Therefore, A and B are logically equivalent. ▮

The concept of logical equivalence is very important. Whenever two compound statements are logically equivalent, in some sense they are logically "equal." Of course, we do not mean to suggest that logically equivalent statements are *identical*. Clearly, for example, the statements $\sim p \lor q$ and $p \to q$ are not identical. Nevertheless, when it comes to the properties of compound statements with which logic is concerned, namely, when statements are true and when they are false, there is no way to distinguish between logically equivalent statements. In this sense, they are "equal."

The concept of logical equivalence, or logical equality, allows us to express some very useful properties of the connectives. For example, as you can easily check for yourself, the statements $p \land q$ and $q \land p$ are logically equivalent. (This seems quite reasonable since, from the point of view of truth values, there should be no difference between the statements "p and q" and "q and p.") Thus, we can write

$$p \land q \equiv q \land p \tag{2.3.1}$$

Let us consider some of the other properties of the connectives that can be expressed in terms of logical equivalence. We will restrict our attention here to properties of conjunction, disjunction, and negation.

It is interesting to compare the following list of properties with those of the union, intersection, and complement given in Section 1.1. In comparing these lists, you should notice that conjunction compares with *intersection*, disjunction with *union*, and negation with *complement*. (To help compare this list with the one in Section 1.1, we have given the properties of disjunction before those of conjunction.) Also, you should notice that the role of the universal set U is taken by any tautology and that the role of the empty set $\varnothing$ is taken by any contradiction.

Properties of Conjunction, Disjunction, and Negation

Let p, q, and r be statement variables, let t denote any tautology, and let c denote any contradiction.

Property 1 (Properties of c and t)

$$p \vee c \equiv p \quad , \quad p \wedge t \equiv p$$

$$p \vee t \equiv t \quad , \quad p \wedge c \equiv c$$

Property 2 (The idempotent properties)

$$p \vee p \equiv p \quad , \quad p \wedge p \equiv p$$

Property 3 (The commutative properties)

$$p \vee q \equiv q \vee p \quad , \quad p \wedge q \equiv q \wedge p$$

Property 4 (The associative properties)

$$p \vee (q \vee r) \equiv (p \vee q) \vee r$$

$$p \wedge (q \wedge r) \equiv (p \wedge q) \wedge r$$

Property 5 (The distributive properties)

$$p \vee (q \wedge r) \equiv (p \vee q) \wedge (p \vee r)$$

$$p \wedge (q \vee r) \equiv (p \wedge q) \vee (p \wedge r)$$

Property 6 (Properties of negation)

$$\sim t \equiv c \quad , \quad \sim c \equiv t$$

$$p \vee \sim p \equiv t \quad , \quad p \wedge \sim p \equiv c$$

$$\sim(\sim p) \equiv p$$

Property 7 (De Morgan's laws)

$$\sim(p \vee q) \equiv \sim p \wedge \sim q \quad , \quad \sim(p \wedge q) \equiv \sim p \vee \sim q$$

We will leave a proof of these properties for the exercises. However, we do want to make a few comments. The associative properties are very important, since they allow us to write expressions such as $p \vee q \vee r$ and $p \wedge q \wedge r$ without having to use parentheses. After all, if $(p \vee q) \vee r$ was not the same as $(p \vee q) \vee r$, then the expression $p \vee q \vee r$ would be meaningless, since we wouldn't know which disjunction to perform first.

De Morgan's laws can clear up a point of confusion that arises in everyday English usage. According to these laws, the negation of the statement "p or q" is "not p *and* not q". Similarly, the negation of the statement "p and q" is "not p *or* not q". Thus, for example, the negation of the statement

It is raining or it is snowing

is

It is not raining *and* it is not snowing,

and the negation of the statement

It is raining and it is snowing

is

It is not raining *or* it is not snowing.

The concept of logical equivalence can help us clear up other confusing points as well. Consider the statement

$$p \to q \tag{2.3.2}$$

There are three other statements that are related to, and often confused with, Statement 2.3.2. These are

$$q \to p$$

which is called the **converse** of Statement 2.3.2,

$$\sim p \to \sim q$$

which is called the **inverse** of Statement 2.3.2, and

$$\sim q \to \sim p$$

which is called the **contrapositive** of Statement 2.3.2.

Let us have an example of these statements.

Example 2.3.1

Consider the statement

If the sun is shining then I go to the store.

The converse of this statement is

If I go to the store then the sun is shining.

The inverse is

If the sun is not shining then I do not go to the store;

and the contrapositive is

If I do not go to the store then the sun is not shining.

Most of the confusion that arises concerning these four statements is whether or not the truth of one of them implies the truth of any of the others. We can settle this question once and for all by looking at the truth tables. For convenience, let us put them all into one large table.

				Statement	Converse	Inverse	Contrapositive
p	q	$\sim p$	$\sim q$	$p \rightarrow q$	$q \rightarrow p$	$\sim p \rightarrow \sim q$	$\sim q \rightarrow \sim p$
T	T	F	F	T	T	T	T
T	F	F	T	F	T	T	F
F	T	T	F	T	F	F	T
F	F	T	T	T	T	T	T

As you can see, this table shows that Statement 2.3.2 is logically equivalent to its *contrapositive*, but not to either its converse or its inverse. However, the converse and inverse of Statement 2.3.2 are logically equivalent.

These facts are very important, and so we should put them into a theorem.

Theorem 2.3.2

Among the four statements

$p \rightarrow q$	$q \rightarrow p$	$\sim p \rightarrow \sim q$	$\sim q \rightarrow \sim p$
statement	converse	inverse	contrapositive

1) The statement and its contrapositive are logically equivalent,
2) The converse and the inverse are logically equivalent,
3) No other pairs of statements are logically equivalent; in particular, the statement and its converse are *not* logically equivalent.

Thus, according to this theorem, we can say that *a statement of the form 2.3.2 is true if and only if its contrapositive is true, but we cannot conclude anything about whether the statement is true from a knowledge of whether or not its converse or inverse is true.*

Let us have an example of Theorem 2.3.2.

Example 2.3.2

Consider the statement

$$\text{If } x = 2, \text{ then } x^2 = 4 \qquad (2.3.4)$$

This statement is certainly true, and therefore, according to Theorem 2.3.2, so is its contrapositive, which is

$$\text{If } x^2 \neq 4, \text{ then } x \neq 2$$

On the other hand, the converse of Statement 2.3.4 is

$$\text{If } x^2 = 4, \text{ then } x = 2$$

which is not true (x could be equal to -2). Similarly, the inverse of Statement 2.3.4 is

If $x \neq 2$, then $x^2 \neq 4$

and this is also not true. □

Example 2.3.3

The statement

If x is positive and $x^2 = 4$, then $x = 2$

is certainly true. Therefore, according to Theorem 2.3.2, so is its contrapositive, which is

If $x \neq 2$, then x is not positive *or* $x^2 \neq 4$ □

In mathematics and science we are often called upon to prove a statement of the form 2.3.2, and it frequently happens that it is much easier to prove the contrapositive of the statement, rather than the statement itself. Theorem 2.3.2 tells us that proving the contrapositive is just as good as proving the original statement. On the other hand, it also tells us that proving the converse or inverse of the statement says nothing at all about whether or not the original statement can be proved. (You might be surprised at how many people think that by proving the converse of a statement, they have also proven the statement.)

As a simple example of this, suppose that we want to prove the following theorem.

Theorem 2.3.3

If n^2 is odd, then n is odd.

PROOF In this case, it is easier to prove the contrapositive, which is

If n is not odd, then n^2 is not odd.

Of course, this can be rephrased as

If n is even, then n^2 is even.

Now, if n is even, then it must have the form $n = 2k$, for some integer k. Therefore, $n^2 = (2k)^2 = 4k^2 = 2(2k^2)$, and so n^2 is even. This concludes the proof. ■

The concept of logical equivalence can clear up another point of confusion involving proofs. Often in mathematics and science we are called upon to prove a statement of the form $p \leftrightarrow q$.

Now, we leave it as an exercise to show that the statements

$$p \leftrightarrow q$$

and

$$(p \rightarrow q) \wedge (q \rightarrow p)$$

are logically equivalent. Therefore, according to the logical meaning of the conjunction, we see that the statement $p \leftrightarrow q$ is true if and only if *both* of the statements $p \rightarrow q$ and $q \rightarrow p$ are true.

Thus, in order to prove a statement of the form $p \leftrightarrow q$, we must actually prove both the statement $p \rightarrow q$ and its converse $q \rightarrow p$.

As an example, consider the following theorem.

Theorem 2.3.4

n^2 is odd if and only if n is odd

PROOF In order to prove this theorem, we must prove both the statement

if n^2 is odd then n is odd

and its converse

if n is odd then n^2 is odd.

The first statement is proved in Theorem 2.3.3, where we verified it by proving its contrapositive. As to the converse statement, if n is odd, then it must have the form $n = 2k + 1$, for some integer k. But then $n^2 = (2k + 1)^2 = 4k^2 + 2k + 1$, and so n^2 is odd. This completes the proof of the converse, and so the theorem is proved. ∎

Before concluding this section, we should make one more important observation about logical equivalence. We know from Theorem 2.2.2 that we can replace the statement variables in a tautology by compound statements, and the result will be another tautology. Using this fact, together with Theorem 2.3.1, we can show that if two statements A and B are logically equivalent, and if we replace statement variables in both A and B by compound statements, then the resulting statements are still logically equivalent. Let us make this precise in a theorem, which we call the **substitution theorem for logical equivalence**.

Theorem 2.3.5

Let A and B be logically equivalent statements, and suppose that A and B contain the statement variables $p_1, p_2, \ldots, p_n$ (and possibly others). Suppose also that $C_1, C_2, \ldots, C_n$ are compound statements. Then, if we replace p_1 by C_1, p_2 by C_2, and so on, in *both* A and B, the resulting statements are still logically equivalent.

PROOF Let us denote by A′ and B′ the statements that result from A and B by replacing p_1 by C_1, p_2 by C_2, and so on. We must show that A′ and B′ are logically equivalent.

Now, according to Theorem 2.3.1, A′ and B′ are logically equivalent if and only if the statement $A' \leftrightarrow B'$ is a tautology. But, the statement $A' \leftrightarrow B'$ comes from *the tautology* $A \leftrightarrow B$ by replacing p_1 by C_1, p_2 by C_2, and so on. (Why is $A \leftrightarrow B$ a tautology?) Hence, according to Theorem 2.2.2, $A' \leftrightarrow B'$ must indeed be a tautology. This completes the proof. ∎

Theorem 2.3.5 is very powerful when it is applied to the list of properties of conjunction, disjunction, and negation. As an example, we know from this list that

$$p \wedge q \equiv q \wedge p$$

and so if, for instance, we substitute $p \rightarrow (q \wedge \sim r)$ for p and $q \leftrightarrow r$ for q, Theorem 2.3.5 tells us that

$$(p \rightarrow (q \wedge \sim r)) \wedge (q \leftrightarrow r) \equiv (q \leftrightarrow r) \wedge (p \rightarrow (q \wedge \sim r))$$

Exercises

1. Verify the properties of c and t given in Property 1.
2. Verify the idempotent properties given in Property 2.
3. Verify the commutative properties given in Property 3.
4. Verify the associative properties given in Property 4.
5. Verify the distributive properties given in Property 5.
6. Verify the properties of negation given in Property 6.
7. Verify De Morgan's laws, given in Property 7.
8. Verify the **absorption laws**, which are

$$p \vee (p \wedge q) \equiv p \qquad \text{and} \qquad p \wedge (p \vee q) \equiv p$$

9. Prove that

$$p \leftrightarrow q \equiv (p \rightarrow q) \wedge (q \rightarrow p)$$

10. Prove that the statement $p \rightarrow (q \rightarrow r)$ is *not* logically equivalent to the statement $(p \rightarrow q) \rightarrow r$. How would you describe this in words? *Hint*, use the word "associative."
11. Prove that if A is logically equivalent to B, and B is logically equivalent to C, then A is logically equivalent to C.
12. Prove that any two of the following three statements are logically equivalent. (*Hint*, it might help to use the results of the previous exercise.)
    a) $p \rightarrow q$ b) $(p \wedge \sim q) \rightarrow \sim p$ c) $(p \wedge \sim q) \rightarrow q$
13. Use Theorem 2.3.5 to prove that
    a) $\sim[(p \leftrightarrow q) \vee (p \wedge \sim q)] \equiv \sim(p \leftrightarrow q) \wedge \sim(p \wedge \sim q)$
    b) $(p \wedge r) \wedge ((p \rightarrow q) \vee r) \equiv [(p \wedge r) \wedge (p \rightarrow q)] \vee [(p \wedge r) \wedge r]$
    c) $(r \wedge \sim p) \wedge ((r \wedge \sim p) \vee (p \wedge \sim q)) \equiv r \wedge \sim p$
    d) $(p \vee q) \rightarrow (p \wedge q) \equiv [((p \vee q) \wedge \sim(p \wedge q)) \rightarrow \sim(p \vee q)]$
    (*Hint*, use the results of the last exercise.)

    Hint, look for recurring compound statements and replace them by new statement variables.
14. Write down the negation of the following statements in grammatically correct English.
    a) I am sick and the weather is bad.
    b) Today is Friday or Saturday.
    c) It is raining or it is not snowing.
    d) John does not love Mary and Mary loves Frank.
    e) It is not raining and it is not snowing.
15. Write down the negation of the following statements in grammatically correct English.
    a) It is not true that it is raining or it is snowing.
    b) It is not true that John loves Mary and Mary does not love Frank.
    c) If it is raining, then we will stay home, and if it is snowing, then we will go.

16. Give the converse, contrapositive, and inverse of the following statements. Use correct grammar.
    a) If today is Friday, then I will go to work.
    b) If John is unhappy, then he cries.
    c) If Mary is happy, then she does not cry.
    d) It is raining only if the sky is cloudy.
17. Give the converse, contrapositive, and inverse of the following statements.
    a) If $x \neq 0$, then $x^2 \neq 0$
    b) If $xy = 0$, then $x = 0$
    c) If x is a rational number, then x is a real number.
    d) If x is negative and $x^2 = 4$, then $x = -2$
    e) If $x \neq 2$ and $x^2 + 3x + 1 = 0$, then $x = 1$
    f) If n is an odd integer or a negative integer, then n is not an even nonnegative integer.
18. Prove, by mathematical induction, that if $p_1, p_2, \ldots, p_n$ are statement variables, then
    a) $\sim(p_1 \wedge p_2 \wedge \cdots \wedge p_n) \equiv \sim p_1 \vee \sim p_2 \vee \cdots \vee \sim p_n$
    b) $\sim(p_1 \vee p_2 \vee \cdots \vee p_n) \equiv \sim p_1 \wedge \sim p_2 \wedge \cdots \wedge \sim p_n$
19. Let A be a compound statement, and suppose that it contains the statement variable p.
    a) Prove that the statement $A \wedge (p \vee \sim p)$ has the same truth table as A.
    b) Use mathematical induction to show that the statement

    $$A \wedge (p \vee \sim p) \wedge (p \vee \sim p) \wedge \cdots \wedge (p \vee \sim p)$$

    where $(p \vee \sim p)$ is repeated n times, for any $n > 0$, has the same truth table as A. This shows that, for any compound statement A, there are an infinite number of other compound statements that have the same truth table.
20. Prove the following theorem.
    Theorem n^2 is even if and only if n is even.
21. Prove the following theorem.
    Theorem $2x + 1 \leq 5$ if and only if $x \leq 2$.
22. Prove the following theorem.
    Theorem The product mn of two integers is odd if and only if both of the integers m and n are odd.
23. Prove the following theorem by proving its contrapositive.
    Theorem If $a \neq b$ and $x \neq 0$, then $ax \neq bx$.
24. Let A and B be compound statements. Then we say that A **logically implies** B if B is true whenever A is true.
    a) Can you interpret logical implication in terms of truth tables?
    b) Prove that two statements A and B are logically equivalent if and only if A logically implies B and B logically implies A.
    c) Prove that A logically implies B if and only if the statement $A \rightarrow B$ is a tautology.

2.4 Valid Arguments

Now that we have the necessary background, we can discuss the question of what constitutes a logically valid argument. First of all, we must decide exactly what we mean by an argument.

It seems reasonable to say that an argument should consist of a collection of statements, which are the *premises* of the argument, followed by another statement, which is the *conclusion* of the argument. So, let us make this a formal definition.

Definition

An **argument** is a finite collection $A_1, A_2, \ldots, A_n$ of statements (simple or compound), followed by a statement A. We denote an argument by the notation

$$A_1, A_2, \ldots, A_n \therefore A$$

or

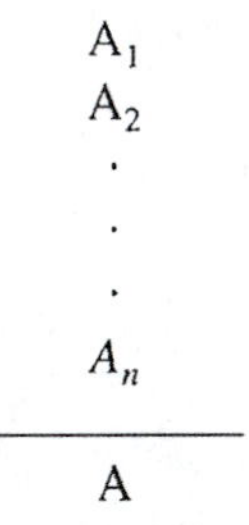

(The symbol ∴ and the horizontal bar stand for the word "therefore.") Each of the statements $A_1, A_2, \ldots, A_n$ is called a **premise** of the argument, and the statement A is called the **conclusion** of the argument. □

This definition is actually very general, since it does not necessarily require that the premises be true. There is really no point to making this requirement, since in practice we are only interested in arguments for which the premises are true.

Now that we know what an argument is, let us decide what it means to say that an argument is valid.

Definition

An argument $A_1, A_2, \ldots, A_n \therefore A$ is **valid** if whenever an assignment of truth values to the statement variables makes the premises $A_1, A_2, \ldots, A_n$ true, then it also makes the conclusion A true. An argument that is not valid is said to be **invalid**. □

Arguments can be tested for validity by constructing a truth table that gives the possible truth values for each of the premises, as well as for the conclusion. Let us consider some examples.

Example 2.4.1

Consider the argument

$$\begin{array}{l} p \to q \\ p \\ \hline q \end{array}$$

An example of such an argument is

If today is Saturday, then I do not have to go to school.

Today is Saturday.

Therefore, I do not have to go to school.

In order to decide whether this argument is valid, we consider the following truth table.

p	q	$p \to q$
T	T	T
T	F	F
F	T	T
F	F	T

As we can see from this table, the only assignment of truth values to the statement variables p and q that make both of the premises p and $p \to q$ true is the assignment in the first row, that is, p true and q true. But in this case, the conclusion q is also true. Hence, according to the definition, the argument is valid.

Incidentally, this simple but important argument is sometimes referred to as **modus ponens.** ☐

■ Example 2.4.2

Consider the argument

$$\begin{array}{l} p \to q \\ q \\ \hline p \end{array}$$

An example of such an argument is

If today is Saturday, then I do not have to go to school.

I do not have to go to school.

Therefore, today is Saturday.

In order to decide whether this argument is valid, we consider the same truth table as in the previous example.

p	q	$p \to q$
T	T	T
T	F	F
F	T	T
F	F	T

In this case, both the first row and the third row correspond to assignments of truth values to the statement variables that make the premises $p \to q$ and q true. But, if we assign the truth values from the third row, namely p false and q true, then the conclusion p is false. Hence, according to the definition, this argument is not valid. □

Example 2.4.3

Consider the argument

$$\begin{array}{l} p \to q \\ q \to r \\ \hline p \to r \end{array}$$

An example of such an argument is

If today is Saturday, then the library is open.

If the library is open, then I must study at the library.

Therefore, if today is Saturday, then I must study at the library.

To test this argument for validity, we consider the following truth table.

p	q	r	$p \to q$	$q \to r$	$p \to r$
T	T	T	T	T	T ←
T	T	F	T	F	F
T	F	T	F	T	T
T	F	F	F	T	F
F	T	T	T	T	T ←
F	T	F	T	F	T
F	F	T	T	T	T ←
F	F	F	T	T	T ←

The rows for which the premises $p \to q$ and $q \to r$ are true are marked with an arrow. Since in each of these cases the conclusion $p \to r$ is also true, the argument is valid.

This important argument is known as the **law of syllogism** and was one of Aristotle's (384–322 B.C.) main contributions to logic. □

Example 2.4.4

Consider the argument

$$\begin{array}{l} (p \wedge q) \to r \\ p \\ \hline r \end{array}$$

To test this argument for validity, we construct the following truth table.

p	q	r	$p \wedge q$	$(p \wedge q) \rightarrow r$
T	T	T	T	T
T	T	F	T	F
T	F	T	F	T
T	F	F	F	T ←
F	T	T	F	T
F	T	F	F	T
F	F	T	F	T
F	F	F	F	T

Now, the row marked with an arrow shows that it is possible for the premises to be true, and at the same time for the conclusion to be false. Hence, this argument is invalid. □

Sometimes it is possible to decide the validity of an argument without constructing a truth table, as the following two examples shows.

Example 2.4.5

Consider the argument

$$\begin{array}{l} p \rightarrow (q \rightarrow r) \\ q \\ \hline p \rightarrow r \end{array}$$

We can test the validity of this argument by trying to find an assignment of truth values that makes the conclusion false, but all of the premises true. If this can be done, then the argument is invalid, but if it cannot be done, then the argument is valid.

Now, according to the logical meaning of the conditional, in order for the conclusion to be false, p must be true and r must be false. Also, in order for the second premise to be true, q must be true.

Therefore, if we want the conclusion to be false and the second premise to be true, we must have p true, q true, and r false. But then the first premise $p \rightarrow (q \rightarrow r)$ is false. Hence, there is no way to make the conclusion false and both of the premises true, and so the argument is valid. □

Example 2.4.6

Consider the argument

$$\begin{array}{l} r \leftrightarrow s \\ p \vee \sim s \\ r \rightarrow q \\ \hline p \vee q \end{array}$$

In order for the conclusion $p \vee q$ to be false, p must be false and q must be false. Let us see if, under these conditions, all of the premises can be true.

Since p is false, in order for the second premise to be true, $\sim s$ must be true; that is, s must be false. Also, since q is false, in order for the third premise to be true, r must be false.

Thus, if we want the conclusion to be false and the second and third premises to be true, we must have p false, q false, r false, and s false. But this assignment makes the first premise true, and so we have found an assignment of truth values that makes all of the premises true and the conclusion false. Hence, this argument is not valid. □

Example 2.4.7

Consider the following argument.

If 2 + 2 = 5, then the moon is made of green cheese.

2 + 2 = 5.

Therefore, the moon is made of green cheese.

In order to determine the validity of this argument, we must first translate it into symbolic form.

So, let p be the statement "2 + 2 = 5," and let q be the statement "the moon is made of green cheese." Then this argument has the form

$$\begin{array}{l} p \rightarrow q \\ p \\ \hline q \end{array}$$

As we saw in Example 2.4.1, this argument is indeed valid. □

The previous example provides a good illustration of the fact that an argument can be valid even if the conclusion is false! According to the definition, all that is required for an argument to be valid is that *if* the premises are true, then so is the conclusion. This says nothing about what happens when some of the premises are not true. However, as we mentioned earlier, this does not really cause any problems since, in practice, we are only interested in arguments for which the premises are true.

Example 2.4.8

Consider the following argument.

If Mr. Johnson resigns the presidency of his company, then Mr. Smith becomes president.

If Mr. Smith is president, then prices will rise or productivity will fall.

Mr. Johnson has resigned and productivity increased. Therefore, prices will rise.

In order to decide whether this argument is valid, we must first put it into symbolic form. So let

p be the statement "Mr. Johnson resigns the presidency of his company";
q be the statement "Mr. Smith is president";
r be the statement "prices will rise";
s be the statement "productivity will rise."

Then this argument has the symbolic form

$$\begin{array}{l} p \to q \\ q \to (r \vee \sim s) \\ p \wedge s \\ \hline r \end{array}$$

(Of course, this is not a literal translation, but it does capture the meaning of the argument.)

Now, we could test the validity of this argument with a truth table, which would have $2^4 = 16$ rows, or we could proceed as in Examples 2.4.5 and 2.4.6. In this case, the second course is a bit easier, so let us take it.

In order for the conclusion of this argument to be false, r must be false. Let us see if, under this circumstance, all of the premises can be true.

In order for the third premise $p \wedge s$ to be true, we must have p true and s true. Then, in order for the first premise $p \to q$ to be true, q must be true. Thus, in order for the conclusion to be false and the first and third premises to be true, we must have r false and p, q, and s true.

However, in this case the second premise $q \to (r \vee \sim s)$ is false. Therefore, it is not possible to make the conclusion false and all of the premises true, and so the argument is valid. □

Before considering any more examples, we should make an important observation. In Example 2.2.1 we showed that the argument

$$\begin{array}{l} p \to q \\ p \\ \hline q \end{array}$$

is valid. Here, of course, p and q are statement variables, and so they represent simple statements. However, since this argument is valid, we can conclude that *any* argument of the form

$$\begin{array}{l} \mathrm{A} \to \mathrm{B} \\ \mathrm{A} \\ \hline \mathrm{B} \end{array}$$

where A and B are any statements, must also be valid. After all, it is clear that any assignment of truth values to the statement variables in A and B that make the premises A → B and A true, will also make the conclusion B true.

More generally, we can state the following theorem, similar to Theorems 2.2.2 and 2.3.5, which we will call **the substitution theorem for valid arguments**.

Theorem 2.4.1

Let $A_1, A_2, \ldots, A_n \therefore A$ be a valid argument, and suppose that the compound statements $A_1, A_2, \ldots, A_n$ and A involve the statement variables $p_1, p_2, \ldots, p_n$ (and perhaps others). Suppose also that $B_1, B_2, \ldots, B_n$ are compound statements. Then the argument obtained by replacing p_1 by B_1, p_2 by B_2, and so on, in *all* of the statements $A_1, A_2, \ldots, A_n$ and A, is also a valid argument.

We will leave a proof of this theorem for the exercises (see Exercise 29). Instead let us have some examples.

Example 2.4.9

The argument

$$\begin{array}{l} (p \rightarrow \sim q) \rightarrow (q \wedge \sim r) \\ p \rightarrow \sim q \\ \hline q \wedge \sim r \end{array}$$

is valid, since it comes from the argument

$$\begin{array}{l} p \rightarrow q \\ p \\ \hline q \end{array}$$

by replacing p by $(p \rightarrow \sim q)$ and q by $(q \wedge \sim r)$.

The argument

$$\begin{array}{l} (r \vee s) \rightarrow (s \leftrightarrow \sim q) \\ (s \leftrightarrow \sim q) \rightarrow (\sim r \rightarrow \sim s) \\ \hline (r \vee s) \rightarrow (\sim r \rightarrow \sim s) \end{array}$$

is valid, since it comes from the law of syllogism

$$\begin{array}{l} p \rightarrow q \\ q \rightarrow r \\ \hline p \rightarrow r \end{array}$$

by replacing p by $(r \vee s)$, q by $(s \leftrightarrow \sim q)$ and r by $(\sim r \rightarrow \sim s)$. □

The law of syllogism is one of the most important forms of argument. By using the principle of mathematical induction, it is possible to extend it to more than two premises.

Theorem 2.4.2

For any integer $n \geq 2$, the argument

$$p_1 \rightarrow p_2, p_2 \rightarrow p_3, \ldots, p_{n-1} \rightarrow p_n \therefore p_1 \rightarrow p_n \qquad (2.4.1)$$

is valid.

PROOF Let us prove this by induction. Let P(n) be the proposition that the argument 2.4.1 is valid. We will leave it to you to check that P(2) is valid. Then P(3) is the proposition that the argument $p_1 \rightarrow p_2, p_2 \rightarrow p_3 \therefore p_1 \rightarrow p_3$ is valid. But this is just the law of syllogism, which we proved is valid in Example 2.4.3.

Now suppose that P(k) is true; that is, suppose that the argument

$$p_1 \rightarrow p_2, p_2 \rightarrow p_3, \ldots, p_{k-1} \rightarrow p_k \therefore p_1 \rightarrow p_k \tag{2.4.2}$$

is valid. Then we must show that $P(k + 1)$ is also true; that is, we must show that the argument

$$p_1 \rightarrow p_2, p_2 \rightarrow p_3, \ldots, p_k \rightarrow p_{k+1} \therefore p_1 \rightarrow p_{k+1} \tag{2.4.3}$$

is also valid.

Consider any assignment of truth values to the statement variables $p_1, p_2, \ldots, p_{k+1}$ that makes the premises in 2.4.3 true. We want to show that this assignment also makes the conclusion $p_1 \rightarrow p_{k+1}$ true.

But since this assignment of truth values makes the first $k - 1$ premises $p_1 \rightarrow p_2, p_2 \rightarrow p_3, \ldots, p_{k-1} \rightarrow p_k$ true, we may use the valid argument 2.4.2 to conclude that the statement $p_1 \rightarrow p_k$ must be true.

Then, since the premise $p_k \rightarrow p_{k+1}$ is also true, we can apply the law of syllogism

$$\begin{array}{l} p_1 \rightarrow p_k \\ p_k \rightarrow p_{k+1} \\ \hline p_1 \rightarrow p_{k+1} \end{array}$$

to conclude that $p_1 \rightarrow p_{k+1}$ must be true. This is what we wanted to show, and so $p(k + 1)$ is true. This completes the proof. ∎

Theorem 2.4.2 can be combined with Theorem 2.2.1 to conclude that if $B_1, B_2, \ldots, B_n$ are any statements, then the argument

$$B_1 \rightarrow B_2, B_2 \rightarrow B_3, \ldots, B_{n-1} \rightarrow B_n \therefore B_1 \rightarrow B_n \tag{2.4.4}$$

is valid. Such an argument is sometimes called a **sorites** (pronounced with a long i and a long e).

Lewis Carroll, the well-known author of *Alice in Wonderland*, wrote a fascinating little book called *Symbolic Logic*, in which he gave, among other things, several examples of sorites. Let us give one of his examples. (Some others appear in the exercises.)

Example 2.4.10 (Lewis Carroll)

Consider the argument

Babies are illogical

Nobody is despised who can manage a crocodile

Illogical persons are despised

Therefore, babies cannot manage crocodiles

In order to check the validity of this argument, we let

p be the statement "he is a baby";
q be the statement "he is illogical";
r be the statement "he manages crocodiles";
s be the statement "he is despised."

Then Lewis Carroll's argument can be written in the symbolic form

$$\begin{array}{l} p \rightarrow q \\ r \rightarrow \sim s \\ q \rightarrow s \\ \hline p \rightarrow \sim r \end{array}$$

Now, recalling that the statement $r \rightarrow \sim s$ is logically equivalent to its contrapositive, which is $s \rightarrow \sim r$, we see that this argument is valid if and only if the argument

$$\begin{array}{l} p \rightarrow q \\ s \rightarrow \sim r \\ q \rightarrow s \\ \hline p \rightarrow \sim r \end{array}$$

is valid (see Exercise 30). But, by rearranging the premises in this argument, we get

$$\begin{array}{l} p \rightarrow q \\ q \rightarrow s \\ s \rightarrow \sim r \\ \hline p \rightarrow \sim r \end{array}$$

which is just a special case of the sorites 2.4.4, and so it is indeed valid. □

□ □ Exercises

In Exercises 1 through 14, determine whether or not the argument is valid. If it is invalid, find an assignment of truth values to the statement variables that makes the premises true and the conclusion false.

1. $\begin{array}{l} p \rightarrow q \\ \sim q \\ \hline \sim p \end{array}$

2. $\begin{array}{l} p \wedge q \\ p \\ \hline q \end{array}$

3. $\begin{array}{l} p \vee q \\ p \\ \hline q \end{array}$

4. $\begin{array}{l} p \vee q \\ \sim p \\ \hline q \end{array}$

5. $\begin{array}{l} p \vee q \\ p \\ \hline \sim q \end{array}$

6. $\begin{array}{l} p \rightarrow q \\ p \rightarrow r \\ \hline q \rightarrow r \end{array}$

7. $\begin{array}{l} p \rightarrow q \\ q \vee r \\ \hline r \rightarrow \sim q \end{array}$

8. $\begin{array}{l} p \rightarrow q \\ \sim p \vee q \\ \hline q \rightarrow p \end{array}$

9. $\begin{array}{l} q \rightarrow p \\ \sim q \leftrightarrow p \\ \hline p \end{array}$

10. $(p \vee q) \rightarrow r$
p

---
r

11. $p \rightarrow q$
$\sim q \rightarrow r$
r

---
p

12. $\sim p \vee q$
$p \rightarrow (r \wedge s)$
$s \rightarrow q$

$q \vee r$

13. $q \rightarrow s$
$(s \wedge r) \rightarrow (p \wedge q)$

$p \rightarrow (r \vee s)$

14. $(p \wedge \sim q) \rightarrow r$
$p \wedge \sim r$

---
q

In Exercises 15 through 18, use Theorem 2.4.1 to show that the given argument is valid.

15. $p \rightarrow (q \rightarrow r)$
p

$q \rightarrow r$

16. $p \rightarrow (q \wedge r)$
$\sim q \vee \sim r$

$\sim p$

17. $(p \wedge q) \rightarrow (r \rightarrow s)$
$(r \rightarrow s) \rightarrow \sim s$

$(p \wedge q) \rightarrow \sim s$

18. $q \wedge (s \vee q)$
$\sim s \wedge \sim q$

---
q

In Exercises 19 through 27, translate the argument into symbolic form and determine whether or not it is valid.

19. It is raining or snowing.
It is not snowing.
Therefore, it is raining.

20. It is raining or it is not snowing.
It is not raining.
Therefore, it is snowing.

21. It is raining or it is not snowing.
It is snowing.
Therefore, it is not raining.

22. Only students who study pass this course.
John is very smart.
Even smart people must study.
Therefore, John will pass this course.

23. If we have a bigger bomb, then democracy is safe.
We do not have a bigger bomb.
Therefore, democracy is not safe.

24. If Mr. Johnson has bought a house, then either he has sold his car or he has borrowed money from a bank.
Mr. Johnson has not borrowed money from a bank.
Therefore, if Mr. Johnson has not sold his car, then he has not bought a house.

25. If it rains today, then either the newspaper is right or the radio is wrong.
It did not rain today, or the newspaper is wrong.
Therefore, the radio is right.

26. (Lewis Carroll)
No ducks waltz.

No officers ever decline to waltz.
All my poultry are ducks.
Therefore, my poultry are not officers.

27. (Lewis Carroll)
No birds, except ostriches, are 9 feet high.
There are no birds in this aviary that belong to anyone but me.
No ostrich lives on mince-pies.
I have no birds less than 9 feet high.
Therefore, no bird in this aviary lives on mince-pie.

28. Prove that an argument $A_1, A_2, \ldots, A_n \therefore A$ is valid if and only if the statement $A_1 \wedge A_2 \wedge \ldots \wedge A_n \rightarrow A$ is a tautology.

29. Use the result of the previous exercise to prove Theorem 2.4.1.

30. Prove that if the argument $A_1, A_2, \ldots, A_n \therefore A$ is valid, and if the statement B_1 is logically equivalent to A_1, B_2 is logically equivalent to $A_2, \ldots,$ and B_n is logically equivalent to A_n, then the argument $B_1, B_2, \ldots, B_n \therefore A$ is also valid.

31. Prove that the argument

$$p_1 \rightarrow (q_1 \rightarrow r_1), p_2 \rightarrow (q_2 \rightarrow r_2), \ldots, p_n \rightarrow (q_n \rightarrow r_n)$$

$$q_1 \wedge q_2 \wedge \cdots \wedge q_n$$

$$\therefore (p_1 \rightarrow r_1) \wedge (p_2 \rightarrow r_2) \wedge \cdots \wedge (p_n \rightarrow r_n)$$

is valid.

2.5 Boolean Functions and Disjunctive Normal Form

Logic has a great many practical uses. One of the most important of these is to the design of so-called *logic circuits*, which are used in computers. The idea rests on the fact that transistors can be made to function as very fast binary switches. As a result, it is possible to design simple circuits that can perform the functions "and," "or," and "not." Such circuits are called **AND gates, OR gates,** and **inverters**, respectively. For example, an AND gate takes two inputs, p and q, in the form of high voltages (denoted by a 1 or a T) or low voltages (denoted by a 0 or an F) and outputs the conjunction $p \wedge q$.

Using combinations of AND gates, OR gates, and inverters (as well as other types of gates), circuits can be designed that will add binary numbers, compare binary numbers, and perform other functions relevant to the operation of a computer. Combinations of these and other gates make up what are known as **integrated circuits,** or simply **chips**.

The study of how to design integrated circuits is taken up in a course in computer logic and design. Our plan in the next few sections is to discuss the connection between logic circuits and compound statements. This will enable us to design simple circuits that perform in a prescribed way.

In order to give a meaningful discussion of logic circuits, we must first discuss certain related concepts. We will do that in this section and turn to logic circuits themselves in the next section.

For the remainder of this chapter, we will use the symbol $'$ to denote the negation. Thus x' is read "not x." This change in notation will save us from having to write some rather messy expressions involving parentheses.

Compound statements will play a key role in our discussion, but in the present context, they generally go by a different name. Also, different symbols are used for the statement variables than the ones we have been using up to now. Let us have a definition.

Definition

Let $x_1, x_2, x_3, \ldots$ be a sequence of statement variables. Then any compound statement in these variables will be called a **Boolean polynomial**. We will use the notation $p(x_1, x_2, \ldots, x_n)$ for a Boolean polynomial in the variables $x_1, x_2, \ldots, x_n$. If there are four or fewer statement variables involved in the discussion, we will generally use the letters x, y, z, and w, rather than x_1, x_2, x_3, and x_4. We also include the special polynomials $\mathbf{0}(x_1, x_2, \ldots, x_n) = 0$ and $\mathbf{1}(x_1, x_2, \ldots, x_n) = 1$ as Boolean polynomials. □

As an example, the following are Boolean polynomials.

$$p_1(x) = x$$

$$p_2(x, y) = x \wedge y$$

$$p_3(x, y, z) = y' \wedge (x \vee z')$$

$$p_4(x, y, z) = (x \wedge y \wedge x) \vee (z \vee z')$$

$$p_5(x, y, z) = x \vee y$$

Boolean polynomials are very similar to ordinary polynomials, and as you know, ordinary polynomials are generally thought of as *functions*, whose variables can take on real values. For example, the polynomial

$$p(x, y) = x^2 + xy + xy^2$$

is a function $p:\mathbf{R} \times \mathbf{R} \rightarrow \mathbf{R}$ of two variables, whose values are obtained by substituting real numbers for these variables. For instance, we have

$$p(3,2) = 3^2 + 3 \cdot 2 + 3 \cdot 2^2 = 27$$

Similarly, Boolean polynomials are generally thought of as functions. However, the variables in a Boolean polynomial can take on only the values 0 and 1. For example, the Boolean polynomial

$$p(x, y) = x \wedge y'$$

is a function $p:\mathbf{B} \times \mathbf{B} \rightarrow \mathbf{B}$ of two variables, where $\mathbf{B} = \{0, 1\}$. The values of this function are

$$p(0, 0) = 0 \wedge 0' = 0 \wedge 1 = 0$$

$$p(0, 1) = 0 \wedge 1' = 0 \wedge 0 = 0$$

$$p(1, 0) = 1 \wedge 0' = 1 \wedge 1 = 1$$

$$p(1, 1) = 1 \wedge 1' = 1 \wedge 0 = 0$$

The Boolean function $\mathbf{0}(x_1, x_2, \ldots, x_n)$ always takes the value 0, and the Boolean function $\mathbf{1}(x_1, x_2, \ldots, x_n)$ always takes the value 1.

In general, if $p(x_1, x_2, \ldots, x_n)$ is a Boolean polynomial in the n variables $x_1, x_2, \ldots, x_n$, then it defines a function of n variables; that is, it defines a function from the cartesian product $\mathbf{B} \times \mathbf{B} \times \cdots \times \mathbf{B}$ to the set $\mathbf{B}$.

In view of this discussion, Boolean polynomials are also called **Boolean functions**, or **truth functions**. We discussed truth functions very briefly in Section 2.2. In that section we used the symbols T and F, rather than 1 and 0, but this is only a matter of notation. To emphasize their role as functions, from now on we will use the term Boolean function.

Perhaps the most convenient way to describe a Boolean function is with its truth table, which is simply a list of all of the values of the function. For example, the truth table for the Boolean function

$$p(x, y, z) = y' \wedge (x \vee z)$$

is

x	y	z	y'	$x \vee z$	$p(x, y, z)$
1	1	1	0	1	0
1	1	0	0	1	0
1	0	1	1	1	1
1	0	0	1	1	1
0	1	1	0	1	0
0	1	0	0	0	0
0	0	1	1	1	1
0	0	0	1	0	0

It is important always to list the variables in the same order in all truth tables. In this way, every Boolean function has a *unique* truth table. However, it is quite possible for two *different* Boolean functions to have the same truth table. As a matter of fact, since Boolean functions are just compound statements, and since two compound statements have the same truth table if and only if they are logically equivalent, we have the following theorem.

Theorem 2.5.1

Two Boolean functions (polynomials) have the same truth table if and only if they are logically equivalent.

Given a truth table with 2^n rows, we would like to be able to find a Boolean function of n variables whose truth table is that table. Let us illustrate a procedure for doing so with an example.

Example 2.5.1

Consider the truth table

Table 2.5.1

x	y	z		
1	1	1	1	←
1	1	0	0	
1	0	1	1	←
1	0	0	0	
0	1	1	1	←
0	1	0	0	
0	0	1	0	
0	0	0	1	←

In order to find a Boolean function $p(x, y, z)$ whose truth table is this one, we proceed as follows. First, we locate each row that ends in a 1. These are marked with arrows in Table 2.5.1. For each of these rows, we create a term of the form

$$e_1 \wedge e_2 \wedge e_3 \tag{2.5.1}$$

where $e_1 = x$ if the entry in the first column of that row is a 1 and $e_1 = x'$ if the entry in the first column is a 0. Similarly, $e_2 = y$ if the entry in the second column of that row is a 1, and $e_2 = y'$ if the second entry is a 0. Finally, $e_3 = z$ if the entry in the third column of that row is a 1, and $e_3 = z'$ if the entry in the third column is a 0.

Thus, in this case, the terms of the form 2.5.1 corresponding to the four marked rows are

$$x \wedge y \wedge z$$

$$x \wedge y' \wedge z$$

$$x' \wedge y \wedge z$$

$$x' \wedge y' \wedge z'$$

Finally, we "or" these expressions together to get the Boolean function

$$\begin{aligned} p(x, y, z) = {} & (x \wedge y \wedge z) \vee (x \wedge y' \wedge z) \\ & \vee (x' \wedge y \wedge z) \vee (x' \wedge y' \wedge z') \end{aligned} \tag{2.5.2}$$

We suggest that you check for yourself that this function has Table 2.5.1 as its truth table. □

If the last column of the truth table under consideration consists entirely of 0's, then the method in Example 2.5.1 will not work. However, in that case, the Boolean function

$$\mathbf{0}(x_1, x_2, \ldots, x_n) = 0 \tag{2.5.3}$$

has that table as its truth table. In either case then, we have found a way to construct a Boolean function with a given truth table.

The Boolean function in 2.5.2 has a rather special form, and since it is clear that the procedure we used to obtain this function will always yield a function of a similar form, we are led to make the following definition.

Definition

A Boolean function $p(x_1, x_2, \ldots, x_n)$ is said to be in **disjunctive normal form** if it is the disjunction of a finite number of *distinct* terms, each of which has the form

$$e_1 \wedge e_2 \wedge \cdots \wedge e_n \tag{2.5.4}$$

where $e_i = x_i$ or $e_i = x_i'$ for all $i = 1, 2, \ldots, n$. Terms of the form 2.5.4 are called **minterms**. Also, the Boolean functions $\mathbf{0}(x_1, x_2, \ldots, x_n) = 0$ and $\mathbf{1}(x_1, x_2, \ldots, x_n) = 1$ are in disjunctive normal form.

For example, the following Boolean functions are in disjunctive normal form.

$$p(x) = x$$

$$p(x, y) = x \wedge y$$

$$p(x, y) = (x \wedge y') \vee (x \wedge y)$$

$$p(x, y, z) = (x \wedge y \wedge z') \vee (x' \wedge y' \wedge z) \vee (x \wedge y' \wedge z)$$

However, the Boolean function

$$q(x, y, z) = (x \wedge y \wedge z) \vee (x' \wedge z) \vee (x \wedge y \wedge z')$$

is not in disjunctive normal form.

It is clear that the method of Example 2.5.1 will work on any truth table, with the exception noted immediately following that example. But since the Boolean function in 2.5.3 is in disjunctive normal form, we can state that *every truth table is the truth table of a function in disjunctive normal form.*

Given a Boolean function $p(x_1, x_2, \ldots, x_n)$, we can compute its truth table and then use the method of Example 2.5.1 to compute a Boolean function $q(x_1, x_2, \ldots, x_n)$ in disjunctive normal form with the same truth table. Therefore, the functions p and q will have the same truth table, and so they will be logically equivalent.

Thus, given any Boolean function p, we have a method for finding a Boolean function q in disjunctive normal form that is logically equivalent to p. We call any such function q a **disjunctive normal form of** p.

Let us do an example.

Example 2.5.2

Consider the Boolean function

$$p(x, y, z) = (x \wedge y) \vee z'$$

We can determine a disjunctive normal form for this function by first computing its truth table.

x	y	z	$x \wedge y$	z'	$(x \wedge y) \vee z'$
1	1	1	1	0	1
1	1	0	1	1	1
1	0	1	0	0	0
1	0	0	0	1	1
0	1	1	0	0	0
0	1	0	0	1	1
0	0	1	0	0	0
0	0	0	0	1	1

Using this table, the method of Example 2.5.1 gives us the Boolean function

$$q(x, y, z) = (x \wedge y \wedge z) \vee (x \wedge y \wedge z') \vee (x \wedge y' \wedge z') \vee (x' \wedge y \wedge z') \vee (x' \wedge y' \wedge z')$$

which is therefore a disjunctive normal form of $p(x, y, z)$. □

Exercises

1. Which of the following Boolean functions are in disjunctive normal form?
   a) $p(x) = x'$
   b) $p(x) = x \vee x'$
   c) $p(x, y) = x \vee y \vee y'$
   d) $p(x, y) = (x' \wedge y) \vee (y \wedge x)$
   e) $p(x, y, z) = x' \vee y' \vee z'$
   f) $p(x, y, z) = (x \wedge y \wedge z) \vee (x' \wedge y' \wedge z')$
2. Let $p(x, y, z) = (x' \wedge y)' \vee z$.
   a) Compute $p(0, 0, 1)$
   b) Compute $p(1, 0, 1)$
   c) Compute $p(1, 1, 1)$
3. Let $p(x, y, z, w) = (x \wedge w') \vee (y \vee x)'$
   a) Compute $p(0, 0, 0, 0)$
   b) Compute $p(0, 1, 0, 1)$
   c) Compute $p(1, 1, 1, 1)$
4. Are the Boolean functions $p(x, y) = x \wedge y$ and $q(x, y) = y \wedge x$ logically equivalent? Justify your answer.
5. Are the Boolean functions $p(x, y) = x \wedge y$ and $q(x, y) = x \vee y$ logically equivalent? Justify your answer.
6. Are the Boolean functions $p(x, y, z) = x \wedge (y \vee z)$ and $q(x, y, z) = (x \wedge y) \vee (x \wedge z)$ logically equivalent? Justify your answer.

In Exercises 7 through 12, find the truth table for the given Boolean function.

7. $p(x) = x'$
8. $p(x, y) = (x \vee y) \wedge y'$

9. $p(x, y) = (x \wedge y)' \wedge x'$
10. $p(x, y, z) = x \wedge y \wedge z$
11. $p(x, y, z) = (x \vee y)' \wedge (z \vee x) \wedge y$
12. $p(x, y, z, w) = (x \wedge z') \wedge (z \vee w')'$

In Exercises 13 through 17, find a Boolean function in disjunctive normal form whose truth table is the given table.

13.

x	y	
1	1	1
1	0	1
0	1	1
0	0	0

14.

x	y	
1	1	1
1	0	0
0	1	1
0	0	0

15.

x	y	z	
1	1	1	1
1	1	0	0
1	0	1	0
1	0	0	0
0	1	1	1
0	1	0	0
0	0	1	0
0	0	0	0

16.

x	y	z	
1	1	1	0
1	1	0	0
1	0	1	0
1	0	0	0
0	1	1	0
0	1	0	0
0	0	1	0
0	0	0	0

17.

x	y	z	
1	1	1	1
1	1	0	1
1	0	1	1
1	0	0	0
0	1	1	0
0	1	0	0
0	0	1	1
0	0	0	1

In Exercises 18 through 29, find a disjunctive normal form for the given Boolean function.

18. $p(x, y) = x \wedge y$
19. $p(x, y) = x \vee y$
20. $p(x, y) = x \wedge y'$
21. $p(x, y, z) = (x \vee y) \wedge z$
22. $p(x, y, z) = (x \wedge y)' \vee z$
23. $p(x, y, z) = (x \vee y) \wedge (x \vee y \vee z')$
24. $p(x, y, z) = (x' \vee y \vee z) \wedge (x' \vee y' \vee z')$
25. $p(x, y, z) = [(x \wedge y') \vee (x \wedge z)]' \vee x'$
26. $p(x, y, z) = (x \wedge y') \vee (x \wedge z) \vee (x \wedge y)$
27. $p(x, y, z) = (x \vee y) \wedge (x \vee y') \wedge (x' \vee z)$
28. $p(x, y, z, w) = x \wedge y \wedge z \wedge w$
29. $p(x, y, z, w) = x \vee y \vee z \vee w$
30. Prove by using induction that the truth table of a Boolean function of n variables has exactly 2^n rows.
31. How do you think *conjunctive* normal form should be defined? Give some examples of Boolean functions that are in conjunctive normal form, and some that are not. Can a Boolean function be in both disjunctive normal form and conjunctive normal form?

2.6 Logic Circuits

Now that we have the necessary background, we can turn to a discussion of the design of logic circuits.

The type of logic circuit that we will study is called an **AND-OR circuit**. An AND-OR circuit is an electrical circuit that contains three types of *gates*, which we will describe in a moment. The wires in the circuit each carry either a high voltage, designated by a 1, or a low voltage, designated by a 0.

The first type of gate in an AND-OR circuit is called an **AND gate**, and is pictured below.

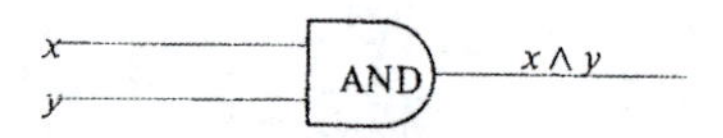

This AND gate has two *inputs*, labeled x and y, and one *output*, labeled $x \wedge y$. As the name implies, the output of an AND gate is 1 (high voltage) if and only if both inputs are equal to 1; that is, $x \wedge y$ is 1 if and only if both x *and* y equal 1. Since the Boolean function $p(x, y) = x \wedge y$ has the value 1 if and only if both x and y equal 1, this accounts for the label $x \wedge y$ on the output wire of the AND gate.

The second type of gate in an AND-OR circuit is the **OR gate**. This is pictured below.

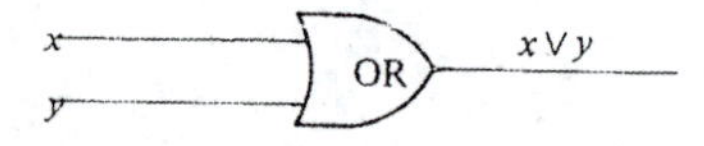

In this case, the output is 1 if and only if either one of the inputs is 1, that is, if and only if either x *or* y is equal to 1. Since the Boolean function $q(x, y) = x \vee y$ has the value 1 if and only if either x or y has value 1, this accounts for the label $x \vee y$ on the output wire of the OR gate.

The third type of gate in an AND-OR circuit is called a **NOT gate**, or an **inverter**. This is pictured below.

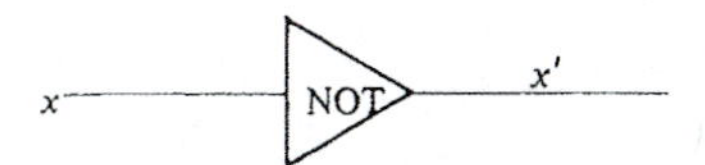

The output of an inverter is the complement of its input. That is, if the input is 0, then the output is 1, and if the input is 1, then the output is 0.

A typical AND-OR circuit is shown in Figure 2.6.1.

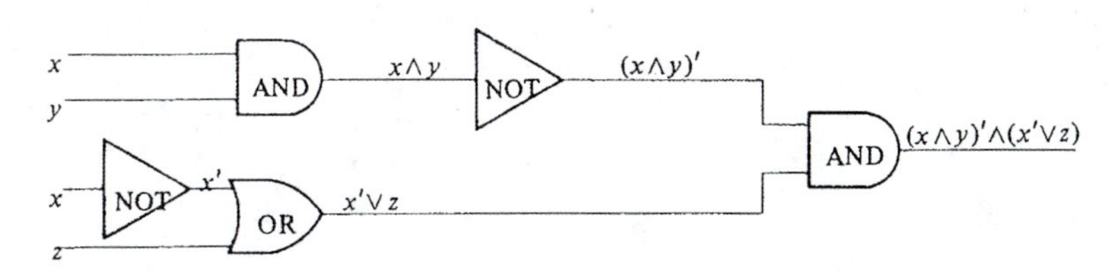

Figure 2.6.1

As this example shows, an AND-OR circuit may have several input wires, and some may have the same label. When two input wires have the same label, they always receive the same input (high or low voltage). On the other hand, we will restrict our attention to AND-OR circuits that have exactly one output wire.

We will usually omit the labels "AND," "OR," and "NOT" in the logic gates. However, the different shapes of these gates will always make it clear what type of gate is under consideration.

As you can see from the circuit in Figure 2.6.1, each wire is labeled with a Boolean function. The label on the output wire is particularly significant, since it describes all of the possible outputs of the circuit.

For example, in the circuit of Figure 2.6.1, the function labeling the output wire is

$$p(x, y, z) = (x \wedge y)' \wedge (x' \vee z)$$

Hence, for instance, if the input to this circuit is

$$x = 0, y = 1, z = 1$$

the output is

$$p(0, 1, 1) = (1 \wedge 0)' \wedge (0' \vee 1) = 1$$

Put another way, if a low voltage is placed at all inputs labeled x, and a high voltage is placed at all inputs labeled y or z, then the output is a high voltage.

It is clear from this example that every AND-OR circuit has a Boolean function associated with it that describes the possible outcomes of the circuit. This function is the one that labels the output wire, and we will call it the **Boolean function of the circuit**.

On the other hand, every Boolean function is the function of a logic circuit. For example, the function

$$q(x, y, z) = (x \wedge y) \vee [(x \wedge y \wedge z') \vee y']'$$

is the function of the circuit shown in Figure 2.6.2.

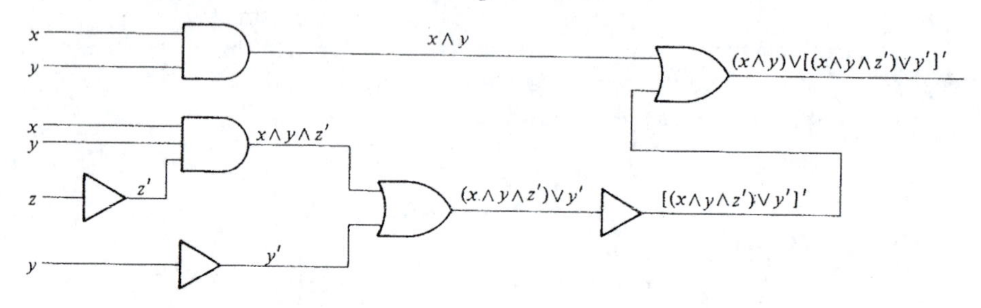

Figure 2.6.2

This example illustrates another feature of AND-OR circuits. Namely, we allow AND and OR gates to have more than two inputs (the NOT gate always has one input, however).

If we are given a logic circuit, then we can determine what the possible outputs of this circuit are by computing its Boolean function. In order to display this information in a form that is easy to access, we can then compute the truth table of the function. For example, the truth table of the function

$$p(x, y, z) = (x \wedge y)' \wedge (x' \vee z)$$

which is the function of the circuit in Figure 2.6.1, is

x	y	z	$p(x, y, z)$
1	1	1	0
1	1	0	0
1	0	1	1
1	0	0	0
0	1	1	1
0	1	0	1
0	0	1	1
0	0	0	1

From this table, we can easily determine the output of the circuit in Figure 2.6.1 for any given input. For example, if x and y are at high voltage ($x = y = 1$), and z is at low voltage ($z = 0$), then the output of the circuit is a low voltage. This comes from the second row of the table.

If p is the Boolean function of a circuit, then we will call the truth table of p the **truth table of the circuit**. The truth table is a very convenient way to describe the possible outcomes of a circuit.

Of course, the main goal in the design of circuits is to be able to design a circuit that behaves in a prescribed manner. As we shall see in the next two examples, frequently the desired behavior can be expressed in terms of a truth table. Thus, the question naturally arises as to whether we can find a circuit with a given truth table.

The answer is yes. All we have to do is find a function with this truth table, by the method of the previous section, for example, and then construct a circuit with this function as its function. The next examples not only illustrate this method, but also illustrate why we might want to find a circuit with a given truth table in the first place.

Example 2.6.1

Let us design a logic circuit that inputs the values from three variables, x, y, and z, and outputs the value 1 if and only if all three variables are equal.

The first step is to construct the truth table that corresponds to this property.

x	y	z	
1	1	1	1
1	1	0	0
1	0	1	0
1	0	0	0
0	1	1	0
0	1	0	0
0	0	1	0
0	0	0	1

As you can see, this table has a 1 in the last column of a given row if and only if the values of the variables are equal. Now we need to find a circuit with this truth table.

In order to accomplish this, we use the method of the previous section to construct a Boolean function with this truth table. We will leave it to you to verify that this gives

$$p(x, y, z) = (x \wedge y \wedge z) \vee (x' \wedge y' \wedge z')$$

Finally, we construct the circuit that has this function.

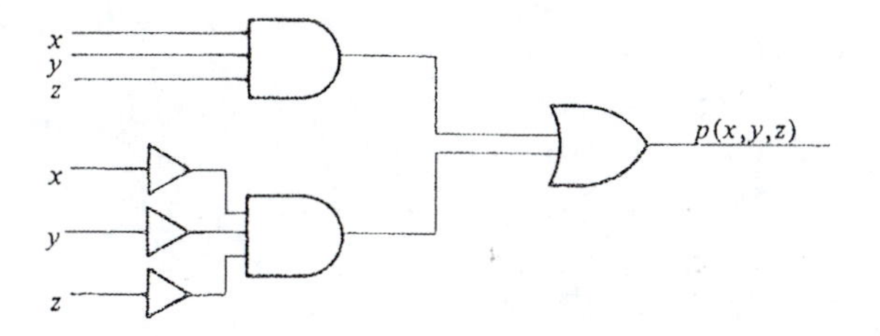

Example 2.6.2

You have probably seen rooms that have a single overhead light that is controlled by three switches, on three different walls of the room. Somehow, these switches seem to know whether or not the light is on at any given moment since, whenever one of the switches is changed, the light goes on if it was off and off if it was on. Have you ever wondered how the switch can possibly know whether the light is on or off?

Actually, it is a simple application of logic circuits and Boolean functions to design a circuit that will behave in this manner.

Let x be the variable that tells whether or not the first switch is closed, and let us assume that $x = 0$ corresponds to an open switch (no current), whereas $x = 1$ corresponds to a closed switch. Similarly, let y and z describe the condition of the other two switches.

Now, we can describe the "state" of the system by an ordered 4-tuple (e_1, e_2, e_3, e_4), where e_1 is the value of x, e_2 is the value of y, e_3 is the value of z, and e_4 is equal to 0 if the light is off and 1 if the light is on.

Using these 4-tuples, we can easily describe the property that we want our circuit to possess. Namely, whenever the circuit is in a particular state and *exactly* one of the first three components is changed, then the fourth component is also changed. This enables us to construct a "state" table for the circuit.

x	y	z	
1	1	1	1
1	1	0	0
1	0	1	0
1	0	0	1
0	1	1	0
0	1	0	1
0	0	1	1
0	0	0	0

We have assumed that when all of the switches are closed ($x = y = z = 1$), the light is on. This gives us the first row of the table. Once this row has been completed, we have no choice as to how to fill in the rest of the table. For example, the second row must end in a 0 because the only difference in the switch positions in the first two rows is that the third switch has changed its state. Hence, the state of the light must be different in these two cases.

Now that we have our table, we can use the method of the previous section to find a function, in disjunctive normal form, with this truth table. We get

$$p(x, y, z) = (x \wedge y \wedge z) \vee (x \wedge y' \wedge z') \vee (x' \wedge y \wedge z') \wedge (x' \wedge y' \wedge z)$$

It is not hard to construct a circuit whose function is this one (see Figure 2.6.3).

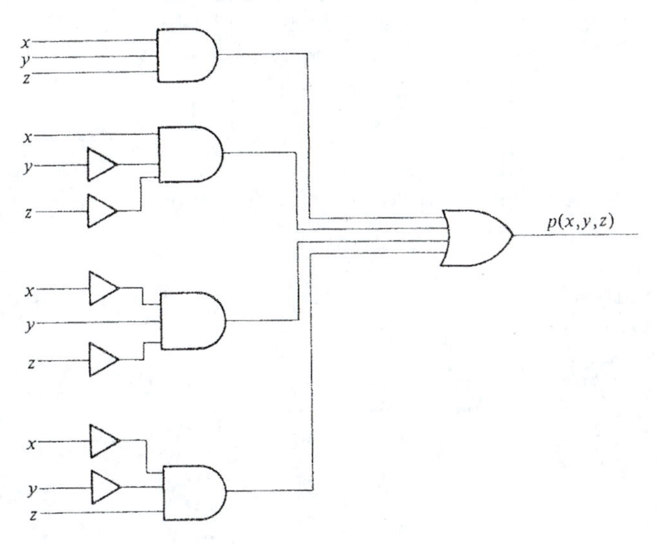

Figure 2.6.3

Finally, we can construct our lighting circuit by connecting a power source to three switches s_1, s_2, and s_3. Switch s_1 is then connected to each of the inputs of the circuit in Figure 2.6.3 that are labeled x, and similarly for the other two switches. Finally, the output of this circuit is connected to the light. (Of course, the circuit must be grounded.) □

As the previous examples illustrate, we often want to construct a circuit with a given truth table. However, there are many different circuits that have a given truth table. In fact, if p is *any* Boolean function whose truth table is the given one, then the circuit whose function is p will also have that truth table. In view of this, it is natural to want to find as simple a circuit as possible with a given truth table.

The method that we used in the previous examples will always give us a circuit with a given truth table. This method can be summarized by the following diagram.

Truth table → Boolean function p in disjunctive normal form with that truth table → circuit whose Boolean function is p

As you can see from this diagram, the method always produces a circuit that comes from a function in disjunctive normal form. Unfortunately, such a circuit is not usually the simplest circuit with the given truth table.

Of course, one way to remedy this situation is to simplify the Boolean polynomial p in disjunctive normal form, that is, to replace it with a simpler, logically equivalent Boolean polynomial before constructing the circuit. This is illustrated in the following diagram.

Truth table →	Boolean function p in disjunctive normal form with that truth table	→	Boolean function q that is logically equivalent to p, but simpler	→	circuit whose Boolean function is q	

(2.6.1)

One way to accomplish this simplification is by using the properties of conjunction, disjunction, and negation given in Section 2.3 before computing the circuit. Let us illustrate this with an example.

Example 2.6.3

Consider the truth table

x	y	z	
1	1	1	1
1	1	0	1
1	0	1	1
1	0	0	0
0	1	1	0
0	1	0	0
0	0	1	1
0	0	0	0

The function in disjunctive normal form with this truth table, obtained by the method of the previous section is

$$p(x, y, z) = (x \wedge y \wedge z) \vee (x \wedge y \wedge z') \vee (x \wedge y' \wedge z) \vee (x' \wedge y' \wedge z)$$

and the circuit with this function is shown below.

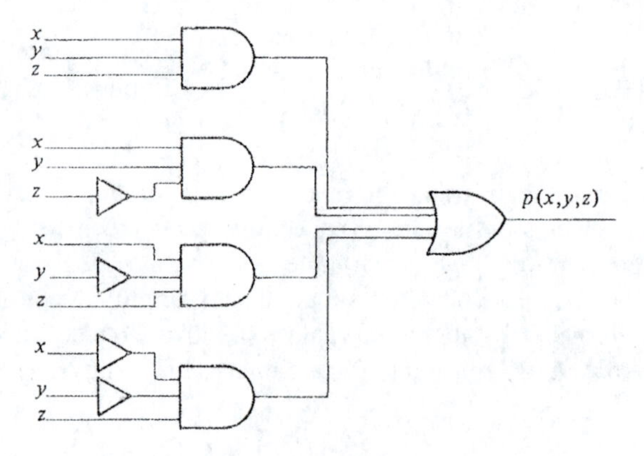

Now let us try to simplify the function $p(x, y, z)$ by using the properties given in Section 2.3. The first two minterms in $p(x, y, z)$ can be simplified using the distributive property. Perhaps the easiest way to see this is to let $a = x \wedge y$, for then we have

$$\begin{aligned}(x \wedge y \wedge z) \vee (x \wedge y \wedge z') &\equiv (a \wedge z) \vee (a \wedge z') \\ &\equiv a \wedge (z \vee z') \\ &\equiv a \wedge t \\ &\equiv a \\ &\equiv x \wedge y\end{aligned}$$

In a similar way, if we let $b = y' \wedge z$, then the last two terms of $p(x, y, z)$ can be simplified as follows

$$\begin{aligned}(x \wedge y' \wedge z) \vee (x' \wedge y' \wedge z) &\equiv (x \wedge b) \vee (x' \wedge b) \\ &\equiv (x \vee x') \wedge b \\ &\equiv t \vee b \\ &\equiv b \\ &\equiv y' \wedge z\end{aligned}$$

Thus, we see that

$$p(x, y, z) \equiv (x \wedge y) \vee (y' \wedge z)$$

That is, the Boolean function

$$q(x, y, z) = (x \wedge y) \vee (y' \wedge z)$$

is logically equivalent to $p(x, y, z)$, and so these two functions have the same truth table. Hence, the circuits of $p(x, y, z)$ and $q(x, y, z)$ have the same truth table. But the circuit of $q(x, y, z)$ is

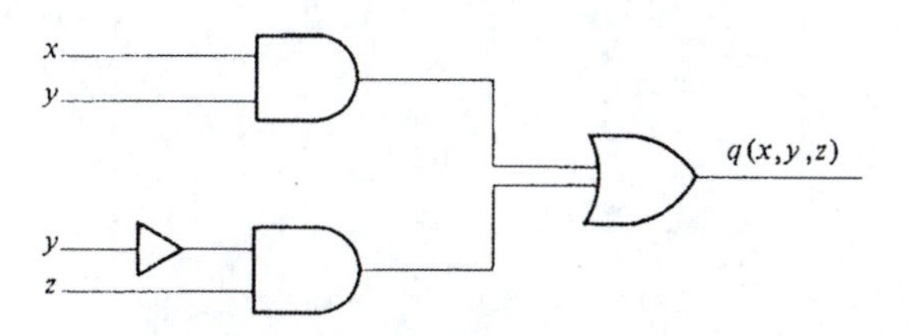

which is much simpler than the circuit of $p(x, y, z)$. □

There are many other methods for performing the simplification step in Diagram 2.6.1. However, none of them are entirely satisfactory. For example, the method that we used in the previous example relies on the ability to spot simplifications through the use of the properties given in Section 2.3. But, when the truth table is long, the disjunctive normal form can be quite complicated, and in this case such a method is very unreliable. In the next section, we will discuss another method for achieving this simplification.

Exercises

In Exercises 1 through 8, label all the wires in the given circuit. Then find the Boolean function of the circuit and compute the truth table of the circuit.

1.

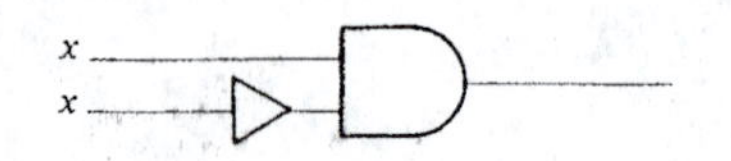

2.

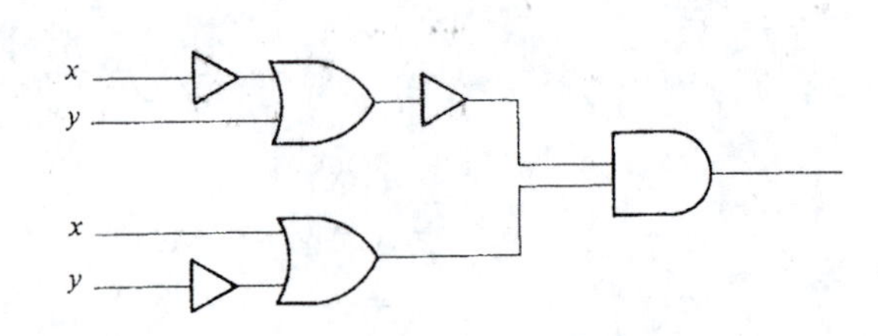

3.

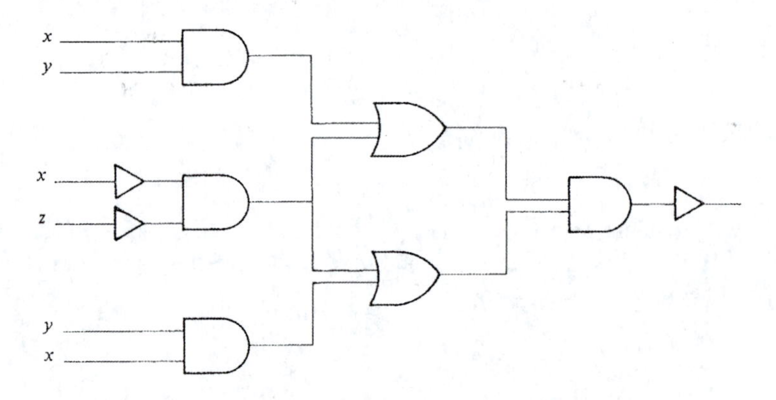

4.

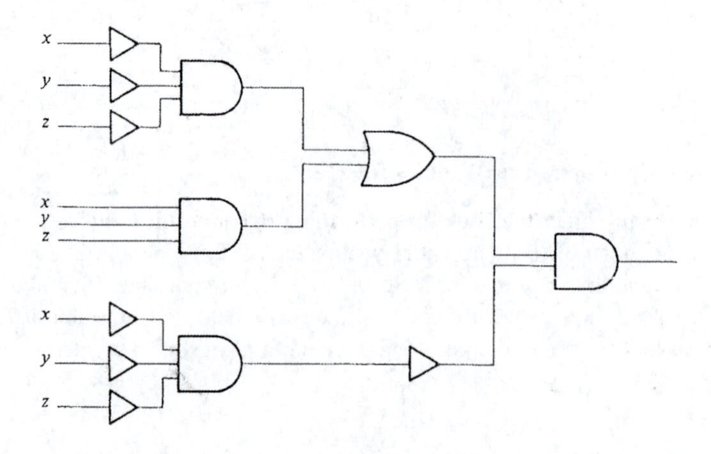

5.

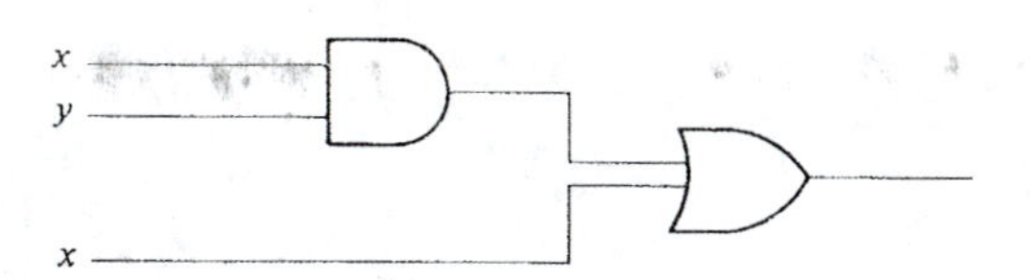

6.

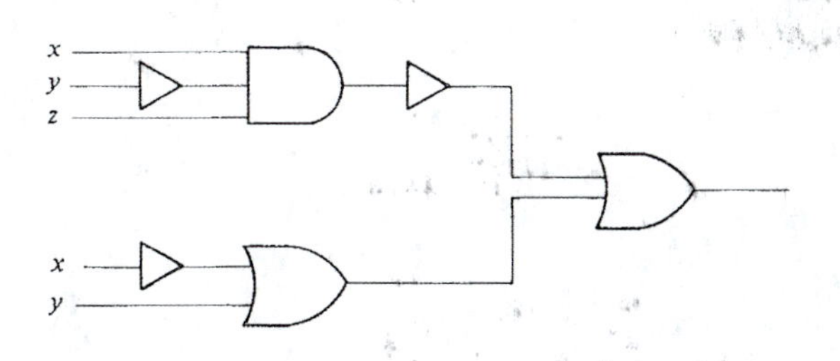

7.

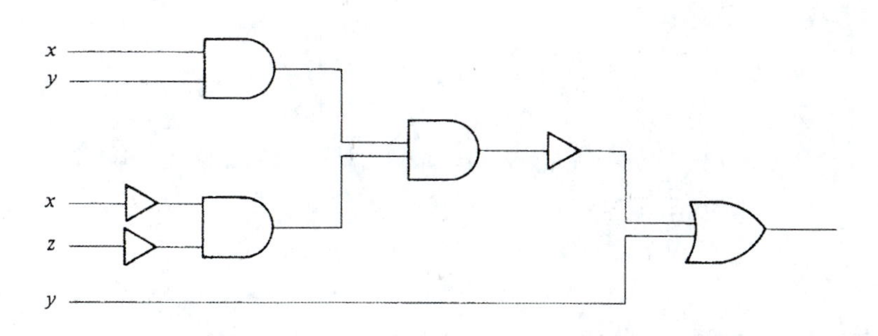

8.

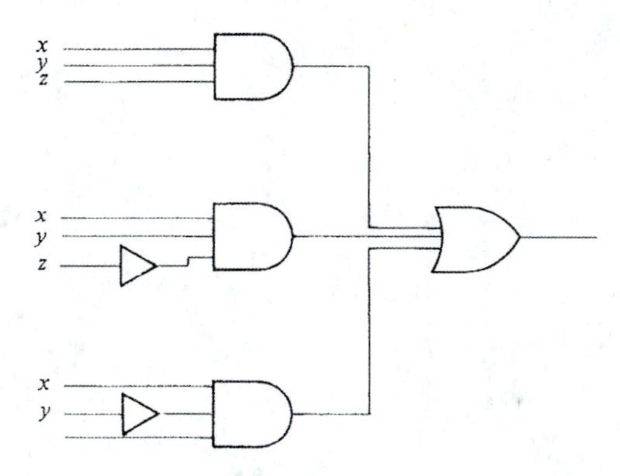

In Exercises 9 through 14, find a circuit with the given truth table.

9.

x	y	
1	1	0
1	0	1
0	1	1
0	0	0

10.

x	y	
1	1	0
1	0	0
0	1	0
0	0	0

11.

x	y	
1	1	0
1	0	0
0	1	1
0	0	0

12.

x	y	z	
1	1	1	0
1	1	0	0
1	0	1	1
1	0	0	0
0	1	1	0
0	1	0	1
0	0	1	0
0	0	0	0

13.

x	y	z	
1	1	1	1
1	1	0	0
1	0	1	1
1	0	0	0
0	1	1	1
0	1	0	0
0	0	1	1
0	0	0	0

14.

x	y	z	
1	1	1	1
1	1	0	0
1	0	1	0
1	0	0	1
0	1	1	0
0	1	0	1
0	0	1	1
0	0	0	0

In Exercises 15 through 18, carry out the procedure in Diagram 2.6.1, simplifying the function p by using the properties of conjunction, disjunction, and negation given in Section 2.3, just as we did in Example 2.6.3.

15.

x	y	
1	1	1
1	0	1
0	1	0
0	0	0

16.

x	y	z	
1	1	1	1
1	1	0	1
1	0	1	0
1	0	0	0
0	1	1	1
0	1	0	0
0	0	1	0
0	0	0	0

17.

x	y	z	
1	1	1	1
1	1	0	1
1	0	1	0
1	0	0	0
0	1	1	1
0	1	0	0
0	0	1	1
0	0	0	0

18.

x	y	z	
1	1	1	1
1	1	0	0
1	0	1	0
1	0	0	0
0	1	1	1
0	1	0	1
0	0	1	1
0	0	0	1

19. Design a logic circuit that inputs the values of three variables x, y, and z, and outputs a 1 if and only if $x = y$, and $x \neq z$.

20. Design a logic circuit that inputs the values of three variables x, y, and z, and outputs a 1 if and only if exactly two of the three variables are equal.

21. Design a logic circuit that inputs the values of three variables x, y, and z, and outputs a 1 if and only if $x < y$.

22. A certain room has an overhead light that is controlled by two switches. Whatever the status of the light, when either of these switches is thrown, the light changes status. Design a logic circuit that will make the switches behave in this manner.

23. A light is to be controlled by three switches, s_1, s_2, and s_3. Switch s_1 is an enabling switch. When it is in the on position, then the other two switches control the light. If the light is on, then flipping either s_2 or s_3 will turn it off, and if it is off, then flipping either s_2 or s_3 will turn it on. However, if s_1 is off, then switch s_3 is deactivated, and only switch s_2 will control the light. Design a logic circuit that will make the switches behave in this manner.

2.7 Karnaugh Maps

As you will recall from our discussion in the previous section, it is common to want to construct a circuit with a given truth table. The procedure that we used in the previous section can be described by the diagram

Truth table	→	Boolean function p in disjunctive normal form with that truth table	→	Boolean function q that is logically equivalent to p, but simpler	→	circuit whose Boolean function is q

As we discussed, the main difficulty that we face in implementing this procedure comes in the second step, namely, in finding a logically equivalent Boolean function that is simpler than the one in disjunctive normal form.

In this section, we will discuss an apprcach to overcoming this difficulty through the use of **Karnaugh maps**. This method is useful only for functions that involve at most four variables, and we will restrict our attention to the cases of two and three variables. (This will be sufficient to give you a feel for the method.)

Karnaugh Maps in Two Variables

A **Karnaugh map in two variables** is simply a square that has been divided into four smaller squares, as shown in Figure 2.7.1(a). We will refer to the smaller squares of the map as **subsquares**. As you can see, each subsquare corresponds to a different minterm, and since there are exactly four different minterms in two variables, these subsquares account for all of the possibilities.

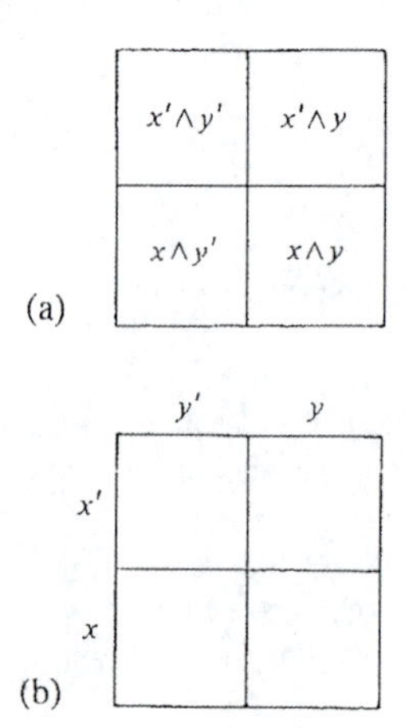

Figure 2.7.1

Since our procedure calls for putting additional information inside the subsquares, it will be more convenient to arrange it so that the minterm labels are not inside the subsquares. Thus, we will work instead with the map as shown in Figure 2.7.1(b), where we have placed the values of x and y at the sides of the square.

We will say that two subsquares are **adjacent** if they share a common side. Just as single subsquares correspond to minterms, which are expressions in two variables, pairs of adjacent subsquares correspond to expressions in only one variable. These correspondences are shown in Figure 2.7.2, where we have indicated the adjacent subsquares by using a pair of 1's.

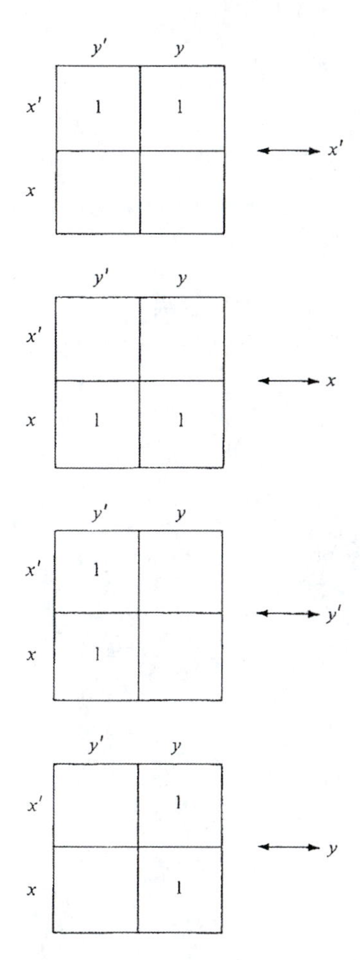

Figure 2.7.2

Now we are ready to outline the simplification procedure with an example.

Example 2.7.1

Consider the Boolean function

$$p(x, y) = (x' \wedge y) \vee (x \wedge y) \vee (x \wedge y')$$

We divide the procedure into three steps.

Step 1.

First we draw a Karnaugh map and place a 1 in each subsquare that is labeled with a minterm in p. Thus, in this case, we get the map

	y'	y
x'		1
x	1	1

Step 2.

Now we circle enough pairs of adjacent 1's so as to "cover" all of the 1's in the map, being careful not to use more circles than is necessary.

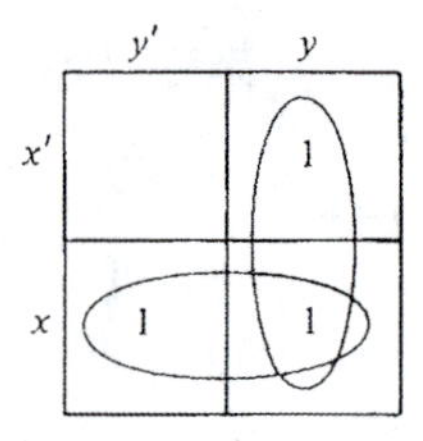

Step 3.

Now we are ready to form our simplified Boolean function $q(x, y)$. For each of the circles obtained in the previous step, we form the single variable expression obtained from Figure 2.7.2, and then "or" these single variable expressions, to get $q(x, y)$. In this case, we get the single variable expressions x and y, and so the simplified expression is

$$q(x, y) = x \vee y$$

We will leave it to you to show that $q(x, y)$ is logically equivalent to $p(x, y)$. There is no question that it is simpler. □

□ □ Karnaugh Maps in Three Variables

Karnaugh maps in three variables work on exactly the same principle as two variable maps. However, since there are more minterms in three variables than in two variables, the maps are a bit more complicated.

A **Karnaugh map in three variables** is a rectangle that has been divided into eight subsquares, as shown in Figure 2.7.3(a). As before, each subsquare is labeled with one of the eight possible minterms in three variables. Since we will also be placing 1's in some of the squares of this map, we will write it in the form shown in Figure 2.7.3(b).

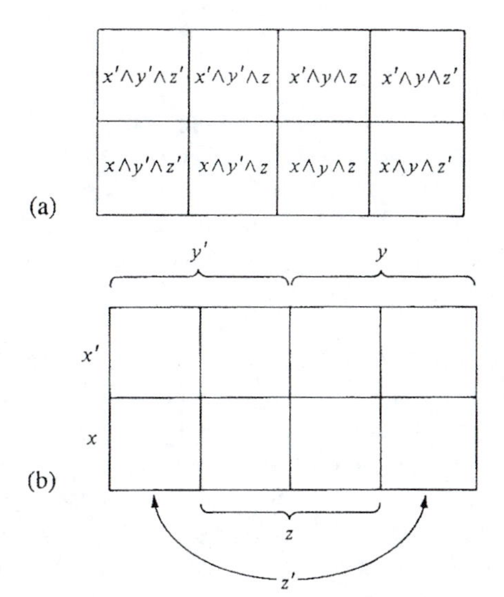

Figure 2.7.3

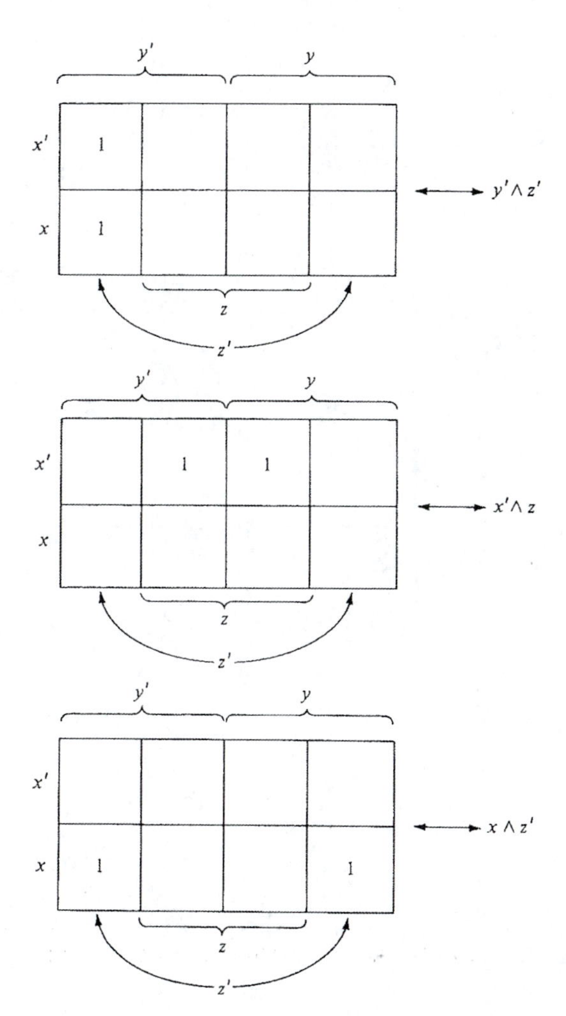

Figure 2.7.4

One of the reasons that the method of Karnaugh maps works is that the labeling in a Karnaugh map is designed so that adjacent subsquares have labels that differ in only one variable. For example, the last two subsquares in the first row have labels $x' \wedge y \wedge z$ and $x' \wedge y \wedge z'$, which differ only in the variable z.

But notice also that the labels in the first and last subsquares in each row differ by only one variable. For example, in the second row, the first and last subsquares are labeled $x \wedge y' \wedge z'$ and $x \wedge y \wedge z'$, and these labels differ only in the variable y. For this reason, we will also say that the first and last subsquares in each row are adjacent. One can think of the map as being wrapped into a cylinder, in which case the first and last columns would in fact be adjacent.

Before illustrating the technique for three variables, we must discuss the counterpart of Figure 2.7.2. In this case, each subsquare corresponds to a minterm, which is an expression involving three variables. However, adjacent pairs of subsquares correspond to expressions involving only two variables. Some examples are given in Figure 2.7.4. Furthermore, adjacent *quadruples* of subsquares correspond to expressions involving only one variable, as illustrated in Figure 2.7.5.

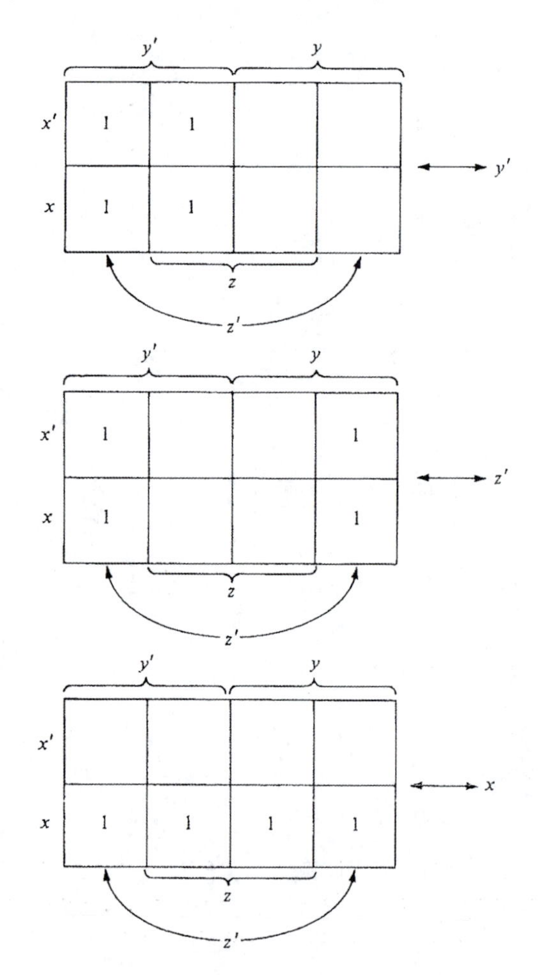

Figure 2.7.5

Now we are ready to illustrate the simplification procedure with some examples.

Example 2.7.2

Consider the Boolean function

$$p(x, y, z) = (x' \wedge y' \wedge z') \vee (x' \wedge y' \wedge z) \vee (x \wedge y' \wedge z') \vee (x \wedge y \wedge z) \vee (x' \wedge y \wedge z)$$

In order to simplify it, we proceed as follows.

Step 1.

First we draw a Karnaugh map and place a 1 in each subsquare that is labeled with a minterm in p. Thus, in this case, we get the map

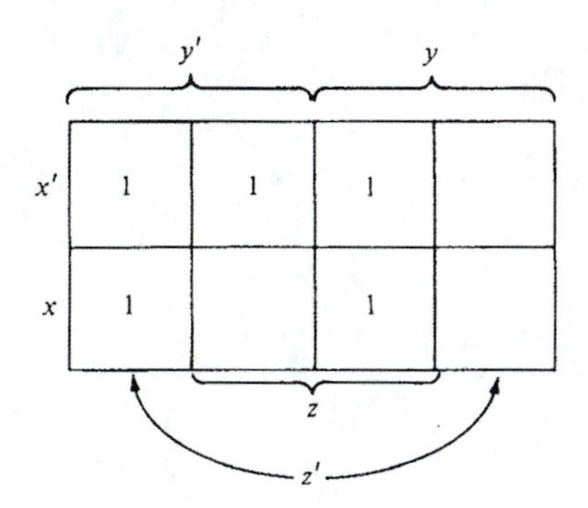

Step 2.

Now we circle enough pairs or quadruples of adjacent 1's so as to "cover" all of the 1's in the map, being careful not to use more circles than is necessary.

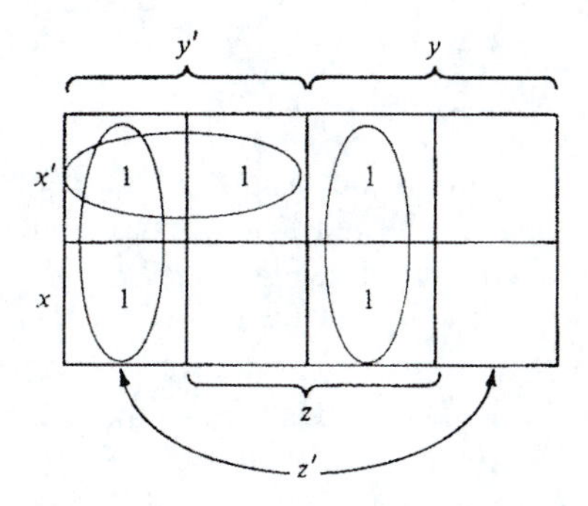

(Generally, when there are quadruples involved, it is best to circle them, since this will tend to make the final expression as simple as possible. Our next example will illustrate this.)

Step 3.

Now we are ready to form the simplified Boolean function $q(x, y, z)$. For each of the circles obtained in the previous step, we form the one- or two-variable expression, as indicated in Figures 2.7.4 and 2.7.5, and then we "or" these expressions to get $q(x, y, z)$. In this case, we get the expressions $y' \wedge z'$, $x' \wedge y'$ and $y \wedge z$, and so the simplified function is

$$q(x, y, z) = (y' \wedge z') \vee (x' \wedge y') \vee (y \wedge z)$$

Again, we will leave it to you to verify that $q(x, y, z)$ is logically equivalent to $p(x, y, z)$. It is certainly simpler. Let us consider one more example. □

Example 2.7.3

Let us simplify the Boolean function

$$p(x, y, z) = (x' \wedge y' \wedge z') \vee (x \wedge y \wedge z) \vee (x \wedge y' \wedge z) \vee (x' \wedge y' \wedge z) \vee (x' \wedge y \wedge z) \vee (x' \wedge y \wedge z')$$

Step 1.

First we draw a Karnaugh map and place a 1 in each subsquare that is labeled with a minterm in p.

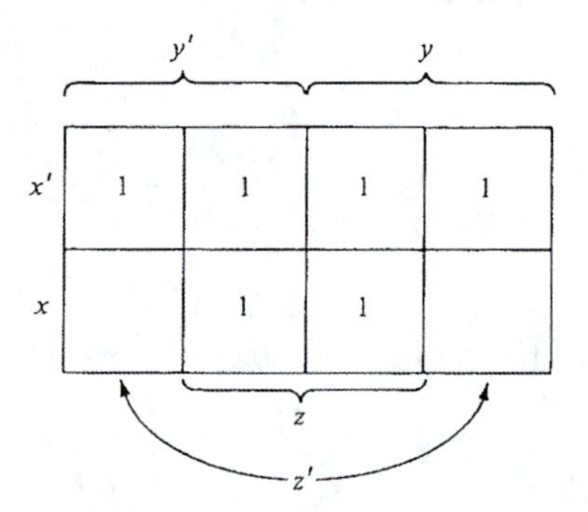

Step 2.

Now we circle enough pairs or quadruples of adjacent 1's so as to "cover" all of the 1's in the map, being careful not to use more circles than is necessary. This can be done in three ways, illustrated in Figure 2.7.6.

Step 3.

Using Figure 2.7.6(a), the two circles give us the two expressions x' and $x \wedge z$. Hence, we get the simplified function

$$q(x, y, z) = x' \vee (x \wedge z)$$

Using Figure 2.7.6(b), we get the two expressions z and $x' \wedge z'$, and so the simplified function is

$$r(x, y, z) = z \vee (x' \wedge z')$$

On the other hand, Figure 2.7.6(c) gives us the two expressions x' and z, and so we get the simplification

$$s(x, y, z) = x' \vee z$$

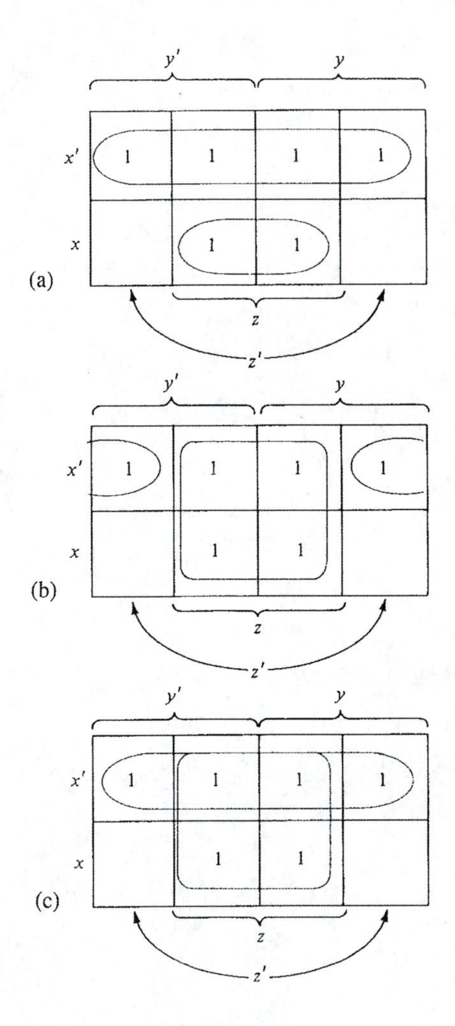

Figure 2.7.6

We will leave it to you to show that all of the functions $p(x, y, z)$, $q(x, y, z)$, $r(x, y, z)$, and $s(x, y, z)$ are logically equivalent. In this case however, the function $s(x, y, z)$ is the simplest of them all, and this supports the comment that we made in step 2 of the previous example that it is better to circle adjacent quadruples than adjacent pairs. □

□ □ Exercises

1. Show that the Boolean functions $p(x, y)$ and $q(x, y)$ in Example 2.7.1 are logically equivalent.
2. Show that the Boolean functions $p(x, y, z)$ and $q(x, y, z)$ in Example 2.7.2 are logically equivalent.
3. Show that the Boolean functions $p(x, y, z)$, $q(x, y, z)$, $r(x, y, z)$, and $s(x, y, z)$ in Example 2.7.3 are logically equivalent.

In Exercises 4 through 16, use the method of Karnaugh maps to simplify the given Boolean function.

4. $(x' \wedge y') \vee (x' \wedge y)$

5. $(x \wedge y) \vee (x' \wedge y)$

6. $(x' \wedge y') \vee (x \wedge y') \vee (x' \wedge y)$

7. $(x \wedge y') \vee (x' \wedge y) \vee (x' \wedge y')$

8. $(x' \wedge y' \wedge z') \vee (x' \wedge y \wedge z) \vee (x' \wedge y' \wedge z) \vee (x' \wedge y \wedge z')$

9. $(x \wedge y' \wedge z) \vee (x' \wedge y' \wedge z') \vee (x' \wedge y' \wedge z) \vee (x \wedge y' \wedge z')$

10. $(x' \wedge y' \wedge z') \vee (x \wedge y \wedge z') \vee (x \wedge y' \wedge z') \vee (x' \wedge y \wedge z')$

11. $(x' \wedge y' \wedge z') \vee (x' \wedge y \wedge z') \vee (x \wedge y' \wedge z) \vee (x \wedge y \wedge z)$

12. $(x' \wedge y' \wedge z) \vee (x \wedge y' \wedge z) \vee (x' \wedge y \wedge z') \vee (x \wedge y \wedge z')$

13. $(x \wedge y' \wedge z') \vee (x \wedge y' \wedge z) \vee (x \wedge y \wedge z)$

14. $(x' \wedge y' \wedge z') \vee (x' \wedge y' \wedge z) \vee (x' \wedge y \wedge z') \vee (x \wedge y' \wedge z) \vee (x \wedge y \wedge z')$

15. $(x' \wedge y \wedge z) \vee (x' \wedge y \wedge z') \vee (x \wedge y' \wedge z') \vee (x \wedge y' \wedge z) \vee (x \wedge y \wedge z)$

16. $(x' \wedge y' \wedge z') \vee (x' \wedge y' \wedge z) \vee (x' \wedge y \wedge z) \vee (x' \wedge y \wedge z')$
$\vee (x \wedge y' \wedge z') \vee (x \wedge y' \wedge z)$

Happy will that house be in which the relations are formed from character.
—Emerson, *Society and Solitude: Domestic Life*

It is a piece of luck to have relations scarce.
—Menander, *Thupopos*

It is a melancholy truth, that even great men have their poor relations.
—Charles Dickens, *Bleak House*

Chapter THREE

Relations on Sets

3.1 Relations

The world is full of relationships. There are father-son relationships and mother-daughter relationships, there are employee-employer relationships, there are student-teacher relationships, and so on. Similarly, the world of mathematics and computer science is full of relationships. For example, there is the "less than" relationship, the "greater than" relationship, the "subset" relationship, the "same size as" relationship, and the "logically equivalent" relationship, to name but a few. In this chapter, we want to study mathematical relationships. In mathematics, however, we use the term *relation*, rather than relationship. Let us begin with a definition.

Definition

Let A and B be nonempty sets. A **binary relation** from A to B is a subset of the cartesian product A × B. A binary relation from A to A is usually called a binary relation *on* A. □

Let R be a binary relation from A to B; that is, let R be a subset of A × B. If the ordered pair (a, b) is in R, then we say that a **is related to** b. On the other hand, if $(a, b) \notin R$, then we say that a **is not related to** b. It is very common to write aRb in place of $(a, b) \in R$ and $a\not R b$ in place of $(a, b) \notin R$. Thus, we can specify a binary relation by specifying the conditions under which aRb.

Let us consider some examples of binary relations.

Example 3.1.1

Let A = {1, 2, 3} and B = $\{a, b, c\}$. Then the set

$$R = \{(1, a), (1, c), (2, c)\}$$

is a binary relation from A to B. In this case, we see that 1 is related to both a and c and that 2 is related to c. In symbols, we have $1Ra$, $1Rc$, and $2Rc$. Also, $1\not R b$, $2\not R a$, $2\not R b$, $3\not R a$, $3\not R b$, and $3\not R c$. □

Example 3.1.2

The set **R** of all real numbers has many important binary relations defined on it. For example, equality is actually a binary relation on **R**. That is, we can define a binary relation R on **R** by specifying that aRb if and only if a is equal to b. The equality relation is denoted by the symbol =, and so we write $a = b$ instead of aRb.

We can also define a binary relation S on the set **R** by specifying that aSb if and only if a is less than or equal to b. As you know, this relation is denoted by the symbol $\leq$, and so we write $a \leq b$ rather than aSb. □

Example 3.1.3

The set **Z** of all integers has some rather interesting binary relations defined on it. For example, we can define a binary relation R on **Z** by saying that mRn if and only if m divides n, that is, if and only if there is an integer k with the property that $n = km$. (In other words, n is a multiple of m.) This relation is generally denoted by a bar, and so we write $m|n$ to denote the fact that m divides n. For example, we have $2|4$ and $3|45$, but $7\nmid 8$ and $4\nmid 2$.

As another example of a binary relation on **Z**, let k be a positive integer. Then we can define a relation S on **Z** by specifying that mSn if and only if k divides $m - n$. This is equivalent to saying that mSn if and only if m and n have the same remainder after being divided by k. (See Exercise 22.) Since this relation is very important, it has a special notation. Namely, if mSn, then we write $m \equiv n \pmod{k}$, and read this as "m is congruent to n modulo k." For example, we have $8 \equiv 5 \pmod{3}$ and $-7 \equiv -15 \pmod{4}$, but $6 \not\equiv 2 \pmod{3}$. We will discuss this relation further in Section 3.3. □

Example 3.1.4

If S is a set, then we can define a relation R on $\mathcal{P}(S)$ by specifying that ARB if and only if A is a subset of B. This relation is denoted by the symbol $\subset$, and so we write A $\subset$ B instead of ARB. □

Example 3.1.5

Let P = $\{p_1, p_2, p_3, \ldots\}$ be a (countable) set of statement variables, and let *Stat*(P) be the set of all statements (both simple and compound) that can be formed using these variables and the connectives $\wedge$, $\vee$, and $\sim$. (See Chapter 2.) Then we can define a binary relation R on *Stat*(P) by specifying that ARB if and only if statement A is logically equivalent to statement B. In Section 2.3, we used the symbol $\equiv$ to denote the relation of logical equivalence. □

Example 3.1.6

Let Σ be an alphabet. Then there are many important binary relations that we can define on the set Σ^* of all words over Σ. For example, we can define a relation R by specifying that wRu if and only if w has the same length as u, that is, if and only if $\ell(w) = \ell(u)$.

Also, we can define a relation S on Σ^* by specifying that wSu if and only if w is shorter than u, that is, if and only if $\ell(w) < \ell(u)$.

As another example, if a $\in \Sigma$, then we can define a binary relation T on Σ^* by specifying that wTu if and only if w and u have the same number of occurrences of the letter a. As you can imagine, there are many other possibilities for relations on Σ^*. □

Example 3.1.7

When a computer programmer inputs a program, written in a high-level language such as BASIC or Pascal, into a computer, the computer must first translate it into a lower-level language before it can execute the instructions. One of the steps in doing this is to create a **symbol table**, which contains, among other things, the names of any variables that were used in the program, along with their values. For example, consider the following portion of a symbol table:

Name of Variable	*Value*
X	10
Y	12
N	100
SUM	200
Q	TRUE
C$	"GO"

This table is nothing but a binary relation from the set A of all variable names to the set B of all possible values of the variables, where each row of the table corresponds to an ordered pair in the relation. Thus, in this example, we have (X)R(10), (Y)R(12), (N)R(100), (SUM)R(200), (Q)R(TRUE), and (C$)$R$("GO"). As you can easily see, the table format is far simpler in this case than the ordered pair format. □

Example 3.1.8

A computer program is frequently made up of several different parts, known as *subroutines*. Generally speaking, the control of the program can be passed from certain subroutines to certain others. This leads us to define a binary relation on the set of all subroutines of a program. In particular, we say that $s_1 R s_2$ if control of the program can pass *directly* from subroutine s_1 to subroutine s_2. The relation R might be called the *flow of control relation*.

In a similar way, data can be passed from certain subroutines to certain others, by means of parameters, for instance. Thus, we can define the *flow of data relation* by saying that $s_1 S s_2$ if and only if subroutine s_1 can pass data to subroutine s_2. □

This short list of examples should begin to convince you that binary relations are very common. In order to continue our discussion of binary relations, we need to make a definition. By a **directed graph**, we mean a set of points, together with a set of directed arcs connecting some of these points. Figures 3.1.1 and 3.1.2 are examples of directed graphs.

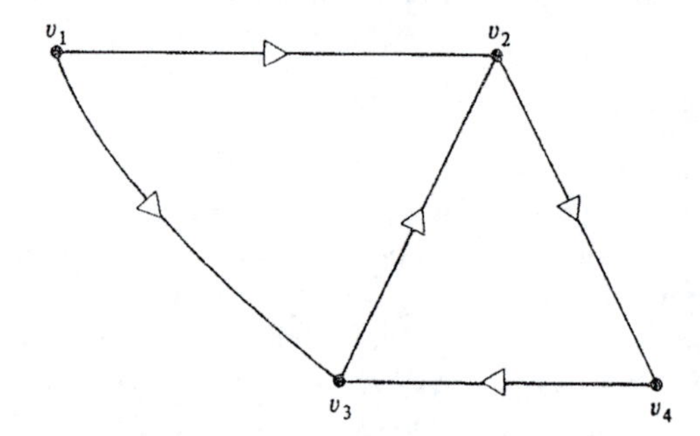

Figure 3.1.1

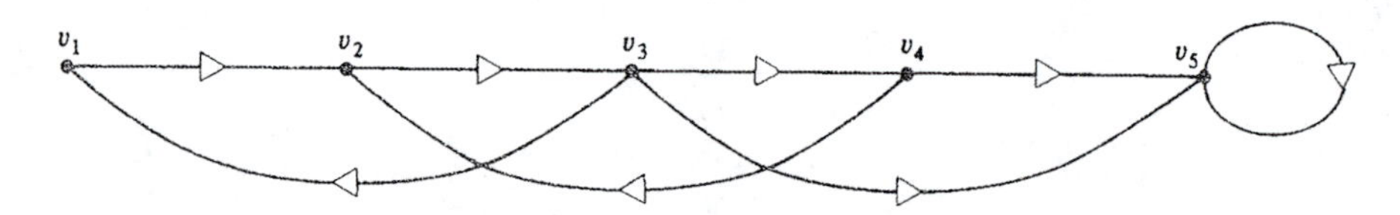

Figure 3.1.2

The points of a directed graph are usually called the **vertices** (singular: vertex) or the **nodes** of the graph. As you can see, we have labeled each vertex for the purposes of identification. If D is a directed graph, then we will let $\mathcal{V}(D)$ denote the set of vertices of D.

Notice also that we have denoted the direction of each arc by a small arrow. Thus, for example, in Figure 3.1.1, there is an arc *from* vertex v_1 *to* vertex v_2, but no arc from vertex v_2 to vertex v_1. The arcs of a directed graph are usually denoted by ordered pairs of vertices. For example, the arcs of the directed graph in Figure 3.1.1 are

$$(v_1, v_2), (v_3, v_1), (v_4, v_2), (v_4, v_3), (v_2, v_3)$$

Notice that the directed graph in Figure 3.1.2 has an arc from vertex v_5 to itself. Thus, the ordered pair (v_5, v_5) is an arc of this graph. An arc of this type is called a **loop**.

Directed graphs are one of the most useful tools in the applications of mathematics, especially to computer science, and we will devote all of Chapter 6 to studying graphs (both undirected and directed). Incidentally, do not confuse this concept of a graph with the graph of a function. The two types of graphs are not the same. (However, see Exercise 16.)

Directed graphs are very closely connected with binary relations. It is easy to see that the arcs of a directed graph D define a binary relation R on the vertex set $\mathcal{V}(D)$. We simply define R by saying that uRv if and only if (u, v) is an arc in D. As an example, the directed graph

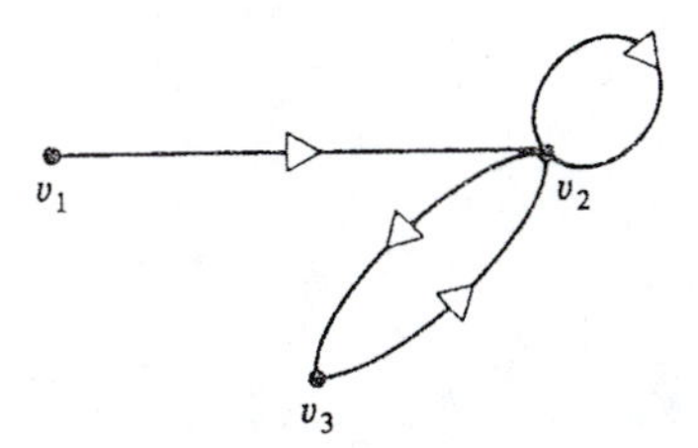

defines the binary relation

$$R = \{(v_1, v_2), (v_2, v_2), (v_2, v_3), (v_3, v_2)\}$$

However, there is much more to the story than this. It so happens that any binary relation on a *finite* set A can be represented by a directed graph D. For if R is a binary relation on A, then we can define a directed graph D by taking the vertices of D to be the elements of A and by letting (u, v) be an arc in D if and only if uRv. We call the directed graph D the **graph of the relation** R.

Example 3.1.9

Let A = {1, 2, 3, 4}. Then the equality relation on A has graph

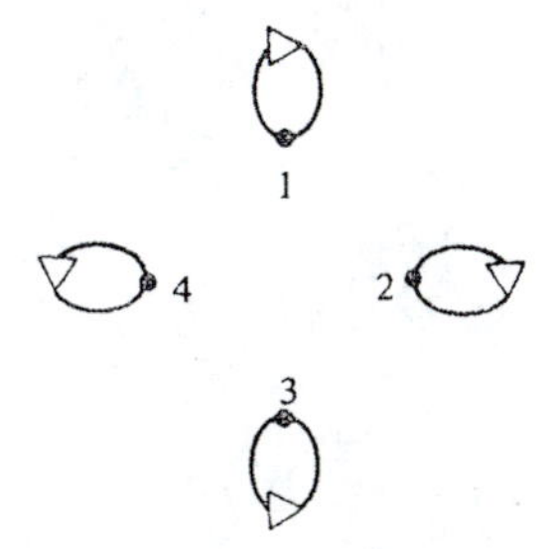

Example 3.1.10

Let S = $\{a, b\}$. Then the subset relation on $\mathcal{P}$(S) has graph

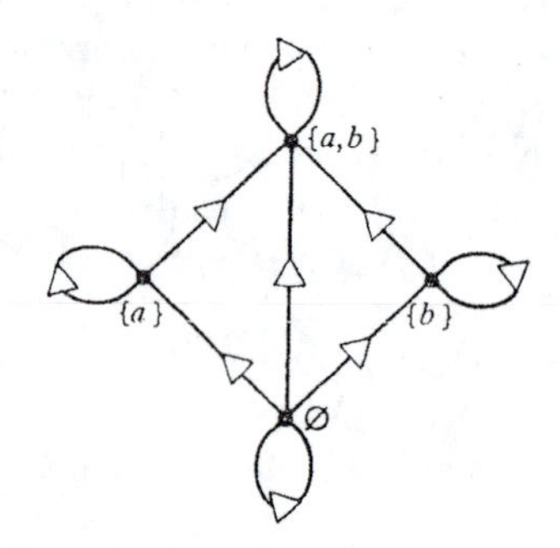

□

■ Example 3.1.11

Let A = {1, 2, 3, 4, 5, 6}. Then the relation "n divides m" on A has graph

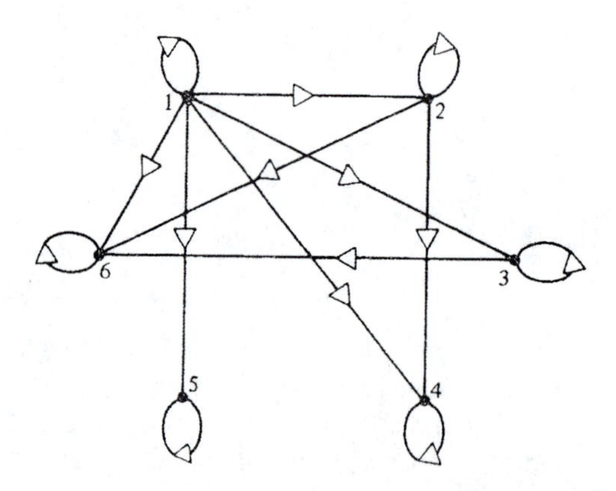

□

We should emphasize that a relation on a set A has a graph if and only if A is a *finite* set. Thus, from now on, whenever we mention the graph of a relation, we will assume that the relation is defined on a finite set.

If R is a relation from A to B, then we can define a relation from B to A, denoted by R^{-1} and called the **inverse** of R, by specifying that $bR^{-1}a$ if and only if aRb. Thus, we have

$$R^{-1} = \{(b, a) \mid (a, b) \in R\}$$

If R is a binary relation on a finite set A, then the graph of R^{-1} can be obtained from the graph of R simply by reversing the directions of all of the arcs.

■ Example 3.1.12

The inverse of the relation R of Example 3.1.1 is the relation

$$R^{-1} = \{(a, 1), (c, 1), (c, 2)\}$$

□

■ Example 3.1.13

The inverse of the subset relation R on $\mathcal{P}(S)$ is the superset relation on $\mathcal{P}(S)$. This follows from the fact that $AR^{-1}B$ if and only if BRA, that is, if and only if $B \subset A$. In other words, $AR^{-1}B$ if and only if $A \supset B$. □

Example 3.1.14

Let Σ be an alphabet, and let R be the relation "is shorter than" defined on the set Σ^*. (See Example 3.1.6.) Then the inverse of R is the relation "is longer than." □

Example 3.1.15

Let *Stat*(P) be the set of all statements, and let $\equiv$ be the relation of logical equivalence. (See Example 3.1.5.) Then since $A \equiv B$ if and only if $B \equiv A$, we see that the relation $\equiv$ is its own inverse! □

Let us conclude this section by mentioning that there are other types of relations on sets besides binary relations, although binary relations are by far the most useful. We will not study other types of relations, but we should at least give the definitions, and a simple example. (We will include the definition of a binary relation for completeness.)

Definition

Let A be a nonempty set.

a) Then a **unary relation** on A is simply a subset of A.

b) A **binary relation** on A is a subset of the cartesian product $A \times A$.

c) A **ternary relation** on A is a subset of the cartesian product $A \times A \times A$.

d) In general, if $n \geq 1$, then an ***n*-ary relation** on A is a subset of the cartesian product $A \times A \times \cdots \times A$, where the product has n factors. □

Example 3.1.16

a) Let $A = \{1, 2, 3, 4, 5\}$. Then the set $R = \{1, 3\}$, being a subset of A, is a unary relation on A.

b) Let $A = \{1, 2, 3, 4, 5\}$. Then the set

$$R = \{(1, 1, 1), (1, 2, 3), (1, 4, 5), (2, 2, 3)\}$$

being a subset of the cartesian product $A \times A \times A$, is a ternary relation on A.

c) Let A be any set. Then the set

$$R = \{(a, a, \ldots, a) \mid a \in A\}$$

of all n-tuples of the form $(a, a, \ldots, a)$ being a subset of the cartesian product $A \times A \times \cdots \times A = A^n$ is an n-ary relation on A. It is called the **diagonal relation** on A. □

From now on, whenever we use the term "relation" without modification, we will mean binary relation.

Exercises

In Exercises 1 through 13, find the graph of the relation R on the set A.

1. Let $A = \mathcal{P}(\{1, 2, 3\})$ and define R by BRC if and only if B is a proper subset of C.
2. Let $A = \mathcal{P}(\{1, 2, 3\})$ and let R be the subset relation $\subset$.
3. Let $A = \{1, 2, 3, 4\}$ and let R be the relation $<$.

4. Let A $= \{1, 2, 3, 4, 5, 6, 7, 8, 9\}$ and let R be defined by aRb if and only if $a|b$.
5. Let A $= \{1, 2, 3, 4, 5\}$ and let R be defined by aRb if and only if a divides $b + 1$.
6. Let A $= \{1, 2, 3, 4, 5, 6\}$ and let R be the relation of congruence modulo 2. (This is the relation defined in Example 3.1.3, for $k = 2$.)
7. Let A $= \{1, 2, 3, 4, 5, 6\}$ and let R be the relation defined by specifying that aRb if and only if a and b have the same parity, that is, if and only if a and b are either both even or both odd. Does this graph resemble the graph of the relation in Exercise 6? What can you conclude from this?
8. Let A $= \{1, 2, 3, 4, 5, 6\}$ and let R be the relation of congruence modulo 3. (See Example 3.1.3.)
9. Let A $= \{-2, -1, 0, 1, 2\}$. Define a relation on A by setting aRa' if and only if $f(a) = f(a')$, where $f(x) = x^2$.
10. Let $\Sigma = \{0, 1\}$. Define a relation R on the set A $= \Sigma_3$ of all binary words of length 3 by saying that wRu if and only if w and u have the same number of 0's.
11. Let $\Sigma = \{0, 1\}$. Define a relation R on the set A $= \Sigma_4$ of all binary words of length 4 by saying that wRu if and only if w and u have the same number of 0's.
12. Let $\Sigma = \{0, 1, 2\}$. Define a relation R on the set A $= \Sigma_2$ of all ternary words of length 2 by saying that wRu if and only if w and u have the same number of 0's.
13. Let $\Sigma = \{0, 1\}$. Define a relation R on the set A $= \Gamma_3$ of all binary words of length at most 3 by saying that wRu if and only if $\ell(w) = \ell(u)$.

A binary relation on a subset A of real numbers can be defined by means of an equation in two variables. For example, if A $= \{-2, -1, 0, 1, 2\}$, *then the equation* $x + y = 0$ *defines the relation R on A given by aRb if and only if* $a + b = 0$. *Thus in this case,* $R = \{(-2, 2), (-1, 1), (0, 0), (1, -1), (2, -2)\}$. *Exercises 14 through 16 concern this type of relation.*

14. Let A $= \{-2, 0, 7\}$ and let R be the relation obtained from the equation $x^2 + y = 7$. Find the graph of R.
15. Let A $= \{0, 1, 2, 3, 4, 5\}$ and let R be the relation obtained from the equation $x^2 + y^2 = 100$. Find the graph of R.
16. The equation $x^2 + y^2 = 4$ defines a binary relation on the set **R** of all real numbers. However, since **R** is an infinite set, it does not have a directed graph, as defined in this section. Do you think that this relation can be described by any other type of graph? If so, sketch it. Do you now see a resemblance between the two different concepts of a graph? Explain.
17. Let R be the relation of congruence modulo 5 on **Z**. (See Example 3.1.3.) Answer true or false.
    a) $6 \equiv 6 \pmod 5$
    b) $a \equiv a \pmod 5$ for all $a \in \mathbf{Z}$
    c) $3 \equiv 8 \pmod 5$
    d) $a \equiv a + 1 \pmod 5$ for some $a \in \mathbf{Z}$
18. What is the inverse of the relation $\leq$ on the set of all real numbers?
19. Find the inverse of the relation R (on the set of all real numbers) defined by the equation $2x + 3y = 0$. (That is, R is defined by aRb if and only if $2a + 3b = 0$.)
20. What is the inverse of the relation "is older than" defined on the set A of all people?
21. Find the inverse of the relation ARB if and only if A logically implies B.
22. Referring to the relation R of congruence modulo k, defined in Example 3.1.3, show that k divides $m - n$ if and only if m and n have the same remainder after dividing by k.
23. a) Let R be the relation of equality, defined on a set A. What is R^{-1}?
    b) Find a relation on a set A, other than the relation of equality, that has the property of being equal to its own inverse.
    c) How can you describe the fact that a relation R has the property that $R = R^{-1}$ in terms of the graph of R?

24. Consider the equation $x^2 + y^2 + z^2 = 9$. How could you use this equation to define a ternary relation on the set of all real numbers? Describe the ternary relation that this equation defines on the set of all *positive* integers.

3.2 Properties of Relations

In this section, we want to continue our study of binary relations on sets. There are several important properties that a binary relation on a set can satisfy. The most important of these are given in the next definition.

Definition

Let R be a binary relation on a set A. Then

1) R is said to be **reflexive** if

$$aRa$$

for all a in A.

2) R is said to be **symmetric** if

$$aRb \text{ implies that } bRa$$

3) R is said to be **antisymmetric** if

$$aRb \text{ and } bRa \text{ imply that } a = b$$

4) R is said to be **transitive** if

$$aRb \text{ and } bRc \text{ imply that } aRc$$

□

One of the best ways to get a feel for these properties is to consider their effect on the graph of a relation. (Of course, we are now referring to relations on a *finite* set A.)

It is easy to see that a relation is reflexive if and only if its graph has a loop at each vertex. [A *loop* is an arc of the form (v, v).] A relation is symmetric if and only if whenever its graph contains the arc (a, b), then it also contains the arc (b, a). A relation is antisymmetric if and only if its graph cannot contain both of the arcs (a, b) and (b, a) for any pair of *distinct* vertices a and b. Finally, a relation is transitive if and only if whenever its graph contains arcs of the form (a, b) and (b, c), then it also contains the arc (a, c), completing the "triangle" with vertices a, b, and c.

Using this information about the graph of a relation, it is easy to construct examples of relations that satisfy some of these properties but not others. For example, the graph

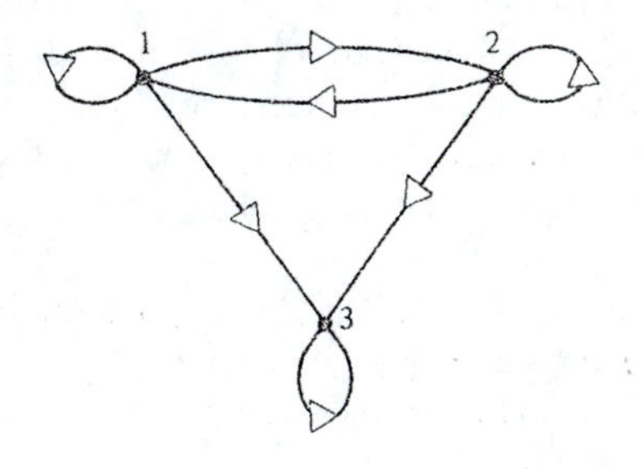

is the graph of a relation, on the set A $= \{1, 2, 3\}$, that is reflexive and transitive, but neither symmetric nor antisymmetric. This relation, call it R, is reflexive since $1R1$, $2R2$, and $3R3$. However, R is not symmetric since, for example, $1R3$ but $3 \not R 1$, and R is not antisymmetric since $1R2$ and $2R1$, but 1 is not equal to 2. Finally, to see that R is transitive, we must check that whenever aRb and bRc, we also have aRc. Making a list of the cases where aRb and bRc, we get

$$1R2 \quad \text{and} \quad 2R3$$

and

$$2R1 \quad \text{and} \quad 1R3$$

But since we also have $1R3$ and $2R3$, we see that R is transitive.

Let us consider what happens if a relation R is both symmetric and antisymmetric. If aRb, then the symmetry of R implies that bRa, and the antisymmetry of R then implies that $a = b$. Thus, in order for a relation R to be both symmetric and antisymmetric, it must have the property that no two distinct elements of A can be related. If A is a finite set, then the graph of R must contain no edges other than loops. There are many such relations (see Exercise 18), but they are not really very useful.

By far the most important types of binary relations are those that are either reflexive, symmetric and transitive, or else reflexive, antisymmetric, and transitive. Accordingly, we will devote the next two sections to a discussion of these two types of relations. For now, let us continue our discussion of the properties of relations in general.

Example 3.2.1

The equality relation on a set A is reflexive since $a = a$ for all a in A. Also, it is symmetric since $a = b$ implies that $b = a$, and it is transitive since $a = b$ and $b = c$ imply that $a = c$. □

Example 3.2.2

The relation $\leq$ on the set of all real numbers is reflexive since $a \leq a$ for all real numbers a. Also, it is antisymmetric since if $a \leq b$ and $b \leq a$ then we must have $a = b$. Finally, it is transitive since if $a \leq b$ and $b \leq c$ then $a \leq c$.

On the other hand, the relation $<$ on the set of all real numbers is neither reflexive nor symmetric, but it is transitive and antisymmetric. (See Exercise 1.) □

Example 3.2.3

Consider the relation of congruence modulo k, defined on the set $\mathbf{Z}$ of all integers. Recall that $m \equiv n \pmod{k}$ if and only if k divides $m - n$. This relation is reflexive, since $n \equiv n \pmod{k}$. Also, it is symmetric since if $m \equiv n \pmod{k}$ then k divides $m - n$, and so k also divides $n - m$, that is, $n \equiv m \pmod{k}$. We leave it as an exercise to show that congruence modulo k is transitive. □

Example 3.2.4

Let Σ be an alphabet. Then the relation R defined by wRu if and only if $\ell(w) = \ell(u)$ is reflexive, symmetric, and transitive. However, the relation S defined by wSu if and only if w is shorter than u is neither reflexive nor symmetric, but it is transitive. (See Exercises 8 and 9.) ▯

■ Example 3.2.5

Each of the properties that we have been discussing has a special meaning for the flow of control relation that we discussed in Example 3.1.8. To say that a flow of control relation R is reflexive is to say that a subroutine can pass control to itself. Of course, this depends on how we interpret the term "pass control." If we want to think of a subroutine as being able to pass control to itself, then the relation is reflexive. Otherwise, it is not. In any case, this is not a crucial issue.

To say that a flow of control relation R is symmetric is to say that if subroutine s_1 can pass control to subroutine s_2, then s_2 can also pass control to s_1. To say that a flow of control relation is antisymmetric is to say that if s_1 and s_2 can pass control to each other, then they must be the same subroutine. Finally, to say that a flow of control relation is transitive is to say that if subroutine s_1 can pass control (directly) to subroutine s_2, and if s_2 can pass control (directly) to s_3, then s_1 can pass control (directly) to s_3.

Whether any of these properties holds depends, of course, on the particular program. It is not hard to imagine that some flow of control relations have these properties and some do not. ▯

Let us continue our discussion of the flow of control relation that we began in the previous example. Given a program that contains subroutines, we can define another binary relation R' by saying that $s_1R's_2$ if and only if control of the program can pass *eventually* from subroutine s_1 to subroutine s_2, perhaps making "intermediate stops" at other subroutines. This is different from the flow of control relation R, which has the property that s_1Rs_2 if and only if control can pass *directly* from s_1 to s_2.

As we pointed out in the previous example, the relation R need not be transitive. However, the relation R' is always transitive. For if $s_1R's_2$ and $s_2R's_3$, then control can pass eventually from s_1 to s_2, and also pass eventually from s_2 to s_3. Therefore, control can pass eventually from s_1 to s_3.

The relationship between the relations R and R' can be summarized in the following list.

1) The relation R' *contains* the relation R, in the sense that if wRu, then $wR'u$. In terms of sets of ordered pairs, the relation R is a *subset* of the relation R'.
2) The relation R' is transitive.
3) The relation R' is the *smallest* relation that contains R and is transitive. That is, if S is another relation that contains R and is transitive, then S is larger than R'.

Because R' has these properties, it is called the **transitive closure** of the relation R.

Any binary relation, on any set, has a transitive closure. In simple terms, the transitive closure of a relation R is the smallest relation that contains R and is transitive. Loosely speaking, the transitive closure of a relation R is obtained by adding only as many ordered pairs as is necessary to make the relation transitive. Of course, if R

itself is transitive, then it is its own transitive closure, in symbols, $R^t = R$. However, if the relation is not transitive, then its transitive closure R^t is bigger than R.

As you can see from the previous example, the concept of a transitive closure can be very useful. In fact, we are frequently more interested in the transitive closure of a relation than in the relation itself.

The transitive closure of a relation on a finite set can be described in terms of the graph of the relation. Suppose that R is a relation with graph D. Then we know that aRb if and only if there is an arc in D going from vertex a to vertex b. The transitive closure can be described by saying that aR^tb if and only if there is a *path* in D going from vertex a to vertex b. Of course, such a path must have all of its arrows pointing in the proper direction. As an example, consider the directed graph in Figure 3.2.1, which we assume to be the graph of a relation R.

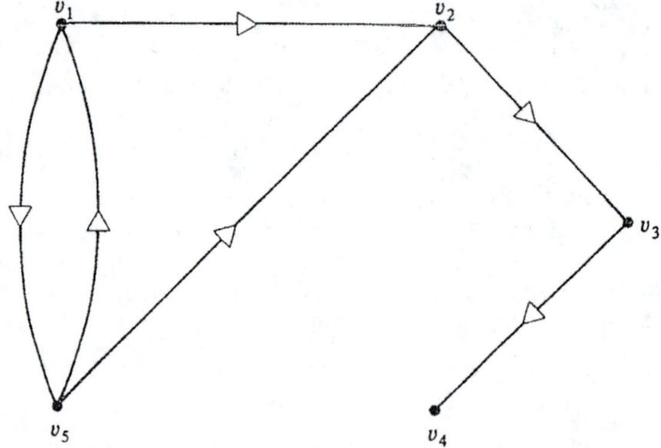

Figure 3.2.1

Then R is not transitive, since v_1Rv_2 and v_2Rv_3, but $v_1\not{R}v_3$. On the other hand, since there is a path from v_1 to v_3, we have $v_1R^tv_3$. Also, since there is a path from v_1 to v_4, we have $v_1R^tv_4$. However, there is no path from v_4 to v_1, and so we have $v_4\not{R}^tv_1$.

As you know, a binary relation R on a set A is just a set of ordered pairs, that is, a subset of A $\times$ A. Therefore, we can take the complement of a binary relation, as well as the union and intersection of two binary relations. Let us examine these concepts.

The complement of a binary relation R is denoted by R^c. According to the definition of complement, we have $(a, b) \in R$ if and only if $(a, b) \notin R^c$. In other words,

$$aR^cb \text{ if and only if } a\not{R}b$$

■ Example 3.2.6

The complement of the relation "is equal to" is the relation "is not equal to." The equality relation is denoted by the symbol $=$, and its complement is denoted by the symbol $\neq$.

The complement of the relation "less than or equal to" on the set **R** of all real numbers is the relation "not less than or equal to," or "greater than." In symbols, "less than or equal to" is denoted by $\leq$, and there are two common notations for the complement of this relation, namely, $\not\leq$ and $>$. □

Now let R and S be binary relations *on the same set* A. Then according to the definition of union, we have $(a, b) \in R \cup S$ if and only if either $(a, b) \in R$ or $(a, b) \in S$ (or both). In other symbols, we have

$$a(R \cup S)b \text{ if and only if either } aRb \text{ or } aSb$$

Similarly, we have

$$a(R \cap S)b \text{ if and only if } aRb \text{ and } aSb$$

Example 3.2.7

Let A = {1, 2, 3, 4, 5}, and let R and S be binary relations defined by

$$R = \{(1, 1), (1, 4), (2, 3), (2, 5), (4, 4), (4, 5)\}$$

and

$$S = \{(1, 1), (1, 2), (1, 4), (3, 3), (4, 5)\}$$

Then we have

$$R \cup S = \{(1, 1), (1, 2), (1, 4), (2, 3), (2, 5), (3, 3), (4, 4), (4, 5)\}$$

and

$$R \cap S = \{(1, 1), (1, 4), (4, 5)\}$$ □

Example 3.2.8

The union of the relations "less than" and "equal to" on the set **R** of all real numbers is the relation "less than or equal to."

Similarly, if Σ is an alphabet, then the union of the relations "same length as" and "shorter than" is the relation "same length or shorter than." □

Example 3.2.9

Let R be the relation of congruence modulo 3 and let S be the relation of congruence modulo 5, both on the set **Z** of all integers. Then, $m(R \cap S)n$ if and only if $m \equiv n \pmod 3$ and $m \equiv n \pmod 5$. Thus, $m(R \cap S)n$ if and only if $m - n$ is divisible by *both* 3 and 5. But this can happen if and only if $m - n$ is divisible by 15, and so $m(R \cap S)n$ if and only if $m \equiv n \pmod{15}$. In other words, the intersection of the relation of congruence modulo 3 and congruence modulo 5 is the relation of congruence modulo 15.

On the other hand, we have $m(R \cup S)n$ if and only if either $m \equiv n \pmod 3$ or $m \equiv n \pmod 5$. That is, $m(R \cup S)n$ if and only if $m - n$ is divisible by *either* 3 or 5 (or both.) □

Exercises

In Exercises 1 through 12, determine whether or not the given relation R, on the set A, is reflexive, symmetric, antisymmetric, or transitive.

1. Let A be the set of all real numbers, and let R be the relation "less than."
2. Let A be the set of all integers, and let R be the relation "greater than or equal to."
3. Let A be the set of all integers, and let R be the relation "m divides n."
4. Let A = $\mathcal{P}$(S), where S is a set, and let R be the subset relation.

5. Let A be the set of all people, and let R be the relation "is taller than."
6. Let A be the set of all straight lines in the plane, and let R be the relation "is perpendicular to."
7. Let A be the set of all straight lines in the plane, and let R be the relation "is parallel to."
8. Let A $= \Sigma^*$, and let R be the relation "same length as."
9. Let A $= \Sigma^*$, and let R be the relation "shorter than."
10. Let A $= \Sigma^*$, where a $\in \Sigma$. Let R be the relation "has the same number of occurrences of the letter a as."
11. Let A be the set of all compound statements, and let R be the relation of logical equivalence.
12. Let f:A $\rightarrow$ B be a function from the set A to the set B, and let R be the relation defined by aRa' if and only if $f(a) = f(a')$.

In Exercises 13 through 17, draw the graph of a relation on the set A = {1, 2, 3, 4} that satisfies the given properties.

13. Not reflexive, symmetric, not antisymmetric, transitive.
14. Not reflexive, not symmetric, antisymmetric, not transitive.
15. Reflexive, not symmetric, not antisymmetric, not transitive.
16. Reflexive, not symmetric, antisymmetric, transitive.
17. Not reflexive, not symmetric, not antisymmetric, transitive.
18. Describe the graphs of all possible binary relations on the set A $= \{1, 2, \ldots, n\}$ that are both symmetric and antisymmetric.
19. Let A be a set, and let R be a relation on A that is reflexive, symmetric, and antisymmetric. How would you describe the relation R?
20. Is a relation that is both symmetric and antisymmetric also transitive? Justify your answer.
21. Prove that the relation of congruence modulo k is transitive.

In Exercises 22 through 26, assume that the directed graph shown in Figure 3.2.2 is the graph of a relation R, and determine whether or not the given statements about the transitive closure R' are true. Justify your answer.

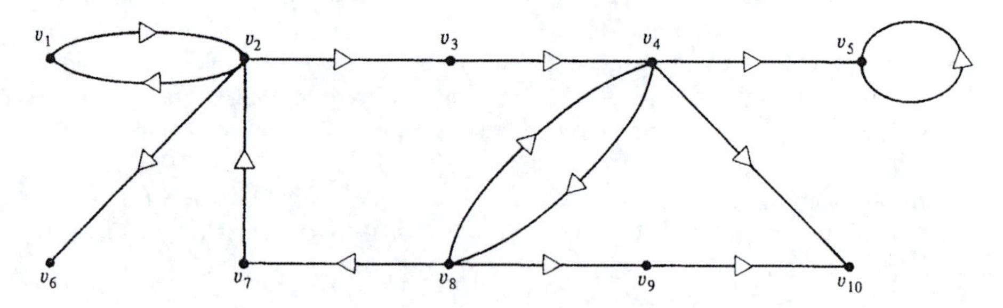

Figure 3.2.2

22. a) $v_1R'v_6$ b) $v_6R'v_1$
23. a) $v_1R'v_8$ b) $v_8R'v_1$
24. $v_{10}R'a$ for some vertex a
25. $v_5R'a$ for some vertex a
26. $v_1R'v_1$ (is v_1Rv_1 ?)

27. Describe the union of the relations ''less than'' and ''greater than,'' defined on the set **R** of all real numbers.
28. Describe the union of the relations ''longer than'' and ''shorter than,'' defined on the set Σ^* of all words over an alphabet Σ.
29. Describe the union of the relations ''longer than'' and ''shorter than or the same length as,'' defined on the set Σ^* over all words over an alphabet Σ.
30. Describe the intersection of the relations ''is a subset of'' and ''is a proper subset of,'' defined on the power set $\mathcal{P}(S)$ of a set S.
31. Describe the intersection of the relations of congruence modulo 5 and congruence modulo 7, on the set **Z**.
32. Describe the intersection of the relations of congruence modulo 2 and congruence modulo 4, defined on the set **Z**.
33. Let R and S be relations on A.
    a) If R is reflexive, must R^c also be reflexive?
    b) If R and S are reflexive, must $R \cap S$ be reflexive?
    c) If R and S are reflexive, must $R \cup S$ be reflexive?
34. Let R and S be relations on A.
    a) If R is symmetric, must R^c also be symmetric?
    b) If R and S are symmetric, must $R \cap S$ also be symmetric?
    c) If R and S are symmetric, must $R \cup S$ also be symmetric?
35. Let R and S be relations on A.
    a) If R is antisymmetric, must R^c also be antisymmetric?
    b) If R and S are antisymmetric, must $R \cap S$ also be antisymmetric?
    c) If R and S are antisymmetric, must $R \cup S$ also be antisymmetric?
36. Let R and S be relations on A.
    a) If R is transitive, must R^c also be transitive?
    b) If R and S are transitive, must $R \cap S$ also be transitive?
    c) If R and S are transitive, must $R \cup S$ also be transitive?
37. Let R be a relation on A that is *not* reflexive. Define a new relation R' on A by specifying that $aR'b$ if and only if either $a = b$ or aRb. What property does R' have? What term would you use to describe the relation R' and why?

3.3 Equivalence Relations

In this section, we want to study binary relations that are reflexive, symmetric, and transitive.

Definition

Let A be a nonempty set. A binary relation E on A that is reflexive, symmetric, and transitive is called an **equivalence relation** on A. If aEb we say that a is **equivalent to** b. □

Let us consider some examples of equivalence relations.

Example 3.3.1

The relation of equality on a set A is an equivalence relation. We actually showed this in Example 3.2.1. □

Example 3.3.2

The relation of logical equivalence is an equivalence relation on the set *Stat*(P). To be specific, the relation is reflexive since A ≡ A for all statements A, it is symmetric since A ≡ B implies B ≡ A, and it is transitive since A ≡ B and B ≡ C imply that A ≡ C. □

Example 3.3.3

Let Σ be an alphabet. Then the relation "has the same length" is an equivalence relation on Σ^*. After all, a word has the same length as itself, and if w has the same length as u, then u has the same length as w. Finally, if w has the same length as u and u has the same length as v, then w has the same length as v. □

Example 3.3.4

Let C be the set of all logic circuits that have a fixed number n of input variables and one output. (See Section 2.6 for a discussion of logic circuits.) Then we can define a binary relation on C by saying that c_1Ec_2 if and only if circuit c_1 has the same truth table as circuit c_2. We will leave it as an exercise to show that this relation is an equivalence relation on C. □

Example 3.3.5

The relation of congruence modulo k is an equivalence relation on the set **Z** of all integers. We showed that this relation is reflexive and symmetric in Example 3.2.3 and left verification of transitivity to Exercise 21 of Section 3.2. □

Now let us suppose that A is a nonempty set, and that P is a partition of A. Recall from Section 1.1 that a partition of a set A is a collection of disjoint, nonempty subsets of A, called the blocks of the partition, with the property that every element of A is in one of the blocks.

Then we can use the partition P to define a binary relation on A, denoted by $\mathcal{E}(P)$, by specifying that $a\mathcal{E}(P)b$ if and only if a and b are in the same block of the partition P. (See Figure 3.3.1.)

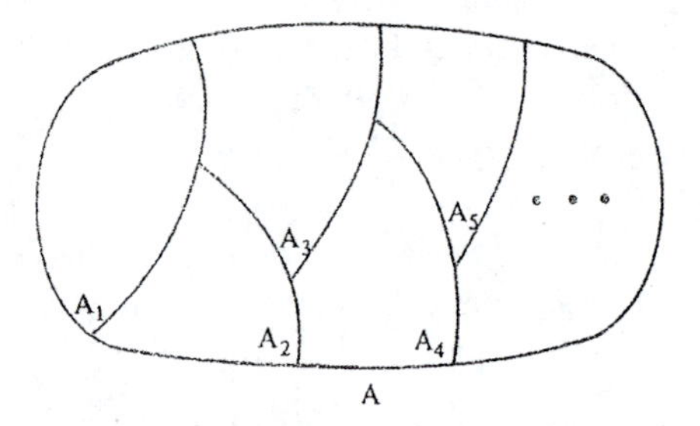

Figure 3.3.1
Two elements a and b of A satisfy $a\mathcal{E}(P)b$ if and only if a and b lie in the same block A_i.

The relation $\mathcal{E}(P)$ is easily seen to be reflexive, symmetric, and transitive (see Exercise 20), and so it is an equivalence relation. We call $\mathcal{E}(P)$ the **equivalence relation associated to the partition** P. Let us have some examples.

Example 3.3.6

Let A be the set $A = \{1, 2, 3, 4, 5, 6\}$. Then the partition $P = \{A_1, A_2, A_3\}$ where

$$A_1 = \{1, 4\} \quad , \quad A_2 = \{2, 3, 5\} \quad , \quad A_3 = \{6\}$$

defines an equivalence relation $\mathcal{E}(P)$ on A. For example, we have $1\mathcal{E}(P)4$ since 1 and 4 are in the same block A_1, and $2\mathcal{E}(P)5$ since 2 and 5 are in the same block A_2. However, $2\not\mathcal{E}(P)6$ since 2 and 6 are not in the same block. □

Example 3.3.7

Consider the set **Z** of all integers, and the partition $P = \{A_1, A_2\}$ where A_1 is the set of all even integers and A_2 is the set of all odd integers. Then the equivalence relation $\mathcal{E}(P)$ has the property that $m\mathcal{E}(P)n$ if and only if m and n have the same parity, that is, if and only if either m and n are both even or both odd. Put another way, $m\mathcal{E}(P)n$ if and only if $m - n$ is even. But since an integer is even if and only if it is divisible by 2, we see that $m\mathcal{E}(P)n$ if and only if $m \equiv n \pmod 2$. Hence, the equivalence relation associated to the partitioning of **Z** into even and odd integers is just the relation of congruence modulo 2. □

Example 3.3.8

Consider the partition $P = \{A_1, A_2, A_3\}$ of **Z** where A_1 is the set of all integers that are divisible by 3,

$$A_1 = \{\ldots, -9, -6, -3, 0, 3, 6, 9, \ldots\}$$

A_2 is the set of all integers whose remainder, after dividing by 3, is equal to 1,

$$A_2 = \{\ldots, -8, -5, -2, 1, 4, 7, 10, \ldots\}$$

and A_3 is the set of all integers whose remainder, after dividing by 3, is equal to 2,

$$A_3 = \{\ldots, -7, -4, -1, 2, 5, 8, 11, \ldots\}$$

(Since the remainder upon dividing any integer by 3 is either 0, 1, or 2, the set P is a partition of **Z**.) Then the equivalence relation $\mathcal{E}(P)$ satisfies $m\mathcal{E}(P)n$ if and only if m and n have the same remainder after dividing by 3. In other words, $m\mathcal{E}(P)n$ if and only if $m - n$ is divisible by 3. (See Exercise 21 of Section 3.1.) Thus, the equivalence relation $\mathcal{E}(P)$ is just congruence modulo 3. In the exercises, we ask you to generalize this result to congruence modulo k. (See Exercise 18.) □

The fact that every partition of a set A leads to an equivalence relation on A is extremely important. As a matter of fact, it turns out that *all* equivalence relations come from partitions in this way. That is, given any equivalence relation E on a set A, there is a partition P of A with the property that E is the equivalence relation associated to P, in symbols $E = \mathcal{E}(P)$. Before we can prove this, we need another definition.

Definition

Let E be an equivalence relation on a set A. For each element a in A, we define the **equivalence class of** A **associated to** a, to be the set

$$[a] = \{x \in A \mid xEa\}$$

In words, the equivalence class $[a]$ associated to a is the set of all elements of A that are equivalent to a. □

Of course, we could define the notion of equivalence class for any binary relation R, but it is mainly useful in the case of an equivalence relation. The following theorem gives some important properties of equivalence classes.

Theorem 3.3.1

Let E be an equivalence relation on a set A. Then the following properties hold.

a) Every element of A is in its own equivalence class; that is, $a \in [a]$ for all a in A.

b) If a and b are elements of A, then $[a] = [b]$ if and only if aEb.

c) If $[a]$ and $[b]$ are equivalence classes, then either $[a] = [b]$ or else $[a] \cap [b] = \varnothing$. In words, any two equivalence classes are either identical or else disjoint.

PROOF The proof of part a is very simple. Since E is reflexive, we have aEa for all a in A, and so according to the definition of $[a]$, we must have $a \in [a]$ for all a in A.

As for part b, let us first suppose that $[a] = [b]$, and show that aEb. This follows easily from the first part of this theorem. For if $[a] = [b]$, then since $a \in [a]$, we must also have $a \in [b]$. Hence aEb.

Now let us assume that aEb and show that $[a] = [b]$. We will do this by first showing that $[a] \subset [b]$. This follows from the fact that if $x \in [a]$, then xEa, and since aEb, the transitivity of E implies that xEb, and so $x \in [b]$. Thus, any element of $[a]$ is also an element of $[b]$, that is $[a] \subset [b]$. By a similar reasoning (which you should go through in detail yourself), we can show that $[b] \subset [a]$. Putting these two pieces together, we see that $[a] = [b]$. This proves part b.

Finally, let us prove part c. Our plan is to show that if $[a]$ and $[b]$ are *not* disjoint, then they must be equal. If $[a]$ and $[b]$ are not disjoint, then there must exist an element c in A with the property that $c \in [a]$ and $c \in [b]$. Now, since $c \in [a]$, we have aEc (why?), and since $c \in [b]$, we have cEb. Hence, by the transitivity of E, we conclude that aEb. Now we can use the second part of this theorem to deduce that $[a] = [b]$. This completes the proof of part c. ∎

If E is an equivalence relation on a set A, and P is the set of all equivalence classes of A, then we can describe P by writing

$$P = \{[a] \mid a \in A\}$$

Actually, there is a great deal of duplication in this description. In fact, whenever aEb, the equivalence classes $[a]$ and $[b]$ are identical, and so $[a]$ and $[b]$ are actually the same element of P. Of course, since P is a *set*, we ignore this duplication. In other words, we think of P as the set of all *distinct* equivalence classes of A. (As a simple example of this, if we were to define a *set* S by saying that S = $\{1, 2, 2, 2, 3, 3\}$, then S is really the set $\{1, 2, 3,\}$.)

Now, according to the last part of Theorem 3.3.1, distinct equivalence classes are disjoint, and so P consists of disjoint subsets of A. Furthermore, according to the first part of Theorem 3.3.1, each element a in A is in its own equivalence class $[a]$. This shows not only that each equivalence class is nonempty, but also that every element of A is in some equivalence class. Hence, the set P is actually a partition of A! It is called the **partition of A associated to the equivalence relation E** and is denoted by $\mathcal{P}(E)$.

Let us have some examples of the partition $\mathcal{P}(E)$.

Example 3.3.9

Let E be the equivalence relation of equality on the set A. Then since an element in A is equivalent only to itself, the equivalence classes of A are simply the one-element subsets of A. In symbols, if a is in A, then $[a] = \{a\}$, and so

$$\mathcal{P}(=) = \{\, \{a\} \mid a \in A\}$$

The partition $\mathcal{P}(=)$ is the "finest" partition of A possible, in that it has the smallest blocks possible (and the largest number of blocks possible). □

Example 3.3.10

Let A consist of the set of all straight lines in the plane, and let E be defined by specifying that $l_1 E l_2$ if and only if line l_1 is parallel to line l_2. We will leave it as an exercise to show that E is an equivalence relation on A. An equivalence class of A consists simply of the set of all straight lines that point in a certain direction. Thus, the partition $\mathcal{P}(E)$ partitions the set A into blocks with the property that two lines are in the same block if and only if they are parallel. In a sense, the blocks of this partition correspond to the different directions in the plane. □

Example 3.3.11

Let Σ be an alphabet, and let E be the equivalence relation on Σ^* defined by wEu if and only if $\ell(w) = \ell(u)$. Then the equivalence classes of Σ^* are just the sets Σ_n, for $n = 0, 1, 2, \ldots$, where Σ_n is the set of all words of length n. □

Let us pause to see where we stand. We know that any partition P defines an equivalence relation $\mathcal{E}(P)$, and that any equivalence relation E defines a partition $\mathcal{P}(E)$ (into equivalence classes). Thus, if A is a set, we have a way of going from partitions of A to equivalence relations on A, which we can picture as follows

$$\begin{array}{l} P_1 \xrightarrow{\ \mathcal{E}\ } \mathcal{E}(P_1) \\ P_2 \xrightarrow{\ \mathcal{E}\ } \mathcal{E}(P_2) \\ P_3 \xrightarrow{\ \mathcal{E}\ } \mathcal{E}(P_3) \\ \quad\vdots \end{array} \tag{3.3.1}$$

Also, we have a way of going from equivalence relations on A to partitions of A, which can be pictured by a similar diagram

$$
\begin{aligned}
E_1 &\xrightarrow{\mathcal{P}} \mathcal{P}(E_1)\\
E_2 &\xrightarrow{\mathcal{P}} \mathcal{P}(E_2)\\
E_3 &\xrightarrow{\mathcal{P}} \mathcal{P}(E_3)\\
&\vdots
\end{aligned}
\tag{3.3.2}
$$

Now, it so happens that the association $P \rightarrow \mathcal{E}(P)$ is a *one-to-one correspondence* between the set of all partitions of A and the set of all equivalence classes on A. Also, the association $E \rightarrow \mathcal{P}(E)$ is a one-to-one correspondence between the set of all equivalence classes on A and the set of all partitions of A.

Furthermore, these two one-to-one correspondences are inverses of each other. That is, if we start with a partition P, then form its equivalence relation $\mathcal{E}(P)$, and then form the partition associated to this relation, we get back to P. In symbols,

$$\mathcal{P}(\mathcal{E}(P)) = P \tag{3.3.3}$$

for all partitions P. Similarly, we have

$$\mathcal{E}(\mathcal{P}(E)) = E \tag{3.3.4}$$

for all equivalence relations E.

We will prove Equation 3.3.3, and the fact that the association in 3.3.2 is a one-to-one correspondence, and leave for the exercises the proof of Equation 3.3.4 and the fact that the association in 3.3.1 is a one-to-one correspondence.

Theorem 3.3.2

Let A be a nonempty set. Then for all partitions P of A we have

$$\mathcal{P}(\mathcal{E}(P)) = P$$

PROOF In order to show that two partitions are equal, we need to show that the blocks in one partition are the same as the blocks in the other. But, according to the definition of $\mathcal{E}$, two elements a and b of A are in the same block of P if and only if $a\mathcal{E}(P)b$. Furthermore, according to the definition of $\mathcal{P}$, we have $a\mathcal{E}(P)b$ if and only if a and b are in the same block of the partition $\mathcal{P}(\mathcal{E}(P))$. Thus, we see that a and b are in the same block of P if and only if they are in the same block of $\mathcal{P}(\mathcal{E}(P))$. Therefore, the blocks of these two partitions are the same, and so $P = \mathcal{P}(\mathcal{E}(P))$. This completes the proof. ∎

Theorem 3.3.3

Let A be a nonempty set. Then the association

$$\mathrm{E} \rightarrow \mathcal{P}(E)$$

described by 3.3.2 is a one-to-one correspondence between the set of all equivalence relations on A and the set of all partitions of A.

PROOF Imagine that we have made a complete list of all the equivalence relations on A and associated them with the corresponding partitions, as in 3.3.2. Thus, each equivalence relation on A appears exactly once on the left side of 3.3.2. Our task is to show that each partition of A appears exactly once on the right side. This will prove that the association is a one-to-one correspondence.

First, we must show that every partition of A does appear on the right side of 3.3.2. But if P is a partition of A, then $\mathcal{E}(P)$ is an equivalence relation on A, and so $\mathcal{E}(P)$ appears on the left side of 3.3.2,

$$\mathcal{E}(P) \longrightarrow \mathcal{P}(\mathcal{E}(P))$$

Now we can use the previous theorem to conclude that $\mathcal{P}(\mathcal{E}(P)) = P$, and so we have

$$\mathcal{E}(P) \longrightarrow \mathcal{P}(\mathcal{E}(P)) = P$$

This shows that P does indeed appear on the right side of 3.3.2. All that remains is to show that no partition P of A appears more than once on the right side of 3.3.2. But in order for a partition to appear more than once, it would have to come from two different equivalence relations on A, by means of the association 3.3.2. By thinking about the definition of $\mathcal{P}$,. we can see that this is not possible. (After all, different equivalence relations must produce different equivalence classes.) Hence, each partition P appears once and only once on the right side of 3.3.2, and so this association is a one-to-one correspondence. This completes the proof. ▮

The concepts expressed in Theorems 3.3.2 and 3.3.3 are not particularly difficult, but they do take a while to get used to. The important thing to remember is that there is a one-to-one correspondence between partitions of a set A and equivalence relations on A. Thus, whenever we are given an equivalence relation, we can obtain a partition, and vice versa. In a sense, partitions and equivalence relations are just different ways to express the same concept.

□ □ Exercises

In Exercises 1 through 17, determine whether or not the given relation R is an equivalence relation on the set A. For those that are equivalence relations, describe the equivalence classes.

1. $A = \{1, 2, 3\}$, $R = \{(1, 1), (2, 2), (3, 3), (1, 2), (2, 1)\}$
2. $A = \{1, 2, 3, 4, 5\}$, $R = \{(1, 1)\}$
3. $A = \{1, 2, 3\}$, $R = \{(1, 1), (2, 2), (3, 3)\}$
4. $A = \{-1, 0, 1\}$, $R = \{(x, y) \mid x^2 = y^2\}$
5. Let R be the subset relation on the power set $A = \mathcal{P}(S)$.
6. Let A be the set of all statements, and let R be the relation of logical equivalence.
7. Let $f{:}A \to B$ be a function, and let R be the relation defined by aRa' if and only if $f(a) = f(a')$.
8. Let A be the set of all expressions of the form a/b, where a and b are integers and $b \neq 0$. Let R be defined by $(a/b)R(c/d)$ if and only if $ad = bc$.

9. Let S be a finite set, and let R be the relation defined on the power set A = $\mathcal{P}(S)$ by specifying that S_1RS_2 if and only if $|S_1| = |S_2|$, where $|X|$ stands for the number of elements in X.
10. A is the set of all words in the English language, and xRy if and only if x and y start with the same letter.
11. A is the set of all words in the English language, and xRy if and only if x precedes y in alphabetical order.
12. Let Σ be an alphabet, and let R be the relation defined by wRu if and only if $\ell(w) \leq \ell(u)$.
13. Let A = **Z**, and let R be the relation "m divides n."
14. Let A be the set of all straight lines in the plane, and let R be defined by l_1Rl_2 if and only if l_1 is parallel to l_2.
15. Let A be the set of all straight lines in the plane, and let R be defined by l_1Rl_2 if and only if l_1 is perpendicular to l_2. (It is interesting to compare this exercise with the previous one.)
16. Let A be the set of all ordered pairs of real numbers, and let R be defined by $(a, b)R(c, d)$ if and only if $a^2 + b^2 = c^2 + d^2$.
17. Let A be the set of all logic circuits with a fixed number n of input variables and only one output. Let R be defined by "has the same truth table." (See Example 3.3.4.)
18. Let **Z** be the set of integers. Find a partition P of **Z** for which $\mathcal{E}(P)$ is the equivalence relation of congruence modulo k. *Hint*, think in terms of the possible remainders after dividing by k.
19. Let f:A $\rightarrow$ B be a function from A to B, and let E be the equivalence relation defined by specifying that aEa' if and only if $f(a) = f(a')$. Describe the partition $\mathcal{P}(E)$ associated to E. What does this partition look like if f is the function, from real numbers to real numbers, defined by $f(x) = x^2$?
20. Let P be a partition of the set A. Show that the relation $\mathcal{E}(P)$ is an equivalence relation on A.
21. Prove Equation 3.3.4.
22. Prove that the association $P \rightarrow \mathcal{E}(P)$ described in 3.3.1 is a one-to-one correspondence. *Hint*, you will probably want to use Equation 3.3.4.
23. Let E_1 and E_2 be equivalence relations on A.
    a) Is E_1^c an equivalence relation on A?
    b) Is $E_1 \cup E_2$ an equivalence relation on A?
    c) Is $E_1 \cap E_2$ an equivalence relation on A?

3.4 Partially Ordered Sets

If you were asked to pick the most common binary relation that is reflexive, antisymmetric, and transitive, you would probably choose the *order* relation $\leq$ on the set **R** of real numbers. In this section, we want to discuss the concept of order in some detail.

Actually, the order relation $\leq$ satisfies *four* properties that give it its characteristic behavior. Three of these properties are the reflexive, antisymmetric, and transitive properties. The fourth property is the fact that every pair of the elements in **R** is *comparable*. In other words, if a and b are real numbers, then we have either $a \leq b$ or $b \leq a$ (or perhaps both.)

As we shall see, the concept of order is very common throughout mathematics and its applications. However, it frequently happens that the binary relations that express order do *not* satisfy this fourth property. As an example, let S = $\{a, b, c, d\}$

and consider the power set A $= \mathcal{P}(S)$. Of course, $\mathcal{P}(S)$ is just the set of all subsets of S. Now, there is a kind of "order" on the elements of $\mathcal{P}(S)$, that is, on the subsets of S, given by set inclusion. For instance, we have $\{a, b\} \subset \{a, b, d\}$, and so, in some sense, the set $\{a, b\}$ is "less than" the set $\{a, b, d\}$.

The reason that we might want to think of set inclusion as a kind of order is that it has many properties in common with ordinary order $\leq$ on the set of real numbers. In fact, both these relations are reflexive, antisymmetric, and transitive. However, set inclusion does *not* have the fourth property that ordinary order $\leq$ has, namely, not all elements of $\mathcal{P}(S)$ are comparable. For example, the elements $\{a, b\}$ and $\{a, c\}$ are not comparable, since neither one is a subset of the other.

The example of set inclusion, along with many other examples, leads us to make the following definitions. If a binary relation satisfies the reflexive, antisymmetric, and transitive properties, then we call it a *partial order*. Of course, it is possible for a partial order to satisfy the fourth property, namely, that every pair of elements be comparable. If that is the case, then we call it a *total order*. Thus, set inclusion is a partial order, and $\leq$ is a total order. We should emphasize that partial orders are much more common, and every bit as important as total orders.

Now let us give a formal definition of a partial order. We will define and discuss total order a bit later in the section.

Definition

Let P be a nonempty set. A binary relation R on P that is reflexive, antisymmetric, and transitive is called a **partial order**. We will denote a partial order by the symbol $\preceq$. If a set P has a partial order defined on it, we call P a **partially ordered set**, and say that $\preceq$ **orders** P.

If a and b are elements of P, and if $a \preceq b$, then we say that a **is less than or equal to** b, or that b **is greater than or equal to** a. If $a \preceq b$ but $a \neq b$, then we write $a < b$, and say that a **is less than** b, or that b **is greater than** a. □

Let us make a few remarks about this definition. As we will see in a moment, it is possible to define many different partial orders on a nonempty set P, and each of these partial orders makes P into a *different* partially ordered set.

Notice also that we use phrases such as "a is less than or equal to b" for *any* partially ordered set, regardless of whether or not P is a set of real numbers. As an example, suppose that P is a set of words in the English language, and that $\preceq$ is defined by specifying that $w_1 \preceq w_2$ if and only if word w_1 precedes word w_2 in alphabetical order. Then $\preceq$ is a partial order (in fact a total order) and if $w_1 \preceq w_2$, we can say that w_1 is *less than or equal to* w_2.

Example 3.4.1

Let P $= \{1, 2, 3, a, b, c\}$ and let $\preceq$ be defined by

$$1 \preceq 1 \quad , \quad 2 \preceq 2 \quad , \quad 3 \preceq 3$$

$$a \preceq a \quad , \quad b \preceq b \quad , \quad c \preceq c$$

$$1 \preceq 2 \quad , \quad 2 \preceq 3 \quad , \quad 1 \preceq 3$$

$$a \preceq b \quad , \quad a \preceq c \quad , \quad b \preceq c$$

The first two rows of this definition insure that the relation $\preceq$ is reflexive, and it is not hard to see that it is also antisymmetric and transitive. Hence, the relation $\preceq$ is a partial order on P, and P is a partially ordered set with this relation.

Now, let us consider the relation $\preceq'$ on P defined by

$$1 \preceq' 1 \quad , \quad 2 \preceq' 2 \quad , \quad 3 \preceq' 3$$

$$a \preceq' a \quad , \quad b \preceq' b \quad , \quad c \preceq' c$$

$$1 \preceq' a \quad , \quad 2 \preceq' b \quad , \quad 3 \preceq' c$$

This relation is also reflexive, antisymmetric, and transitive, and so it is also a partial order on P. Thus, P is a partially ordered set under this relation as well. However, it is very important to realize that P is a *different* partially ordered set under $\preceq'$ than under $\preceq$. Of course, it is the same *set*, but there is more to a partially ordered set than just its elements. We must also take into account the order relation. (There are many other ways to make P into a partially ordered set. See Exercise 11.) □

Example 3.4.2

The "less than or equal to" relation on **R** is a partial order, since it is reflexive, antisymmetric, and transitive. Thus, in this case, P = **R** and $\preceq$ = $\leq$.

Also, the relation of equality is a partial order, since it is reflexive, antisymmetric, and transitive. □

Example 3.4.3

If S is a set, then since the subset relation on $\mathcal{P}$(S) is reflexive, antisymmetric, and transitive, it is a partial order. Hence, this relation makes $\mathcal{P}$(S) into a partially ordered set. □

Example 3.4.4

Let **N** be the set of non-negative integers, and let $\preceq$ = $|$ be the relation "m divides n." Then this relation is a partial order on **N**, and so it makes **N** into a partially ordered set. (You should verify this for yourself.) □

Example 3.4.5

Let Σ be an alphabet, and suppose that Σ is a partially ordered set, with partial order denoted by $\preceq$. Then we can use the relation $\preceq$ to define a partial order on the set Σ^* of all words over Σ, which we will also denote by the symbol $\preceq$. The idea is very simple—we define the order on Σ^* to be "alphabetical order"!

Thus, to compare two words

$$w = a_1a_2 \cdots a_n \quad \text{and} \quad u = b_1b_2 \cdots b_k$$

we first compare their first "letters." If $a_1 \prec b_1$ (notice that this is "less than" and not "less than or equal to"), then we have $w \preceq u$. However, if $a_1 = b_1$, then we compare the second letters in each word. If $a_1 = b_1$ and $a_2 \prec b_2$, then we have $w \preceq u$. If $a_1 = b_1$ and $a_2 = b_2$, then we compare the third letters in each word, and so on.

Just as we do with ordinary alphabetical order, if one of the words runs out of letters before we get to a place where the letters in the two words differ, then we say that the shorter word is less than or equal to the longer word. Put another way, if w is a *prefix* of u, then $w \leq u$. Also, we must include the fact that if $w = u$, then $w \leq u$. (This possibility has not been covered so far.) Finally, we can agree to let the empty word θ precede every other word in order. Thus, $\theta \leq w$ for all words w in Σ^*.

The partial order $\leq$ defined on Σ^* is called **lexicographic order**, or **dictionary order**. Of course, the reason that it is called dictionary order is that when Σ is the ordinary Roman alphabet, lexicographic order is just ordinary alphabetical order, as used in a dictionary. □

Example 3.4.6

Lexicographic order can be used to order the set $\mathbf{R}^2$ of all ordered pairs in the plane. All we have to do is think of an ordered pair (a, b) as a word ab of length 2.

In terms of ordered pairs, lexicographic order is defined as follows:

$$(a, b) \leq (c, d)$$

$$\text{if and only if either } a < c \text{ or else } a = c \text{ and } b \leq d$$

Perhaps the best way to get a feel for this definition is to look at Figure 3.4.1, which shows the set of all ordered pairs that are less than or equal to the ordered pair (2, 4).

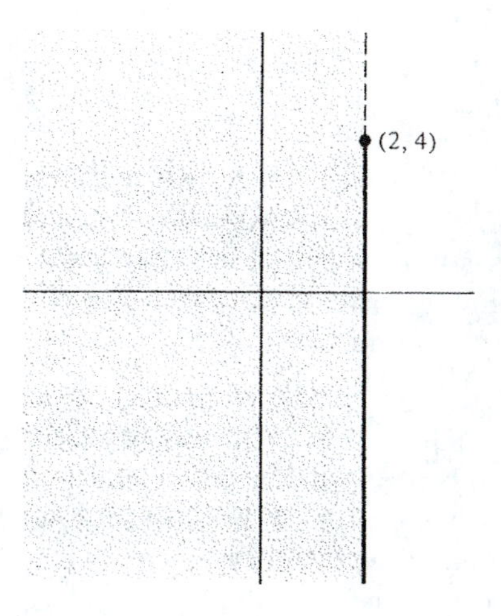

Figure 3.4.1 □

As we discussed at the beginning of this section, one of the most important characteristics of a partially ordered set P is that not all of the elements of P need be comparable.

Thus, if $\leq$ is a partial order on P, then just because $a \not\leq b$, it does not necessarily follow that $b \leq a$. (It is tempting to think that $a \not\leq b$ does imply $b \leq a$, since this is the case in the most familiar partially ordered set **R**, with the usual order $\leq$.) This leads us to make the following definition.

Definition

Let $\preceq$ be a partial order on a set P. Then we say that two elements a and b in P are **comparable** if either $a \preceq b$ or $b \preceq a$ (or both). Otherwise, we say that a and b are **incomparable**, or **not comparable**. □

Example 3.4.7

Let S = {1, 2, 3}, and consider the partially ordered set $\mathcal{P}(S)$, ordered by set inclusion $\subset$. Then, according to the definition, two elements A and B of $\mathcal{P}(S)$ are comparable if and only if one is a subset of the other. For example, A = {1, 3} and B = {1} are comparable, since B $\subset$ A. On the other hand, C = {1, 2} and D = {2, 3} are not comparable, since C $\not\subset$ D and D $\not\subset$ C. □

Example 3.4.8

Consider the partially ordered set P = {2, 3, 5, 6, 30}, ordered by the relation "m divides n." Two integers in P are comparable if and only if one divides the other. For example, the integers 2 and 6 are comparable since $2|6$, but the integers 3 and 5 are not comparable, since $3 \nmid 5$ and $5 \nmid 3$. □

Of course, it is true that if a and b are *comparable* elements in a partially ordered set P, then if $a \npreceq b$, we must have $b \preceq a$.

Partially ordered sets in which all elements are comparable are very special, and they deserve a special name.

Definition

If $\preceq$ is a partial order on P, and if every pair of elements of P is comparable, then we say that $\preceq$ is a **total order** on P and that P is a **totally ordered set**. (The terms **linear order** and **linearly ordered set** are also used.)

More specifically, a partial order $\preceq$ is a total order if for all a and b in P, we must have either $a \preceq b$ or $b \preceq a$ (or both). □

Of course, the most common example of a totally ordered set is the set **R** of all real numbers, with the total order $\leq$. If Σ is totally ordered by a relation $\preceq$, then the lexicographic order on Σ^* obtained from $\preceq$ is also a total order. However, the subset relation and the relation "m divides n" are in general not total orders. (We will ask you to verify these statements in the exercises.)

There are several important definitions associated with partially ordered sets, and we should discuss some of them now.

Let $\preceq$ be a partial order on a set P. If a, b, and c are in P, and if $a \preceq b$ and $b \preceq c$, we sometimes abbreviate this by writing $a \preceq b \preceq c$. Similarly, if $a_1, a_2, \ldots, a_n$ are elements of P and if $a_i \preceq a_{i+1}$ for all $i = 1, 2, \ldots, n - 1$ then we abbreviate this by writing $a_1 \preceq a_2 \preceq \cdots \preceq a_n$.

We say that an element b in P **covers** an element a in P if $a < b$ and if whenever $a \preceq c \preceq b$, we must have either $c = a$ or $c = b$. Loosely speaking, we say that the element b covers the element a if b is greater than a, and if there are no elements of P "between" a and b.

■ Example 3.4.9

Let S be any nonempty set, and consider the partially ordered set $\mathcal{P}(S)$, ordered by set inclusion. Then an element B of $\mathcal{P}(S)$ covers an element A of $\mathcal{P}(S)$ if $A \subset B$ and if B contains exactly one additional element that is not in A. □

In Section 3.1, we learned that binary relations on a finite set can be described by means of directed graphs. When the relation is a partial order, we can use the concept of covering to improve upon this description. In effect, we show only those arcs (u, v) in the graph for which v covers u. However, it is customary to make one other change in the graph.

As before, we let the vertices of the graph represent the elements of P. But in this case we place vertex v higher than vertex u whenever v is greater than u. Then we draw an arc between two vertices if and only if one of them covers the other. Furthermore, since we have placed some vertices higher than others, we do not need to assign a direction to each arc. (In a sense, all directions are "up.") Such a graph is called the **Hasse diagram** of the partially ordered set P. Also, the arcs of a Hasse diagram are usually called **edges**. Some examples may help clarify the definition.

■ Example 3.4.10

Referring to Example 3.4.1, the Hasse diagram of the partially ordered set P, with partial order $\preceq$ is

and the Hasse diagram of the partially ordered set P, with partial order $\preceq'$ is

a b c

1 2 3

□

■ Example 3.4.11

Let S = {1, 2, 3}. Then the Hasse diagram for the partially ordered set $\mathcal{P}(S)$ ordered by set inclusion, is

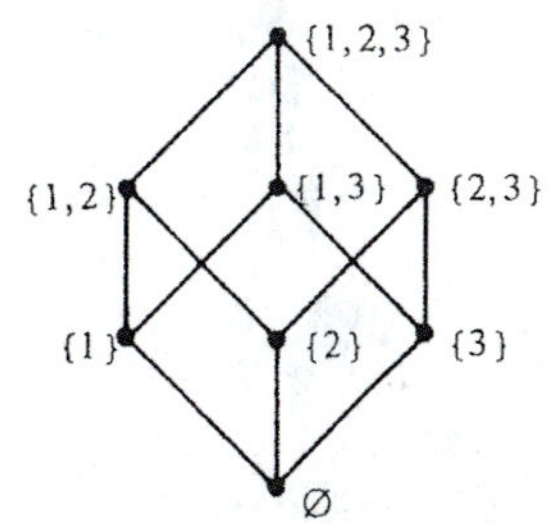

It is interesting to notice that, in this case, the vertices and edges of the Hasse diagram form the vertices and edges of a cube. □

Example 3.4.12

Let P = {1, 2, 3, 4, 6, 8, 9}. Then the Hasse diagram for the partially ordered set P, with partial order "*m* divides *n*," is

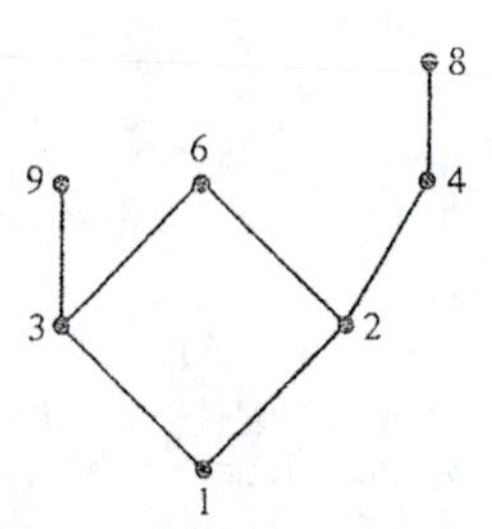

In a Hasse diagram, there is an edge between two elements in P if and only if one of them covers the other. However, we can still tell from the diagram whether or not one element of P is less than or equal to another element. For if a and b are distinct elements of P, then $a \leq b$ if and only if there is a *path* in the Hasse diagram that goes *up* from a to b. Thus, the Hasse diagram of a partially ordered set still completely describes the partially ordered set.

If P is a partially ordered set, and if S is a subset of P, then S is also a partially ordered set, *using the same partial order*. In a sense, subsets of P "inherit" the partial order from P, and so become partially ordered sets themselves. If S is a *totally* ordered set, then we say that S is a **chain** in P. In other words, a *chain* in a partially ordered set P is a totally ordered subset of P.

Example 3.4.13

Consider the partially ordered set P = {2, 3, 4, 6, 8, 14}, ordered by the partial order "*m* divides *n*." If S_1 = {2, 4, 8}, then since all of the elements of S_1 are comparable, the set S_1 is a totally ordered set. Hence, S_1 is a chain in P.

If S_2 = {2, 3, 4}, then the set S_2 is a partially ordered set, but it is not a chain, since the integers 2 and 3 are not comparable. (That is, 2 does not divide 3, nor does 3 divide 2.)

Exercises

1. Let P be the set of all students in your mathematics class, and let R be the relation defined by saying that aRb if and only if student a got a higher grade on the first quiz than student b. Is R a partial order? Justify your answer.
2. Let $\Sigma = \{0, 1\}$ and let n be a positive integer. Define a relation R on Σ_n by wRu if and only if the number of occurrences of 0 in w is less than or equal to the number of occurrences of 0 in u.
   a) Is R a partial order on Σ_n? Justify your answer.
   b) Let $A \subset \Sigma_n$ be the set of all words in Σ_n of the form $0 \cdots 01 \cdots 1$. Is R a partial order on A? Justify your answer.
3. Let S = {1, 2, 3}. Define a relation on $\mathcal{P}(S)$ by saying that ARB if and only if $|A| \leq |B|$. ($|X|$ is the size of the set X.) Is R a partial order on $\mathcal{P}(S)$? Justify your answer.
4. a) Let P = {2, 6, −1, 10} and let R be defined by aRb if and only if $|a| \leq |b|$. ($|a|$ is the absolute value of a.) Is R a partial order on P? Justify your answer.

b) Let Q = $\{2, -2, 6, -1, 10\}$ and let R be defined by aRb if and only if $|a| \leq |b|$. Is R a partial order on Q? Justify your answer.

5. Consider the relation R defined on the set $\mathbf{Z} \times \mathbf{Z}$ by saying that $(a, b)\ R\ (c, d)$ if and only if $a \leq c$ and $b \leq d$. Is R a partial order on $\mathbf{Z} \times \mathbf{Z}$? Justify your answer.

In Exercises 6 through 10, draw the Hasse diagram for the given partially ordered set P, ordered by the relation $\preceq$.

6. P = $\mathcal{P}(S)$, $\preceq$ is set inclusion, S = $\{a, b, c, d\}$
7. P = $\{2, 3, 5, 7, 11, 13, 17\}$, $\preceq$ is "m divides n."
8. P = $\{2, 3, 5, 7, 10, 21, 30, 35, 316\}$, $\preceq$ is "m divides n."
9. Let $\Sigma = \{0, 1\}$. Let P = Γ_3 be the set of all binary words of length at most 3, ordered by lexicographic order.
10. Let $\Sigma = \{0, 1\}$. Let P = Σ_4 be the set of all binary words of length 4, ordered by lexicographic order.
11. Find three more distinct partial orders on the set P in Example 3.4.1. Draw the Hasse diagrams for each of the resulting partially ordered sets.
12. a) How many distinct partial orders are there on the set P = $\{1, 2\}$? Draw the Hasse diagrams for each of the resulting partially ordered sets.
    b) How many distinct partial orders are there on the set Q = $\{1, 2, 3\}$? Draw the Hasse diagrams for each of the resulting partially ordered sets.
13. Suppose that P is a partially ordered set and that the relation $\preceq$ is also an equivalence relation. What can you say about $\preceq$? Explain.
14. Let Σ be an alphabet and define a relation $\preceq$ on Σ^* by $w \preceq u$ if and only if $\ell(w) \leq \ell(u)$. Is $\preceq$ a partial order? Justify your answer.
15. Show that if Σ is a totally ordered set, and if we put lexicographic order on Σ^*, then Σ^* is also totally ordered.
16. Are there any conditions under which set inclusion, defined on the power set $\mathcal{P}(S)$, is a total order? If so, describe them and if not explain why not.
17. a) Find an example of a set A of integers for which the relation "m divides n" is a partial order, but not a total order.
    b) Find an example of an *infinite* set A of integers for which the relation "m divides n" is a total order.
18. Verify the statements made in Example 3.4.9 concerning the covering property in $\mathcal{P}(S)$, ordered by set inclusion.
19. Let P be a partially ordered set. Show that if $x_1, x_2, \ldots, x_n$ are elements of P with the property that $x_1 \preceq x_2 \preceq \cdots \preceq x_n \preceq x_1$, then $x_1 = x_2 = \cdots = x_n$.
20. Let P be a finite partially ordered set, and suppose a and b are elements of P with the property that $a < b$. Show that there must exist elements $a_0, a_1, a_2, \ldots, a_n$ in P for which $a = a_0 < a_1 < a_2 < \cdots < a_n = b$ where a_i covers a_{i-1} for all $i = 1, 2, \ldots, n$.

3.5 More on Partially Ordered Sets; Maximal and Minimal Elements and Topological Sorting

An element m in a partially ordered set P is called a **maximal element** of P if it has the property that no other element of P is greater than m. Put another way, an element m in P is a maximal element if it has the property that all elements in P are either less than or equal to m or else not comparable to m.

The concept of a *minimal element* is defined similarly. An element n in P is called a **minimal element** of P if it has the property that no other element of P is less than n. Put another way, an element n in P is a minimal element if it has the property that all elements in P are either greater than or equal to n or else not comparable to n.

Let us have some examples.

Example 3.5.1

Consider the partially ordered set P $= \{1, 2, 3, a, b, c\}$ whose Hasse diagram is shown below.

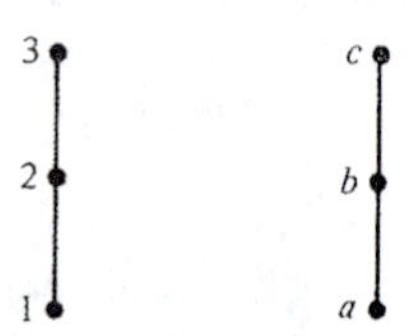

Then the elements 3 and c are both maximal elements. After all, there is no element of P that is greater than 3, and similarly for c. Also, the elements 1 and a are minimal elements. This example shows that a partially ordered set may have more than one maximal, or minimal, element. The next example shows that a partially ordered set may have no maximal or minimal elements. □

Example 3.5.2

The set **Z** of integers, ordered by $\leq$, has no maximal elements, since there is no integer m with the property that no other integer is larger than m. (After all, $m + 1$ is larger than m.) Similarly, **Z** has no minimal elements. □

We should be very careful to distinguish between a *maximal* element of P and the *maximum*, or *largest*, element of P, if it exists. An element M in P is called the **maximum**, or **largest**, element of P if it has the property that *all* elements of P are less than or equal to M; in symbols, $p \leq M$, for all p in P. This implies, of course, that all elements of P must be comparable to M, and that is the key difference between a maximal element and the maximum element.

Of course, the maximum element of P, if it exists, must be a maximal element, but a maximal element need not be the maximum element. (The next example will illustrate this.)

The concept of the *minimum* element of P is defined similarly. An element N in P is called the **minimum**, or **smallest**, element of P, if it has the property that *all* elements of P are greater than or equal to N, in symbols, $N \leq p$, for all p in P.

Perhaps the easiest way to see the difference between the concepts of "maximal" and "maximum" is to consider some examples.

Example 3.5.3

The partially ordered set in Example 3.5.1 has no maximum element, since there is no element M in P for which $p \leq M$ for *all* p in P. Similarly, it has no minimum element. This shows that a partially ordered set can have maximal elements without having a maximum element.

On the other hand, if we order the set $P = \{1, 2, 3, a, b, c\}$ as shown in the following Hasse diagram,

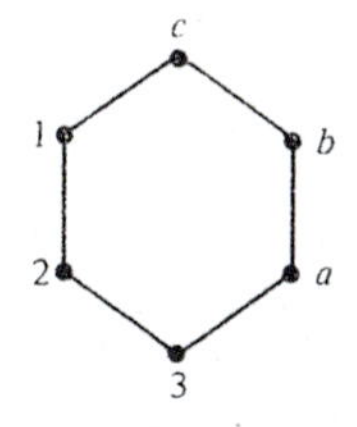

then the element c is the maximum element, for it is greater than or equal to all elements of P. Also, the element 3 is the minimum. □

■ Example 3.5.4

Let $P = \{1, 2, 3, 5, 6, 15\}$. The partially ordered set P, ordered by "m divides n" has two maximal elements, namely, 6 and 15. This is easy to see by considering the Hasse diagram for P.

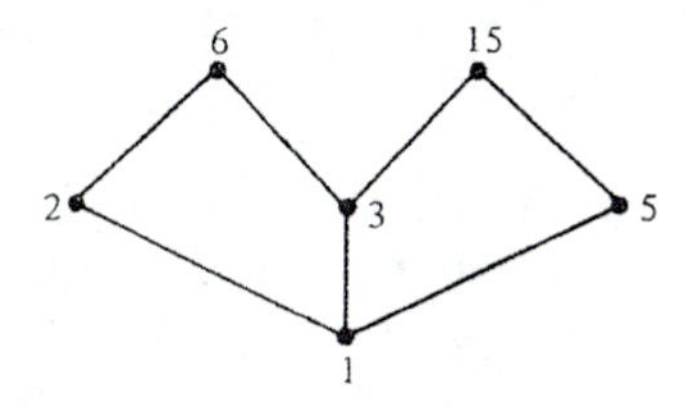

The element 6 is maximal since no element of P is greater than 6. Also, the element 15 is maximal since no element of P is greater than 15. (Remember that "greater than" in this case means "is a multiple of.")

Notice, however, that the partially ordered set P has no *maximum* element. For example, in order for 15 to be the maximum element of P, it would have to be true that $p|15$ for *all* p in P. But 6 does not divide 15, and so 15 cannot be the maximum element of P.

Of course, if we order the set P with the partial order "less than or equal to," then 15 is indeed the maximum element. This points out the fact that the concepts of maximal and maximum depend very much on the partial order in question and not just on the set itself. In other words, which elements of a partially ordered set P are maximal and which are maximum (if any) depends not just on the set P, but also on the particular partial order $\leq$. □

■ Example 3.5.5

If $S = \{a, b, c, d\}$, then the partially ordered set $\mathcal{P}(S)$, ordered by set inclusion, has one maximal element, namely the set S. This set also happens to be the maximum element of P as well. (Does this partially ordered set have a minimum element, and if so, what is it?) □

If P is a partially ordered set, ordered by $\leq$, and if P has a minimum element N, then we say that an element a in P is an **atom** of P if a covers N. In other words, an atom a is an element of P that is greater than N but that is so "small" that there are no elements of P "between" N and a.

For example, in the partially ordered set $\mathcal{P}(S)$, ordered by set inclusion, the minimum element is the empty set $N = \varnothing$. Hence, the atoms of $\mathcal{P}(S)$ are the subsets of S that contain exactly one element.

Now let us turn to another topic. Frequently, we are called upon to input the elements of a *finite* partially ordered set into a computer. (By *finite* we mean, of course, that P has a finite number of elements.) For example, suppose that we want to input the elements of $\mathcal{P}(S)$ into a computer, where $S = \{1, 2\}$. (Let us assume that we have a way to input subsets of a given set into the computer.) Thus, we want to input the subsets

$$\varnothing, \{1\}, \{2\}, \{1, 2\} \tag{3.5.1}$$

Now the very process of inputting a set into the computer imposes an order on its elements, namely, the order in which the elements are input. In fact, this order is a *total* order. Let us refer to it as the **input order**.

Of course, the input order need not have any connection with the original partial order on the set. For example, if we input the elements in 3.5.1 in the order

$$\{1, 2\}, \{1\}, \varnothing, \{2\} \tag{3.5.2}$$

then this order has little to do with the partial order of set inclusion. However, it is often desirable for the input order to be **compatible** with the partial order $\leq$ on the set. By this we mean that if $p \leq q$, then the element p should be input into the computer before the element q. In other words, if p is less than q in the partial order $\leq$, then p should also be less than q in the input order.

As an example, the input order shown in 3.5.2 is *not* compatible with set inclusion, since $\{1\}$ is a subset of $\{1, 2\}$, but $\{1\}$ is not input before $\{1, 2\}$. On the other hand, the input order

$$\varnothing, \{1\}, \{2\}, \{1, 2\} \tag{3.5.3}$$

is compatible with set inclusion, as is the input order

$$\varnothing, \{2\}, \{1\}, \{1, 2\} \tag{3.5.4}$$

(In fact, these are the only two orders that are compatible with set inclusion. See Exercise 11.)

The process of finding an input order that is compatible with a given partial order amounts to finding any *total* order that is compatible with the partial order. For then we simply input the elements according to that total order. Thus, we are left with the following question. Given a partial order $\leq$ on P, how can we find a total order $\ll$ on P that is compatible with $\leq$, in the sense that

$$\text{if} \quad p \leq q \quad \text{then} \quad p \ll q$$

The process of finding a compatible total order is sometimes called **topological sorting**. Fortunately, there is a very simple algorithm for this type of sorting, which we will give for *finite* partially ordered sets.

Algorithm for Topological Sorting

Let P be a *finite* partially ordered set, ordered by $\preceq$. Then we can define an input order $\ll$ on P that is compatible with the partial order $\preceq$ as follows.

1) Choose an element n of P that is minimal with respect to the partial order $\preceq$. (Any *finite* partially ordered set must have a minimal element. See Exercise 24.)
2) The element n comes next in the input order.
3) Remove n from the set P.
4) Repeat steps 1 through 3 until there are no elements left.

Let us try an example of this algorithm.

Example 3.5.6

Consider the partially ordered set P = {2, 3, 4, 8, 9, 36, 288}, ordered by the relation "m divides n." The Hasse diagram of P is shown below.

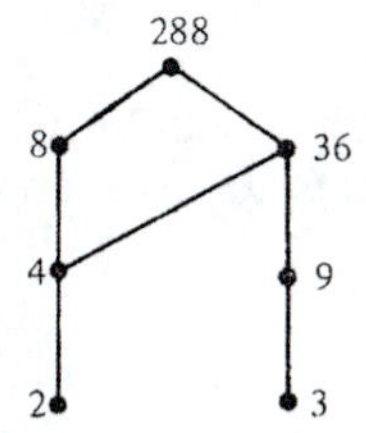

(It is a good idea to use the Hasse diagram to help with the sorting algorithm.) The first step in the algorithm is to pick a minimal element. We could choose either 2 or 3. Let us choose 3. Let us also keep track of our input order:

input order: 3

Then we remove this element, which we can do simply by crossing it out in the Hasse diagram.

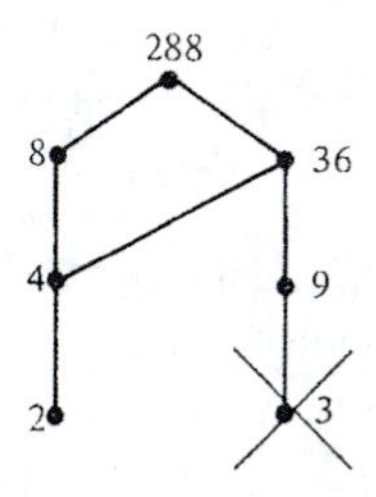

Then we pick a minimal element of the remaining set. Our choices now are either 2 or 9. Let us choose 2. Now we have

input order: 3, 2

and the Hasse diagram becomes

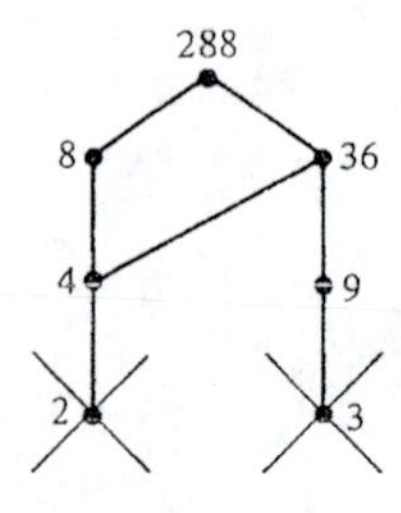

Next we choose 9, which makes the input order

input order: 3, 2, 9

Repeating the process of choosing a minimal element and crossing it off in the Hasse diagram will lead us to a compatible input order, for example,

input order: 3, 2, 9, 4, 8, 36, 288

We will leave it to you to check that if $p|q$, then p precedes q in the input order. Hence, this order is indeed compatible with the partial order $|$. (Incidentally, it is *not* true that if p precedes q in the input order, then p *must* divide q.) □

□ □ Exercises

In Exercises 1 through 10, for the given partially ordered set P, find all maximal elements, all minimal elements, and the maximum and minimum elements, if they exist. In case P is finite, draw the Hasse diagram of P.

1. $P = \mathcal{P}(S)$, $\leq$ is set inclusion, $S = \{a, b, c, d\}$
2. $P = \mathbf{Z}$, $\leq$ is "less than or equal to."
3. P is the set of all positive integers, $\leq$ is "less than or equal to."
4. P is the set of all positive integers, $\leq$ is "m divides n."
5. $P = \{2, 3, 4, \ldots\}$, $\leq$ is "m divides n."
6. $P = \{2, 3, 5, 7, 10, 21, 30, 35, 316\}$, $\leq$ is "m divides n."
7. $P = \{2, 3, 5, 7, 11, 13, 17\}$, $\leq$ is "m divides n."
8. $P = \{r \in \mathbf{R} \mid 0 \leq r \leq 1\}$, $\leq$ is $\leq$.
9. $P = \{r \in \mathbf{R} \mid 0 < r < 1\}$, $\leq$ is $\leq$.
10. P is the set of all positive rational numbers, and $\leq$ is the relation "less than or equal to."
11. a) Verify that the input orders given in 3.5.3 and 3.5.4 are compatible with set inclusion.
b) Show that these input orders are the *only* ones compatible with set inclusion.
12. Verify that the input order in Example 3.5.6 is compatible with the partial order given in that example.

In Exercises 13 through 16, apply the algorithm for topological sorting to obtain a total order that is compatible with the given partial order $\leq$ on P. Then verify that it is indeed compatible.

13. $P = \mathcal{P}(S)$, $S = \{1, 2, 3\}$, $\leq$ is set inclusion.
14. $P = \mathcal{P}(S)$, $S = \{1, 2, 3, 4\}$, $\leq$ is set inclusion.
15. $P = \{a, b, c, d, 1, 2, 3, 4\}$, where $\leq$ is given by the following Hasse diagram.

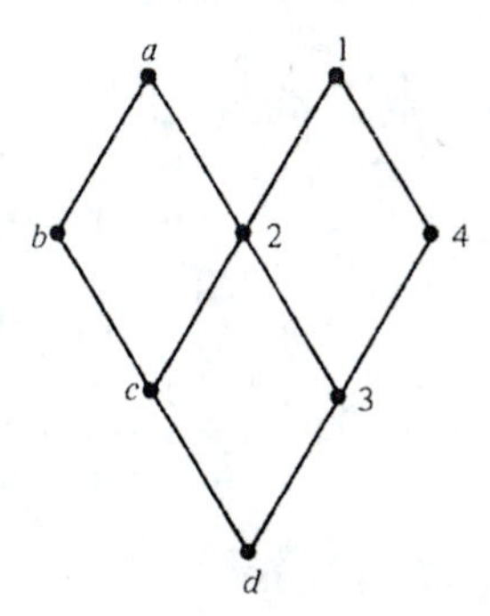

16. P = {1, 2, 3, 4, 5, 6, 7, 8, 9, 10, 11}, where $\preceq$ is given by the following Hasse diagram.

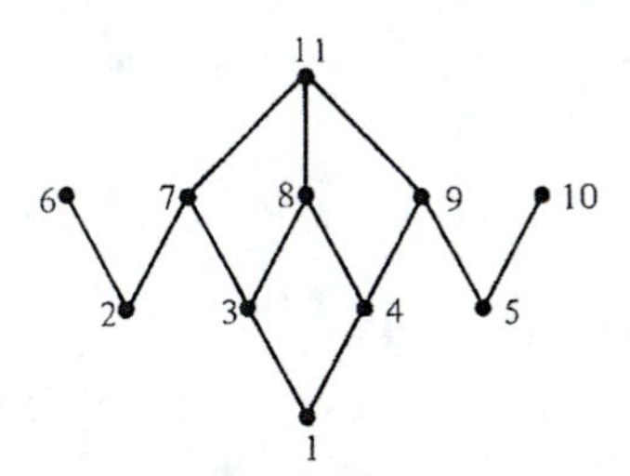

17. If $\preceq$ is a total order on P, what can you say about the input order obtained by applying the algorithm for topological sorting? Justify your answer.
18. Find all atoms in the following partially ordered sets P.
    a) $\mathbf{N} - \{0\}$, ordered by "m divides n"
    b) $P = \{2, 4, 6\}$, ordered by "m divides n"
    c) $\mathbf{R}$ ordered by $\leq$
    d) $P = \{12, -3, 14, -20\}$, ordered by the relation $a \preceq b$ if and only if $|a| \leq |b|$
19. a) Find a partially ordered set that has no maximal elements.
    b) For any positive integer n, find a partially ordered set that has exactly n maximal elements.
    c) Find a partially ordered set that has an infinite number of different maximal elements. *Hint*, you can use the integers for all three parts.
20. a) Prove that a partially ordered set can have at most one maximum element.
    b) Prove that a partially ordered set can have at most one minimum element.
21. a) Prove that if a partially ordered set has a maximum element, then it has exactly one maximal element.
    b) Is it true that if a partially ordered set has exactly one maximal element, then it must have a maximum element? *Hint*, try the set S consisting of all integers, together with the element a. Define an appropriate partial order on this set.
22. Prove that any *finite* partially ordered set P must have a maximal element, and a minimal element.
23. a) Prove that any finite totally ordered set has a maximum and a minimum element.
    b) Find an example of a totally ordered (infinite) set with no maximum or minimum element.

*The **length** of a finite chain is defined to be one less than the number of elements in the chain. If P is a partially ordered set, then a chain in P is said to be **maximal** if it is not contained in any chains with greater length.*

For example, consider the partially ordered set P = {2, 3, 4, 5, 9, 8, 27, 81}, ordered by the relation "m divides n." Then the chains C_1 = {5}, C_2 = {2, 4, 8}, and C_3 = {3, 9, 27, 81} have length 0, 2, and 3 respectively, and they are all maximal chains, since none of them is contained in a longer chain. On the other hand, the chain D = {3, 27}, for example, is not maximal since it is contained in the longer chain C_2.

24. Draw the Hasse diagrams for each of the following partially ordered sets P, ordered by "*m* divides *n*," and find all of the maximal chains in P.
   a) P = {2, 3, 5, 6, 36}
   b) P = {2, 3, 5, 7, 11}
   c) P = {1, 2, 3, 5, 6, 30, 90, 180, 900}

25. Find all maximal chains in the partially ordered set $\mathcal{P}$(S), ordered by set inclusion, where S = {1, 2, 3, 4}.

26. Does the partially ordered set **R**, ordered by the relation "less than or equal to," have a maximal chain? Explain your answer.

3.6 Order Isomorphisms

Let us consider the partially ordered set $\mathcal{P}$(S), where S = {1, 2, 3}, ordered by set inclusion, and the partially ordered set P = {1, 2, 3, 5, 6, 10, 15, 30}, ordered by the relation "*m* divides *n*." The Hasse diagrams for these partially ordered sets are shown in Figure 3.6.1.

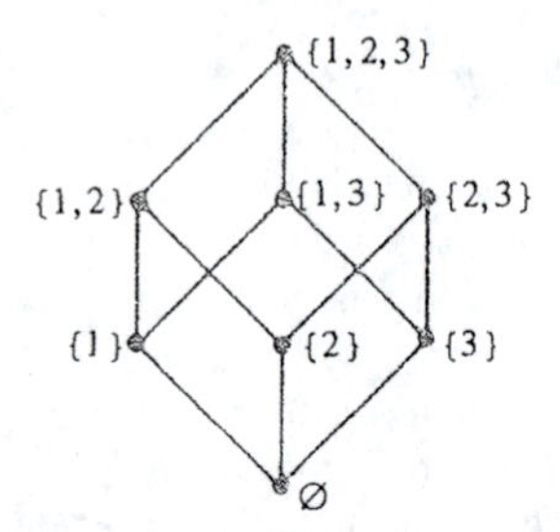

The Hasse diagram for ($\mathcal{P}$(S), ⊂)

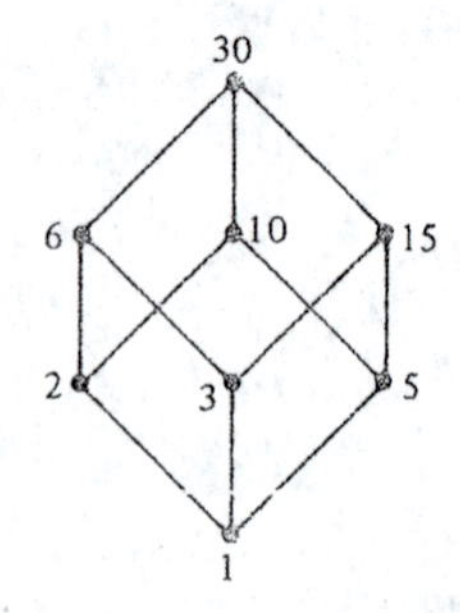

The Hasse diagram for (P,|)

Figure 3.6.1

As you can see, these diagrams are very similar. We can describe this similarity by observing that there is a one-to-one correspondence between the elements of $\mathcal{P}(S)$ and the elements of P that *preserves order*. We want to make this more precise in a definition, but first we need a bit of notation.

If P is a partially ordered set, under the partial order $\preceq$, then it is quite common to denote this by writing the ordered pair $(P, \preceq)$ and referring to it as a partially ordered set. Thus, for example, $(\mathcal{P}(S), \subset)$ and $(\mathbf{R}, \leq)$ are partially ordered sets.

If $(P_1, \preceq_1)$ and $(P_2, \preceq_2)$ are partially ordered sets, and if $f{:}P_1 \rightarrow P_2$ is a function from P_1 to P_2, then whenever we wish to emphasize the order relations, we will write

$$f{:}(P_1, \preceq_1) \rightarrow (P_2, \preceq_2)$$

and say that f is a function from $(P_1, \preceq_1)$ to $(P_2, \preceq_2)$.

Definition

Let $f{:}(P_1, \preceq_1) \rightarrow (P_2, \preceq_2)$ be a function from $(P_1, \preceq_1)$ to $(P_2, \preceq_2)$. Then we say that f is an **order isomorphism** from $(P_1, \preceq_1)$ to $(P_2, \preceq_2)$ if it satisfies the following two conditions.

1) The function f is a bijection (that is, f is one-to-one and onto P_2).

2) The function f preserves order in the sense that

$$a \preceq_1 b \text{ if and only if } f(a) \preceq_2 f(b)$$

Whenever there is an order isomorphism $f{:}(P_1, \preceq_1) \rightarrow (P_2, \preceq_2)$, from $(P_1, \preceq_1)$ to $(P_2, \preceq_2)$, we say that the partially ordered sets $(P_1, \preceq_1)$ and $(P_2, \preceq_2)$ are **order isomorphic**. □

Of course, the fact that f is a bijection implies that f defines a one-to-one correspondence between the elements of P_1 and the elements of P_2.

Incidentally, the terms *isomorphism* and *isomorphic* come from the Greek. The prefix *iso* comes from the Greek word *isos*, meaning *same*, and the words *morphism* and *morphic* come from the Greek word *morphe*, meaning *form*. Hence, *isomorphic* means "of the same form." We will encounter these terms again in Chapter 6. Let us now turn to some examples of order isomorphisms.

Example 3.6.1

Consider the partially ordered sets $(\mathcal{P}(S), \subset)$ and $(P, |)$, whose Hasse diagrams are given in Figure 3.6.1. The function $f{:}(\mathcal{P}(S), \subset) \rightarrow (P, |)$ defined by

$$f(\{1, 2, 3\}) = 30$$

$$f(\{1, 2\}) = 6 \quad , \quad f(\{1, 3\}) = 10 \quad , \quad f(\{2, 3\}) = 15$$

$$f(\{1\}) = 2 \quad , \quad f(\{2\}) = 3 \quad , \quad f(\{3\}) = 5$$

$$f(\emptyset) = 1$$

is an order isomorphism from $(\mathcal{P}(S), \subset)$ to $(P, |)$. To verify this, it is necessary to check that f is a bijection and that $A \subset B$ if and only if $f(A)|f(B)$. We will leave this as an exercise.

The function $g:(\mathcal{P}(S), \subset) \rightarrow (P, |)$ defined by

$$g(\{1, 2, 3\}) = 30$$

$$g(\{1, 2\}) = 15 \quad , \quad g(\{1, 3\}) = 6 \quad , \quad g(\{2, 3\}) = 10$$

$$g(\{1\}) = 3 \quad , \quad g(\{2\}) = 5 \quad , \quad g(\{3\}) = 2$$

$$g(\varnothing) - 1$$

is also an order isomorphism from $(\mathcal{P}(S), \subset)$ to $(P, |)$, but the function $h:(\mathcal{P}(S), \subset) \rightarrow (P, |)$ defined by

$$h(\{1, 2, 3\}) = 30$$

$$h(\{1, 2\}) = 15 \quad , \quad h(\{1, 3\}) = 10 \quad , \quad h(\{2, 3\}) = 6$$

$$h(\{1\}) = 2 \quad , \quad h(\{2\}) = 3 \quad , \quad h(\{3\}) = 5$$

$$h(\varnothing) = 1$$

is not an order isomorphism, since $\{1\} \subset \{1, 2\}$ but $h(\{1\}) = 2$ does not divide $h(\{1, 2\}) = 15$. □

The last example makes an important point, namely, that according to the definition, two partially ordered sets are order isomorphic if there is *at least one* order isomorphism between them. However, just because two partially ordered sets happen to be order isomorphic does not mean that *all* bijective functions between them are order isomorphisms.

■ Example 3.6.2

Let P_1 be the set of all subsets of $\{a, b, c, d\}$ that do not contain the element a, and let P_2 be the set of all subsets of $\{a, b, c, d\}$ that do contain the element a. Then $(P_1, \subset)$ and $(P_2, \subset)$ are both partially ordered sets. If we define $f:(P_1, \subset) \rightarrow (P_2, \subset)$ by

$$f(A) = A \cup \{a\}$$

then f is an order isomorphism from $(P_1, \subset)$ to $(P_2, \subset)$. The function f preserves order since $A \subset B$ if and only if $(A \cup \{a\}) \subset (B \cup \{a\})$. In other words, $A \subset B$ if and only if $f(A) \subset f(B)$. We will leave it as an exercise to show that f is a bijection. Hence, the partially ordered sets $(P_1, \subset)$ and $(P_2, \subset)$ are isomorphic. We suggest that you draw the Hasse diagrams for each of these partially ordered sets. □

Before considering other examples of order isomorphisms, let us discuss some of their properties. We will leave the proof of the following simple theorem as an exercise.

■ Theorem 3.6.1

Let $f:(P_1, \preceq_1) \rightarrow (P_2, \preceq_2)$ be a function from $(P_1, \preceq_1)$ to $(P_2, \preceq_2)$. Then f is an order isomorphism if and only if

1) f is a bijection, and

2) $a <_1 b$ if and only if $f(a) <_2 f(b)$

As we learned in Section 1.4, any bijective function f has an inverse function f^{-1}. In particular, if $f:(P_1, \preceq_1) \rightarrow (P_2, \preceq_2)$ is an order isomorphism, then since it is a bijection, the inverse function $f^{-1}:(P_2, \preceq_2) \rightarrow (P_1, \preceq_1)$ exists. Of course, we have no right to *assume* that f^{-1} is an order isomorphism. Fortunately, however, this does turn out to be the case, as we see in the next theorem.

Theorem 3.6.2

Let $f:(P_1, \preceq_1) \rightarrow (P_2, \preceq_2)$ be an order isomorphism from $(P_1, \preceq_1)$ to $(P_2, \preceq_2)$. Then $f^{-1}:(P_2, \preceq_2) \rightarrow (P_1, \preceq_1)$ is an order isomorphism from $(P_2, \preceq_2)$ to $(P_1, \preceq_1)$.

PROOF First we observe that since f is a bijection, so is f^{-1}. (We asked you to show this in Exercise 36 of Section 1.4.) Our task now is to show that f^{-1} preserves order; that is, we want to show that $c \preceq_2 d$ if and only $f^{-1}(c) \preceq_1 f^{-1}(d)$.

So suppose that $c \preceq_2 d$. Then if we let $a = f^{-1}(c)$ and $b = f^{-1}(d)$, we have $f(a) = c$ and $f(b) = d$. Hence, $c \preceq_2 d$ is the same as $f(a) \preceq_2 f(b)$. But, since f is an order isomorphism, we know that $f(a) \preceq_2 f(b)$ if and only if $a \preceq_1 b$, that is, if and only if $f^{-1}(c) \preceq_1 f^{-1}(d)$. Thus, $c \preceq_2 d$ if and only if $f^{-1}(c) \preceq_1 f^{-1}(d)$. This shows that f^{-1} preserves order and completes the proof. ∎

In general, to determine whether or not a function $f:(P_1, \preceq_1) \rightarrow (P_2, \preceq_2)$ is an order isomorphism, it is necessary to verify the conditions of the definition. However, if P_1 and P_2 are *finite* sets, then there is a simpler way, which we describe in the next theorem.

Theorem 3.6.3

Let $f:(P_1, \preceq_1) \rightarrow (P_2, \preceq_2)$ be a function from the partially ordered set $(P_1, \preceq_1)$ to the partially ordered set $(P_2, \preceq_2)$. Suppose that both P_1 and P_2 are finite sets. Then f is an order isomorphism from $(P_1, \preceq_1)$ to $(P_2, \preceq_2)$ if and only if

1) f is a bijection
2) f preserves the covering property, in the sense that an element b covers an element a if and only if $f(b)$ covers $f(a)$

PROOF Let us first assume that $f:(P_1, \preceq_1) \rightarrow (P_2, \preceq_2)$ is an order isomorphism and show that it must satisfy the two conditions of this theorem.

According to the definition of order isomorphism, f is a bijection. Now suppose that b covers a. Then of course $a \preceq_1 b$, and so $f(a) \preceq_2 f(b)$. We need to show that $f(b)$ actually covers $f(a)$.

We can do this by contradiction. For if there was an element p in P_2 between $f(a)$ and $f(b)$, then since f is surjective, we can write $p = f(c)$, and we would have $f(a) <_2 f(c) <_2 f(b)$. Now we can apply the order isomorphism f^{-1} to this, which according to Theorem 3.6.1, gives $a <_1 c <_1 b$. But this contradicts the fact that b covers a, and so there can be no element of P_2 between $f(a)$ and $f(b)$. Hence $f(b)$ does cover $f(a)$.

So far, we have shown that if b covers a, then $f(b)$ covers $f(a)$. Now we must show that if $f(b)$ covers $f(a)$, then b covers a. So suppose that $f(b)$ covers $f(a)$. If we apply what we have just proved to the inverse function f^{-1} (rather than to f), we can conclude that $f^{-1}(f(b))$ covers $f^{-1}(f(a))$. But since $f^{-1}(f(b)) = b$ and $f^{-1}(f(a)) = a$, we see that b covers a. Thus, we have shown that b covers a if and only if $f(b)$ covers $f(a)$.

Now let us turn to the converse. We must prove that if a function $f:(P_1, \leq_1) \to (P_2, \leq_2)$ satisfies the two conditions of this theorem, then it is an order isomorphism. According to the first condition, it is a bijection, so we need only show that it preserves order.

To show that f preserves order, let us suppose that $a <_1 b$. We would like to conclude that $f(a) <_2 f(b)$. Of course, if b covers a, then we know from condition 2 that $f(b)$ covers $f(a)$, and so we would have $f(a) <_2 f(b)$. But what if b does not cover a?

This is where we use the fact that P_1 is a finite set. As we asked you to show in Exercise 20 of Section 3.4, there must exist elements $a_1, a_2, \ldots, a_n$ in P_1 with the property that

$$a <_1 a_1 <_1 a_2 <_1 \cdots <_1 a_n <_1 b$$

and that each element in this sequence covers the one immediately preceding it. Hence, we can apply condition 2 to the elements in this sequence to conclude that

$$f(a) <_2 f(a_1) <_2 f(a_2) <_2 \cdots <_2 f(a_n) <_2 f(b)$$

where each element covers the one immediately preceding it. In particular then, we must have $f(a) <_2 f(b)$, which is what we wanted to show.

This proves that if $a <_1 b$, then $f(a) <_2 f(b)$. In order to prove that if $f(a) <_2 f(b)$ then $a <_1 b$, we again use the inverse function f^{-1}. If $f(a) <_2 f(b)$, then by applying what we have just proved to f^{-1} (rather than to f), we may conclude that $f^{-1}(f(a)) <_1 f^{-1}(f(b))$, that is, $a <_1 b$. Thus, we have shown that $a <_1 b$ if and only if $f(a) <_2 f(b)$. Since f is a bijection, we may use Theorem 3.6.1 to conclude that f is an order isomorphism. This completes the proof. ∎

Let us now consider a few more examples of order isomorphisms.

■ Example 3.6.3

Let $S = \{a_1, a_2, a_3, a_4, a_5\}$ be a set with 5 elements and let us associate the elements of A with the first 5 prime numbers 2, 3, 5, 7, and 11. (A **prime** number is an integer $p \geq 2$ that is not divisible by any positive integers other than 1 and itself.) Thus, our association is

$$a_1 \leftrightarrow 2$$

$$a_2 \leftrightarrow 3$$

$$a_3 \leftrightarrow 5$$

$$a_4 \leftrightarrow 7$$

$$a_5 \leftrightarrow 11$$

Now consider the partially ordered set $(\mathcal{P}(S), \subset)$ and the partially ordered set $(P, |)$, where P is the set of all possible products of distinct prime numbers from the list 2, 3, 5, 7, and 11. We also include 1 in the set P. Let us define a function $f:(\mathcal{P}(S), \subset) \to (P, |)$ as follows. If A is in $\mathcal{P}(S)$ then we define $f(A)$ to be the product of those prime numbers that are associated to the elements of A. For example, we have

$$f(\{a_1\}) = 2$$

$$f(\{a_2, a_3, a_5\}) = 3 \cdot 5 \cdot 11 = 165$$

$$f(\{a_1, a_2, a_3, a_4, a_5\}) = 2 \cdot 3 \cdot 5 \cdot 7 \cdot 11 = 2310$$

Also, we set $f(\emptyset) = 1$.

Then the function f is an order isomorphism from $(\mathcal{P}(S), \subset)$ to $(P, |)$. We will show this by using Theorem 3.6.3. First, we must show that f is a bijection. The function f is a surjection simply because of the way in which we defined P. After all, P is precisely the set of all products of distinct primes from the list 2, 3, 5, 7, and 11, and any such product comes from an element of $\mathcal{P}(S)$ under the function f. For instance, the product $2 \cdot 3 \cdot 5$ comes from the set $\{a_1, a_2, a_3\}$, since $f(\{a_1, a_2, a_3\}) = 2 \cdot 3 \cdot 5$.

One way to show that f is injective is to notice that the sets $\mathcal{P}(S)$ and P have the same size (why?) and therefore, since f is surjective, it must also be injective. (See Exercise 35 of Section 1.4). Thus, f is a bijection.

Now we must show that f preserves the covering property. So suppose that A and B are elements of $\mathcal{P}(S)$, and that B covers A. This means that B contains all of the elements of A, and *one* additional element, call it x. Hence, $f(B)$ is equal to the product of $f(A)$ and the particular prime number that corresponds to x. This implies that $f(B)$ covers $f(A)$. (See Exercise 13.)

On the other hand, if $f(B)$ covers $f(A)$, this means that $f(B)$ must be equal to $f(A)$ times some prime number, call it p. Now, p must be one of the primes 2, 3, 5, 7, or 11. Thus, B must come from A by adding whichever element of S corresponds to p. Hence, B covers A. This proves that f preserves the covering property, and so f is an order isomorphism. □

Example 3.6.4

Consider the partially ordered set $(P, |)$, where $P = \{2, 4, 8, 3, 9, 27\}$ and the partially ordered set $(Q, \leq)$, where $Q = \{1, 2, 3, 4, 5, 6\}$. The Hasse diagrams of these partially ordered sets are

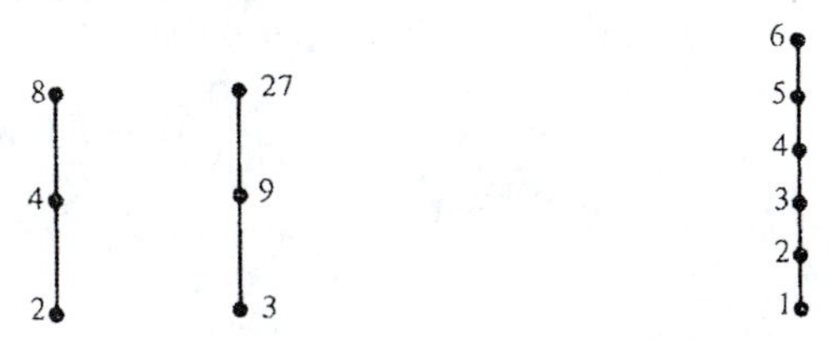

The Hasse diagram for (P, |) The Hasse diagram for $(Q, \leq)$

Let us define a function $f:(P, |) \to (Q, <)$ by

$$f(2) = 1 \quad , \quad f(4) = 2 \quad , \quad f(8) = 3$$
$$f(3) = 4 \quad , \quad f(9) = 5 \quad , \quad f(27) = 6$$

It is not hard to see that f is a bijection. Therefore, in order to show that f is an order isomorphism, we would need to show that $a|b$ if and only if $f(a) < f(b)$.

Now, f does have the property that if $a|b$ then $f(a) \leq f(b)$. However, it does *not* have the property that if $f(a) \leq f(b)$ then $a|b$. For example, $f(2) \leq f(9)$, but 2 does not divide 5. Hence, f is *not* an order isomorphism.

This example illustrates how important it is to carefully check all of the conditions in the definition of an order isomorphism, or in Theorem 3.6.3. □

□ □ Exercises

1. Consider the partially ordered sets $P = \{1, 2, 3, 4\}$, ordered by $\leq$, and $Q = \{1, 4, 9, 16\}$, also ordered by $\leq$. Is the function $f:(P, \leq) \to (Q, \leq)$ defined by $f(x) = x^2$ an order isomorphism? Justify your answer.
2. Show that the partially ordered sets $P = \{1, 2, 3, 4\}$ and $Q = \{5, 6, 7, 8\}$, both ordered by the relation $\leq$, are order isomorphic. *Hint*, first draw their Hasse diagrams.
3. Let $P = \{2, 3, 6, 9\}$ and $Q = \{ \{a\}, \{b\}, \{a, b\}, \{a, c\} \}$. Show that the partially ordered sets $(P, |)$ and $(Q, \subset)$ are order isomorphic. *Hint*, first draw the Hasse diagrams of these partially ordered sets.
4. Consider the partially ordered set $(P, |)$, where $P = \{2, 3, 4, 6, 24\}$. Find a partially ordered set of the form $(Q, \leq)$ that is order isomorphic to $(P, |)$. Explain how you arrived at your answer.
5. Let $\mathbf{N}$ be the set of all nonnegative integers, and let $\mathbf{E}$ be the set of all nonnegative even integers.
   a) Show that the partially ordered sets $(\mathbf{N}, \leq)$ and $(\mathbf{E}, \leq)$ are order isomorphic. *Hint*, consider the function $f(x) = 2x$.
   b) Find a bijection $g:(\mathbf{N}, \leq) \to (\mathbf{E}, \leq)$ that is *not* an order isomorphism.
6. Referring to Example 3.6.1, prove that the function f is an order isomorphism.
7. Referring to Example 3.6.2, show that the function f is a bijection.
8. Prove Theorem 3.6.1.
9. Let $f:(P_1, \preceq_1) \to (P_2, \preceq_2)$ and $g:(P_2, \preceq_2) \to (P_3, \preceq_3)$ be order isomorphisms. Prove that the composition $g \circ f:(P_1, \preceq_1) \to (P_3, \preceq_3)$ is an order isomorphism.
10. Let $(T, \preceq)$ be a totally ordered set, where $|T| = n$. Prove that $(T, \preceq)$ is order isomorphic to the totally ordered set $(P, \leq)$, where $P = \{1, 2, \ldots, n\}$. *Hint*, use induction.
11. Let $f:(P_1, \preceq_1) \to (P_2, \preceq_2)$ be an order isomorphism.
    a) Prove that m is maximal in $(P_1, \preceq_1)$ if and only if $f(m)$ is maximal in $(P_2, \preceq_2)$.
    b) Prove that M is the maximum element in $(P_1, \preceq_1)$ if and only if $f(M)$ is the maximum element in $(P_2, \preceq_2)$.
    c) State and prove results similar to those of parts a and b for minimal elements and minimum elements.
12. Consider the partially ordered set $(\mathcal{P}(S), \subset)$, where $S = \{a_1, a_2, \ldots, a_n\}$. Generalize the method of Example 3.6.3 to find a partially ordered set of the form $(P, |)$ that is order isomorphic to $(\mathcal{P}(S), \subset)$.
13. Consider the partially ordered set $(\mathbf{N}, |)$, where $\mathbf{N}$ is the set of all integers, and $|$ is the relation "m divides n." Show that an integer b covers an integer a if and only if $b = pa$, for some prime number p.

If, my dear, you seek to slumber,
Count of stars an endless number;
If you still continue wakeful,
Count the drops that make a lakeful;
Then, if vigilance yet above you
Hover, count the times I love you;
And if slumber still repel you,
Count the times I did not tell you.
—Franklin P. Adams, *Lullaby*

Chapter FOUR

Combinatorics—The Art of Counting

4.1 Introduction

In this chapter and the next, we are going to study various techniques for solving counting problems. These techniques are part of a branch of mathematics known as **combinatorics**, and mathematicians who specialize in this particular branch of mathematics are called **combinatorialists**. Combinatorics is one of the most important branches of mathematics when it comes to applications to computer science.

Perhaps the best way to get a feel for counting problems is to consider several examples. During the course of our studies, we will learn techniques for solving all of these problems.

Example 4.1.1

A certain company manufactures computer chips. Each type of chip must be given a name, so that it can be identified by customers. The company has decided that it is more impressive if the names consist of 3 letters followed by 3 digits, rather than simply naming the chips Chip 1, Chip 2, Chip 3, and so on. For example, one possible name is Chip CDX154. Our first counting problem is to count the number of possible names for computer chips. □

Example 4.1.2

Every morning, a computer operator receives one computer program from each of 9 students and must decide in which order to run these programs. In order to be fair, the operator wants to assign a different order to the programs each day. For how many days can he do this? In other words, how many different orderings are there of 9 programs? □

Example 4.1.3

If Σ is an alphabet and $|\Sigma| = m$, how many elements are there in Σ_n? That is, how many words are there over Σ that have length n? How many elements are there in Γ_n? That is, how many words are there over Σ that have length at most n? □

Example 4.1.4

Suppose that a bookstore is having a sale on books. You can purchase any 5 books from a collection of 50 different books for only one dollar. How many different choices of 5 books are there? □

Example 4.1.5

Consider the map pictured in Figure 4.1.1. In order to get from point A to point B by the shortest possible route, traveling on the streets only, you must cover a total of 11 blocks. But, of course, there are many possible paths that you can take. Two of these are illustrated in Figure 4.1.1. The problem is to count the total number of ways to travel from A to B, covering exactly 11 blocks.

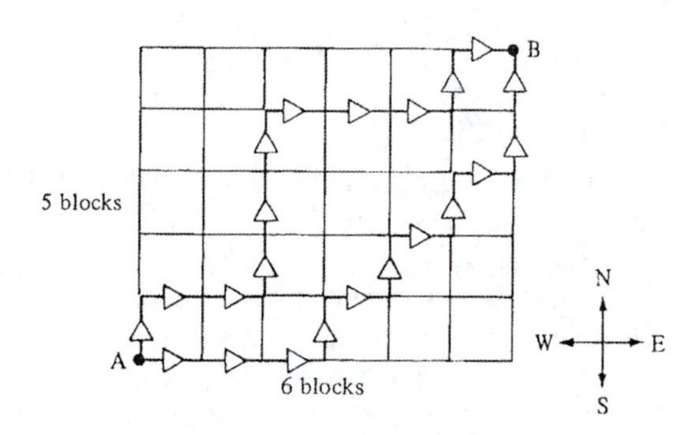

Figure 4.1.1 □

Example 4.1.6

Suppose that your mathematics class has 12 students. On the day that your midterm exams are returned, your professor makes a total mess of things and returns the exams in such a way that every student gets someone else's exam, instead of his or her own. In how many ways can the exams be returned in such a way that every student gets someone else's exam? □

Example 4.1.7

A game known as the Towers of Hanoi is pictured in Figures 4.1.2 and 4.1.3. There are 3 pegs and 8 circular rings of increasing outside diameter. At first, all of the rings are placed on the first peg, as shown in Figure 4.1.2. The object of the game is to transfer all of the rings onto the third peg, so that they appear as in Figure 4.1.3. However, you are allowed to move only one ring at a time, and you cannot place a ring on top of another ring that has a *smaller* outside diameter. The problem in this case is to count the *minimum* number of moves required to transfer this "tower" of rings to the third peg. □

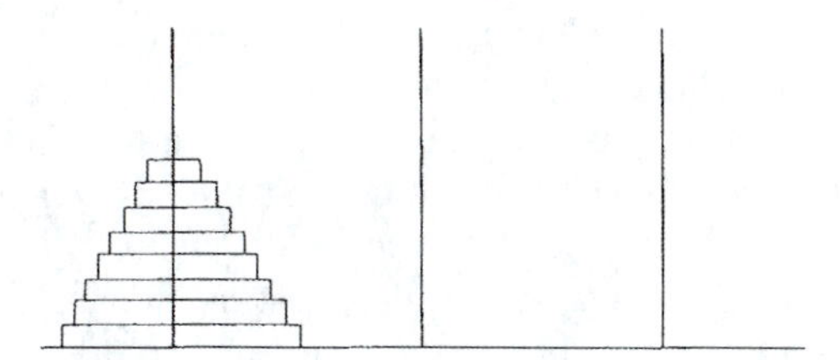

Figure 4.1.2
The Towers of Hanoi puzzle: starting position.

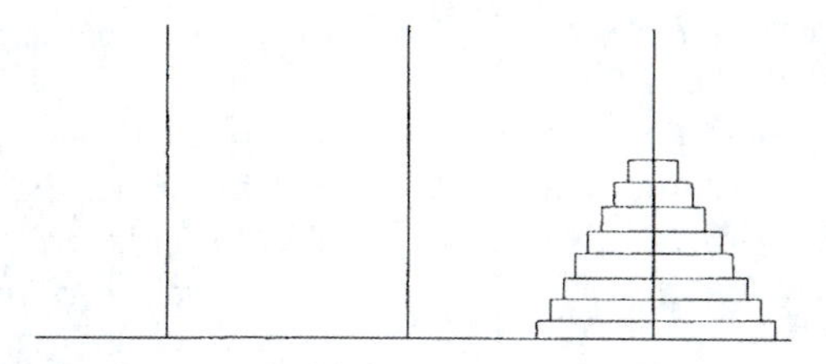

Figure 4.1.3
The Towers of Hanoi puzzle: ending position.

Example 4.1.8

How many ways are there to give change for a dollar bill using pennies, nickels, dimes, and quarters? For example, two possible ways are

1) 5 pennies, 2 dimes, and 3 quarters
2) 10 pennies, 1 nickel, 6 dimes, and 1 quarter □

Example 4.1.9

Memory for a microcomputer comes in blocks of 64K bytes ($64K = 64 \times 1024$). If you have a total of 50 blocks of memory, how many ways can this memory be distributed between 4 microcomputers if computer 1 must receive at least 2 blocks of memory, computer 2 must receive at least 3 blocks of memory, computer 3 must receive at least 1 block of memory, and computer 4 must receive at least 5 blocks of memory? □

Hopefully, these few examples of counting problems have shown you that such problems do occur frequently in various contexts. Now let us proceed to a discussion of how to solve counting problems.

4.2 The Multiplication Rule

We will begin our study of counting techniques with two very simple, yet very powerful rules, which we will discuss in this and the next section. In fact, much of what we will do in this chapter is based on these two rules.

It turns out that many counting problems amount to counting the number of ways to perform a certain sequence of tasks, where each task can be performed in several different ways.

As a simple example, consider the problem of counting names for computer chips in Example 4.1.1. Deciding on a name that consists of 3 letters followed by 3 digits amounts to performing a sequence of 6 tasks. The first task is to choose the first letter in the name, and this task can be performed in 26 different ways, since there are 26 letters in the alphabet. The second task is to choose the second letter, and it too can be performed in 26 different ways. The third task is to choose the third letter, the fourth task is to choose the first digit in the name, and so on. Clearly, the number of possible names is simply the number of ways to perform this sequence of 6 tasks.

Our first counting rule, called the **multiplication rule**, will tell us the number of ways to perform such a sequence of tasks, say $T_1, T_2, \ldots, T_k$. By the phrase "performing a sequence $T_1, T_2, \ldots, T_k$ of tasks," we mean of course, that task T_1 is performed first, then task T_2, and so on.

Before we can state the multiplication rule, however, we must discuss one very important point. The multiplication rule applies only under the assumption that the number of ways to perform any one of the tasks in the sequence, say T_i, does not depend on how the previous tasks $T_1, T_2, \ldots, T_{i-1}$ were performed. In a sense, this is an assumption about the *independence* of the tasks in the sequence. If a sequence of tasks does not satisfy this type of independence condition, then we cannot apply the multiplication rule.

With this in mind, let us now state the multiplication rule.

The Multiplication Rule Suppose that $T_1, T_2, \ldots, T_k$ is a sequence of tasks with the property that the number of ways to perform any task in the sequence does not depend on how the *previous* tasks in the sequence were performed. Then, if there are n_i ways to perform the i-th task T_i, for all $i = 1, 2, \ldots, k$, the number of ways to perform the entire sequence of tasks is the product $n_1 n_2 \cdots n_k$.

As a simple example of the multiplication rule, suppose that we want to perform the following sequence of two tasks. Task T_1 is to choose an integer between 1 and 10 (inclusive), and task T_2 is to choose a letter from the alphabet. How many ways are there to perform this sequence of tasks?

Of course, there are 10 ways to perform task T_1, since there are 10 integers between 1 and 10. Also, there are 26 ways to perform task T_2, since there are 26 letters in the alphabet. Furthermore, this sequence of tasks satisfies the independence condition required in order to apply the multiplication rule. That is, the number of ways to select a letter does not depend on which integer was chosen.

Therefore, we can apply the multiplication rule, which tells us that the number of ways to perform the sequence T_1, T_2 of tasks is equal to the product $10 \cdot 26 = 260$. In other words, there are 260 ways to choose an integer between 1 and 10 and a letter from the alphabet.

You may be wondering why the multiplication rule tells us to multiply the numbers $n_1, n_2, \ldots, n_k$ together, rather than add them, for instance. To see why, let us look again at our simple example. In this case, we take the product $10 \cdot 26$ because for *each* of the 10 different ways to perform the first task, there are 26 ways to perform the second task.

Put another way, if we choose the number 1, then we can choose the letter in 26 different ways; if we choose the number 2, then we can choose the letter in 26 different ways; and so on. Thus, we get a total of

$$\underbrace{26 + 26 + \cdots + 26}_{10 \text{ terms}} = 10 \cdot 26 = 260$$

different ways to perform this sequence of tasks.

This same reasoning applies, of course, to the case of more than two tasks (see also Exercise 25).

Let us consider some additional examples of the multiplication rule.

Example 4.2.1

A computer store sells 7 types of microcomputers, 5 types of monitors, and 12 types of printers. How many different system configurations are possible with this hardware? (A system configuration consists of 1 microcomputer, 1 monitor, and 1 printer.)

This is a problem for the multiplication rule. In particular, if we let T_1 be the task of choosing the microcomputer, T_2 be the task of choosing a monitor, and T_3 be the task of choosing a printer, then this problem becomes one of determining the number of ways to perform the sequence of tasks T_1, T_2, T_3.

Furthermore, these tasks do satisfy the independence condition needed in order to apply the multiplication rule. Namely, the number of ways to choose a monitor does not depend on which microcomputer is chosen, and the number of ways to choose a printer does not depend on which microcomputer or which monitor is chosen. (Alas, this assumption is not always true in real life, but we will assume for the purposes of illustration that all of the components are compatible.)

Since there are 7 ways to perform task T_1, 5 ways to perform task T_2, and 12 ways to perform task T_3, the multiplication rule tells us that there are $7 \cdot 5 \cdot 12 = 420$ ways to perform this sequence of tasks. That is, there are 420 possible system configurations. □

Example 4.2.2

The multiplication rule is just what we need to solve the problem of counting the number of names for computer chips in Example 4.1.1. Referring to the discussion at the beginning of this section, we see that there are

$$26 \cdot 26 \cdot 26 \cdot 10 \cdot 10 \cdot 10 = 17{,}576{,}000$$

different possibilities for a name that consists of 3 letters followed by 3 digits. □

Example 4.2.3

The multiplication rule can be used to count the number of words of a given length. More precisely, suppose that Σ is an alphabet, and that $|\Sigma| = m$. Then forming a word of length n over Σ amounts to performing a sequence of n tasks. Task 1 is to choose the first "letter" for the word, task 2 is to choose the second "letter" for the word, and so on. Since each task can be performed in m different ways, and since these tasks are independent, the multiplication rule tells us that there are

$$\underbrace{m \cdot m \cdots m}_{n\ factors} = m^n$$

different ways to perform the entire sequence of n tasks. That is, there are m^n words of length n over the set Σ. In summary,

$$\text{if } |\Sigma| = m \qquad \text{then} \qquad |\Sigma_n| = m^n$$

As a special example, the number of binary words of length n is 2^n.

We can use this information to compute the size of the set Γ_n of all words of length at most n. First we observe that

$$\Gamma_n = \Sigma_0 \cup \Sigma_1 \cup \cdots \cup \Sigma_n$$

and since the sets $\Sigma_0, \Sigma_1, \ldots, \Sigma_n$ are disjoint, we conclude that

$$\begin{aligned} |\Gamma_n| &= |\Sigma_0| + |\Sigma_1| + \cdots + |\Sigma_n| \\ &= 1 + m + m^2 + \cdots + m^n \end{aligned}$$

(This would not be true if the sets were not disjoint, for then we would be counting some words more than once.) Now, we will leave it as an exercise to show by induction that, for $m \neq 1$,

$$1 + m + m^2 + \cdots + m^n = \frac{m^{n+1} - 1}{m - 1}$$

Of course, for $m = 1$, we have $1 + m + m^2 + \cdots + m^n = n + 1$. Therefore, we can state that

$$\text{if } |\Sigma| = m \text{ then } |\Gamma_n| = \begin{cases} \dfrac{m^{n+1} - 1}{m - 1} & \text{when } m \neq 1 \\ n + 1 & \text{when } m = 1 \end{cases}$$

□

Example 4.2.4

How many even integers with distinct digits are there between 1 and 99? (By distinct digits, we mean that the two digits must be different. For example, the integer 22 does not have distinct digits. Also, we write the integers between 1 and 9 in the form 01, 02, . . ., 09.)

Let T_1 be the task of choosing the units digit from among the 5 possibilities 0, 2, 4, 6, and 8. (Remember, the number must be even.) Let T_2 be the task of choosing the tens digit. Then there are 5 ways to perform the first task T_1, and *no matter how that task is performed*, there are 9 ways to perform the second task T_2 (since we can choose any one of the 10 digits 0 through 9, except the one that was chosen for the units digit). Thus, according to the multiplication rule, there are $5 \cdot 9 = 45$ even integers with distinct digits between 1 and 99. ☐

The previous example points out a very important fact, namely, the order in which we pick our tasks can be crucial. For if, in the previous example, we had let T_1 be the task of choosing the tens digit, and T_2 be the task of choosing the units digit, then we would have run into a little trouble. There would be 10 ways to perform task T_1, since there are 10 possible digits for the tens place. However, the number of ways to perform task T_2 would then depend on how we performed task T_1. For if we had chosen an odd digit for the tens place, then we would have 5 possibilities for the units digit, but if we had chosen an even number for the tens digit, then we would have only 4 possibilities for the units digit. (Why?)

Thus, if we had chosen our tasks in this order, the independence condition would not hold, and we could not apply the multiplication rule in the same way that we did above.

■ Example 4.2.5

The multiplication rule can be used to determine the number of subsets of a set containing n elements. We proved in Section 1.5, by using mathematical induction, that any set of size n has 2^n subsets. But you might have wondered how we got the number 2^n in the first place. One answer to this is "from the multiplication rule."

Suppose that $S = \{x_1, x_2, \ldots, x_n\}$ is a set with n elements. Then we can think of the subsets of S as being formed by performing a sequence of n tasks. The first task is to decide whether or not to include the first element x_1 in the subset. The second task is to decide whether or not to include the second element x_2 in the subset, and so on. Each subset of S can be obtained in this way, and the total number of subsets of S is equal to the total number of ways of performing this sequence of n tasks.

But, there are exactly 2 ways to perform each of the tasks, regardless of how any of the previous tasks in the sequence were performed, and so, according to the multiplication rule, there are

$$\underbrace{2 \cdot 2 \cdots 2}_{n \textit{ factors}} = 2^n$$

ways to perform all n tasks; that is, there are 2^n subsets of the set S. ☐

The multiplication rule is really a very simple idea—the trick is in recognizing when it applies, that is, in recognizing whether or not a given problem can be thought of as a problem involving a sequence of tasks. As you will see, a great many problems can be expressed in this way, and so this rule does apply in a surprisingly large number of problems.

From now on, whenever we use the multiplication rule, we will assume that the required independence condition has been verified.

Exercises

1. Suppose that you own 3 pairs of shoes, 6 pairs of socks, 4 pairs of pants, and 6 shirts. How many different outfits can you make out of these articles of clothing? (An outfit consists of one pair of shoes, one pair of socks, one pair of pants, and one shirt.)
2. Consider the following roadmap.

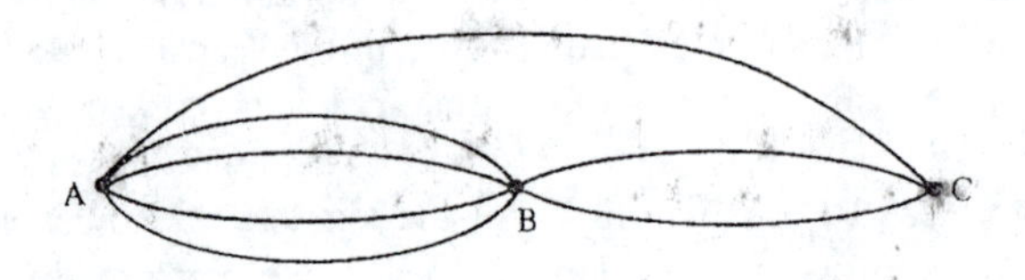

   a) How many ways are there to travel from A to B, and back to A, without going through C?
   b) How many ways are there to go from A to C, stopping once at B?
   c) How many ways are there to go from A to C, making at most one intermediate stop?
3. A certain apartment complex has 26 television antennas. Each pair of apartments shares a common antenna. How many apartments are there in the complex?
4. The computer identification numbers on the computers of a certain company consist of 2 letters followed by 2 digits followed by 2 more letters followed by 4 more digits. How many identification numbers are possible?
5. How many different license plates are possible in your state? (You may ignore the fact that some combinations are not allowed.)
6. Recall that a ternary word is a word over the alphabet $\Sigma = \{0, 1, 2\}$.
   a) How many ternary words are there of length 8? (Such words might be called *ternary bytes*.)
   b) How many ternary words are there of length at most 8?
7. How many binary words are there of length at least 4 and at most 6?
8. How many binary words are there of length at least 3 and at most 10?
9. a) How many words are there over the alphabet $\Sigma = \{0, 1, 2, 3\}$ of length 8?
   b) How many words are there over the alphabet Σ of length at most 8?
10. a) How many words are there over the alphabet $\Sigma = \{0, 1, 2, 3\}$ of length 8 whose first "letter" is a 0?
    b) How many words are there over the alphabet $\Sigma = \{0, 1, 2, 3\}$ of length 8 whose first "letter" is not a 0?
11. A certain type of switch can be in one of 4 possible states, represented by the alphabet $\Sigma = \{\rightarrow, \leftarrow, \uparrow, \downarrow\}$. The state of a circuit consisting of 10 switches of this type is thus given by a word of length 10 over this alphabet. How many possible switch configurations are there for this circuit?
12. Two cards are drawn from a deck of cards, one at a time. How many outcomes are possible
    a) if the order in which the cards are drawn matters?
    b) if the order in which the cards are drawn does not matter?
13. A computer room contains 96 microcomputers and several printers. Each printer is cabled to 3 microcomputers, and each microcomputer is cabled to exactly 1 printer. How many printers are there in the room?
14. A microcomputer store sells 4 types of microcomputers, 3 types of monitors, 6 types of printers, and 8 types of modems. (A modem is a device for communicating data between computers over telephone lines.) How many configurations are possible?
15. Suppose that you flip a coin 5 times and record the sequence of heads and tails. How

many possibilities are there for this sequence? How does this problem relate to alphabets and words?

16. Suppose that you are given 3 boxes. Box 1 contains 5 objects, box 2 contains 4 objects, and box 3 contains 2 objects.
   a) How many ways are there to choose exactly one object?
   b) How many ways are there to choose exactly one object from each box?

17. Suppose that you are given a box that contains 10 balls numbered 1 through 10. You select 3 of these balls and record their numbers, obtaining in this way a sequence of three numbers.
   a) How many such sequences of three numbers are there if you replace each ball as soon as you have recorded its number, thus allowing for the possibility of choosing the same ball more than once?
   b) How many such sequences of three numbers are there if you do not replace a ball once it is chosen?

18. A computer salesman sells 62 boxes of floppy diskettes on a given day. One of his customers bought 1 box, two of his customers each bought 2 boxes, one of his customers bought 5 boxes, and the rest of his customers each bought 4 boxes. How many customers did the salesman have that day?

19. Solve the counting problem in Example 4.1.2 of Section 4.1. Give full details.

20. How many integers between 0 and 1,000,000 contain the digit 9? *Hint*, it is easier to count the number of integers that do *not* contain the digit 9.

21. a) How many integers between 100 and 999 (inclusive) have distinct digits?
   b) How many of these integers are odd?

22. A certain set contains 192 elements. One half of the elements are grouped into groups of size 4, and the rest are grouped into groups of size 2. How many groups are there?

23. A certain set contains n elements. One half of the elements are grouped into groups of size k and the rest of the elements are grouped into groups of size j. How many groups are there? (Assume that n is even and that both k and j divide $n/2$.)

24. a) Prove, without using mathematical induction, that the truth table of a compound statement that involves exactly n different statement variables has 2^n rows.
   b) Prove that the number of possible truth tables for statements involving exactly n statement variables is 2^{2^n}.

25. Assuming that the multiplication rule holds for the case of two tasks T_1, T_2, prove that it holds for any series of tasks $T_1, T_2, \ldots, T_n$.

26. Prove by induction that if $m \neq 1$, then

$$1 + m + m^2 + \cdots + m^n = \frac{m^{n+1} - 1}{m - 1}$$

4.3 The Pigeonhole Principle

Let us now consider the second of our simple, yet very powerful rules, known as the **Dirichlet pigeonhole principle**. This rule is named after the French mathematician Peter Gustav Lejeune Dirichlet (1805–1859), but we will refer to it simply as the pigeonhole principle.

The Pigeonhole Principle If m balls are placed in n boxes, and if m is larger than n, then one of the boxes must receive at least two balls.

Of course, the pigeonhole principle can be worded in many different ways. For example, we could reword it as follows: If m users want to use n computer terminals, and if m is larger than n, then at least two users must share a terminal. (The term *pigeonhole principle* comes from the fact that this rule is sometimes worded in terms of putting pigeons into pigeonholes.)

The pigeonhole principle seems like a very simple rule, almost not even worth mentioning. But, as you will soon see, it can be used in an amazingly large variety of ways to solve complicated counting problems. Let us begin with some simple examples.

Example 4.3.1

If $r_1, r_2, \ldots, r_m$ are integers between 1 and n (inclusive), and if m is larger than n, then at least two of these integers must be equal.

This follows easily from the pigeonhole principle. We simply label m different balls with the integers $r_1, r_2, \ldots, r_m$, and n different boxes with the integers 1, 2, . . ., n. Then, we place the m balls into the n boxes in such a way that each ball goes into the box with the same label. Since m is larger than n, the pigeonhole principle tells us that some box must receive at least two balls. This means that at least two balls must have the same label, and this translates into the fact that at least two of the integers $r_1, r_2, \ldots, r_m$ must be equal. □

Example 4.3.2

In any group of 367 people, at least two people must have the same birthday. In this case, we label 366 boxes with the 366 possible birthdays (including February 29), and label 367 balls with the names of the 367 people. Then we place each ball into the box that bears the birthdate of the person whose name is on the ball. According to the pigeonhole principle, one of the boxes must receive at least two balls; that is, there must be at least two people with the same birthday. □

Example 4.3.3

The population of New York City is at least 7,000,000 people. If we assume that no person has more than 500,000 hairs on his or her head (a reasonable assumption, by the way), then according to the pigeonhole principle, there must be at least two people in New York City with the same number of hairs on their heads. (Can you phrase this in terms of balls into boxes?) □

The pigeonhole principle can be used to solve certain problems from a branch of mathematics known as **number theory**, which can be described as the study of the integers. Let us consider a few examples of this type.

Example 4.3.4

Given any subset of $n + 1$ integers from the set $\{1, 2, \ldots, 2n\}$, there must be two integers from the subset with the property that one of them divides the other. (An integer k *divides* an integer j if, when we divide j by k, the remainder is equal to 0. Put another way, k divides j if j is a multiple of k, that is, if $j = qk$, where q is some integer.)

Let us write each of the $n + 1$ integers in the form of a power of 2 times an odd factor. For example, we would write

$$84 = 2^2 \cdot 21 \quad , \quad 26 = 2^1 \cdot 13, \quad \text{and} \quad 7 = 2^0 \cdot 7.$$

Of course, all of the $n + 1$ odd factors obtained in this way are between 1 and $2n$. But there are only n odd numbers between 1 and $2n$, and so there are only n possibilities for these $n + 1$ odd factors. Therefore, we can apply the pigeonhole principle, which tells us that at least two of the odd factors must be equal. (Here we have n "boxes" and $n + 1$ "balls.") This means that two of the integers in the subset must have the form

$$x = 2^k m \quad \text{and} \quad y = 2^j m$$

where m is the odd factor. But if $k \leq j$, then x divides y, and if $j < k$, then y divides x. (You should check this for yourself by dividing.) In either case, one of the numbers x or y divides the other, which is what we wanted to show. □

The argument in the previous example is rather involved, and you should not feel discouraged if you need to read it through several times in order to follow it. Let us consider another example of the same type as the previous one.

Example 4.3.5

Given any collection of n integers, not necessarily distinct, there is some subcollection of these integers whose sum is divisible by n.

Let us denote the n integers by $a_1, a_2, \ldots, a_n$. We want to consider the sums

$$s_1 = a_1$$

$$s_2 = a_1 + a_2$$

$$s_3 = a_1 + a_2 + a_3$$

$$\vdots$$

$$s_n = a_1 + a_2 + \cdots + a_n$$

(These are not all of the possible sums of subcollections of the integers $a_1, a_2, \ldots, a_n$, but these are the only ones that we need.)

Let us divide each of these sums by n and denote the quotients by $q_1, q_2, \ldots, q_n$ and the remainders by $r_1, r_2, \ldots, r_n$. Thus, we can write

$$s_1 = a_1 = q_1 n + r_1$$

$$s_2 = a_1 + a_2 = q_2 n + r_2$$

$$s_3 = a_1 + a_2 + a_3 = q_3 n + r_3$$

$$\vdots$$

$$s_n = a_1 + a_2 + \cdots + a_n = q_n n + r_n$$

Since each of the n remainders $r_1, r_2, \ldots, r_n$ is a remainder after dividing by n, it must be less than n. In symbols,

$$0 \leq r_i < n$$

for all $i = 1, 2, \ldots, n$.

Now, if one of the remainders, say r_j, is equal to zero, then the corresponding sum s_j would satisfy

$$s_j = q_j n$$

and so it would be divisible by n. Thus we would have found a sum, namely s_j, which is divisible by n.

If none of the remainders are equal to zero, then we can still find a sum that is divisible by n. For then all of the remainders $r_1, r_2, \ldots, r_n$ would be *strictly* between 0 and n, in symbols,

$$0 < r_i < n$$

for all $i = 1, 2, \ldots, n$. But there are only $n - 1$ integers strictly between 0 and n, and so we can use the pigeonhole principle to conclude that at least two of the n remainders must be equal. Let us suppose that

$$r_j = r_k \qquad \textbf{(4.3.1)}$$

where j and k are integers betwen 1 and n, and $j < k$.

Using Equation 4.3.1, we have

$$s_j = q_j n + r_j$$

and

$$s_k = q_k n + r_k = q_k n + r_j$$

Subtracting these two equations gives

$$s_k - s_j = (q_k n + r_j) - (q_j n + r_j) = (q_k - q_j)n \qquad \textbf{(4.3.2)}$$

But, since $j < k$, we also have

$$\begin{aligned} s_k - s_j &= (a_1 + a_2 + \cdots + a_k) - (a_1 + a_2 + \cdots + a_j) \\ &= a_{j+1} + a_{j+2} + \cdots + a_k \end{aligned} \qquad \textbf{(4.3.3)}$$

Comparing Equations 4.3.2 and 4.3.3, we see that

$$a_{j+1} + a_{j+2} + \cdots + a_k = (q_k - q_j)n$$

Hence, in this case, the sum $a_{j+1} + a_{j+2} + \cdots + a_k$ is divisible by n.

In either case, that is, whether some remainder is equal to zero or not, there is a subcollection whose sum is divisible by n, which is what we wanted to show. □

The pigeonhole principle can also be used to solve certain types of geometric problems. Let us consider an example of this type.

■ Example 4.3.6

What is the largest number of points that can be placed in a square whose side has length 2, in such a way that no two points are a distance of $\sqrt{2}$ or less from each other?

Consider the square of Figure 4.3.1, which we have divided into 4 smaller squares, each with side length equal to 1. According to the Pythagorean Theorem, the diagonal of each of the small squares has length $\sqrt{2}$. Thus, if we are going to place points in the large square in such a way that no 2 points are within a distance of $\sqrt{2}$, we cannot place more than 1 point in each of the smaller squares. Therefore, according to the pigeonhole principle, we cannot place *more than* 4 points in the large square.

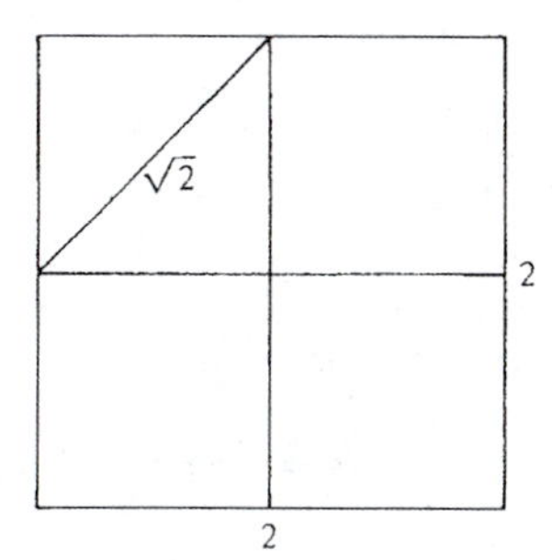

Figure 4.3.1

Notice that the pigeonhole principle does not tell us that we *can* place 4 points in the square—it only tells us that we cannot place *more than* 4 points. However, if we place 1 point in each of the 4 corners of the large square, then no 2 of these points will be within a distance of $\sqrt{2}$ from each other. Hence, the answer to the question is 4. □

□ □ Exercises

In each of the following exercises, if you use the pigeonhole principle, indicate which are the balls and which are the boxes.

1. Suppose that you have 10 pairs of socks in your dresser drawer and each pair is a different color. Assume also that they are loose in the drawer, that is, they are not bound together in matching pairs. The room is completely dark and you cannot see the color of any of the socks. How many socks must you remove in order to be certain that you get a matched pair?
2. Suppose that you have 9 microcomputer chips, each of which has 16 prongs, 7 chips each of which has 32 prongs, and 3 chips each of which has 128 prongs. Will all of these chips fit on a board that has 750 sockets? (A socket accommodates one prong.) Explain your answer.
3. Show that, given any 3 integers, at least 2 of them must have the property that their difference is even.
4. How many integers between 0 and $2n$ must you pick in order to be certain that at least one of them is odd?

5. How many books must you choose from a collection of 7 chemistry books, 18 mathematics books, 12 computer science books, and 11 physics books in order to be certain that you will get at least 5 books of one type?
6. Show that at any party consisting of 10 people at least 2 of the people must have the same number of friends at the party. *Hint*, the possibilities for the number of friends are 0, 1, . . ., 9. Is it possible for one person at the party to have exactly 0 friends, and for another person to have exactly 9 friends?
7. Show that, for any group of people, there must be at least 2 people with the same number of friends in the group.
8. Suppose that 5 microcomputers are connected to 3 printers. How many connections are necessary between the computers and the printers in order to be certain that, whenever any 3 of the computers each require a separate printer, the printers are available?
9. Let $2n + 1$ balls be placed into $n \geq 1$ boxes, and let M be the largest number of balls that fall into a single box. What is the smallest possible value of M?
10. Let $kn + 1$ balls be placed into $n \geq 1$ boxes, and let M be the largest number of balls that fall into a single box. What is the smallest possible value of M?
11. Show that if $a_1 + a_2 + \cdots + a_n - n + 1$ balls are placed into n boxes, then either the first box receives at least a_1 balls, or else the second box receives at least a_2 balls, or else the third box receives at least a_3 balls, and so on. *Hint*, suppose not. Then how many balls can there be?
12. Suppose that m balls are placed in n boxes and that

$$m < \frac{n(n-1)}{2}$$

Show that at least 2 boxes must receive the same number of balls. *Hint*, suppose that each box receives a different number of balls. Can you find a lower estimate on the total number of balls?
13. Let n be a positive odd integer, and let $b_1, b_2, \ldots, b_n$ be a rearrangement of the integers $1, 2, \ldots, n$. Prove that the number

$$(b_1 - 1)(b_2 - 2) \cdots (b_n - n)$$

is even. *Hint*, a product of integers is even if and only if at least one of the factors is even, so you only need to prove that one of the factors must be even. Group the factors into two types, namely,

$$b_i - \text{even} \quad \text{and} \quad b_j - \text{odd}$$

Is it possible for all of the factors b_j − odd to be odd? What would this say about the parity of the numbers b_j?
14. Suppose that you are given n blocks, each of which weighs an integral number of pounds but less than n pounds. Suppose also that the total weight of the n blocks is less than $2n$. Prove that the blocks can be divided into two groups, one of which weighs exactly n pounds. *Hint*, let $w_1, w_2, \ldots, w_n$ be the weights of the n blocks, and consider the sums

$$s_1 = w_1$$

$$s_2 = w_1 + w_2$$

$$\vdots$$

$$s_n = w_1 + w_2 + \cdots + w_n$$

These sums form an increasing sequence of numbers, where $s_1 < n$ and $s_n \geq n$. If one of the sums s_i is equal to n, then you are done. (Why?) But if not, then there is an integer k for which $s_k < n$, but $s_{k+1} > n$. (How do we know this?) Now consider the integers

$$s_1, s_2, \ldots, s_k, s_{k+1} - n, s_{k+2} - n, \ldots, s_n - n$$

Is it possible for all of these integers to be distinct? If not, then two of them must be equal. What does this tell you?

15. What is the largest number of points that can be placed in an equilateral triangle of side length 1 in such a way that no two of the points are within a distance of $\frac{1}{2}$ from each other? *Hint*, use a method similar to the one in Example 4.3.6.

16. a) Show that it is possible to invite 5 people to a party under the restriction that no group of 3 of the people are either total strangers or mutual acquaintances. *Hint*, consider the following diagram

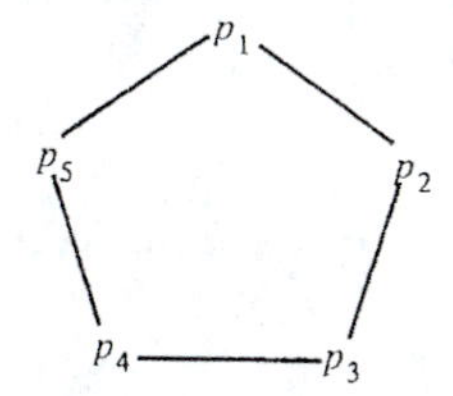

The symbols $p_1, p_2, \ldots, p_5$ represent the 5 people, and there is a line connecting two people if and only if they are acquaintances. Show that this party of 5 people does not violate the previous restriction.

b) Show that the *maximum* number of people that can be invited to a party under the restriction that no group of 3 of the people are either total strangers or mutual acquaintances is 5. *Hint*, you know from part a that it is possible to invite 5 people to such a party. You must show that no more than 5 people can be invited. Put another way, you must show that if 6 people $p_1, p_2, \ldots, p_6$ are at a party, then one of the restrictions is violated. First show that person p_1 must either be acquainted with or else be a stranger to at least 3 of the other 5 people, then explore the relationships among these 3 people.

4.4 Permutations

Let us consider the counting problem in Example 4.1.2. Recall that the problem is to count the number of possible ways in which a computer operator can order 9 computer programs.

Now, if we denote the programs by the symbols $p_1, p_2, \ldots, p_9$, then any order of the programs corresponds to an ordered arrangement of these symbols. For example, the arrangement

$$p_6\, p_2\, p_1\, p_3\, p_9\, p_4\, p_8\, p_7\, p_5$$

is one possible order, and the arrangement

$$p_2\, p_9\, p_1\, p_3\, p_8\, p_4\, p_5\, p_7\, p_6$$

is another possible order.

Ordered arrangements of objects occur very frequently in combinatorics, and they deserve a special name.

Definition

An ordered arrangement of a set of objects is called a **permutation** of the objects. If there are n objects in the permutation, we say that the permutation has **size** n, or is an **n-permutation.** □

Thus, in this case, each order of the programs is simply a permutation of the 9 programs, and the question in Example 4.1.2 is "How many permutations are there of 9 programs?"

It would be nice to have a formula for the number of permutations of any set of n objects, but before deriving such a formula, let us consider a few more examples of permutations.

Example 4.4.1

There are only 2 permutations of the 2 letters a and b, namely

ab and ba □

Example 4.4.2

There are 6 permutations of the 3 letters a, b, and c, namely

abc	acb
bac	bca
cab	cba

□

Example 4.4.3

There are 24 permutations of the 4 objects a, b, c, and 1. These are

abc1	ab1c	acb1	ac1b
a1bc	a1cb	bac1	ba1c
bca1	bc1a	b1ac	b1ca
cab1	ca1b	cba1	cb1a
c1ab	c1ba	1abc	1acb
1bac	1bca	1cab	1cba

□

The number of permutations of a set of objects depends only on the number of objects, and not on what the objects are. Thus, for example, there are 24 permutations of *any* set of 4 objects, regardless of whether the objects are numbers, letters, people, books, programs, or anything else. Put another way, there are 24 permutations of size 4.

It is not hard to derive a formula for the number of permutations of n objects, if we use the multiplication rule.

Theorem 4.4.1

The number of permutations of a set of n objects is the product of the first n positive integers, that is,

$$n(n - 1) \cdots 1$$

(Notice that we have written the factors in this product in descending order.)

PROOF Forming a permutation of n objects can be thought of as performing a sequence of n tasks. The first task is to choose the first object for the permutation, and this can be done in n different ways. The second task is to choose the second object for the permutation, and this can be done in $n - 1$ different ways. (There are only $n - 1$ objects left after the first task has been performed.) The third task is to choose the third object for the permutation, and this can be done in $n - 2$ different ways. This continues until we reach the n-th task, which is to choose the n-th object for the permutation. This can be done in only one way, since there is only one object left at this point.

Now, the *number* of ways to perform the i-th task in the sequence does not depend on how the previous $i - 1$ tasks were performed. We emphasize the word *number* here because the outcome for the i-th task does depend on how the previous tasks were performed—it is only the number of possible outcomes that does not.

Therefore, we can apply the multiplication rule, which tells us that there are $n(n - 1) \cdots 1$ different ways to perform the entire sequence of n tasks; that is, there are $n(n - 1) \cdots 1$ permutations of n objects. This proves the theorem. ▮

Since the product $n(n - 1) \cdots 1$ occurs very frequently in combinatorics, we give it a special name, and a special symbol. Namely, we write

$$n! = n(n - 1) \cdots 1$$

and read $n!$ as "n factorial."

Thus, Theorem 4.4.1 says that there are $n!$ permutations of n objects; that is, there are $n!$ permutations of size n. As we will see later, it is convenient to define $0!$ by setting $0! = 1$. This is only a convention and has no meaning in terms of permutations.

For reference, let us list the first few factorials

$$0! = 1$$
$$1! = 1$$
$$2! = 2 \cdot 1 = 2$$
$$3! = 3 \cdot 2 \cdot 1 = 6$$
$$4! = 4 \cdot 3 \cdot 2 \cdot 1 = 24$$
$$5! = 5 \cdot 4 \cdot 3 \cdot 2 \cdot 1 = 120$$
$$6! = 6 \cdot 5 \cdot 4 \cdot 3 \cdot 2 \cdot 1 = 720$$
$$7! = 7 \cdot 6 \cdot 5 \cdot 4 \cdot 3 \cdot 2 \cdot 1 = 5040$$
$$8! = 8 \cdot 7 \cdot 6 \cdot 5 \cdot 4 \cdot 3 \cdot 2 \cdot 1 = 40{,}320$$
$$9! = 9 \cdot 8 \cdot 7 \cdot 6 \cdot 5 \cdot 4 \cdot 3 \cdot 2 \cdot 1 = 362{,}880$$
$$10! = 10 \cdot 9 \cdot 8 \cdot 7 \cdot 6 \cdot 5 \cdot 4 \cdot 3 \cdot 2 \cdot 1 = 3{,}628{,}800$$

As you can see from this list, the factorials grow very rapidly (see also Exercises 4 and 20).

Using Theorem 4.4.1, we can easily solve the problem of ordering the 9 programs. According to this theorem, there are $9! = 362{,}880$ possible ways to order 9 programs. (Thus, the computer operator will be able to give a different order to the programs each day for approximately 1000 years!) Imagine having to solve this problem by trying to write down all of the possible orders one by one. Even if you could write one order every 5 seconds, 24 hours a day, it would take approximately 3 weeks to write them all down, and how could you be sure that you didn't miss one along the way? This is an excellent demonstration of the power of the multiplication rule.

Since we will be using factorials a great deal, it is important to spend a little time becoming familiar with their arithmetic properties. The next example may help in this regard.

Example 4.4.4

$$\frac{6!}{4!} = \frac{6 \cdot 5 \cdot 4 \cdot 3 \cdot 2 \cdot 1}{4 \cdot 3 \cdot 2 \cdot 1} = 6 \cdot 5 = 30$$

$$\frac{8!}{5!\,3!} = \frac{8 \cdot 7 \cdot 6 \cdot 5 \cdot 4 \cdot 3 \cdot 2 \cdot 1}{5 \cdot 4 \cdot 3 \cdot 2 \cdot 1 \cdot 3 \cdot 2 \cdot 1} = \frac{8 \cdot 7 \cdot 6}{3 \cdot 2 \cdot 1} = 8 \cdot 7 = 56$$

$$\frac{8}{4}! = 2! = 2$$

$$9! - 6! = 6!(9 \cdot 8 \cdot 7 - 1) = 503 \cdot 6! = 362{,}160 \quad \square$$

Some care must be taken in working with factorials. For example, the expression $(n + m)!$ is *not* in general equal to $n! + m!$, nor is $(nm)!$ equal to $n!m!$ (see Exercise 3).

Example 4.4.5

Suppose that we want to arrange 6 different math books, 4 different computer science books, and 3 different chemistry books on a single bookshelf.

a) In how many ways can this be done?

b) In how many ways can this be done if the math books must come first, then the computer science books, and finally the chemistry books?

c) In how many ways can this be done if all books of the same subject must be kept together?

To answer part a, we simply observe that each arrangement of the 13 books on the bookshelf is a permutation of the books. Hence there are $13! = 6{,}227{,}020{,}800$ possible arrangements of the books.

As to part b, there are 6! different ways to arrange the math books, there are 4! different ways to arrange the computer science books, and there are 3! different ways to arrange the chemistry books. Hence, according to the multiplication rule, there are $(6!)(4!)(3!) = 103{,}680$ different ways to arrange the books on the bookshelf, under the conditions given in part b.

Finally, we can use the results of part b, along with the multiplication rule, to solve part c. For if we let T_1 be the task of deciding the order in which the three *subjects*—math, computer science, and chemistry—appear on the shelf, and if we let

T_2 be the task of ordering the actual books, once the order of the subjects has been decided, then arranging the books on the bookshelf is the same as performing the sequence of tasks T_1, T_2. Now, there are $3! = 6$ ways of performing task T_1 and there are 103,680 ways of performing task T_2. (Why?) Hence, according to the multiplication rule, there are $6 \cdot 103{,}680 = 622{,}080$ different ways to arrange the books on the bookshelf, under the conditions given in part c. □

Example 4.4.6

How many ways can we order 9 computer programs $p_1, p_2, \ldots, p_9$, if program p_2 cannot immediately follow program p_1?

This is an example of a type of problem where it is much easier to count what we do *not* want and subtract that from the total. In this case, there are a total of 9! orderings and so if we subtract from 9! the number of orderings in which program p_2 *does* immediately follow program p_1, then we will have the answer to our question.

In order to compute the number of orderings in which program p_2 does follow program p_1, we reason as follows. As long as program p_2 must follow program p_1, we can think of these two programs as "tied together" into one program. In effect then, there are only 8 programs, and so there are 8! orderings in which p_2 follows p_1.

Thus, there are

$$9! - 8! = 362{,}880 - 40{,}320 = 322{,}560$$

orderings in which program p_2 does not follow program p_1. □

The technique that we used in the last example is very important, and so we should discuss it a bit more. It is important to keep in mind, when considering a counting problem, that it is *sometimes* easier to count what we do not want, and then subtract that from the total, rather than to count what we do want directly. Let us have another example of this.

Example 4.4.7

How many bytes are there that contain at least two 1's? (Recall that a byte is a binary word of length 8.)

Rather than try to count this number directly, it is much easier in this case to count the number of bytes that contain *less than* two 1's (these are the bytes we do *not* want) and subtract that number from 2^8, which is the total number of bytes.

Now, it is easy to count the number of bytes that have less than two 1's. After all, there is only one byte that contains no 1's, namely the byte with all 0's. Also, there are 8 bytes that contain exactly one 1—one byte for each of the 8 places to put the 1.

Hence, we see that there are 9 bytes that contain less than two 1's, and so there are $2^8 - 9 = 247$ bytes that contain at least two 1's. □

Example 4.4.8

We know that there are $n!$ different ways to arrange n people in a line, since each arrangement corresponds to a permutation of the n people. But how many ways are there to arrange n people in a circle?

The answer is *not* $n!$. The reason is that two circular arrangements are considered to be the same if one can be obtained from the other by a rotation. For example, the two circular arrangements in Figure 4.4.1 are the same. Our problem is to figure out a way to avoid counting these two arrangements as being different.

Figure 4.4.1

These two circular arrangements of the five people $p_1, \ldots, p_5$ are really the same, since one can be obtained from the other by a rotation.

Let us denote the n people by $p_1, p_2, \ldots, p_n$, and imagine that we are putting these n people in n seats around a round table. Then we can avoid the problem of rotations by simply agreeing to always put the first person, p_1, at the head of the table, as pictured in Figure 4.4.2. Then we can fill the remaining $n - 1$ seats with the remaining $n - 1$ people.

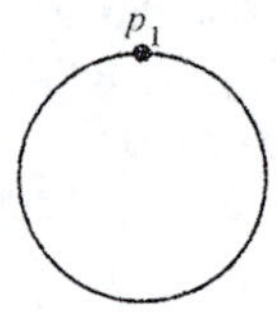

Figure 4.4.2

By always putting person p_1 at the head of the table, we avoid the problem of rotations.

Now, there are $n - 1$ choices for the first seat next to p_1, there are $n - 2$ choices for the next seat, and so on. Hence, according to the multiplication rule, there are $(n - 1)(n - 2) \cdots 1 = (n - 1)!$ ways to fill the remaining $n - 1$ seats; that is, there are $(n - 1)!$ ways to arrange the n people in a circle. □

Circular arrangements such as these are known as **circular permutations**, and the previous example shows that there are $(n - 1)!$ circular permutations of n objects. Let us put this into a theorem.

■ Theorem 4.4.2

The number of circular permutations of n objects is $(n - 1)!$.

For example, there are $2! = 2$ ways to arrange three people in a circle, as pictured in Figure 4.4.3. Also, there are $3! = 6$ ways to arrange four people in a circle, as pictured in Figure 4.4.4.

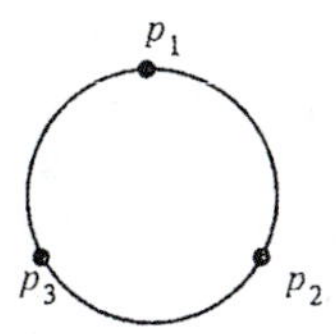

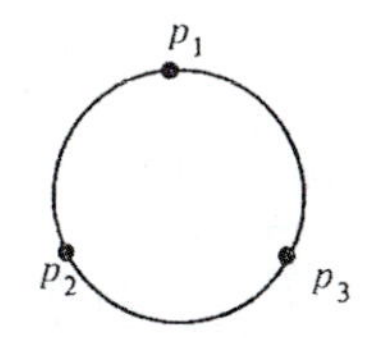

Figure 4.4.3
The two circular arrangements of three people.

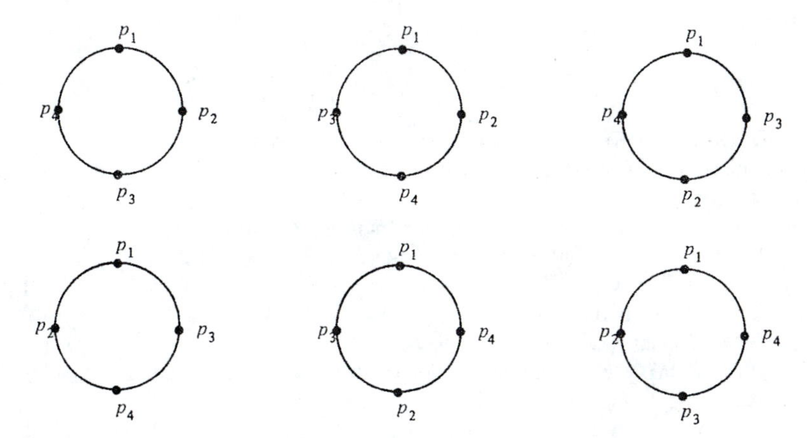

Figure 4.4.4
The six circular arrangements of four people.

Exercises

1. Simplify as much as possible

   a) $\dfrac{7!}{3!6!}$ b) $\dfrac{2!}{1!0!}$ c) $4! - 3!$

   d) $(0!)(1!)(2!)$ e) $\dfrac{9!}{10! - 9!}$ f) $7! - 5! + 6!$

2. Simplify as much as possible

   a) $\dfrac{n!}{(n-1)!}$ b) $\dfrac{(n+2)!}{(n-1)!}$ c) $\dfrac{(2n)!}{n!}$

   d) $n! - (n-1)!$ e) $n[n! + (n-1)!]$

3. a) Find an example to show that $(m + n)!$ is not necessarily equal to $m! + n!$. Are there any values of m and n for which $(m + n)! = m! + n!$?

   b) Find an example to show that $(mn)!$ is not necessarily equal to $m!n!$. Are there any values of m and n for which $(mn)! = m!n!$?

4. Find the smallest value of n for which $n!$ is larger than the population of the world (approximately 4,530,000,000).

5. Verify that, if you were to write down one ordering every 5 seconds, 24 hours a day, it would take at least 3 weeks to write all 9! orderings of 9 programs.

6. How many ways are there to order 9 programs if a certain program must always come last?
7. How many ways are there to order 8 programs if program p_3 must directly follow program p_2, which must directly follow program p_1?
8. How many ways are there to order 7 programs if program p_2 cannot follow program p_1, but program p_4 must follow program p_3?
9. How many binary words are there of length 10 that contain at least two 0's?
10. How many bytes are there that contain at least three 1's?
11. How many ternary words (that is, words over the alphabet $\Sigma = \{0, 1, 2\}$) are there of length 6 that contain at least two 2's?
12. How many ternary words are there of length 5 that contain at least two 1's and at least two 2's?
13. If you toss a coin 10 times, how many possible outcomes are there with at least 2 tails?
14. How many subsets are there of size at least 2, taken from a set of size 10? *Hint,* relate subsets to binary words.
15. How many ways are there to choose three numbers (not necessarily distinct) between 1 and 10 in such a way that the sum of the three numbers is at least 5?
16. Suppose that you wish to arrange 2 math books, 3 computer science books, and 4 physics books on a single bookshelf.
    a) In how many ways can this be done?
    b) In how many ways can this be done if the math books must come first, then the computer science books, and finally the physics books?
    c) In how many ways can this be done if all books of the same subject must be kept together?
17. Suppose that you wish to arrange 7 math books, 7 computer science books, and 7 physics books on a single bookshelf.
    a) In how many ways can this be done if the subjects must alternate—first math, then computer science, and finally physics?
    b) In how many ways can this be done if the subjects must alternate, with no restriction on the order of the subjects?
18. How many sequences of $2n$ integers are there, taken from the set $\{1, 2, \ldots, 2n\}$
    a) if all of the odd numbers must come first?
    b) if the numbers in the sequence must alternate even, odd, even, odd, . . .?
19. Show that, for $1 \le k \le n - 1$,

$$\frac{(n+1)!}{(k+1)!(n-k)!} = \frac{n!}{k!(n-k)!} + \frac{n!}{(k+1)!(n-k-1)!}$$

20. Solve for n.
    a) $(2n)! = 2n!$ b) $(3n)! = 3n!$
21. What value of X is printed when a computer runs the following program segment?

```
X := 1;

FOR I := 1 TO 10 DO

    X := I*X;

PRINT X
```

22. What value of X is printed when a computer runs the following program segment?

```
X := 1;
FOR I := 1 TO 10 DO
    X := 2*I*X;
PRINT X
```

23. What value of X is printed when a computer runs the following program segment?

```
X := 1;
FOR I := 1 TO 10 DO
    IF I <= 5 THEN X := I*I*X
ELSE X := I*X;
PRINT X
```

24. How many necklaces can be made by using 10 round beads, all of a different color?

25. a) How many ways are there to arrange 6 people in a circle?

b) How many ways are there to arrange 6 people in a circle if person p_2 cannot sit next to person p_1?

c) How many ways are there to arrange 6 people in a circle if person p_2 cannot sit to the immediate right of person p_1?

26. a) How many ways are there to arrange 3 men and 3 women in a circle in such a way that the men and women alternate?

b) How many ways are there to arrange n men and n women in a circle in such a way that the men and women alternate?

Hint, there are $(n - 1)!$ ways of arranging the n men in a circle. How many ways are there to insert the women in between the men?

27. Prove by induction that

$$1 \cdot 1! + 2 \cdot 2! + 3 \cdot 3! + \cdots + n \cdot n! = (n + 1)! - 1$$

28. Prove Theorem 4.4.1 by using mathematical induction.

29. Show that $(n!)^2 > n^n$. (Hence, $n! > n^{n/2}$, which shows that $n!$ gets *very* large.) *Hint*, write $(n!)^2$ in the form

$$(n!)^2 = (1 \cdot n)(2 \cdot (n - 1))(3 \cdot (n - 2)) \cdots (n \cdot 1)$$

A typical factor has the form $k \cdot (n - k + 1)$, where k is an integer between 1 and n. Show that the smallest value that these factors can be is when $k = 1$ (or $k = n$). One way to do this is to consider the quadratic function $f(x) = x(n - x + 1)$.

4.5

More on Permutations

It often happens that we have a set S of size n, but that we wish to form permutations using only k of the objects at a time, where $k \leq n$. In this section, we want to find a formula for the number of permutations of size k, *taken from a set of size n*.

Example 4.5.1

The permutations of size 2, taken from the set of 4 letters {a, b, c, d} are

ab	ba	ac
ca	ad	da
bc	cb	bd
db	cd	dc

The permutations of size 3, taken from this same set are

abc	acb	bac	*those 3-permutations involving*
bca	cab	cba	*the letters a, b, and c*
abd	adb	bad	*those 3-permutations involving*
bda	dab	dba	*the letters a, b, and d*
acd	adc	cad	*those 3-permutations involving*
cda	dac	dca	*the letters a, c, and d*
bcd	bdc	cbd	*those 3-permutations involving*
cdb	dbc	dcb	*the letters b, c, and d*

For convenience, we have organized these permutations into four groups according to which letters they involve. □

As you can see from these examples, if Σ is a finite set, then a permutation of size k, taken from the set Σ, is exactly the same as a word of length k over Σ *that has no repeated letters*. Therefore, when we count the number of permutations of size k, we will also be counting the number of words of length k that have no repeated letters.

Using the multiplication rule, we can easily obtain a formula for the number of permutations of size k, taken from a set of size n. Let us denote this number by $P(n, k)$.

Theorem 4.5.1

The number $P(n, k)$ of permutations of size k, taken from a set of size n, is

$$P(n, k) = n(n-1)\cdots(n-k+1)$$

where $1 \leq k \leq n$.

PROOF Forming a permutation of size k, from a set of size n, can be thought of as performing a sequence of k tasks. The first task is to choose the first element for the permutation, and there are n ways to do this. The second task is to choose the second element for the permutation, and there are $n - 1$ ways to do this, and so on. The k-th task is to choose the k-th element for the permutation, and since there are $n - (k - 1)$ objects left to choose from at this point, there are $n - (k - 1) = n - k + 1$ ways to perform this task. Hence, according to the multiplication rule, we have

$$P(n, k) = n(n-1)\cdots(n-k+1) \tag{4.5.1}$$

which proves the theorem. ■

Let us have an example of Theorem 4.5.1.

Example 4.5.2

The number of permutations of size 3, taken from a set of size 4, is

$$P(4, 3) = 4 \cdot 3 \cdot 2 = 24$$

(In this case, $n = 4$ and $k = 3$, and so $n - k + 1 = 2$.) Of course, this agrees with Example 4.4.3.

If Σ is an alphabet of size 8, then the number of words of length 4 over Σ that do not have repeated letters is

$$P(8, 4) = 8 \cdot 7 \cdot 6 \cdot 5 = 1680$$

(In this case, $n = 8$ and $k = 4$, and so $n - k + 1 = 5$.) □

Theorem 4.4.1 is a special case of Theorem 4.5.1. After all, $P(n, n)$ is, *by definition*, the number of permutations of n objects, and if we set $k = n$ in Theorem 4.5.1, we get

$$\begin{aligned} P(n, n) &= n(n-1) \cdots (n-n+1) \\ &= n(n-1) \cdots 1 \\ &= n! \end{aligned}$$

which is exactly what Theorem 4.4.1 says.

Products of the form $n(n - 1) \cdots (n - k + 1)$ occur very frequently in combinatorics, and so there is a special notation for them. Namely, we write

$$(n)_k = n(n-1) \cdots (n-k+1)$$

The symbol $(n)_k$ is read "n lower factorial k." Notice that

$$(n)_n = n(n-1) \cdots 1 = n!$$

and so the factorial $n!$ is just a special case of the lower factorial.

Using the lower factorial notation, the conclusion of Theorem 4.5.1 can be written

$$P(n, k) = (n)_k$$

In words, this says that the number $P(n, k)$ of permutations of size k, taken from a set of size n, is the lower factorial $(n)_k$.

In order to be absolutely clear about this, we should emphasize that $P(n, k)$ stands for the number of permutations of size k, taken from a set of size n. On the other hand, $(n)_k$ stands for the product

$$(n)_k = n(n-1) \cdots (n-k+1)$$

Thus, $P(n, k)$ and $(n)_k$ stand for entirely different things. However, it is a *theorem* that $P(n, k) = (n)_k$.

Let us consider some simple examples of permutations of size k, taken from a set of size n.

Example 4.5.3

In order to send messages from one boat to another, flags are sometimes used. Suppose that a certain boat has 10 different flags and 1 flagpole. If each ordered arrangement of 3 flags on the flagpole represents a different message, how many messages are possible?

Since there is 1 message for each permutation of size 3, taken from the 10 flags, there are

$$P(10, 3) = (10)_3 = 10 \cdot 9 \cdot 8 = 720$$

possible messages. □

Example 4.5.4

Suppose that a certain state has license plates consisting of 3 letters followed by 3 digits. How many license plates are there in this state that do not have a repeated letter or a repeated digit?

Forming such a license plate can be thought of as performing two tasks. The first task is to determine the 3 letters, and there are P(26, 3) ways of doing this, since there are P(26, 3) different permutations of size 3, taken from the 26 letters of the alphabet. The second task is to choose the 3 digits, and there are P(10, 3) ways to do this. Hence, according to the multiplication rule, there are

$$\begin{aligned} P(26, 3)P(10, 3) &= (26)_3(10)_3 \\ &= 26 \cdot 25 \cdot 24 \cdot 10 \cdot 9 \cdot 8 \\ &= 11{,}232{,}000 \end{aligned}$$

such license plates.

It is interesting to compare this number with the total number of license plates possible when repetitions are allowed. According to the multiplication rule, there are

$$26 \cdot 26 \cdot 26 \cdot 10 \cdot 10 \cdot 10 = 17{,}576{,}000$$

license plates consisting of 3 letters followed by 3 digits. Now, since

$$\frac{11{,}232{,}000}{17{,}576{,}000} \approx 0.64$$

we see that approximately 36% of all license plates have either a repeated letter or a repeated digit! □

We should say a few words about how to evaluate the lower factorial $(n)_k$. If k is small, the easiest way is simply to write down the k factors, starting with the factor n. For example, to evaluate $(50)_4$, we write

$$(50)_4 = 50 \cdot 49 \cdot 48 \cdot 47 = 5{,}527{,}200$$

In this case, we do not need to worry about computing $n - k + 1$ in order to determine the last factor.

On the other hand, if k is large, then we must settle for computing $n - k + 1$ and writing down only the first few factors and the last factor. For example, if we want to evaluate $(50)_{35}$, the best we can do is compute $n - k + 1 = 50 - 35 + 1 = 16$, and write

$$(50)_{35} = 50 \cdot 49 \cdots 16$$

The lower factorial $(n)_k$ can be written completely in terms of the ordinary factorials, since

$$\begin{aligned}(n)_k &= n(n-1)\cdots(n-k+1)\\ &= \frac{n(n-1)\cdots(n-k+1)(n-k)\cdots 1}{(n-k)\cdots 1}\\ &= \frac{n!}{(n-k)!}\end{aligned}$$

This gives us the nice formula

$$(n)_k = \frac{n!}{(n-k)!}$$

For example,

$$(50)_{35} = \frac{50!}{(50-35)!} = \frac{50!}{15!}$$

Combining this with Theorem 4.5.1, we can now write

$$P(n, k) = (n)_k = \frac{n!}{(n-k)!} \tag{4.5.2}$$

for $1 \leq k \leq n$, where $(n)_k = n(n-1)\cdots(n-k+1)$.

The last expression in Equation 4.5.2 makes perfectly good sense for $k = 0$, and in fact equals 1. For this reason, we *define* $P(n, 0)$ and $(n)_0$ by setting them both equal to 1,

$$P(n, 0) = 1 \qquad \text{and} \qquad (n)_0 = 1$$

for all integers n (including $n = 0$). Then Equation 4.5.2 holds for all values of k satisfying $0 \leq k \leq n$.

Let us conclude this section by mentioning that there are many useful relationships among the lower factorials. As an example, we have

$$(n)_k = n(n-1)_{k-1}$$

[One of the reasons that such relationships are useful is that, since $(n)_k = P(n, k)$, they tell us things about the numbers $P(n, k)$.]

In order to verify this equation, we simply evaluate both sides. The left side is

$$(n)_k = n(n-1)\cdots(n-k+1) \tag{4.5.3}$$

In order to evaluate the right side, we first evaluate $(n - 1)_{k-1}$,

$$(n - 1)_{k-1} = (n - 1)(n - 2) \cdots ((n - 1) - (k - 1) + 1)$$

$$= (n - 1)(n - 2) \cdots (n - k + 1)$$

(Do you see how we got this?) Hence

$$n(n - 1)_{k-1} = n(n - 1)(n - 2) \cdots (n - k + 1) \tag{4.5.4}$$

Comparing Equations 4.5.3 and 4.5.4, we see that

$$(n)_k = n(n - 1)_{k-1}$$

which is what we wanted to show.

Exercises

1. Let Σ be an alphabet of size 10. How many words of length 5 are there over Σ that have no repeated letters?
2. Let $\Sigma = \{a, b, \ldots, z\}$. How many 5-letter words can be formed over Σ if no letter is allowed to be used more than once in any word?
3. The license plates of a certain state consist of 4 letters followed by 2 digits. How many such license plates are there if
   a) repeated letters are not allowed?
   b) neither repeated letters nor repeated digits are allowed?
   c) neither repeated letters nor repeated digits are allowed and 0 is not allowed to be the first digit?
4. The serial numbers on the back of a certain brand of computer consist of 3 letters followed by 7 digits.
   a) How many possible serial numbers are there?
   b) How many of these serial numbers do not have any repeated letters?
   c) How many of these serial numbers do not have repeated digits?
   d) How many of these serial numbers do not have either repeated letters or repeated digits?
5. How large can the population of a town be if no 2 people are allowed to have the same 3-letter initials?
6. Let Σ be an alphabet for which $|\Sigma| = m$. How many words are there of size m over Σ with no repeated letters?
7. How many ordered triplets of letters are there, taken from the letters A, T, C, and G
   a) if repeated letters are allowed?
   b) if repeated letters are not allowed?
8. How many ways are there to hand out 7 free books among 15 students if no student is to receive more than 1 book?
9. How many ways are there to place k balls into n boxes in such a way that no box receives more than 1 ball?
10. A combination for a combination lock consists of a sequence of 4 integers between 0 and 59 (inclusive).
    a) How many possible combinations are there?
    b) How many possible combinations are there if no number can be repeated?
11. Suppose that you are given 10 red flowers and 10 yellow flowers. (All flowers are distinguishable.)
    a) How many ways are there to arrange 10 of these flowers in a row in a window box?

b) How many ways are there to arrange 10 of these flowers in a row if you want the colors to alternate?

c) How many ways are there to arrange 5 of the red flowers and 5 of the yellow flowers if you want all of the red flowers to be together?

12. Verify that

a) $(n)_k(n-k)_{n-k} = n!$ b) $(n)_k = (n-1)_k + k(n-1)_{k-1}$

13. Verify that

a) $P(n, n) = P(n, n-1)$

b) $P(n, k+1) = nP(n-1, k)$

c) $P(n, k)/k! = P(n, n-k)/(n-k)!$

d) $P(n, k) = P(n-1, k) + kP(n-1, k-1)$

14. Verify that

$$(n)_k = \frac{n!}{(n-k)!}$$

when $k = n$. How does this justify choosing $0! = 1$?

15. Solve for n.

a) $5P(n, 3) = 2P(n-1, 4)$

b) $P(n, 4) = 3024$

c) $10P(n, 1) + 2P(n, 2) = P(n, 3)$

16. Let $p_1, p_2, \ldots, p_n$ denote n people.

a) How many ways are there to arrange k of these people in a line?

b) How many ways are there to arrange k of these people in a line if person p_1 refuses to be next to person p_2 in line? *Hint*, it might help to review Example 4.4.6.

c) How many ways are there to arrange k of these people in a line if person p_1 refuses to be in the same line with person p_2?

17. What value of X is printed by a computer that runs the following program segment?

```
X := N;
FOR I := 1 TO K - 1 DO
    X := X*(N - I);
PRINT X
```

18. What value of X is printed by a computer that runs the following program segment?

```
X := N;
FOR I := 1 TO N - K DO
    X := X*(N - I);
PRINT X
```

Give your answer in terms of lower factorials.

19. Given n people, show that there are

$$\frac{(n)_k}{k}$$

possible ways to arrange any k of these people in a circle. (Each such arrangement is called a **circular permutation of size k**, taken from a set of size n.) *Hint*, one way to do this is to consider the set of all ordinary permutations of size k, taken from a set of size n, and to group this set into groups as follows. Two k-permutations are to be placed in the same group if and only if they represent the same circular permutation when they are formed into a circle by joining the last person to the first one. For example, the permutation $p_1\, p_3\, p_2\, p_4$ becomes the circular permutation

$$\begin{matrix} & p_1 & \\ p_4 & & p_3 \\ & p_2 & \end{matrix}$$

Thus, for example, the two permutations $p_1\, p_3\, p_2\, p_4$ and $p_2\, p_4\, p_1\, p_3$ are in the same group, since they give the same circular permutation, Show that there are k permutations in each group. How many groups are there? How does this solve the problem?

4.6 Combinations

In the previous two sections, we have been studying *ordered* arrangements of objects. Now, we want to study *unordered* arrangements. Actually, we have already studied unordered arrangements in Chapter 1, since an unordered arrangement of the elements of a set S is nothing more than a *subset* of S.

However, in this section, we want to consider certain combinatorial questions relating to subsets that we did not discuss in Chapter 1. For example, we want to determine how many subsets of size k can be formed from a set of size n. (We do know that a set of size n has 2^n subsets of *all* sizes.) Let us begin with a definition.

Definition

Let S be a set. Then an unordered arrangement of k elements of S, that is, a subset of S of size k, is also called a **combination of size k**, or a **k-combination**, taken from S. □

We will denote the number of combinations of size k, taken from a set of size n, by $C(n, k)$. *Thus, $C(n, k)$ is the number of subsets of size k, taken from a set of size n.*

Example 4.6.1

For purposes of comparison, let us make a list of all of the 3-combinations and 3-permutations of the set $\{1, 2, 3, 4\}$.

	3-permutations of {1, 2, 3, 4}			*3-combinations of {1, 2, 3, 4}*
those permutations involving the integers 1, 2, 3	123	132	213	{1, 2, 3}
	231	312	321	
those permutations involving the integers 1, 2, 4	124	142	214	{1, 2, 4}
	241	412	421	
those permutations involving the integers 1, 3, 4	134	143	314	{1, 3, 4}
	341	413	431	
those permutations involving the integers 2, 3, 4	234	243	324	{2, 3, 4}
	342	423	432	

Notice that we have arranged the permutations into four groups, each of which consists of those permutations that use the same three integers. Also, we have placed each combination next to the group of permutations that involves the same three integers. □

As it turns out, there is a simple relationship between the numbers $C(n, k)$ and $P(n, k)$, and since we already have a formula for $P(n, k)$, we can use this relationship to obtain a formula for $C(n, k)$. First, let us give this relationship.

Theorem 4.6.1

The number $C(n, k)$ of combinations of size k, taken from a set of size n, satisfies

$$C(n, k) = \frac{P(n, k)}{k!}$$

for $0 \leq k \leq n$, where $P(n, k)$ is the number of permutations of size k, taken from a set of size n.

PROOF Let us first consider the case $k = 0$. Since $C(n, 0)$ is the number of subsets of size 0 of a set of size n, we know that $C(n, 0) = 1$. On the other hand, we decided in Section 4.5 to let $P(n, 0) = 1$, and since $0! = 1$, we do indeed have

$$C(n, 0) = \frac{P(n, 0)}{0!}$$

Now let us consider the case $1 \leq k \leq n$. To make the argument easier to follow, we will determine the number of k-combinations of the set $S = \{1, 2, \ldots, n\}$. Certainly, this number will be the same as the number of k-combinations of *any* set of size n.

Let us imagine that we have made a list, similar to the one in the previous example, of the $P(n, k)$ permutations and the $C(n, k)$ combinations of size k, taken from the set S. Thus, the permutations are grouped together according to which integers they involve, and each combination is placed next to that group of permutations that involves the same integers.

Now, each group of permutations in this list contains exactly $k!$ permutations, since it contains all of the permutations of a particular choice of k of the integers 1, 2, . . ., n. (For instance, the first group in Example 4.6.1 contains the $3! = 6$ permutations of the three integers 1, 2, and 3 and the second group contains the $3! = 6$ permutations of the three integers 1, 2, and 4.)

Therefore, and this is the key to the proof, each of the $C(n, k)$ combinations in the list corresponds to exactly $k!$ permutations. This means that if we multiply $C(n, k)$ by $k!$, we must get the total number of permutations $P(n, k)$. In symbols,

$$k!C(n, k) = P(n, k)$$

Dividing by $k!$, we arrive at

$$C(n, k) = \frac{P(n, k)}{k!}$$

which is what we wanted to prove. ■

Using Theorem 4.6.1 and the fact that $P(n, k) = (n)_k$, we now get a formula for $C(n, k)$.

$$C(n, k) = \frac{(n)_k}{k!} \tag{4.6.1}$$

for all $0 \le k \le n$. Also, since $(n)_k = n!/(n - k)!$, we have

$$C(n, k) = \frac{n!}{k!(n - k)!} \tag{4.6.2}$$

It is probably safe to say that the expression

$$\frac{n!}{k!(n - k)!}$$

is the most common expression in all of combinatorics, and so it deserves a special symbol. Namely, we set

$$\binom{n}{k} = \frac{n!}{k!(n - k)!}$$

The symbol $\binom{n}{k}$ is read "n choose k" since, according to Equation 4.6.2, it is equal to $C(n, k)$, which can be thought of as the number of ways of *choosing* a subset of size k from a set of size n.

As in the case of permutations, it is important to emphasize that the notation $C(n, k)$ stands for the number of subsets of size k of a set of size n, whereas the notation $\binom{n}{k}$ stands for the expression $n!/k!(n - k)!$. Thus, these two notations stand for different things, but according to Theorem 4.6.1, they happen to be equal.

We can now combine our results into one formula,

$$C(n, k) = \frac{P(n, k)}{k!} = \frac{(n)_k}{k!} = \frac{n!}{k!(n - k)!} = \binom{n}{k}$$

The first equality is a result of Theorem 4.6.1, the second equality is a result of Theorem 4.5.1, the third equality follows from the fact that $(n)_k = n!/(n - k)!$, and the last equality is simply the definition of the symbol $\binom{n}{k}$.

The expression $\binom{n}{k}$ is called a **binomial coefficient**, since it plays an important role in a formula known as the *binomial formula*, which we will discuss in the next section.

■ Example 4.6.2

In evaluating the binomial coefficient $\binom{n}{k}$, we can of course use either of the formulas

$$\binom{n}{k} = \frac{n!}{k!(n-k)!} \quad \text{or} \quad \binom{n}{k} = \frac{(n)_k}{k!}$$

For example,

$$\binom{7}{3} = \frac{(7)_3}{3!} = \frac{7 \cdot 6 \cdot 5}{3 \cdot 2 \cdot 1} = 35$$

$$\binom{10}{5} = \frac{10!}{5!\,5!} = \frac{10 \cdot 9 \cdot 8 \cdot 7 \cdot 6}{5 \cdot 4 \cdot 3 \cdot 2 \cdot 1} = 252$$

$$\binom{100}{2} = \frac{(100)_2}{2!} = \frac{100 \cdot 99}{2} = 4950$$

$$\binom{100}{3} = \frac{(100)_3}{3!} = \frac{100 \cdot 99 \cdot 98}{3 \cdot 2 \cdot 1} = 161{,}700$$

$$\binom{100}{96} = \frac{100!}{96!\,4!} = \frac{100 \cdot 99 \cdot 98 \cdot 97}{4 \cdot 3 \cdot 2 \cdot 1} = 3{,}921{,}225$$

□

In the next section, we will study properties of the binomial coefficients in more detail. For now, let us simply note that

$$\binom{n}{k} = \binom{n}{n-k} \tag{4.6.3}$$

for all values of k such that $0 \leq k \leq n$. (We will leave verification of this as an exercise.)

Equation 4.6.3 can be very useful. For instance, since we showed in the previous example that

$$\binom{100}{2} = 4950$$

Equation 4.6.2 tells us that

$$\binom{100}{98} = 4950$$

(Here $n = 100$ and $k = 2$, and so $n - k = 98$.)

Example 4.6.3

The number of subsets of size 3, taken from a set of size 4, is

$$C(4, 3) = \binom{4}{3} = \frac{4!}{3!\,1!} = 4$$

The number of subsets of size 10, taken from a set of size 15, is

$$C(15, 10) = \binom{15}{10} = \frac{15!}{10!\,5!} = \frac{15 \cdot 14 \cdot 13 \cdot 12 \cdot 11}{5 \cdot 4 \cdot 3 \cdot 2 \cdot 1} = 3003$$

The number of subsets of size 95, taken from a set of size 100, is

$$C(100, 95) = \binom{100}{95} = \frac{100!}{95!\,5!}$$

$$= \frac{100 \cdot 99 \cdot 98 \cdot 97 \cdot 96}{5 \cdot 4 \cdot 3 \cdot 2 \cdot 1} = 75{,}287{,}520$$

Imagine trying to figure this out by actually writing down all of the subsets of size 95 of a set of size 100. This is another example of how powerful our results can be. As this example shows, we can do things with the theorems we have been learning that would have been impossible to do without them. □

Example 4.6.4

How many different 5-card hands can be made from a deck of 52 cards?

Since a 5-card hand is simply a subset of size 5 taken from a set of size 52, there are

$$C(52, 5) = \frac{(52)_5}{5!} = \frac{52 \cdot 51 \cdot 50 \cdot 49 \cdot 48}{5 \cdot 4 \cdot 3 \cdot 2 \cdot 1} = 2{,}598{,}960$$

possible 5-card hands that can be made from a deck of 52 cards. (For those of you familiar with the game of poker, this is the number of possible poker hands.) □

Example 4.6.5

A certain club has 5 male and 7 female members. How many ways are there to form a 7-person committee consisting of 3 men and 4 women?

There are C(5, 3) ways of choosing 3 men to serve on the committee, and there are C(7, 4) ways of choosing 4 women to serve on the committee. Hence, there are

$$C(5, 3)C(7, 4) = \binom{5}{3}\binom{7}{4} = 350$$

ways to form the 7-person committee. Incidentally, what rule did we use in this example? □

Example 4.6.6

The number of ternary words of length 12 that have exactly seven 0's, three 1's, and two 2's is given by

$$\binom{12}{7}\binom{5}{3}\binom{2}{2} = 7920$$

This follows from the multiplication rule, since there are $\binom{12}{7}$ ways to choose 7 of the 12 positions for 0's, there are $\binom{5}{3}$ ways to choose 3 of the remaining 5 positions for the 1's and there are $\binom{2}{2}$ ways to choose 2 of the remaining 2 positions for the 2's. □

Example 4.6.7

How many subsets of size 2 are there of the set $\{1, 2, \ldots, 20\}$ that do not consist of two consecutive integers? (For example, the set $\{5, 6\}$ does consist of two consecutive integers.)

The simplest way to answer this question is to count the number of subsets of size 2 that do consist of two consecutive integers and then subtract that number from C(20, 2), which is the total number of subsets of size 2. Clearly, there are 19 subsets of size 2 consisting of consecutive integers, namely, $\{1, 2\}, \{2, 3\}, \ldots, \{19, 20\}$. Hence there are

$$C(20, 2) - 19 = \binom{20}{2} - 19 = 171$$

subsets of $\{1, 2, \ldots, 20\}$ that do not consist of two consecutive integers. □

Example 4.6.8

For $N \geq 2$, the program segment

```
X := 0;
FOR I := 2 TO N DO
    FOR J := 1 TO I - 1 DO
        X := X + 1;
PRINT X
```

prints the value $X = \binom{N}{2}$. This can be seen by reasoning as follows. The statement $X := X + 1$ is executed once for each value of I and J satisfying $1 \leq J < I \leq N$. Therefore, X is equal to the number of pairs (J, I) of integers that satisfy this inequality. But this is equal to the number of ways to choose two integers from the set $\{1, 2, \ldots, N\}$, the smaller one being J and the larger one being I. □

Example 4.6.9

How many ways can 12 people be partitioned into 3 groups of 4 people each?

Let us denote the people by $p_1, p_2, \ldots, p_{12}$. We might try to reason as follows. The number of ways of choosing 4 people for the first group, call it G_1, is C(12, 4); the number of ways of choosing 4 people for the second group, G_2, is C(8, 4) (there are only 8 people left after the first group is chosen); and the number of ways of choosing 4 people for the last group, G_3, is C(4, 4). Hence, according to the multiplication rule, there are C(12, 4)C(8, 4)C(4, 4) ways of partitioning the 12 people into 3 groups of equal size.

Unfortunately, there is a flaw in this reasoning. (Do you see it?) What we have counted is the number of ways to partition 12 people into 3 *distinct* groups, G_1, G_2, and G_3, of equal size. Thus, for instance, the partition

$$G_1 = \{1, 2, 3, 4\},\ G_2 = \{5, 6, 7, 8\},\ G_3 = \{9, 10, 11, 12\}$$

is counted separately from the partition

$$G_1 = \{5, 6, 7, 8\},\ G_2 = \{1, 2, 3, 4\},\ G_3 = \{9, 10, 11, 12\}$$

even though, from the way that the question is worded, we do not want to count these partitions as being different. (If we did, then the answer would be C(12, 4)C(8, 4)C(4, 4).)

But all is not lost. We have simply overcounted the number we are looking for, and all we have to do is determine how much we have overcounted. To do this, we simply observe that each of the partitions of the 12 people into 3 groups of equal size is counted 3! times in the number C(12, 4)C(8, 4)C(4, 4). For example, the partition

$$\{1, 2, 3, 4\},\quad \{5, 6, 7, 8\},\quad \{9, 10, 11, 12\}$$

is counted 3! (= 6) times as follows

$$G_1 = \{1, 2, 3, 4\},\quad G_2 = \{5, 6, 7, 8\},\quad G_3 = \{9, 10, 11, 12\}$$

$$G_1 = \{1, 2, 3, 4\},\quad G_2 = \{9, 10, 11, 12\},\quad G_3 = \{5, 6, 7, 8\}$$

$$G_1 = \{5, 6, 7, 8\},\quad G_2 = \{1, 2, 3, 4\},\quad G_3 = \{9, 10, 11, 12\}$$

$$G_1 = \{5, 6, 7, 8\},\quad G_2 = \{9, 10, 11, 12\},\quad G_3 = \{1, 2, 3, 4\}$$

$$G_1 = \{9, 10, 11, 12\},\quad G_2 = \{1, 2, 3, 4\},\quad G_3 = \{5, 6, 7, 8\}$$

$$G_1 = \{9, 10, 11, 12\},\quad G_2 = \{5, 6, 7, 8\},\quad G_3 = \{1, 2, 3, 4\}$$

Since each of the ways of partitioning 12 people into 3 equal size groups is counted exactly 3! = 6 times in the number C(12, 4)C(8, 4)C(4, 4), the actual number of ways to partition the 12 people is

$$\frac{C(12, 4)C(8, 4)C(4, 4)}{3!} = \frac{495 \cdot 70 \cdot 1}{6} = 5775$$

□

□ □ Exercises

1. Evaluate the following expressions

a) $\binom{4}{4}$ b) $\binom{13}{5}$ c) $\binom{1{,}000{,}000}{999{,}998}$

d) $\binom{7}{0}$ e) $\binom{10}{6}\binom{6}{3}$ f) $\dfrac{\binom{10}{9}}{\binom{5}{2}}$

2. Simplify as much as you can

a) $\binom{n}{0}$ b) $\binom{n}{1}$ c) $\binom{n}{2}$

d) $\binom{n}{3}$ e) $\binom{n}{n}$ f) $\binom{n}{n-1}$

g) $\binom{n}{n-2}$

3. Verify the following equations

a) $\binom{n}{k} = \frac{n}{k}\binom{n-1}{k-1}$ b) $\binom{n}{k} = \frac{n}{n-k}\binom{n-1}{k}$

c) $\binom{n}{k} = \frac{n-k+1}{k}\binom{n}{k-1}$ d) $\binom{n+k}{n} = \binom{n+k}{k}$

4. Simplify as much as you can

a) $\dfrac{\binom{n}{k+1}}{\binom{n-1}{k}}$ b) $\dfrac{\binom{n}{k+1}}{\binom{n}{k}}$

5. Prove that

$$\binom{n}{k} = \binom{n-1}{k} + \binom{n-1}{k-1}$$

for $1 \leq k \leq n - 1$.

6. Prove that

$$\binom{n}{k}\binom{k}{j} = \binom{n}{j}\binom{n-j}{k-j}$$

for $j \leq k \leq n$.

7. Prove that if $n \geq 1$ then

$$\binom{2n}{n}$$

is even. *Hint*, use part a of Exercise 3.

8. How many different 13-card hands can be made from a deck of 52 cards?

9. A certain club consists of 5 men and 6 women.
- a) How many ways are there to form a committee of 3 people?
- b) How many ways are there to form a committee consisting of 3 men and 4 women?
- c) How many ways are there to form a committee of 6 people if a certain pair of women refuse to serve on the same committee?
- d) How many ways are there to form a committee of 4 men and 3 women if 2 of the men refuse to serve on the same committee?

10. How many bytes are there that contain exactly four 1's?

11. a) How many binary words are there of length 10 that contain exactly six 0's?
- b) How many binary words are there of length 10 that contain at least three 1's?

12. How many binary words are there of length 1000 that contain exactly 998 1's?

13. How many ternary words are there of length 15 that contain exactly six 0's and four 1's?

14. a) How many ternary words are there of length 10 that contain exactly four 2's?
- b) How many ternary words are there of length 10 that contain at most eight 0's?

15. A certain classroom has two rows of seats. The front row contains 8 seats and the back row contains 10 seats. How many ways are there to seat 15 students if a certain group of 4 of them refuses to sit in the back row and if a certain group of 5 others refuses to sit in the front row?

16. How many 2-element subsets of $\{1, 2, \ldots, 2n\}$ are there with the property that the sum of the 2 elements is even?

17. a) A student must choose any 10 questions from a 12-question test. How many ways can she do this?

b) How many ways can she do this if she must choose 7 questions from the first 8 questions and 3 questions from the last 4 questions?

c) How many ways can she do this if she must choose at least 5 questions from the first 8 questions and at least 3 questions from the last 4 questions?

18. A certain party is attended by 8 men and 16 women. How many ways can the 8 men be paired off with the 16 women?

19. A man has n friends. He is able to invite a different combination of 3 of his friends to his home each night for a full year. What is the smallest possible value of n?

20. In how many ways can 16 objects be divided into 4 groups of equal size?

21. If all of the integers from 10,000 to 99,999 are written down, how many 0's will there be?

22. Ten points are placed in the plane in such a way that no 3 of the points are on the same line.

a) How many straight lines can be formed by joining points?

b) How many triangles can be formed by using the points as vertices?

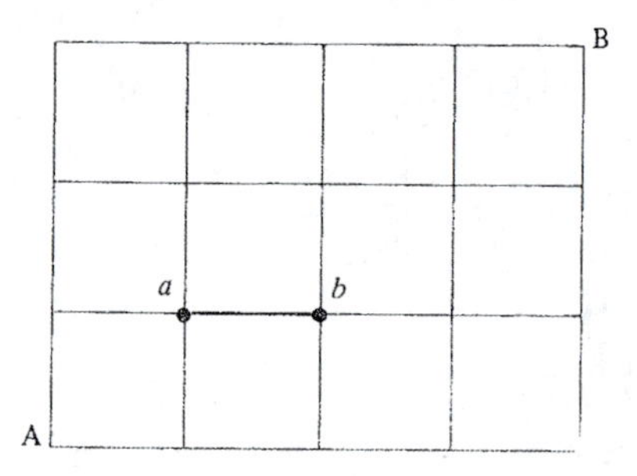

23. a) In the preceding map, how many routes are there from point A to point B? (Assume that you must always travel either north or east.)

b) How many routes are there from point A to point B if it is not possible to travel along the street from point a to point b? (Assume that you must always travel either north or east.) *Hint*, you might try computing the number of routes that include the segment from a to b.

24. What value of X is printed by a computer that runs the following program segment? Justify your answer.

```
X := 0;
FOR I := 1 TO N DO
    FOR J := 1 TO N DO
        IF I < > J THEN X := X + 1;
PRINT X
```

25. What value of X is printed by a computer that runs the following program segment? Justify your answer. (Assume N $\geq$ 3.)

```
X := 0;
FOR I := 3 TO N DO
    FOR J := 2 TO I - 1 DO
        FOR K := 1 TO J - 1 DO
            X := X + 1;
PRINT X
```

26. What value of X is printed by a computer that runs the following program segment? Justify your answer.

```
X := 0;
FOR I1 := 1 TO N DO
    FOR I2 := 1 TO I1 - 1 DO
        FOR I3 := 1 TO I2 - 1 DO
            .
            .
            .
                FOR IK := 1 TO IK-1 - 1 DO
                    X := X + 1;
PRINT X
```

27. Given $2n$ objects, n of which are identical (so that they cannot be distinguished from each other), how many ways are there to choose n of these objects? *Hint*, label the $2n$ objects

$$a, a, \ldots, a, b_1, b_2, \ldots, b_n$$

28. How many 3-element subsets are there of the set $\{1, 2, \ldots, 99\}$ with the property that the sum of the 3 elements is divisible by 3? *Hint*, first write each of the integers in the set $\{1, 2, \ldots, 99\}$ in the form $3q + r$, where $r = 0, 1$, or 2. Then divide the integers into 3 groups, according to the value of r. Decide how the 3-element sets can be chosen from these 3 groups. For example, any 3 elements all of which are from the same group have sum that is divisible by 3. (Why?) Are there any other ways to choose the 3-element sets?

29. How many subsets of size 3 are there from the set $\{1, 2, \ldots, n\}$ with the property that no two of the integers in any of the subsets are consecutive?

30. Show that the number of ways to arrange k 1's and j 0's in such a way that no two 1's are adjacent is

$$\binom{j+1}{k}$$

Hint, first write down j 0's, leaving a space between each of the 0's for a possible 1, as pictured below.

$$\underbrace{\_\,0\,\_\,0\,\_\,0\,\_\ \cdots\ \_\,0\,\_}_{j\ 0\text{'s}}$$

31. How many ways are there to choose k integers from the set $\{1, 2, \ldots, n\}$ in such a way that no two integers in any of the sets are consecutive? *Hint*, use the previous exercise.

4.7 Properties of the Binomial Coefficients

In this section, we want to study the binomial coefficients in more detail. One of the simplest and most useful properties of the binomial coefficients is given in the following theorem.

Theorem 4.7.1

$$\binom{n}{k} = \binom{n-1}{k} + \binom{n-1}{k-1} \tag{4.7.1}$$

for $1 \leq k \leq n - 1$.

PROOF Identities such as this one, involving binomial coefficients, can frequently be proved in two entirely different ways. One way is purely algebraic, using the definition of the binomial coefficient. The other way is combinatorial, by showing that both sides of the equation count the same quantity, and therefore must be equal. Let us prove this identity in both ways.

An algebraic proof might proceed as follows.

$$\begin{aligned}
\binom{n-1}{k} + \binom{n-1}{k-1} &= \frac{(n-1)!}{k!(n-1-k)!} + \frac{(n-1)!}{(k-1)!(n-k)!} \\
&= (n-1)!\left[\frac{1}{k!(n-1-k)!} + \frac{1}{(k-1)!(n-k)!}\right] \\
&= (n-1)!\left[\frac{n-k}{k!(n-k)!} + \frac{k}{k!(n-k)!}\right] \\
&= \frac{(n-1)!}{k!(n-k)!}[n-k+k] \\
&= \frac{n(n-1)!}{k!(n-k)!} \\
&= \binom{n}{k}
\end{aligned}$$

On the other hand, we can give a combinatorial proof as follows. Consider the set $S = \{1, 2, \ldots, n\}$. We know that the number of subsets of S of size k is equal to $\binom{n}{k}$. If we can show that this number is also equal to

$$\binom{n-1}{k} + \binom{n-1}{k-1}$$

then we will have shown that both sides of Equation 4.7.1 count the same thing, and so they must be equal.

Now, the total number of subsets of S of size k is equal to the number of subsets of S of size k that do not contain the integer n, plus the number of subsets of S of size k that do contain the integer n.

The number of subsets of S of size k that do not contain the integer n is

$$\binom{n-1}{k}$$

since these subsets are formed simply by choosing k of the $n - 1$ integers $1, 2, \ldots, n - 1$. Also, the number of subsets of S of size k that do contain the integer n is

$$\binom{n-1}{k-1}$$

since these subsets are formed by choosing $k - 1$ of the $n - 1$ integers $1, 2, \ldots, n - 1$ and then including the integer n.

Hence, the total number of subsets of S of size k is equal to the sum

$$\binom{n-1}{k} + \binom{n-1}{k-1}$$

This shows that both sides of Equation 4.7.1 count the same thing, and so they are equal. This proves the theorem. ∎

Equation 4.7.1 is known as **Pascal's formula**, named after the French mathematician Blaise Pascal (1623–1662). (Pascal invented the first calculating machine, and the computer language "Pascal" is named after him.)

Theorem 4.7.2

For any integer $n \geq 0$, we have

$$\binom{n}{0} + \binom{n}{1} + \binom{n}{2} + \cdots + \binom{n}{n} = 2^n \qquad (4.7.2)$$

or in summation notation

$$\sum_{k=0}^{n} \binom{n}{k} = 2^n$$

PROOF Let us give a combinatorial proof of this formula. Of course, the right side of Equation 4.7.2 counts the total number of subsets, of all sizes, of a set S of size n. But, so does the left side. After all, $\binom{n}{0}$ is the number of subsets of S of size 0, $\binom{n}{1}$ is the number of subsets of S of size 1, $\binom{n}{2}$ is the number of subsets of S of size 2, and so on. Hence the left side of Equation 4.7.2 is also equal to the total number of subsets, of all sizes, of a set S of size n. Since the two sides of Equation 4.7.2 count the same thing, they must be equal. This completes the proof. An algebraic proof of Theorem 4.7.2 can be constructed by using Equation 4.7.1 and mathematical induction. We will leave this for the exercises. ∎

Theorem 4.7.3

$$\sum_{i=k}^{n} \binom{i}{k} = \binom{n+1}{k+1} \qquad (4.7.3)$$

for all $0 \leq k \leq n$. (We suggest that you write out the sum on the left.)

PROOF Let us also give a combinatorial proof of this formula. The right side of Equation 4.7.3 counts the number of ways of choosing $k + 1$ integers from the set $S = \{1, 2, \ldots, n + 1\}$. But the left-hand side also counts this number, by grouping the different choices according to which is the smallest integer chosen.

For example, there are $\binom{n}{k}$ ways to choose $k + 1$ integers from S if the smallest integer chosen is 1, since choosing such a subset amounts to choosing k of the integers $2, 3, \ldots, n + 1$ to include with the integer 1. Similarly, there are $\binom{n-1}{k}$ ways to choose $k + 1$ integers from S if the smallest integer chosen is 2, since choosing such a set amounts to choosing k of the integers $3, 4, \ldots, n + 1$ to include with the integer 2.

Continuing in this way, we obtain all of the terms on the left side of Equation 4.7.3, in reverse order. The first term $\binom{k}{k}$ is the number of ways to choose $k + 1$ integers from S, if the smallest integer chosen is $n - k$.

Hence, the left-hand side of Equation 4.7.3 also counts the number of the ways to choose $k + 1$ integers from the set S, and so it must be equal to the right-hand side. This completes the proof. ∎

The binomial coefficients get their name from the fact that they appear as coefficients in the expansion of the powers of the binomial $x + y$. This expansion is known as the **binomial formula**, and the next theorem is known as the **binomial theorem**.

■ Theorem 4.7.4

$$(x + y)^n = \binom{n}{0} x^n + \binom{n}{1} x^{n-1}y + \binom{n}{2} x^{n-2}y^2 + \cdots + \binom{n}{n-1} xy^{n-1} + \binom{n}{n} y^n$$

Using summation notation, this can be written

$$(x + y)^n = \sum_{k=0}^{n} \binom{n}{k} x^{n-k}y^k$$

(Of course, we could simplify some of the binomial coefficients in this formula, for example, by replacing $\binom{n}{0}$ by 1 and $\binom{n}{1}$ by n, but the pattern is easier to see when the formula is written this way.)

PROOF It is possible to prove the binomial formula by using induction, and we will leave this as an exercise. Instead, let us give a more combinatorial proof.

First, we write

$$(x + y)^n = (x + y)(x + y) \cdots (x + y) \tag{4.7.4}$$

where, of course, there are n factors on the right side.

Now, if we imagine expanding the product on the right completely, then the result will be the sum of several terms, each of the form

$$x^{n-k}y^k$$

where k is an integer between 0 and n. After all, each term in the expansion comes by choosing an x from some of the n factors and a y from the remaining factors and multiplying these variables together. If we happen to choose k y's, then we must also choose $n - k$ x's, and so the resulting term is $x^{n-k}y^k$.

Now, for each value of k between 0 and n, there is one term $x^{n-k}y^k$ for each way of choosing k of the n factors to contribute a y. In other words, for each value of k between 0 and n, there are $\binom{n}{k}$ terms $x^{n-k}y^k$ in the expansion. Therefore, when we collect all of these terms together, we will get

$$\binom{n}{k} x^{n-k}y^k \tag{4.7.5}$$

Thus, if the product on the right side of 4.7.4 is expanded completely, and like terms are collected, the result is a sum of terms of the form 4.7.5, as k ranges over the integers between 0 and n. But this is precisely what the binomial formula says, and so the proof is complete. ∎

Example 4.7.1

Let us expand the expression $(3x + y^2)^4$ using the binomial formula.

$$\begin{aligned}(3x + y^2)^4 &= \binom{4}{0}(3x)^4 + \binom{4}{1}(3x)^3(y^2) + \binom{4}{2}(3x)^2(y^2)^2 + \\ &\quad \binom{4}{3}(3x)(y^2)^3 + \binom{4}{4}(y^2)^4 \\ &= 81x^4 + 108x^3y^2 + 54x^2y^4 + 12xy^6 + y^8\end{aligned}$$

□

Example 4.7.2

Let us find the coefficient of x^4y^3 in the expansion of $(x^2 + 2y)^5$. Making the substitution $z = x^2$, the problem becomes that of finding the coefficient of z^2y^3 in the expansion of $(z + 2y)^5$. According to the binomial formula, the term involving z^2y^3 is

$$\binom{5}{2} z^2(2y)^3$$

and so the coefficient of z^2y^3 is

$$\binom{5}{2} 2^3 = 80$$

□

Example 4.7.3

The binomial formula can be used to give a very simple proof of Theorem 4.7.2. All we have to do is set $x = 1$ and $y = 1$ in the binomial formula to get

$$2^n = \binom{n}{0} + \binom{n}{1} + \binom{n}{2} + \cdots + \binom{n}{n}$$ □

Example 4.7.4

If we set $x = 1$ and $y = -1$ in the binomial formula, we obtain the formula

$$\binom{n}{0} - \binom{n}{1} + \binom{n}{2} - \cdots + (-1)^n\binom{n}{n} = 0$$

where the signs alternate $+, -, +, -, \ldots$. Each term of the form $\binom{n}{k}$, where k is even, occurs with a plus sign, and each term of this form, where k is odd, occurs with a minus sign. (Do you see why?) (The factor $(-1)^n$ appears in the last term to make it positive if n is even and negative if n is odd.)

Incidentally, by moving all of the terms where k is odd to the other side of the equal sign, this equation becomes

$$\binom{n}{0} + \binom{n}{2} + \cdots = \binom{n}{1} + \binom{n}{3} + \cdots \tag{4.7.6}$$

What does this equation tell you about subsets of a set of size n? □

We will explore some additional uses of the binomial formula in the exercises. Let us conclude this section by mentioning that the binomial coefficients are frequently arranged in the form of a triangle, as shown in Figure 4.7.1. This triangle is known

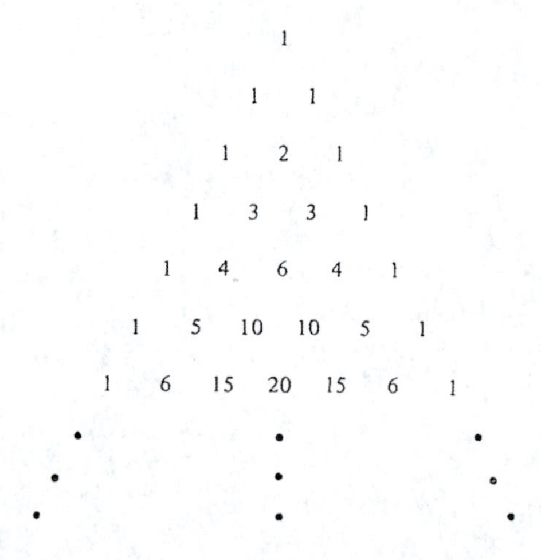

Figure 4.7.1
The first part of Pascal's triangle. The n-th row of this triangle contains the binomial coefficients $\binom{n-1}{0}, \binom{n-1}{1}, \ldots, \binom{n-1}{n-1}$.

as **Pascal's triangle**. (Pascal did not invent this triangle, but he was the first to study its properties in depth.) A typical piece of Pascal's triangle has the form given in Figure 4.7.2.

$$\binom{n-2}{k-1}$$

$$\underbrace{\binom{n-1}{k-1} \quad + \quad \binom{n-1}{k}}$$

$$\binom{n}{k-1} \qquad \binom{n}{k} \qquad \binom{n}{k+1}$$

$$\binom{n+1}{k-1} \qquad \binom{n+1}{k} \qquad \binom{n+1}{k+1} \qquad \binom{n+1}{k+2}$$

Figure 4.7.2
A piece of Pascal's triangle illustrating Pascal's formula.

As you can see, the n-th row of Pascal's triangle consists of the coefficients in the expansion of the n-th power $(x + y)^n$.

Some of the properties of the binomial coefficients can be easily expressed in terms of Pascal's triangle. For example, Pascal's formula says that in order to obtain a particular binomial coefficient, we can simply add the two nearest binomial coefficients in the row immediately above it, as indicated in Figure 4.7.2.

☐☐ Exercises

1. Compute the first six rows of Pascal's triangle (give numerical values).
2. What does Equation 4.7.6 tell you about subsets of a set of size n?
3. Work out the details of Example 4.7.2.
4. Expand the following expressions using the binomial formula.
   a) $(x + y)^3$ b) $(x + y)^4$ c) $(1 + x)^n$
   d) $(x - 3y)^3$ e) $(a^2 + 2b^{-3})^4$ f) $(a + b + c)^2$
5. a) Find the coefficient of a^7b^6 in $(a + b)^{13}$
   b) Find the coefficient of a^4b^2 in $(a^2 + b)^4$
   c) Find the coefficient of $x^{-6}y^3$ in $(2x^{-1} + y)^9$
   d) Find the coefficient of x^{-8} in $(x^3 - 2x^{-2})^4$
6. a) Find the coefficient of x^n in $(x^2 + x^{-2})^n$
   b) Find the coefficient of x^n in $(x^3 - x^{-2})^n$
   c) Find the coefficient of a^6 in $(2a + (a - 1)^2)^9$
7. Prove that

$$\sum_{k=0}^{n} \binom{n}{k} 2^k = 3^n$$

8. Use Equation 4.7.3 to evaluate the sum $1 + 2 + \cdots + n$. *Hint*, choose k wisely.

9. Use the result of the previous exercise, along with Equation 4.7.3 for $k = 2$ and the fact that

$$\binom{j}{2} = \frac{j^2 - j}{2}$$

to evaluate the sum $1^2 + 2^2 + \cdots + n^2$.

10. Prove Theorem 4.7.2 by using mathematical induction.

11. Prove the binomial formula by using mathematical induction.

12. Prove that

$$\binom{k}{0} + \binom{k+1}{1} + \cdots + \binom{n}{n-k} = \binom{n+1}{n-k}$$

Hint, use Equation 4.7.3.

13. Prove that for $n \geq 1$,

$$\binom{n+1}{k} = \binom{n}{k-1} + \binom{n-1}{k} + \binom{n-1}{k-1}$$

a) algebraically.

b) by a combinatorial argument. *Hint*, the left side counts the number of ways to choose k integers from the set $\{1, 2, \ldots, n + 1\}$. So does the right side, according to whether or not the first two integers are involved in the choice.

14. Prove that

$$\sum_{k=0}^{n} \binom{2n+1}{k} = 2^{2n}$$

Hint, use Theorem 4.7.2.

15. Use the result of the previous exercise to solve the following problem. Suppose that you are given $3n + 1$ objects, n of which are identical (so that they cannot be distinguished from each other). How many ways are there to choose n of these objects?

16. Prove that

$$\binom{m}{0} + \binom{m+1}{1} + \cdots + \binom{m+j}{j} = \binom{m+j+1}{j}$$

a) by using Exercise 12 for an appropriate choice of n and k.

b) by a combinatorial argument. *Hint*, the left side counts the number of ways to choose j integers from the set $\{1, 2, \ldots, m + j + 1\}$. So does the right side, by grouping the choices according to which is the smallest integer *not* chosen.

17. Prove that

$$\binom{2n}{2} = 2\binom{n}{2} + n^2$$

a) algebraically.

b) by a combinatorial argument.

18. Prove that

$$\binom{3n}{3} = 3\binom{n}{3} + 6n\binom{n}{2} + n^3$$

a) algebraically.
b) by a combinatorial argument.

19. Prove that

$$\binom{n}{0} + 2\binom{n}{1} + \cdots + (n+1)\binom{n}{n} = 2^n + n2^{n-1}$$

20. Prove that

$$\binom{n+m}{k} = \binom{n}{0}\binom{m}{k} + \binom{n}{1}\binom{m}{k-1} + \cdots + \binom{n}{k}\binom{m}{0}$$

by a combinatorial argument. *Hint*, the left side counts the number of ways to choose k objects from a set of $n + m$ objects. The right side also counts this number, by grouping the $n + m$ objects into one group of n objects and one group of m objects. Each choice is then counted according to how many of the k objects are in the first group and how many are in the second group. Incidentally, this formula is called **Vandermonde's convolution formula**, after the mathematician Alexandre Theophile Vandermonde (1735–1796).

21. Use Vandermonde's convolution formula (Exercise 20) to prove that

$$\binom{n}{0}^2 + \binom{n}{1}^2 + \cdots + \binom{n}{n}^2 = \binom{2n}{n}$$

22. Let n be a fixed integer, and consider the binomial coefficients

$$\binom{n}{0}, \binom{n}{1}, \binom{n}{2}, \ldots, \binom{n}{n}$$

In this exercise, we are interested in how these binomial coefficients grow.

a) If n is even, show that these binomial coefficients grow larger until we reach the middle one

$$\binom{n}{\frac{n}{2}}$$

and then they grow smaller.

b) If n is odd, show that these binomial coefficients grow larger until we reach the two middle ones

$$\binom{n}{\frac{(n-1)}{2}} \quad \text{and} \quad \binom{n}{\frac{(n+1)}{2}}$$

which are equal. Then they grow smaller.

(You can see that this is true for $0 \le n \le 5$ if you did Exercise 1.) *Hint*, in case n is even, it might be easier to replace n by $2m$. In case n is odd, it might be easier to replace n by $2m + 1$.

4.8 The Multinomial Coefficient

The binomial formula tells us how to expand the powers of the *binomial* $x + y$. In this section, we want to develop a similar formula for expanding the powers of any *multinomial*

$$x_1 + x_2 + \cdots + x_k$$

Let us begin with the case of a *trinomial* $x + y + z$.

As we did in the previous section, the first step is to write out the product

$$(x + y + z)^n = (x + y + z)(x + y + z) \cdots (x + y + z)$$

where there are n factors on the right side.

Now, if we imagine that the product on the right is completely expanded, then all of the terms will have the form

$$x^i y^j z^k$$

where i, j, and k are nonnegative integers with the property that $i + j + k = n$. After all, each term in this expansion is formed by choosing either an x, y, or z from each of the n factors and multiplying these variables together. Whenever we choose i x's, j y's, and k z's (where $i + j + k = n$), then the resulting term is $x^i y^j z^k$.

The problem we must face now is to determine, for each possible choice of i, j, and k, the number of times the term $x^i y^j z^k$ appears in the expansion. But this is not hard to do. For there are $\binom{n}{i}$ ways to choose i of the n factors to contribute an x, and once this has been done, there are $\binom{n-i}{j}$ ways to choose j of the remaining factors to contribute a y, and finally, there are $\binom{n-i-j}{k}$ ways to choose k of the remaining factors to contribute a z. Thus, according to the multiplication rule, the number of times that the term $x^i y^j z^k$ appears in the expansion of $(x + y + z)^n$ is

$$\binom{n}{i}\binom{n-i}{j}\binom{n-i-j}{k} = \frac{n!}{i!(n-i)!}\frac{(n-i)!}{j!(n-i-j)!}\frac{(n-i-j)!}{k!(n-i-j-k)!}$$

$$= \frac{n!}{i!\,j!\,k!}$$

(Remember that $i + j + k = n$, and so $(n - i - j - k)! = 0! = 1$.) Hence, if we collect like terms in the expansion of $(x + y + z)^n$, we will get the term

$$\frac{n!}{i!\,j!\,k!} x^i y^j z^k$$

Therefore, the expansion of $(x + y + z)^n$ is just the sum of terms of this form, as i, j, and k range over all nonnegative integers with $i + j + k = n$. Let us put this into a theorem, which is known as the **trinomial theorem**.

Theorem 4.8.1

The expansion of $(x + y + z)^n$ consists of the sum of all possible terms of the form

$$\frac{n!}{i!\,j!\,k!}\,x^i y^j z^k$$

where i, j, and k are nonnegative integers with the property that $i + j + k = n$.

Example 4.8.1

According to the trinomial theorem, the expansion of $(x + y + z)^2$ is the sum of all possible terms of the form

$$\frac{2!}{i!\,j!\,k!}\,x^i y^j z^k$$

where i, j, and k are nonnegative integers with the property $i + j + k = 2$. Let us make a list of the possibilities for i, j, and k, together with the corresponding term in the expansion.

i	j	k	*term in the expansion*
2	0	0	x^2
0	2	0	y^2
0	0	2	z^2
1	1	0	$2xy$
1	0	1	$2xz$
0	1	1	$2yz$

(You should verify each entry in the last column for yourself.) Now, the expansion of $(x + y + z)^2$ is the sum of the terms in the last column of this list, and so

$$(x + y + z)^2 = x^2 + y^2 + z^2 + 2xy + 2xz + 2yz$$

In this case, of course, we can verify this simply by performing the multiplication. (In more complicated examples, this would not be so easy to do.) □

Example 4.8.2

According to the trinomial theorem, the expansion of $(x + y + z)^3$ is the sum of all possible terms

$$\frac{3!}{i!\,j!\,k!}\,x^i y^j z^k$$

where i, j, and k are nonnegative integers with the property that $i + j + k = 3$. Let us make a list of the possibilities.

i	j	k	*term in the expansion*
3	0	0	x^3
0	3	0	y^3
0	0	3	z^3
2	1	0	$3x^2y$
2	0	1	$3x^2z$
0	2	1	$3y^2z$
1	2	0	$3xy^2$
1	0	2	$3xz^2$
0	1	2	$3yz^2$
1	1	1	$6xyz$

The expansion of $(x + y + z)^3$ is the sum of the terms in the last column of this list, and so

$$(x + y + z)^3 = x^3 + y^3 + z^3 + 3x^2y + 3x^2z + 3y^2z + 3xy^2 + 3xz^2 + 3yz^2 + 6xyz$$

This can also be verified by performing the multiplication.

Example 4.8.3

According to the trinomial theorem, the coefficient of the term $x^3y^5z^2$ in the expansion of $(x + y + z)^{10}$ is

$$\frac{10!}{3!\ 5!\ 2!} = 2520$$

In other words, the expansion of $(x + y + z)^{10}$ contains the term $2520x^3y^5z^2$. This is not so easy to verify by performing the multiplication.

The binomial theorem and the trinomial theorem are actually quite similar to each other, even though it may not seem like it from the way that they are worded. In order to see the resemblance more clearly, we observe that the binomial theorem can be restated as follows (see Exercise 15).

Alternate Form of the Binomial Theorem The expansion of $(x + y)^n$ consists of the sum of all possible terms of the form

$$\frac{n!}{i!\ j!} x^i y^j$$

where i and j are nonnegative integers with the property that $i + j = n$.

We can use the same approach as we did for the expansion of the powers of a trinomial to obtain the expansion of the powers of any multinomial. The next theorem is called the **multinomial theorem**.

Theorem 4.8.2

The expansion of $(x_1 + x_2 + \cdots + x_k)^n$ consists of the sum of all possible terms of the form

$$\frac{n!}{i_1!\, i_2! \cdots i_k!} x_1^{i_1} x_2^{i_2} \cdots x_k^{i_k}$$

where $i_1, i_2, \ldots, i_k$ are nonnegative integers with the property that $i_1 + i_2 + \cdots + i_k = n$.

PROOF The proof follows the same lines as the argument we used to obtain Theorem 4.8.1, and so we will only outline it. We first write out the product

$$(x_1 + x_2 + \cdots + x_k)^n$$
$$= (x_1 + x_2 + \cdots + x_k)(x_1 + x_2 + \cdots + x_k) \cdots (x_1 + x_2 + \cdots + x_k)$$

where there are n factors on the right.

Now, each term in the expansion of this product has the form $x_1^{i_1} x_2^{i_2} \cdots x_k^{i_k}$ where $i_1, i_2, \ldots, i_k$ are nonnegative integers whose sum is equal to n. All that we need to do is determine the number of times each of the terms appears in the expansion. We will leave it as an exercise to verify that this number is

$$\binom{n}{i_1}\binom{n-i_1}{i_2}\binom{n-i_1-i_2}{i_3} \cdots \binom{n-i_1-i_2-\cdots-i_{k-1}}{i_k}$$
$$= \frac{n!}{i_1!\, i_2! \cdots i_k!}$$

In other words, after all like terms are collected, the coefficient of $x_1^{i_1} x_2^{i_2} \cdots x_k^{i_k}$ in the expansion of $(x_1 + x_2 + \cdots + x_k)^n$ is

$$\frac{n!}{i_1!\, i_2! \cdots i_k!}$$

and so the expansion of $(x_1 + x_2 + \cdots + x_k)^n$ consists of the sum of all terms of the form

$$\frac{n!}{i_1!\, i_2! \cdots i_k!} x_1^{i_1} x_2^{i_2} \cdots x_k^{i_k}$$

where $i_1 + i_2 + \cdots + i_k = n$. This proves the theorem. ∎

Example 4.8.4

The coefficient of $x_1^2 x_2 x_3^2 x_4$ in the expansion of $(x_1 + x_2 + x_3 + x_4)^6$ is

$$\frac{6!}{2!\, 1!\, 2!\, 1!} = 180$$

The coefficient of $x_1x_2x_5^2$ in the expansion of $(x_1 + x_2 + x_3 + x_4 + x_5)^4$ is

$$\frac{4!}{1!\,1!\,0!\,0!\,2!} = 12$$

The coefficient of $x_1^2x_2^3x_3x_4^4$ in the expansion of $(x_1 + x_2 + x_3 + x_4)^{10}$ is

$$\frac{10!}{2!\,3!\,1!\,4!} = 12{,}600$$

The expression

$$\frac{n!}{i_1!\,i_2!\cdots i_k!}$$

where $i_1, i_2, \ldots, i_k$ are nonnegative integers with the property that $i_1 + i_2 + \cdots + i_k = n$, is important enough to deserve a special name, and a special notation. Namely, we set

$$\binom{n}{i_1, i_2, \ldots, i_k} = \frac{n!}{i_1!\,i_2!\cdots i_k!}$$

and refer to this as a **multinomial coefficient**.

We will not make as thorough a study of multinomial coefficients as we did of binomial coefficients. However, we do want to discuss two important points.

First, a binomial coefficient is a special case of a multinomial coefficient, where $k = 2$. In particular, we have

$$\binom{n}{k} = \frac{n!}{k!(n - k)!} = \binom{n}{k, n - k}$$

The second point that we want to discuss is the fact that multinomial coefficients count something, just as binomial coefficients do. As we know from Section 4.6, the binomial coefficient $\binom{n}{k}$ counts the number of subsets of size k, taken from a set of size n.

In order to get a feeling for what multinomial coefficients count, let us look at what binomial coefficients count in a slightly different way. The binomial coefficient $\binom{n}{k}$ can be thought of as counting the number of ways of dividing a set S of size n into two *disjoint* subsets A_1 and A_2, where A_1 has size k and A_2 has size $n - k$. After all, choosing a subset of S amounts to the same thing as dividing S into two disjoint subsets—the subset of "chosen" elements and the subset of "leftover" elements.

Now, multinomial coefficients also count the number of ways of dividing a set into disjoint subsets, but in general more than just two subsets. Let us be more specific in a theorem.

Theorem 4.8.3

The multinomial coefficient

$$\binom{n}{i_1, i_2, \ldots, i_k}$$

where $i_1, i_2, \ldots, i_k$ are nonnegative integers satisfying $i_1 + i_2 + \cdots + i_k = n$, counts the number of ways of dividing a set S of size n into k *mutually disjoint* subsets $A_1, A_2, \ldots, A_k$, where A_j has size i_j, for all $j = 1, 2, \ldots, k$.

PROOF Dividing a set S of size n into k mutually disjoint subsets A_1, $A_2, \ldots, A_k$, where A_j has size i_j for all $j = 1, 2, \ldots, k$, amounts to performing a sequence of k tasks. The first task is to choose i_1 of the elements of S to form the subset A_1, the second task is to choose i_2 of the remaining $n - i_1$ elements of S to form the second subset A_2, and so on. The k-th task is to choose i_k of the remaining $n - i_1 - i_2 - \cdots - i_{k-1}$ elements of S to form the set A_k.

According to the multiplication rule, the number of ways to perform this sequence of tasks is

$$\binom{n}{i_1}\binom{n - i_1}{i_2}\binom{n - i_1 - i_2}{i_3} \cdots \binom{n - i_1 - i_2 - \cdots - i_{k-1}}{i_k}$$

$$= \frac{n!}{i_1!\, i_2! \cdots i_k!}$$

$$= \binom{n}{i_1, i_2, \ldots, i_k}$$

In other words, there are

$$\binom{n}{i_1, i_2, \ldots, i_k}$$

ways to divide the set S into subsets $A_1, A_2, \ldots, A_k$ of the desired size. This completes the proof. ∎

Example 4.8.5

There are

$$\binom{4}{2, 1, 1} = \frac{4!}{2!\ 1!\ 1!} = 12$$

ways to divide the set $S = \{a, b, c, d\}$ into 3 mutually disjoint subsets A_1, A_2, and A_3, where A_1 has size 2, A_2 has size 1, and A_3 has size 1. Let us make a list of the different ways.

A_1	A_2	A_3
$\{a, b\}$	$\{c\}$	$\{d\}$
$\{a, b\}$	$\{d\}$	$\{c\}$
$\{a, c\}$	$\{b\}$	$\{d\}$
$\{a, c\}$	$\{d\}$	$\{b\}$
$\{a, d\}$	$\{b\}$	$\{c\}$
$\{a, d\}$	$\{c\}$	$\{b\}$
$\{b, c\}$	$\{a\}$	$\{d\}$
$\{b, c\}$	$\{d\}$	$\{a\}$
$\{b, d\}$	$\{a\}$	$\{c\}$
$\{b, d\}$	$\{c\}$	$\{a\}$
$\{c, d\}$	$\{a\}$	$\{b\}$
$\{c, d\}$	$\{b\}$	$\{a\}$

□

Example 4.8.6

There are

$$\binom{7}{2, 1, 3} = \frac{7!}{2!\,1!\,3!} = 420$$

ways to divide the set $S = \{a, b, c, d, e, f, g\}$ into three mutually disjoint subsets A_1, A_2, and A_3, where A_1 has size 2, A_2 has size 1, and A_3 has size 3.

There are

$$\binom{8}{5, 0, 2, 1} = \frac{8!}{5!0!2!1!} = 168$$

ways to divide the set $S = \{1, 2, \ldots, 8\}$ into 4 subsets A_1, A_2, A_3, and A_4, where A_1 has size 5, A_2 has size 0 (that is, A_2 is empty), A_3 has size 2, and A_4 has size 1. □

As you can see from the list that we made in Example 4.8.5, the division

$$A_1 = \{a, b\} \quad , \quad A_2 = \{c\} \quad , \quad A_3 = \{d\}$$

is different from the division

$$A_1 = \{a, b\} \quad , \quad A_2 = \{d\} \quad , \quad A_3 = \{c\}$$

even though they both involve exactly the same sets $\{a, b\}$, $\{c\}$, and $\{d\}$. Thus, this type of division is different from the concept of a partition that we discussed in Section 1.1.

Example 4.8.7

Let $\Sigma = \{a_1, a_2, \ldots, a_k\}$ be an alphabet. We will leave it as an exercise for you to show that there are

$$\binom{n}{i_1, i_2, \ldots, i_k}$$

words of length n that contain exactly i_1 a_1's, i_2 a_2's, . . ., i_k a_k's. □

Exercises

1. Use the trinomial theorem to expand $(x + y + z)^4$
2. Use the multinomial theorem to expand $(x + y + z + w)^2$
3. Use the multinomial theorem to expand $(x + y + z + w)^3$
4. Find the coefficient of x^2y^3z in the expansion of $(x + y + z)^6$
5. Find the coefficient of $x_1^2x_2^2x_3^2x_4^2$ in the expansion of $(x_1 + x_2 + x_3 + x_4)^8$
6. Find the coefficient of x_1^{12} in the expansion of $(x_1 + x_2 + \cdots + x_k)^{12}$
7. Find the coefficient of x^5z^4 in the expansion of $(x + y + z + w)^9$
8. How many ways are there to divide the set $S = \{a, b, c, d, e, f\}$ into three mutually disjoint subsets A_1, A_2, and A_3, where A_1 has size 3, A_2 has size 1, and A_3 has size 2?
9. How many ways are there to divide the set $S = \{1, 2, \ldots, 10\}$ into five mutually disjoint subsets, each of which has size 2?
10. How many ways are there to divide the set $S = \{1, 2, \ldots, 8\}$ into four mutually disjoint subsets A_1, A_2, A_3, and A_4, where A_1 and A_2 have size 4, and A_3 and A_4 have size 0?
11. How many ways are there to divide the set $S = \{1, 2, \ldots, 2n\}$ into n mutually disjoint subsets of equal size?
12. Let $\Sigma = \{a_1, a_2, \ldots, a_k\}$. Show that there are

$$\binom{n}{i_1, i_2, \ldots, i_k}$$

words of length n that contain exactly i_1 a_1's, i_2 a_2's, . . ., i_k a_k's.
13. How many ternary words are there of length 8 that contain exactly two 0's, three 1's and three 2's?
14. How many ternary words are there of length 12 that contain exactly six 0's and three 1's?
15. Let $\Sigma = \{a, b, c, d\}$. How many words of length 10 are there over Σ that have exactly 3 a's, 3 b's, 2 c's, and 2 d's?
16. Compute the sum of all of the coefficients in the expansion of $(x + y + z)^8$.
17. Compute the sum of all of the multinomial coefficients

$$\binom{n}{i_1, i_2, \ldots, i_k}$$

where $i_1, i_2, \ldots, i_k$ are nonnegative integers whose sum is equal to n.
18. Complete the proof of Theorem 4.8.2.
19. Show that the binomial theorem can be restated as done in this section.
20. Prove that

$$\binom{n+1}{i, j, k} = \binom{n}{i-1, j, k} + \binom{n}{i, j-1, k} + \binom{n}{i, j, k-1}$$

   a) by an algebraic argument, using the definition of the multinomial coefficient.
   b) by a combinatorial argument, using Theorem 4.8.3.

4.9 An Introduction to Recurrence Relations

In the year 1202, a mathematician named Leonardo Fibonacci (1170?–1250?) posed the following simple counting problem.

Let us assume that pairs of rabbits do not produce offspring during their first month of life, but after their first month, they produce a new pair of offspring each month. If we start with one pair of newborn rabbits, and if we assume that no rabbits die, how many pairs of rabbits will there be after n months?

To solve this problem, let us denote the number of pairs of rabbits at the end of the n-th month by s_n. Then, of course,

$$s_0 = 1 \quad \text{and} \quad s_1 = 1 \tag{4.9.1}$$

After 2 months, the first pair of rabbits produces a pair of offspring, and so $s_2 = 2$. At the end of the third month, there will be 3 pairs of rabbits, that is, $s_3 = 3$.

In general, we can obtain information about the number s_n by reasoning as follows. At the end of the n-th month, the s_{n-1} pairs of rabbits that were alive at the end of the previous month will still be alive, since we are assuming that no rabbits die. This contributes s_{n-1} pairs of rabbits to the total number of pairs for the n-th month. But, there will also be some newborn pairs. In fact, each of the s_{n-2} pairs of rabbits that were alive 2 *months* prior to the n-th month, being at least 2 months old themselves, will bear a new pair of rabbits. This contributes s_{n-2} additional pairs of rabbits to the total for the n-th month. Hence, we have

$$s_n = s_{n-1} + s_{n-2} \tag{4.9.2}$$

for all values of $n \geq 2$. In words, Equation 4.9.2 says that any term in the sequence

$$s_0, s_1, s_2, s_3, \ldots$$

(from the third one on) is equal to the sum of the two preceding terms.

This information, together with the knowledge that $s_0 = 1$ and $s_1 = 1$, completely determines the entire sequence. By this we mean that, given this information, and enough time, we can find any term in the sequence. For example, we can easily compute the first few terms in the sequence, which are

$$1, 1, 2, 3, 5, 8, 13, 21, 34, 55, 89, 144, \ldots$$

However, finding a general formula for the n-th term s_n of this sequence is another matter!

Equation 4.9.2 is an example of what is known as a **recurrence relation** for the sequence $s_0, s_1, s_2, s_3, \ldots$. It is a formula that describes each member of a sequence in terms of previous members. The equations in 4.9.1 are called the **initial conditions** of the recurrence relation 4.9.2.

Recurrence relations such as 4.9.2 occur frequently in many different branches of mathematics and the sciences. For example, they occur in such diverse areas as physics, computer science, statistics, genetics, botany, economics, psychology, sociology, and many others. Recurrence relations are also known as **difference equations**.

In this section, we will consider some other examples of counting problems that lead to recurrence relations, and we will discuss a very simple method for solving certain recurrence relations. By *solving a recurrence relation*, we mean finding a general formula for the n-th term s_n. In the next two sections, we will consider two other methods for solving recurrence relations.

The sequence $s_0, s_1, s_2, s_3, \ldots$, described by Equations 4.9.1 and 4.9.2, is called the **Fibonacci sequence**, and the numbers in the sequence are called the **Fibonacci numbers**. We will see the Fibonacci numbers again when we count certain types of binary words. Also, these numbers occur in the most remarkable places in nature. For example, on some plants, thorns and leaves grow in a spiral pattern, and the number of thorns per revolution about the stalk is a ratio of two Fibonacci numbers. For instance, the apple or oak tree has 5 growths for every 2 turns around the stalk, the pear tree has 8 growths for every 3 turns, and the willow tree has 13 growths for every 5 turns. (If you are interested in exploring these matters further, a good reference is Peter Stevens' book *Patterns in Nature*, published by Little, Brown and Co., Boston.)

Let us now turn to some examples of counting problems that can lead to recurrence relations. We will not attempt to solve any of these recurrence relations until after we have discussed some methods for obtaining a solution.

Example 4.9.1

One of the simplest recurrence relations arises from the problem of determining the number of subsets of a set of size n. Let s_n be this number.

For the sake of argument, consider the set $S = \{1, 2, \ldots, n\}$. Then the subsets of S can be divided into two groups—those that contain the element n and those that do not.

Clearly, there are s_{n-1} subsets of S that do not contain the element n, since these are just the subsets of the set $\{1, 2, \ldots, n - 1\}$, which has size $n - 1$. But, there are also s_{n-1} subsets of S that do contain n, since these subsets can be formed by taking each of the s_{n-1} subsets of $\{1, 2, \ldots, n - 1\}$ and including the element n.

Thus, the total number of subsets of S is equal to $s_{n-1} + s_{n-1} = 2s_{n-1}$, and we get the recurrence relation

$$s_n = 2s_{n-1} \qquad (4.9.3)$$

valid for all $n \geq 1$. As for initial conditions, since a set of size 0 has exactly 1 subset (what is it?), we have

$$s_0 = 1$$

Notice that, in this case, we need only one initial condition, rather than two. The reason is that only the first term, s_0, in the sequence is needed in order to be able to compute the others from Equation 4.9.3. On the other hand, in the Fibonacci example, the first two terms, s_0 and s_1, were needed in order to use Equation 4.9.2. □

Example 4.9.2

The Towers of Hanoi game, which we discussed in Example 4.1.7, is a counting problem that naturally leads to a recurrence relation.

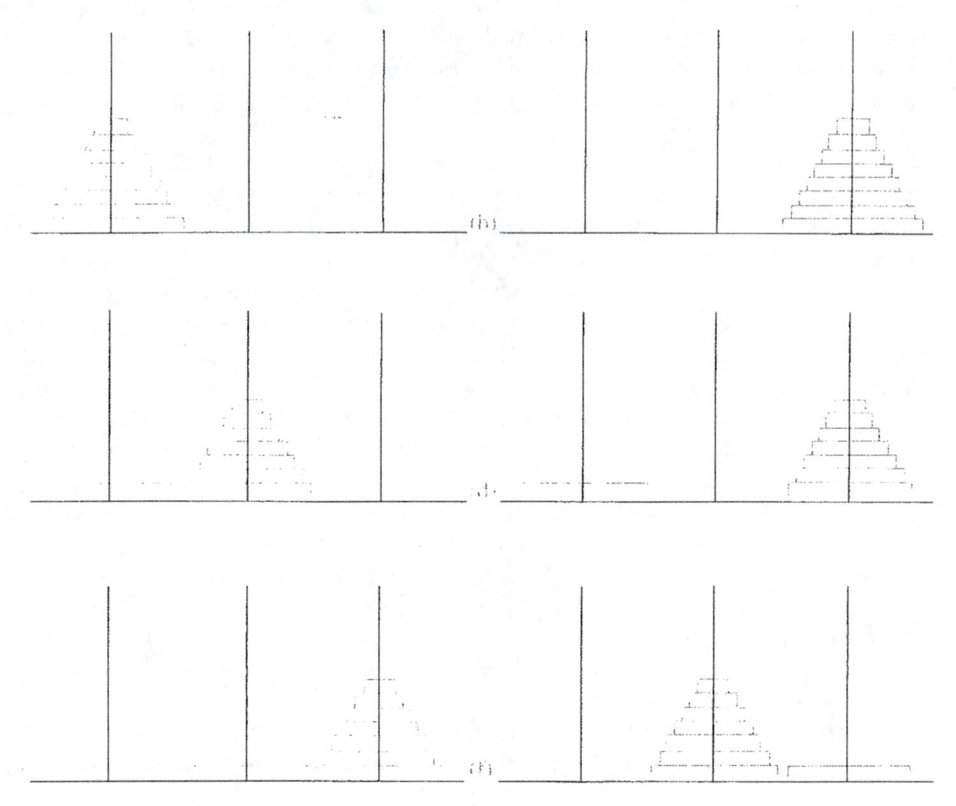

Figure 4.9.1

Recall that the object of this game is to transfer a tower of eight rings from the first peg, as pictured in Figure 4.9.1(a), to the third peg, as pictured in Figure 4.9.1(b), under the conditions that only one ring can be moved at a time and the no ring can be placed on top of another ring of *smaller* outside diameter. The problem is to determine the *minimum* number of moves required to perform this operation.

Suppose that we let s_n be the minimum number of moves required to transfer a tower of n rings from the first peg to the third peg under these conditions. The actual problem is to find s_8, since there are eight rings in this version of the game. Of course, it would be even better if we could find s_n for all values of $n \geq 1$.

Notice that in this case, we start the sequence with s_1, rather than s_0, since there is no such thing as a tower of zero rings. This is typical of recurrence relations. Sometimes it is more convenient to start with the term s_0, and other times it is more convenient to start with the term s_1.

Let us see if we can find a recurrence relation for the sequence $s_1, s_2, s_3, \ldots,$ that is, an expression for s_n in terms of the previous terms $s_1, s_2, \ldots, s_{n-1}$.

Imagine moving a tower of n rings from the first peg to the third peg in as few moves as possible. At some point during this operation, the largest ring must be taken off the first peg and placed on one of the other two pegs. Let us ask ourselves the question, "At this point, where must all of the other rings be?"

Actually, there are only two possibilities. The other rings are either on the second peg, as pictured in Figure 4.9.1(c), or else they are on the third peg, as pictured in Figure 4.9.1(d). For if there were rings on both the second and the third pegs, then we could not move the largest ring.

But, in order to keep the total number of moves to a minimum, it would be better if the other rings were on the second peg, rather than on the third.

To see this, suppose that the other rings were on the third peg. Then, of course, the largest ring would have to be placed on the second peg, and we would be in the position illustrated in Figure 4.9.1(e). From this point, we would have to remove all of the rings from the third peg, put the largest ring on that peg, and then move the other rings back onto the third peg.

However, if the other rings were on the second peg, then the largest ring can be moved directly to the third peg, putting us in the position illustrated in Figure 4.9.1(f). From this point, it would only be necessary to move the other rings from the second peg to the third peg, on top of the largest ring.

Therefore, we can see that, in performing the entire transfer in as few moves as possible, we must arrive at the position illustrated in Figure 4.9.1(f). All we have to do now is determine the minimum number of moves it takes to arrive at this position, as well as the minimum number of moves it would take to finish the transfer from this position, and add these two numbers together. The sum will be equal to s_n.

Getting to the position in Figure 4.9.1(f) in as few moves as possible is accomplished by first moving the tower of $n - 1$ rings (all but the largest ring) from the first peg to the second peg, which takes s_{n-1} moves, and then moving the largest ring from the first peg to the third, which takes one additional move. Thus, getting to the position in Figure 4.9.1(f), in as few moves as possible, requires $s_{n-1} + 1$ moves.

On the other hand, completing the transfer from the position in Figure 4.9.1(f) in as few moves as possible amounts to moving the tower of $n - 1$ rings that are on the second peg, onto the third peg, and this also takes s_{n-1} moves.

Hence, the minimum number of moves required to complete the entire transfer is $s_{n-1} + 1 + s_{n-1} = 2s_{n-1} + 1$, and this gives us the recurrence relation

$$s_n = 2s_{n-1} + 1 \tag{4.9.4}$$

valid for all $n \geq 2$. As for initial conditions, since it takes only one move to transfer a tower consisting of one ring, we have $s_1 = 1$.

Example 4.9.3

A set of straight lines in the plane is said to be in **general position** if none of the lines are parallel, and if no three of the lines go through the same point. Into how many distinct regions does a set of n lines in general position divide the plane?

To get a feel for this problem, let us consider a few simple cases. Of course, a single line divides the plane into two distinct regions. This is pictured in Figure

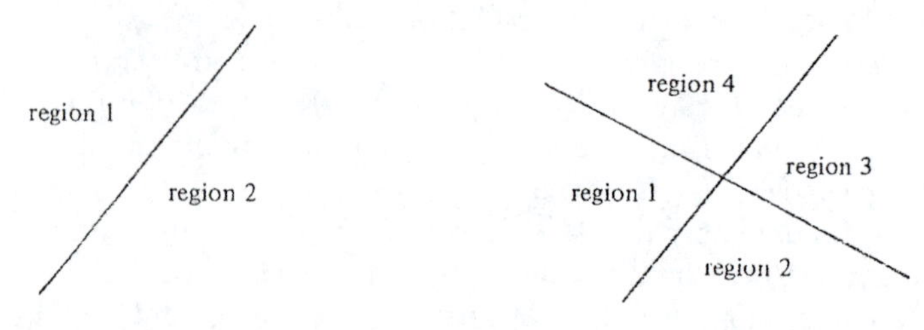

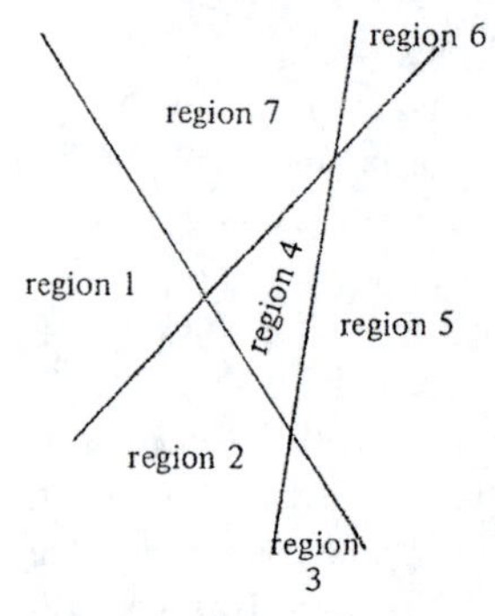

Figure 4.9.2
(a) A single line divides the plane into two regions. (b) Two lines in general position divide the plane into four regions. (c) Three lines in general position divide the plane into seven regions.

4.9.2(a). Two lines in general position divide the plane into four regions, as we see in Figure 4.9.2(b), and three lines in general position divide the plane into seven distinct regions, as in Figure 4.9.2(c). (Incidentally, what about three lines that are *not* in general position?)

Now, if we let p_n be the number of distinct regions into which n lines in general position divide the plane, then we have

$$p_1 = 2, \qquad p_2 = 4, \qquad \text{and} \qquad p_3 = 7$$

Let us try to find a recurrence relation for the sequence $p_1, p_2, p_3, \ldots$.

Consider a set of $n - 1$ lines in general position. These lines divide the plane into p_{n-1} regions. Now let us imagine adding another line to these in such a way that the new set of lines is still in general position. Thus, the new line must intersect all of the old lines, but not at any points where two of the old lines intersect. This is illustrated in Figure 4.9.3.

Our plan is to determine what happens to the p_{n-1} old regions, that is, the regions formed by the old lines, when the new line is added.

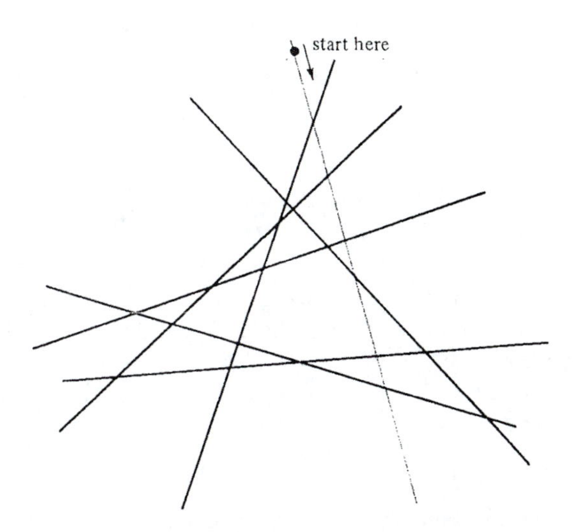

Figure 4.9.3
A set of n lines in general position (in this case, $n = 7$). The colored line is the new one. It goes through $n = 7$ regions, dividing each one into two new regions, for a net gain of $n = 7$ regions.

Perhaps the easiest way to see what happens is to imagine moving along the new line, starting at a point so far out on the line that it has not yet met any of the old lines. (See Figure 4.9.3.) Every time the new line crosses an old line, it enters into a new region and divides that region into two parts. Since the new line meets $n - 1$ old lines, no two at a time, it actually goes through n regions (counting the ones before the new line meets the first old line, and after it has met the last old line).

Thus, the new line divides n of the old regions into two regions each, for a net gain of n regions, and so, after the new line is added, there are a total of $p_{n-1} + n$ regions. This gives us the recurrence relation

$$p_n = p_{n-1} + n \tag{4.9.5}$$

valid for all $n \geq 2$.

Example 4.9.4

Recurrence relations occur naturally in trying to count words that have some restrictions placed on the pattern of letters. For example, suppose that we want to count the number of binary words that do not contain two (or more) consecutive 0's. In other words, we want to count binary words that do not contain the bit pattern 00. (A *bit* is a binary digit, thus it is either a 0 or a 1.)

If s_n is the number of binary words of length n that do not contain the bit pattern 00, then we can obtain a recurrence relation for s_n as follows. Such binary words can be classified according to their first letter. Those that start with a 1 are of the form

$$1x \cdot \cdots \cdot x \tag{4.9.6}$$

where $x \cdots\cdots x$ is any binary word of length $n - 1$ that does not contain the bit pattern 00, and those that start with a 0 are of the form

$$01x \cdots\cdots x \tag{4.9.7}$$

where $x \cdots\cdots x$ is a binary word of length $n - 2$ that does not contain the bit pattern 00. (Such a word cannot start with 00, for that is the forbidden bit pattern.)

Now, there are precisely s_{n-1} binary words of length $n - 1$ that do not contain the bit pattern 00, and so there are s_{n-1} binary words of the type in 4.9.6. Also, there are precisely s_{n-2} binary words of length $n - 2$ that do not contain the bit pattern 00. Hence, there are s_{n-2} binary words of the type in 4.9.7.

Therefore, there are $s_{n-1} + s_{n-2}$ binary words of length n that do not contain the bit pattern 00, and so we have

$$s_n = s_{n-1} + s_{n-2}$$

which happens to be the Fibonacci recurrence relation. Also, it is easy to see that $s_0 = 1$ and $s_1 = 2$. (Notice that the initial conditions in this case are a bit different than in the case of Fibonacci's rabbits. This simply means that the sequence starts with 1, 2, 3, 5, 8, . . ., rather than 1, 1, 2, 3, 5, 8, ...) □

Example 4.9.5

Let us find a recurrence relation for the number s_n of binary words of length n that do not contain the bit pattern 001. Such words are of one of the following three types. Those that begin with a 1 are of the form

$$1x \cdots\cdots x \tag{4.9.8}$$

where $x \cdots\cdots x$ is a binary word of length $n - 1$ that does not contain the bit pattern 001. Those that begin with a 0 are of two possible types. Either they begin with 01, in which case they are of the form

$$01x \cdots\cdots x \tag{4.9.9}$$

where $x \cdots\cdots x$ is a binary word of length $n - 2$ that does not contain the bit pattern 001, or they begin with 00. But, there is only one word of length n that begins with 00 and does not contain the bit pattern 001, namely, the word

$$00 \cdots\cdots 0 \qquad (n \text{ 0's}) \tag{4.9.10}$$

Now we can reason as in the previous example. Since there are s_{n-1} words of the type 4.9.8, s_{n-2} words of the type 4.9.9, and 1 word of the type 4.9.10, we get the recurrence relation

$$s_n = s_{n-1} + s_{n-2} + 1$$

Also, we have the initial conditions $s_0 = 1$ and $s_1 = 2$. □

Now that we have considered some examples of recurrence relations, let us discuss how we might solve them. One of the simplest methods available to us is called **iteration**. The idea behind this method is simply to use the given recurrence relation over and over again, each time for a different value of n, in the hopes of seeing a pattern. A few examples will make the method clear.

Example 4.9.6

Let us begin with the recurrence relation of Example 4.9.1

$$s_n = 2s_{n-1} \tag{4.9.11}$$

with initial condition $s_0 = 1$.

Replacing n by $n - 1$ in Equation 4.9.11 gives

$$s_{n-1} = 2s_{n-2}$$

and substituting this into Equation 4.9.11, we get

$$\begin{aligned} s_n &= 2s_{n-1} \\ &= 2(2s_{n-2}) \\ &= 2^2 s_{n-2} \end{aligned} \tag{4.9.12}$$

Replacing n by $n - 2$ in 4.9.11 gives

$$s_{n-2} = 2s_{n-3}$$

and using this in Equation 4.9.12 gives

$$\begin{aligned} s_n &= 2^2 s_{n-2} \\ &= 2^2(2s_{n-3}) \\ &= 2^3 s_{n-3} \end{aligned}$$

Continuing to substitute in this manner, we get

$$\begin{aligned} s_n &= 2s_{n-1} \\ &= 2^2 s_{n-2} \\ &= 2^3 s_{n-3} \\ &= 2^4 s_{n-4} \\ &= \cdots \\ &= 2^n s_0 \end{aligned}$$

Finally, we use the initial condition $s_0 = 1$ to obtain the solution,

$$s_n = 2^n$$

which, as we already know, is the number of subsets of a set of size n. □

Each of the substitutions that we made in the last example is called an **iteration**, hence the name of the method. Let us try another example of iteration.

Example 4.9.7

The recurrence relation of Example 4.9.2

$$s_n = 2s_{n-1} + 1 \tag{4.9.13}$$

can also be solved by iteration. Replacing n by $n - 1$ in Equation 4.9.13 gives $s_{n-1} = 2s_{n-2} + 1$, and substituting this into Equation 4.9.13 gives

$$\begin{aligned} s_n &= 2s_{n-1} + 1 \\ &= 2(2s_{n-2} + 1) + 1 \end{aligned}$$

Before doing another iteration, we should try to make this a bit more presentable. (Remember, we are looking for a pattern.)

$$s_n = 4s_{n-2} + 2 + 1 \tag{4.9.14}$$

Now we replace n by $n - 2$ in Equation 4.9.13, to get $s_{n-2} = 2s_{n-3} + 1$. Substituting this into 4.9.14 gives

$$\begin{aligned} s_n &= 4(2s_{n-3} + 1) + 2 + 1 \\ &= 8s_{n-3} + 4 + 2 + 1 \end{aligned}$$

At this point, it appears as though we should write this in the form

$$s_n = 2^3 s_{n-3} + 1 + 2 + 2^2$$

and now the pattern is clear. From this we can guess that

$$s_n = 2^{n-1}s_1 + 1 + 2^2 + \cdots + 2^{n-2}$$

But we learned in Section 1.5 that $1 + 2 + 2^2 + \cdots + 2^{n-2} = 2^{n-1} - 1$, and since $s_1 = 1$, we have

$$\begin{aligned} s_n &= 2^{n-1} + 2^{n-1} - 1 \\ &= 2^n - 1 \end{aligned}$$

Of course, this value for s_n is only a guess, and so it is important to make sure that it really is a solution to the recurrence relation by substituting it into Equation 4.9.13. We will leave this for you to do. Don't forget to check the initial conditions as well!

Thus, we see that the minimum number of moves required to complete the Towers of Hanoi puzzle, with a tower consisting of n rings, is $2^n - 1$. In particular, with 8 rings, the minimum number of moves is $2^8 - 1 = 255$.

Example 4.9.8

The recurrence relation

$$p_n = p_{n-1} + n$$

with initial condition $p_1 = 2$ of Example 4.9.3 can also be solved by iteration, and we have

$$\begin{aligned}
p_n &= p_{n-1} + n \\
&= (p_{n-2} + n - 1) + n \\
&= p_{n-2} + n + (n - 1) \\
&= (p_{n-3} + n - 2) + n + (n - 1) \\
&= p_{n-3} + n + (n - 1) + (n - 2) \\
&\ \ \vdots \\
&= p_{n-(n-1)} + n + (n - 1) + (n - 2) + \cdots + (n - (n - 2)) \\
&= p_1 + n + (n - 1) + (n - 2) + \cdots + 2
\end{aligned}$$

Using the fact that $1 + 2 + \cdots + n = n(n + 1)/2$ (see Section 1.5), we get

$$\begin{aligned}
p_n &= 2 + \frac{n(n + 1)}{2} - 1 \\
&= \binom{n + 1}{2} + 1
\end{aligned}$$

Having solved this recurrence relation, we can now answer the question posed in Example 4.9.3, namely, n lines in general position divide the plane into

$$\binom{n + 1}{2} + 1$$

distinct regions. □

Exercises

1. Let $s_n = 3s_{n-1} + ns_{n-2} + 4$, for $n \geq 2$, $s_0 = 1$, $s_1 = 2$. Compute s_3, s_4, s_5, s_6, and s_7.

2. Let $s_n = ns_{n-1} + s_{n-2}^2 + s_{n-3}$, for $n \geq 3$, $s_0 = 1$, $s_1 = 2$, $s_3 = 3$. Compute s_3, s_4, s_5, and s_6.

3. Let $s_n = (s_{n-1} + s_{n-2})^{1/2}$, for $n \geq 2$, $s_0 = 3$, $s_1 = 6$. Compute s_2, s_3, s_4, s_5, and s_6.

4. Let $s_n = as_{n-1} + ns_{n-2} + 3$ for $n \geq 2$. Express s_{n-1} in terms of previous terms. Express s_{n-k} in terms of previous terms.

In Exercises 5 through 19, solve the given recurrence relation. You may need to use some of the exercises in Section 1.5.

5. $s_n = as_{n-1}$, $s_0 = b$

6. $s_n = ns_{n-1}$, $s_0 = 1$

7. $p_n = 2np_{n-1}$, $p_0 = 1$

8. $s_n = (n + 1)s_{n-1}$, $s_0 = a$

9. $p_n = (2n - 1)p_{n-1}$, $p_0 = 1$

10. $s_n = s_{n-1} + 2^n$, $s_0 = 1$

11. $r_n = r_{n-1} + 3^{n-1}$, $r_0 = 1$

12. $s_n = 3s_{n-1} + 1,\ s_0 = 1$
13. $q_n = 4q_{n-1} - 1,\ q_0 = 1$
14. $s_n = as_{n-1} + b$, where $a \neq 0$ and $a \neq 1$, $s_0 = c$
15. $s_n = s_{n-1} + n - 1,\ s_0 = 1$
16. $s_n = s_{n-1} + 3n + 1,\ s_0 = 1$
17. $a_n = a_{n-1} + n^2,\ a_0 = 1$
18. $b_n = b_{n-1} + (n + 1)^3,\ b_0 = 1$
19. $s_n = s_{n-1} + n + n^2,\ s_0 = 1$
20. Solve the recurrence relation

$$s_n = s_{n-1} + \binom{n}{2}$$

with initial condition $s_0 = 1$.

21. Use the recurrence relation derived in Example 4.9.5 to compute the number of binary words of length 8 that do not have the bit pattern 001.

22. Find a recurrence relation for the number s_n of binary words of length n that do not contain the bit pattern 111. Then use this relation to find s_6.

23. Find a recurrence relation for the number s_n of binary words of length n that do not contain the bit pattern 0000. Then use this relation to find s_7.

24. Find a recurrence relation for the number t_n of ternary words of length n that do not contain the bit pattern 00. Then use this relation to find t_6.

25. Find a recurrence relation for the number t_n of ternary words of length n that do not contain the bit pattern 222. Then use this relation to find t_8.

26. Let $\Sigma = \{a, b, c, d\}$. Find a recurrence relation for the number s_n of words over Σ of length n that do not contain the pattern aa. Then use this relation to find s_6.

27. Solve the recurrence relation

$$s_n = 4s_{n-2}$$

for all $n \geq 2$, with initial conditions $s_0 = 1$, $s_1 = 2$. *Hint*, consider separately the cases n even and n odd.

28. Suppose that $2n$ people want to compete in a doubles tennis tournament. Find a recurrence relation for the number p_n of ways to group these people into teams of size 2.

29. a) Suppose that you deposit d dollars in a bank account, at an annual interest rate equal to i. Find a recurrence relation for the amount p_n of money in your account after n years. Solve this recurrence relation.

b) Suppose that *each year* you deposit d dollars into the same bank account, at an annual interest rate equal to i. Find a recurrence relation for the amount s_n in your account at the end of n years. Solve this recurrence relation.

30. Consider the recurrence relation

$$s_n = s_{n-1}s_{n-2}$$

for all $n \geq 2$, with initial conditions $s_1 = s_2 = a$. Compute the first few terms of the sequence and guess a general formula for s_n. Prove that your guess is correct.

4.10

Second Order Linear Homogeneous Recurrence Relations with Constant Coefficients

Unfortunately, there is no general method that can be used to solve all types of recurrence relations. However, there is a very important class of recurrence relations that we can solve. (This class includes the Fibonacci recurrence relation.) Let us begin with a definition.

Definition

A **second order linear homogeneous recurrence relation with constant coefficients** is a recurrence relation of the form

$$s_n = as_{n-1} + bs_{n-2} \tag{4.10.1}$$

where a and b are constants.

Let us make a comment about the terms used in this definition. *Second order* refers to the fact that there are two terms on the right side of the recurrence. The recurrence relation is *linear* because it does not involve any powers or products of members of the sequence s_n. For example, the recurrence relation

$$s_n = s_{n-1}^2 + s_{n-2}$$

is not linear. Equation 4.10.1 is *homogeneous* because it does not have a constant term. For example, the recurrence relation

$$s_n = s_{n-1} + s_{n-2} + 1$$

is not homogeneous.

The first thing that we should notice is that a recurrence relation of the form of Equation 4.10.1, together with initial conditions of the form $s_0 = u$ and $s_1 = v$ (where u and v are constants) completely determines a sequence s_n. Therefore, once we find a solution to Equation 4.10.1 that satisfies the given initial conditions, we will know that our solution is the *only* one possible.

As you can see, the Fibonacci sequence

$$s_n = s_{n-1} + s_{n-2}$$

is a second order linear recurrence relation with constant coefficients, where $a = b = 1$.

Let us begin our study of this type of recurrence relation by making the substitution $s_n = r^n$ in Equation 4.10.1, where r is a constant. This gives

$$r^n = ar^{n-1} + br^{n-2}$$

Cancelling a factor of r^{n-2} gives

$$r^2 = ar + b$$

or

$$r^2 - ar - b = 0$$

Thus, we can see (by reasoning backwards) that if r is any solution to the quadratic equation

$$x^2 - ax - b = 0 \tag{4.10.2}$$

then $s_n = r^n$ is a solution to the recurrence relation 4.10.1.

From this discussion, you can see that the quadratic equation 4.10.2 plays a key role in finding solutions to the recurrence relation 4.10.1. For this reason, Equation 4.10.2 is called the **characteristic equation** for the recurrence 4.10.2, and its solutions are called the **characteristic roots** of 4.10.1.

Actually, we can say a great deal more about solutions to Equation 4.10.1 and characteristic roots. Let us put the information into a theorem.

Theorem 4.10.1

Consider the second order linear homogeneous recurrence relation with constant coefficients

$$s_n = as_{n-1} + bs_{n-2} \tag{4.10.3}$$

with initial conditions $s_0 = u$ and $s_1 = v$, and characteristic equation

$$x^2 - ax - b = 0$$

1) If the characteristic equation has two distinct roots r_1 and r_2, then the solution to the recurrence relation 4.10.3 is given by

$$s_n = c_1r_1^n + c_2r_2^n$$

where the coefficients c_1 and c_2 are determined by the initial conditions.

2) If the characteristic equation has only one root r, then the solution to the recurrence relation 4.10.3 is given by

$$s_n = c_1r^n + c_2nr^n$$

where the coefficients c_1 and c_2 are determined by the initial conditions.

Before proving this theorem, we should make one remark. Theorem 4.10.1 applies equally well to the case where the characteristic equation has only complex (nonreal) roots. (As you may know, every quadratic equation has either one or two roots if we allow complex numbers.) However, since we will not assume any familiarity with complex numbers in this book, we have omitted the case of nonreal roots.

Now let us prove the theorem.

PROOF In order to prove part 1, we assume that the characteristic equation has two roots r_1 and r_2. As we have already seen, both r_1^n and r_2^n are solutions to 4.10.1. We need only show that the sequence $s_n = c_1r_1^n + c_2r_2^n$ is also a solution for any constants c_1 and c_2. But if we make the substitutions $s_{n-1} = c_1r_1^{n-1} + c_2r_2^{n-1}$ and $s_{n-2} = c_1r_1^{n-2} + c_2r_2^{n-2}$ in the right-hand side of 4.10.3 and use the fact that both r_1^n and r_2^n satisfy the recurrence relation, we get

$$
\begin{aligned}
as_{n-1} + bs_{n-2} &= a(c_1r_1^{n-1} + c_2r_2^{n-1}) + b(c_1r_1^{n-2} + c_2r_2^{n-2}) \\
&= c_1(ar_1^{n-1} + br_1^{n-2}) + c_2(ar_2^{n-1} + br_2^{n-2}) \\
&= c_1r_1^n + c_2r_2^n \\
&= s_n
\end{aligned}
$$

This shows that $s_n = c_1r_1^n + c_2r_2^n$ is indeed a solution to 4.10.3 and completes the proof of part 1.

In order to prove part 2, we must show that

$$s_n = c_1r^n + c_2nr^n \tag{4.10.4}$$

is a solution to 4.10.3. Again according to our discussion before the statement of the theorem, we know that r^n is a solution to this recurrence.

Now, in this case, the fact that the characteristic equation has only one solution tells us something about the coefficients a and b. In fact, since r is the only solution to the characteristic equation, it must have the form $(x - r)^2$. In other words, we must have

$$x^2 - ax - b = (x - r)^2$$

Multiplying out the right side of this equation gives

$$x^2 - ax - b = x^2 - 2rx + r^2$$

and so we must have

$$a = 2r \qquad \text{and} \qquad b = -r^2$$

Substituting these values for the coefficients into Equation 4.10.3 gives

$$s_n = 2rs_{n-1} - r^2s_{n-2}$$

Now we can easily show that 4.10.4 is a solution to this recurrence relation. Substituting 4.10.4 into the right-hand side, and using the fact that r^n is a solution gives

$$
\begin{aligned}
&2r[c_1r^{n-1} + c_2(n-1)r^{n-1}] - r^2[c_1r^{n-2} + c_2(n - 2)r^{n-2}] \\
&= 2c_1r^n + 2c_2(n-1)r^n - c_1r^n - c_2(n-2)r^n \\
&= 2c_1r^n + 2c_2nr^n - 2c_2r^n - c_1r^n - c_2nr^n + 2c_2r^n \\
&= c_1r^n + c_2nr^n \\
&= s_n
\end{aligned}
$$

which shows that 4.10.4 is indeed a solution to the recurrence relation 4.10.3. This completes the proof of part 2. ∎

Now let us turn to some examples.

Example 4.10.1

Let us solve the recurrence relation

$$s_n = 2s_{n-1} + 3s_{n-2}$$

with initial conditions $s_0 = 0$ and $s_1 = 8$. The characteristic equation for this recurrence is

$$x^2 - 2x - 3 = 0$$

whose roots are $r_1 = -1$, $r_2 = 3$. Hence, the solution to the recurrence must be of the form

$$s_n = c_1(-1)^n + c_2 3^n \tag{4.10.5}$$

In order to determine the coefficients c_1 and c_2, we use the initial conditions in Equation 4.10.5. When $n = 0$, we get

$$0 = s_0 = c_1 + c_2 \tag{4.10.6}$$

and when $n = 1$, we get

$$8 = s_1 = -c_1 + 3c_2 \tag{4.10.7}$$

From Equation 4.10.6, we see that $c_1 = -c_2$. Substituting this into 4.10.7 gives

$$8 = c_2 + 3c_2$$

or $c_2 = 2$. Hence, $c_1 = -2$, and the solution to the recurrence relation is

$$s_n = -2(-1)^n + 2 \cdot 3^n$$

We suggest that you verify this solution by substitution. (Don't forget to verify the initial conditions as well.) □

Example 4.10.2

Let us solve the recurrence relation

$$s_n = 4s_{n-1} - 4s_{n-2}$$

with initial conditions $s_0 = 1$ and $s_1 = 1$. The characteristic equation for this recurrence is

$$x^2 - 4x + 4 = 0$$

or

$$(x - 2)^2 = 0$$

whose only root is $r = 2$. Hence, the solution to the recurrence must be of the form

$$s_n = c_1 2^n + c_2 n 2^n \tag{4.10.8}$$

In order to determine the coefficients c_1 and c_2, we use the initial conditions in Equation 4.10.8. When $n = 0$, we get

$$1 = s_0 = c_1$$

and when $n = 1$, we get

$$1 = s_1 = 2c_1 + 2c_2 \tag{4.10.9}$$

Substituting $c_1 = 1$ into 4.10.9 gives $c_2 = -\frac{1}{2}$, and so the solution to our recurrence relation is

$$s_n = 2^n - (\tfrac{1}{2})n2^n = 2^n\left[1 - \frac{n}{2}\right]$$

Again we suggest that you verify this by substitution. □

Example 4.10.3

Let us now consider the Fibonacci recurrence relation

$$s_n = s_{n-1} + s_{n-2}$$

with the initial conditions $s_0 = 1$, $s_1 = 1$. In this case, the characteristic equation is

$$x^2 - x - 1 = 0$$

whose solutions are found by the quadratic formula to be

$$r_1 = \frac{1 + \sqrt{5}}{2} \quad \text{and} \quad r_2 = \frac{1 - \sqrt{5}}{2} \tag{4.10.10}$$

Thus, the solution to the Fibonacci recurrence has the form

$$s_n = c_1 r_1^n + c_2 r_2^n \tag{4.10.11}$$

where r_1 and r_2 are given by 4.10.10. Using the initial conditions, we get the equations

$$1 = s_0 = c_1 + c_2$$

and

$$1 = s_1 = c_1 r_1 + c_2 r_2$$

From the first of these equations, we get $c_2 = 1 - c_1$. Substituting this into the second equation gives

$$1 = c_1 r_1 + (1 - c_1)r_2 = r_2 + c_1(r_1 - r_2)$$

and so

$$c_1 = \frac{1 - r_2}{r_1 - r_2} = \frac{r_1}{\sqrt{5}}$$

and

$$c_2 = 1 - \frac{r_1}{\sqrt{5}} = -\frac{r_2}{\sqrt{5}}$$

Putting these values of c_1 and c_2 into 4.10.11 gives the final solution

$$\begin{aligned} s_n &= \frac{r_1}{\sqrt{5}} r_1^n - \frac{r_2}{\sqrt{5}} r_2^n \\ &= \frac{1}{\sqrt{5}} (r_1^{n+1} - r_2^{n+1}) \end{aligned}$$

where r_1 and r_2 are given by 4.10.10. Let us put this important result into a theorem.

Theorem 4.10.2

The Fibonacci recurrence relation

$$s_n = s_{n-1} + s_{n-2} \tag{4.10.12}$$

with initial conditions $s_0 = 1$ and $s_1 = 1$, has solution given by

$$s_n = \frac{1}{\sqrt{5}}\left[\left(\frac{1+\sqrt{5}}{2}\right)^{n+1} - \left(\frac{1-\sqrt{5}}{2}\right)^{n+1}\right] \tag{4.10.13}$$

There is something very remarkable about Theorem 4.10.2. We know that, for all positive integers n, the number s_n, being the number of pairs of rabbits alive at the end of n months, must be an integer. This means that, for all positive integers n, the right side of 4.10.13 must also be an integer! This is quite remarkable in view of the fact that it involves the irrational number $\sqrt{5}$.

Before concluding this section, we should make one more remark about recurrence relations. There are many other more sophisticated methods for solving recurrence relations, and we will briefly discuss one of them in the next section. However, as we mentioned at the beginning of this section, there is no general method that will always work to solve any recurrence relation, and there are many important recurrence relations that have never been solved.

Actually, this situation is not as bad as it may seem. For from a strictly computational point of view, the recurrence relation itself may turn out to be much more valuable than its solution! As an example, if we wanted to compute the first 1000 Fibonacci numbers, with the aid of a computer, it might be preferable to have the computer calculate these numbers directly from the recurrence relation 4.10.12, rather than from Equation 4.10.13.

Exercises

In Exercises 1 through 16, solve the given recurrence relation.

1. $s_n = 3s_{n-1} - 2s_{n-2}$, $s_0 = 0$, $s_1 = 1$
2. $s_n = 2s_{n-1} + 3s_{n-2}$, $s_0 = 1$, $s_1 = 5$
3. $s_n = 2s_{n-1} - s_{n-2}$, $s_0 = a$, $s_1 = b$
4. $s_n = -4s_{n-1} + 4s_{n-2}$, $s_0 = -2$, $s_1 = 2$
5. $s_n = -3s_{n-2}$, $s_0 = 1$, $s_1 = 1$
6. $s_n = 2s_{n-2}$, $s_0 = 0$, $s_1 = 5$
7. $s_n = 16s_{n-2}$, $s_0 = 1$, $s_1 = 1$
8. $s_n = 2s_{n-2}$, $s_0 = a$, $s_1 = b$
9. $s_n = 3s_{n-2}$, $s_0 = 0$, $s_1 = 0$
10. $s_n = 2\sqrt{2}s_{n-1} - 2s_{n-2}$, $s_0 = \sqrt{2}$, $s_1 = \sqrt{2}$
11. $s_n = \pi^2 s_{n-2}$, $s_0 = a$, $s_1 = b$
12. $s_n = 2\pi s_{n-1} - \pi^2 s_{n-2}$, $s_0 = \pi$, $s_1 = \pi$
13. $s_n = -3s_{n-1} - s_{n-2}$, $s_0 = 0$, $s_1 = 1$
14. $s_n = -4s_{n-1} - s_{n-2}$, $s_0 = 1$, $s_1 = 1$
15. $s_n = -5s_{n-1} - 2s_{n-2}$, $s_0 = a$, $s_1 = b$
16. $s_n = 5s_{n-1} - 3s_{n-2}$, $s_0 = a$, $s_1 = b$

17. Find a recurrence relation for the number s_n of ternary words of length n that do not contain the pattern 11. Solve the recurrence relation, and if you have a calculator, approximate the number s_{100}. Compare it with the total number of ternary words of length 100.

18. Let $\Sigma = \{a, b, c, d\}$. Find a recurrence relation for the number s_n of words of length n over Σ that do not contain the pattern dd. Solve the recurrence relation and find s_{10}.

19. Suppose that a person can climb a ladder, taking steps of either one rung at a time or two rungs at a time. Find a recurrence relation for the number s_n of ways he can climb a ladder that has n rungs. Does this recurrence relation look familiar? Find s_{10}.

20. In the design of a certain computer, it is necessary to stack 3 types of circuit boards. Boards of type 1 require a 2-inch clearance, boards of type 2 and type 3 require a 1-inch clearance. Find a recurrence relation for the number s_n of ways to arrange the 3 types of circuit boards in a case that has a total clearance of n inches. (The number of each type of board used does not matter.) Solve the recurrence relation and compute s_{10}.

4.11 Second Order Linear Nonhomogeneous Recurrence Relations with Constant Coefficients

In the previous section, we discussed a special type of homogeneous recurrence relation. In this section, we want to remove the restriction that the recurrence relation be homogeneous. Let us begin with a definition.

Definition

A **second order linear recurrence relation with constant coefficients** is a recurrence relation of the form

$$s_n = as_{n-1} + bs_{n-2} + f(n) \tag{4.11.1}$$

where a and b are constants and $f(n)$ is an expression that depends only on n, and not on the sequence s_n. □

When $f(n) = 0$ for all n, Equation 4.11.1 is a homogeneous recurrence relation of the type that we discussed in the previous section. However, when $f(n)$ is not always equal to 0, then 4.11.1 is called a **nonhomogeneous** recurrence relation. For example, the recurrence relations

$$s_n = s_{n-1} + 2n$$

$$s_n = s_{n-2} + 1$$

and

$$s_n = 2s_{n-1} + 3s_{n-2} + n - 1 \tag{4.11.2}$$

are nonhomogeneous.

If the recurrence relation

$$s_n = as_{n-1} + bs_{n-2} + f(n) \tag{4.11.3}$$

is nonhomogeneous, that is, if $f(n) \neq 0$ for some n, then the recurrence relation obtained from 4.11.3 by dropping the term $f(n)$,

$$s_n = as_{n-1} + bs_{n-2} \tag{4.11.4}$$

is called the **associated homogeneous recurrence relation**. For example, the associated homogeneous recurrence relation for the nonhomogeneous relation (4.11.2) is

$$s_n = 2s_{n-1} + 3s_{n-2}$$

In general, there is no method for solving nonhomogeneous recurrence relations of the form 4.11.1. However, we can take a big step forward by relating the solutions of 4.11.1 to the solutions of the associated homogeneous recurrence relation (4.11.4), which we can solve. Let us see how this can be done.

Suppose that q_n is a solution to the associated homogeneous recurrence relation 4.11.4, and suppose that p_n is a particular solution to the nonhomogeneous relation 4.11.1. Then it is easy to see that the sum

$$t_n = q_n + p_n$$

is also a solution to the nonhomogeneous relation, for we have

$$\begin{aligned} at_{n-1} + bt_{n-2} + f(n) &= a[q_{n-1} + p_{n-1}] + b[q_{n-2} + p_{n-2}] + f(n) \\ &= [aq_{n-1} + bq_{n-2}] + [ap_{n-1} + bp_{n-2} + f(n)] \\ &= q_n + p_n \\ &= t_n \end{aligned}$$

Thus, t_n is a solution to 4.11.1. However, we can say much more than this. Namely, *all* solutions to 4.11.1 are of this form. Let us be more precise in a theorem.

Theorem 4.11.1

Let p_n be a particular solution to the recurrence relation

$$s_n = as_{n-1} + bs_{n-2} + f(n) \tag{4.11.5}$$

1) If q_n is any solution to the associated homogeneous recurrence relation 4.11.4, then $t_n = q_n + p_n$ is a solution to 4.11.5.

2) All solutions to 4.11.5 are given by

$$s_n = q_n + p_n$$

where q_n is a solution to the associated homogeneous recurrence relation 4.11.4.

Theorem 4.11.1 tells us that if we can find just one solution to the recurrence relation 4.11.5, then we can find all solutions. For we already know how to find all solutions to the associated homogeneous relation 4.11.4. The problem, of course, is to find one solution to 4.11.5. Unfortunately, there is no known general method for doing this. However, in many special cases, we can actually guess at the form of a solution, and then verify our guess. The reason that this works is that, frequently, one of the solutions to 4.11.5 has the same form as the term $f(n)$. Let us illustrate this with some examples.

Example 4.11.1

Consider the nonhomogeneous recurrence relation

$$s_n = -2s_{n-1} + n - 4 \tag{4.11.6}$$

The associated homogeneous relation is

$$s_n = -2s_{n-1}$$

whose solutions are given by

$$q_n = c(-2)^n$$

where c is any constant. Thus, we need only find a particular solution to 4.11.6. In order to do this, we look at the term $f(n) = n - 4$. Since this is a linear polynomial in n, we try a solution of this same form. In other words, we try a solution of the form

$$p_n = an + b \tag{4.11.7}$$

where a and b are constants that are not yet determined. Substituting 4.11.7 into 4.11.6, we get

$$\begin{aligned} an + b &= -2[a(n - 1) + b] + n - 4 \\ &= (-2a + 1)n + (2a - 2b - 4) \end{aligned} \tag{4.11.8}$$

Now, if we can find values of a and b satisfying the equations

$$a = -2a + 1$$

$$b = 2a - 2b - 4$$

then 4.11.8 will also hold, and so $p_n = an + b$ will be a particular solution to 4.11.6. Solving the first of these equations for a gives $a = 1/3$, and then solving the second equation for b gives $b = -10/9$. Thus,

$$p_n = \frac{1}{3}n - \frac{10}{9}$$

is a particular solution to 4.11.6, as you should check by substitution.

Now we can apply Theorem 4.11.1 to conclude that *all* solutions to 4.11.6 are given by

$$s_n = c(-2)^n + \frac{1}{3}n - \frac{10}{9} \tag{4.11.9}$$

where c is any constant. If the recurrence relation has an initial condition, this will determine a particular value for c. For example, if we have the initial condition $s_0 = -1/9$, then from 4.11.9 we get

$$-\frac{1}{9} = s_0 = c - \frac{10}{9}$$

and so $c = 1$. Hence the solution to 4.11.6, with this initial condition, is

$$s_n = (-2)^n + \frac{1}{3}n - \frac{10}{9}$$ □

Example 4.11.2

Consider the nonhomogeneous recurrence relation

$$s_n = -s_{n-1} + n^2 + 1 \tag{4.11.10}$$

whose associated homogeneous recurrence relation

$$s_n = -s_{n-1}$$

has solutions $q_n = d(-1)^n$, where d is any constant. In order to find a particular solution to 4.11.10, we try a solution of the same form as $f(n)$, that is, we try a quadratic polynomial in n,

$$p_n = an^2 + bn + c \tag{4.11.11}$$

where a, b and c are constants that are not yet determined. Substituting 4.11.11 into 4.11.10 and simplifying, we get

$$\begin{aligned} an^2 + bn + c &= -[a(n-1)^2 + b(n-1) + c] + n^2 + 1 \\ &= (-a+1)n^2 + (2a-b)n + (-a+b-c+1) \end{aligned}$$

Thus, if we can find values of a, b and c for which

$$\begin{aligned} a &= -a + 1 \\ b &= 2a - b \\ c &= -a + b - c + 1 \end{aligned}$$

then $p_n = an^2 + bn + c$ will be a particular solution to 4.11.10. Solving the first of these equations for a gives $a = 1/2$. Using this information in the second equation gives $b = 1/2$. Finally, the third equation tells us that $c = 1/2$, and so

$$p_n = \frac{1}{2}n^2 + \frac{1}{2}n + \frac{1}{2}$$

is a particular solution to 4.11.10. Now we can use Theorem 4.11.1 to deduce that all solutions to 4.11.10 are given by

$$s_n = d(-1)^n + \frac{1}{2}n^2 + \frac{1}{2}n + \frac{1}{2}$$

where d is any constant. (As in the previous example, any initial condition will determine a particular value for d.) □

Example 4.11.3

Consider the nonhomogeneous recurrence relation

$$s_n = s_{n-1} + 6s_{n-2} + 3 \cdot 2^n \tag{4.11.12}$$

The associated homogeneous recurrence relation is

$$s_n = s_{n-1} + 6s_{n-2} \tag{4.11.13}$$

Now, the characteristic equation of this recurrence relation is $x^2 - x - 6 = 0$, whose solutions are $x = -2$ and $x = 3$. Therefore, the solutions to 4.11.13 are given by

$$q_n = c_1(-2)^n + c_2 3^n$$

where c_1 and c_2 are constants. In order to find a particular solution to 4.11.12, we try a solution of the same form as $f(n)$, that is, we try a solution of the form

$$p_n = a \cdot 2^n$$

where a is a constant. Substituting this into 4.11.12 gives

$$a \cdot 2^n = a \cdot 2^{n-1} + 6a \cdot 2^{n-2} + 3 \cdot 2^n$$

Dividing both sides of this equation by 2^{n-2} gives

$$4a = 2a + 6a + 12$$

whose solution is $a = -3$. Thus $p_n = -3 \cdot 2^n$ is a particular solution to 4.11.12, and so all solutions to 4.11.12 are given by

$$s_n = c_1(-2)^n + c_2 3^n - 3 \cdot 2^n$$

where c_1 and c_2 are constants.

Exercises

In Exercises 1 through 18, solve the given recurrence relation.

1. $s_n = 4s_{n-1} + 1$
2. $s_n = 3s_{n-1} + n - 1$
3. $s_n = 4s_{n-1} + 2n - 5$
4. $s_n = -s_{n-1} + 3n$
5. $s_n = 2s_{n-1} - n$
6. $s_n = -5s_{n-1} + 4s_{n-2} + n + 2$
7. $s_n = -s_{n-1} + 6s_{n-2} + 3n$
8. $s_n = 2s_{n-1} + n^2 + 2$
9. $s_n = -s_{n-1} - n^2 - n - 1$
10. $s_n = 3s_{n-1} + n^2$
11. $s_n = 4s_{n-2} + 2n^2 + 3n - 1$
12. $s_n = 2s_{n-1} + 4s_{n-2} + n^2 - 4n$
13. $s_n = 2s_{n-1} + (-1)^n$
14. $s_n = 2s_{n-1} + s_{n-2} + 5 \cdot 2^n$
15. $s_n = 6s_{n-1} + 2^n + 3^n$
16. $s_n = -2s_{n-1} + 3^{n+1}$
17. $s_n = 2s_{n-1} + 3s_{n-2} + \dfrac{1}{2^n}$
18. $s_n = s_{n-1} + \sin n$
19. Using the method of this section, solve the Tower of Hanoi recurrence relation in Example 4.9.2.
20. Using the method of this section, solve the recurrence relation in Example 4.9.3.
21. Suppose that a certain bank account pays 10% simple interest. If you deposit \$100 at the beginning of every year, find a recurrence relation for the total amount s_n of money in

the account after n years. Solve the recurrence relation. How much money is in the account after 10 years?

22. Find a recurrence relation for the number of ternary words of length n that contain an even number of 0's. Then solve the recurrence relation. How many ternary words of length 10 are there with an even number of 0's? *Hint,* in looking for the recurrence relation, consider the possibilities for the last digit in a word.

23. A **quaternary word** is a word over the alphabet $\Sigma = \{0,1,2,3\}$. Find a recurrence relation for the number of quaternary words of length n that contain an even number of 0's. Then solve the recurrence relation. How many quaternary words of length 10 are there with an even number of 0's? *Hint,* in looking for the recurrence relation, consider the possibilities for the last digit in a word.

4.12 Generating Functions and Recurrence Relations

In this section, we want to discuss another method for solving recurrence relations that rests on a concept known as a *generating function*. Let us define this concept.

Definition

Let $s_0, s_1, s_2, \ldots$ be a sequence of numbers. Then the **generating function** for this sequence is the expression

$$g(x) = s_0 + s_1x + s_2x^2 + \cdots + s_nx^n + \cdots \qquad (4.12.1)$$

For example, the generating function for the sequence 1, 2, 3, . . . is

$$g(x) = 1 + 2x + 3x^2 + 4x^3 + \cdots + (n+1)x^n + \cdots$$

Let us make a few remarks about this definition. The expression on the right side of 4.12.1 is known as a **formal power series**. It is simply a notational device for keeping track of the numbers $s_0, s_1, s_2, \ldots$. In a sense, the powers x^n are just place keepers, used to keep the different members of the sequence separate from each other.

As we will see in the coming examples, the main reason that generating functions are important is that we can perform simple algebraic operations on them. In particular, we can add, subtract, and multiply them.

Before considering examples, we must state a fact that we will need. If m is any positive integer, then

$$\frac{1}{(1-x)^m} = 1 + \binom{m}{1}x + \binom{m+1}{2}x^2 + \binom{m+2}{3}x^3 + \cdots + \binom{m+n-1}{n}x^n + \cdots \qquad (4.12.2)$$

This formula is usually proved in a course in calculus, and so we will not do so now. Instead, let us give some specific examples of this formula. If $m = 1$, we get

$$\frac{1}{1-x} = 1 + x + x^2 + x^3 + \cdots + x^n + \cdots \tag{4.12.3}$$

if $m = 2$, we get

$$\frac{1}{(1-x)^2} = 1 + 2x + 3x^2 + 4x^3 + \cdots + (n+1)x^n + \cdots \tag{4.12.4}$$

and if $m = 3$, then we get

$$\frac{1}{(1-x)^3} = 1 + 3x + 6x^2 + 10x^3 + \cdots + \frac{(n+2)(n+1)}{2}x^n + \cdots \tag{4.12.5}$$

Now let us consider some examples of how to use generating functions to solve recurrence relations.

Example 4.12.1

Let us solve the recurrence relation

$$s_n = 2s_{n-1} \tag{4.12.6}$$

where $s_0 = 1$, by using generating functions. The first step is to write the generating function for the sequence $s_0, s_1, s_2, \ldots$

$$g(x) = s_0 + s_1x + s_2x^2 + \cdots + s_nx^n + \cdots$$

Now we multiply both sides of this by $2x$ and place the result underneath the original generating function $g(x)$, lining up like powers of x,

$$\begin{aligned} g(x) &= s_0 + s_1x + s_2x^2 + \cdots + s_nx^n + \cdots \\ 2xg(x) &= \phantom{s_0 +{}} 2s_0x + 2s_1x^2 + \cdots + 2s_{n-1}x^n + \cdots \end{aligned}$$

If we subtract the second equation from the first, the result is

$$g(x) - 2xg(x) = s_0 + (s_1 - 2s_0)x + (s_2 - 2s_1)x^2 + \cdots + (s_n - 2s_{n-1})x^n + \cdots$$

Now, the coefficient of x is equal to $s_1 - 2s_0$, which, according to the recurrence relation 4.12.6, is equal to 0. Similarly, the coefficients of x^2, x^3, and in general x^n are all equal to 0. Hence, we get

$$g(x) - 2xg(x) = s_0 = 1$$

Solving this for $g(x)$ gives

$$(1 - 2x)g(x) = 1$$

or

$$g(x) = \frac{1}{1-2x}$$

Now we can use Equation 4.12.3, with x replaced by $2x$, to get

$$\begin{aligned} g(x) &= 1 + 2x + (2x)^2 + (2x)^3 + \cdots + (2x)^n + \cdots \\ &= 1 + 2x + 4x^2 + 8x^3 + \cdots + 2^n x^n + \cdots \end{aligned}$$

But since the coefficient of x^n in the generating function $g(x)$ must be s_n, we see that

$$s_n = 2^n$$

Of course, this agrees with our previous results. □

Example 4.12.2

Let us solve the recurrence relation

$$s_n = s_{n-1} + 2 \tag{4.12.7}$$

where $s_0 = 1$, by using generating functions. Again we write the generating function

$$g(x) = s_0 + s_1 x + s_2 x^2 + \cdots + s_n x^n + \cdots$$

Now, the recurrence relation 4.12.7 can be written in the form $s_n - s_{n-1} - 2 = 0$. With this in mind, we write

$$\begin{aligned} g(x) &= s_0 + s_1 x + s_2 x^2 + \cdots + s_n x^n + \cdots \\ -xg(x) &= \quad - s_0 x - s_1 x^2 - \cdots - s_{n-1} x^n - \cdots \\ -\frac{2}{1-x} &= -2 - 2x - 2x^2 - \cdots - 2x^n - \cdots \end{aligned}$$

where the last equation came from 4.12.3. Adding these three equations gives

$$\begin{aligned} &g(x) - xg(x) - \frac{2}{1-x} \\ &= (s_0 - 2) + (s_1 - s_0 - 2)x + (s_2 - s_1 - 2)x^2 + \cdots + (s_n - s_{n-1} - 2)x^n + \cdots . \\ &= s_0 - 2 \\ &= -1 \end{aligned}$$

where we have used the fact that $s_n - s_{n-1} - 2 = 0$. Solving this for $g(x)$ gives

$$\begin{aligned} (1-x)g(x) &= \frac{2}{1-x} - 1 \\ (1-x)g(x) &= \frac{1+x}{1-x} \\ g(x) &= \frac{1+x}{(1-x)^2} \end{aligned}$$

and so, using Equation 4.12.4

$$\begin{aligned} g(x) &= \frac{1 + x}{(1 - x)^2} \\ &= \frac{1}{(1 - x)^2} + \frac{x}{(1 - x)^2} \\ &= [1 + 2x + 3x^2 + \cdots + (n + 1)x^n + \cdots] + \\ &\qquad [x + 2x^2 + 3x^3 + \cdots + nx^n + \cdots] \\ &= 1 + 3x + 5x^2 + \cdots + (2n + 1)x^n + \cdots \end{aligned}$$

Therefore, just as in the previous example, since the coefficient of x^n in $g(x)$ must be s_n, we see that

$$s_n = 2n + 1$$

We will leave it to you to check that this is indeed a solution to the recurrence relation. □

Example 4.12.3

Let us solve the recurrence relation

$$s_n = -2s_{n-1} - s_{n-2} \tag{4.12.8}$$

where $s_0 = 0$ and $s_1 = 1$. Of course, this can also be written $s_n + 2s_{n-1} + s_{n-2} = 0$, and this motivates us to write

$$\begin{aligned} g(x) &= s_0 + s_1x + s_2x^2 + \cdots + s_nx^n + \cdots \\ 2xg(x) &= 2s_0x + 2s_1x^2 + \cdots + 2s_{n-1}x^n + \cdots \\ x^2g(x) &= s_0x^2 + \cdots + s_{n-2}x^n + \cdots \end{aligned}$$

Adding these equations gives

$$\begin{aligned} &g(x) + 2xg(x) + x^2g(x) \\ &\quad = s_0 + (s_1 + 2s_0)x + (s_2 + 2s_1 + s_0)x^2 + \cdots + (s_n + 2s_{n-1} + s_{n-2})x^n + \cdots \\ &\quad = s_0 + (s_1 + 2s_0)x \\ &\quad = x \end{aligned}$$

Solving for $g(x)$ gives

$$(1 + 2x + x^2)g(x) = x$$

or

$$g(x) = \frac{x}{1 + 2x + x^2} = \frac{x}{(1 + x)^2}$$

Using Equation 4.12.4, with x replaced by $-x$, we see that

$$\begin{aligned} g(x) &= x[1 + 2(-x) + 3(-x)^2 + 4(-x)^3 + \cdots + (n + 1)(-x)^n + \cdots] \\ &= x[1 - 2x + 3x^2 - 4x^3 + \cdots + (-1)^n(n + 1)x^n + \cdots] \\ &= x - 2x^2 + 3x^3 - 4x^4 + \cdots + (-1)^{n-1}nx^n + \cdots \end{aligned}$$

As in the other examples, since the coefficient of x^n in $g(x)$ must be s_n, we see that

$$s_n = (-1)^{n-1}n$$

Again we leave it to you to check that this really is the solution to 4.12.8. □

□ □ Exercises

In Exercises 1 through 15, find the generating function for the sequence $s_0, s_1, s_2, \ldots$. Solve the given recurrence relation by using this generating function.

1. $s_n = 3s_{n-1}, s_0 = 1$

2. $s_n = 5s_{n-1}, s_0 = 2$

3. $s_n = -2s_{n-1}, s_0 = 1$

4. $s_n = -4s_{n-1}, s_0 = -1$

5. $s_n = s_{n-1} + 3, s_0 = 1$

6. $s_n = s_{n-1} + 5, s_0 = -1$

7. $s_n = s_{n-1} - 1, s_0 = 1$

8. $s_n = s_{n-1} - 2, s_0 = 2$

9. $s_n = 2s_{n-1} - s_{n-2}, s_0 = 0, s_1 = 1$

10. $s_n = 4s_{n-1} - 4s_{n-2}, s_0 = 0, s_1 = 1$

11. $s_n = -2s_{n-1} - s_{n-2}, s_0 = 1, s_1 = -2$

12. $s_n = 2s_{n-1} - s_{n-2}, s_0 = 1, s_1 = 1$

13. $s_n = 3s_{n-1} + 3s_{n-2} + s_{n-3}, s_0 = s_1 = 0, s_2 = 1$

14. $s_n = s_{n-1} + 2(n - 1), s_0 = 2$

15. $s_n = 2s_{n-1} + 4^{n-1}, s_0 = 1$. *Hint,*

$$\frac{1 - 3x}{(1 - 2x)(1 - 4x)} = \frac{\frac{1}{2}}{1 - 2x} + \frac{\frac{1}{2}}{1 - 4x}$$

When angry, count ten before you speak;
if very angry, an hundred.
—Thomas Jefferson, Writings

More on Combinatorics

5.1
Permutations with Repetitions

In our study of permutations and combinations in the previous chapter, we did not allow any of the objects to be used more than once. In this section, we want to consider permutations where repetitions are allowed. In the next section, we will consider combinations with repetitions.

It is convenient to divide permutations with repetitions into three types, depending on whether there are any restrictions on the number of times an element can be repeated in the permutation.

Definition

Let $S = \{a_1, a_2, \ldots, a_n\}$ be a set with n elements.

a) Let $r_1, r_2, \ldots, r_n$ be nonnegative integers. An ordered arrangement of k of the elements of S, where the element a_i is repeated *exactly* r_i times, for all $i = 1, 2, \ldots, n$, is called a **permutation of size k with fixed repetitions**, taken from the set S. The number r_i is called the **repetition number** of the element a_i. Of course, in order for such a permutation to exist, we must have

$$k = r_1 + r_2 + \cdots + r_n$$

b) Let $s_1, s_2, \ldots, s_n$ and $t_1, t_2, \ldots, t_n$ be nonnegative integers with the property that $s_i \leq t_i$ for all $i = 1, 2, \ldots, n$. An ordered arrangement of k of the elements of S, where the element a_i is repeated *at least* s_i times and *at most* t_i times, for all $i = 1, 2, \ldots, n$, is called a **permutation of size *k* with restricted repetitions**, taken from the set S. We will also allow the possibility that any of the numbers t_i can be equal to infinity. This amounts to saying that there is no restriction on the maximum number of times the corresponding element a_i can be repeated.

c) An ordered arrangement of k of the elements of S, where any element of S can be repeated any number of times in the arrangement, is called a **permutation of size *k* with unrestricted repetitions**, taken from the set S. ▯

Actually, the first and third types of permutations are really just special cases of the second type. A permutation with unrestricted repetitions is just a permutation with restricted repetitions, where s_i is equal to 0 and t_i is equal to infinity, for all $i = 1, 2, \ldots, n$. Also, a permutation with fixed repetitions is just a permutation with restricted repetitions, where $s_i = t_i = r_i$ for all $i = 1, 2, \ldots, n$. (What type are the permutations that we studied in Section 4.4?)

The definitions just given can be rephrased in terms of words over an alphabet. For example, a permutation of size k, taken over a set S, with unlimited repetitions is nothing more than a word of length k over the alphabet S. The other types of permutations are just words with certain restrictions placed on the number of times each letter can be used in each word. We have phrased the definitions in terms of permutations, rather than words, because the concept of a permutation is more common in combinatorics than the concept of a word. In any case, we will leave it to you to make a complete translation of our definitions into ones using the concept of a word.

In this section, we will derive formulas for the number of permutations with unrestricted repetitions, and for the number of permutations with fixed repetitions. Unfortunately, no one has ever been able to find a simple general formula for the number of permutations with restricted repetitions.

Let us first consider permutations with unrestricted repetitions. We begin with some examples.

■ Example 5.1.1

Consider the set S = {0, 1, 2}. Then the permutations of size 3 with unrestricted repetitions, taken from S, are

000	001	002	010	011	012
020	021	022	100	101	102
110	111	112	120	121	122
200	201	202	210	211	212
220	221	222			

These are just the ternary words of length 3. ▯

Example 5.1.2

The permutations of size 4 with unrestricted repetitions, taken from the set $S = \{0, 1\}$, are

0000	0001	0010	0100
1000	0011	0101	1001
0110	1010	1100	0111
1011	1101	1110	1111

These are just the binary words of length 4. □

It is not hard to find a formula for the number of permutations with unrestricted repetitions. We will leave the proof of the next theorem as an exercise.

Theorem 5.1.1

Let S be a set with n elements. Then the number of permutations of size k with unrestricted repetitions, taken from the set S, is n^k.

Example 5.1.3

There are $3^3 = 27$ permutations of size 3 with unrestricted repetitions taken from a set of size 3. This agrees with Example 5.1.1.

Also, there are $2^4 = 16$ permutations of size 4 with unrestricted repetitions taken from a set of size 2. This agrees with Example 5.1.2. □

Let us now turn to permutations with a fixed number of repetitions.

Theorem 5.1.2

Let $S = \{a_1, a_2, \ldots, a_n\}$ be a set with n elements. Then the number of permutations of size k with fixed repetitions, taken from the set S, where a_i has repetition number r_i for all $i = 1, 2, \ldots, n$, is the multinomial coefficient

$$\binom{k}{r_1, r_2, \ldots, r_n}$$

PROOF Let us consider a set of k boxes, numbered 1 through k,

$$\underset{1}{\square}\ \underset{2}{\square} \cdots \underset{k}{\square}$$

Then, forming a permutation of size k, with fixed repetitions of the type specified in the theorem, amounts to filling these boxes with the elements of S in such a way that r_1 of the boxes receive the element a_1; r_2 of the boxes receive the element a_2; and so on.

But, this is the same as dividing the set of k boxes into n mutually disjoint subsets. The first subset A_1 consists of those boxes that receive the element a_1, the second subset A_2 consists of those boxes that receive the element a_2, and so on.

Now, since $|A_i| = r_i$ for all $i = 1, 2, \ldots, n$, Theorem 4.8.3 tells us that there are exactly

$$\binom{k}{r_1, r_2, \ldots, r_n}$$

ways to divide the set of k boxes into n mutually disjoint subsets $A_1, A_2, \ldots, A_n$. Hence, this is also the number of permutations of size k, with fixed repetitions, of the type described in the theorem. This completes the proof. ∎

Example 5.1.4

According to Theorem 5.1.2, there are

$$\binom{11}{2, 4, 4, 1} = \frac{11!}{2!\,4!\,4!\,1!} = 34{,}650$$

permutations of size 11, with fixed repetitions, of the letters

a, b, e, g

where a has repetition number 2, b and e have repetition number 4, and g has repetition number 1. □

Example 5.1.5

How many words of length 9 can be formed by using 4 a's, 3 b's, and 2 c's? Such a word is simply a permutation of size 9, taken from the set

$$S = \{a, b, c\}$$

where a has repetition number 4, b has repetition number 3, and c has repetition number 2. Hence, according to Theorem 5.1.2, there are

$$\binom{9}{4, 3, 2} = \frac{9!}{4!\,3!\,2!} = 630$$

such words. □

Example 5.1.6

How many permutations are there of the letters in the word SEERESS?

In this case, we want to count permutations of size 7, with fixed repetitions, taken from the set

$$S = \{e, r, s\}$$

where e has repetition number 3, r has repetition number 1, and s has repetition number 3. Since $k = 7$, Theorem 5.1.2 tells us that there are

$$\binom{7}{3, 3, 1} = \frac{7!}{3!\,3!\,1!} = 140$$

such permutations. (This compares with $7! = 5040$ permutations of 7 *distinct* letters.) □

Exercises

1. What type are the permutations that we studied in Section 4.4? Explain your answer.
2. How many permutations of size 3 with unrestricted repetitions are there, taken from the set $\{a, b, c, d, e\}$?
3. How many ternary words are there of length 10 that contain exactly four 0's, four 1's and two 2's?

4. Make a list of all the permutations of size 2 with unrestricted repetitions, taken from the set $\{1, a, 2, b\}$.
5. How many permutations of size 4 with unrestricted repetitions are there, taken from the objects

$$a, a, b, c$$

6. How many permutations are there of the letters

a, c, a, b, a, d

7. How many permutations are there of the letters

a, a, a, a

8. How many permutations are there of the letters in the word SWOONS?
9. How many permutations are there of the letters in the word MATHEMATICAL?
10. How many permutations are there of the letters in the word VICISSITUDES?
11. How many three-letter words can be formed from the alphabet S = {1, a,), *}?
12. Prove Theorem 5.1.1.
13. How many integers between 1000 and 100,000 have no digits other than 2, 3, and 8?
14. How many permutations are there of the letters

a, a, a, a, a, b, d, c

with the property that no two a's are adjacent?
15. How many permutations with fixed repetitions are there of the numbers

1, 2, 3, 4

if the number 1 must be repeated three times, the number 2 must be repeated five times, the number 3 must be repeated four times, and the number 4 must be repeated once?
16. How many binary words of length n are there in which 0 and 1 occur the same number of times and in which no two 0's are adjacent?
17. How many ways are there to order the letters

u, u, u, u, v, v, v, w, w, w, w, w

18. Let Σ be an alphabet of size k.
   a) How many words are there over Σ of length n?
   b) How many words are there over Σ of length n if a certain set of m of the k letters can only be used for the first or last letter of a word?
19. How many permutations of size 5 are there of the letters

a, a, a, b, b, b

20. Suppose that you have 6 different flashlights, each of which gives out a different color of light. You wish to send messages with these flashlights. A message consists of a sequence of 3 flashes from the flashlights, at 2-second intervals. How many possible messages can you send?
21. A song of length k consists of a sequence of k notes, sounded at 1-second intervals. If there are n possible notes, how many possible songs are there of length k?
22. Suppose that you have 3 different ribbons available for a certain printer. One ribbon is red and will last for 10 hours of continuous printing, one ribbon is blue and will last for 20 hours of continuous printing, and one ribbon is black and will last for 30 hours of continuous printing. You wish to run the printer as long as possible, perhaps changing colors every hour. How many ways are there to do this? How many ways are there to do this if you must change colors every hour?

5.2
Combinations with Repetitions

In the previous section, we discussed permutations with repetitions. In this section, we want to discuss combinations with repetitions. We will divide combinations with repetitions into only two types, depending on whether or not there are any restrictions on the number of times an element can be repeated in the combination.

Definition

Let $S = \{a_1, a_2, \ldots, a_n\}$ be a set with n elements.

a) Let $s_1, s_2, \ldots, s_n$ and $t_1, t_2, \ldots, t_n$ be nonnegative integers with the property that $s_i \leq t_i$ for all $i = 1, 2, \ldots, n$. An unordered selection of k of the elements of S, where the element a_i appears *at least* s_i times and *at most* t_i times, for all $i = 1, 2, \ldots, n$, is called a **combination of size *k* with restricted repetitions**, taken from the set S. We will also allow the possibility that any of the numbers t_i can be equal to infinity. This amounts to saying that there is no restriction on the maximum number of times the corresponding element a_i can appear.

b) An unordered selection of k of the elements of S, where any element of S can appear any number of times in the selection, is called a **combination of size *k* with unrestricted repetitions**, taken from the set S. □

We have not included combinations with fixed repetitions as a separate type because there is not much to say about them (see Exercise 9).

Our plan in this section is to derive a formula for the number of combinations of size k, with unrestricted repetitions, taken from a set of size n. Unfortunately, there is no *simple* general formula for the number of combinations with restricted repetitions. (There is a formula, but it is rather complicated.) However, we will obtain a nice formula in an important special case.

Let us begin with some examples of combinations with unrestricted repetitions. It might be helpful to compare the next two examples with Examples 4.1.1 and 4.1.2.

Example 5.2.1

Consider the set $S = \{0, 1, 2\}$. Then the combinations of size 2 with unrestricted repetitions, taken from S, are

$$\{0, 0\} \quad \{0, 1\} \quad \{0, 2\}$$
$$\{1, 1\} \quad \{1, 2\} \quad \{2, 2\}$$

The combinations of size 3 with unrestricted repetitions, taken from S, are

$$\{0, 0, 0\} \quad \{0, 0, 1\} \quad \{0, 0, 2\} \quad \{0, 1, 1\}$$
$$\{0, 1, 2\} \quad \{0, 2, 2\} \quad \{1, 1, 1\} \quad \{1, 1, 2\}$$
$$\{1, 2, 2\} \quad \{2, 2, 2\}$$

□

Example 5.2.2

The combinations of size 4 with unrestricted repetitions, taken from the set $S = \{0, 2\}$, are

$$\{0, 0, 0, 0\} \quad \{0, 0, 0, 2\} \quad \{0, 0, 2, 2\} \quad \{0, 2, 2, 2\} \quad \{2, 2, 2, 2\}$$

□

As you can see from these examples, we have used set notation to denote combinations with repetitions. In fact, combinations with repetitions are very similar to ordinary sets and are frequently called **multisets**.

Let us now turn to the problem of finding a formula for the number of combinations of size k with unrestricted repetitions, taken from a set of size n.

As is usually the case, this not a hard problem to solve if we look at it in the right way. For the sake of illustration, let us compute the number of combinations of size 5 with unrestricted repetitions, taken from the set S = {a, b, c, d} of size 4.

The first step is to find a simple way to represent all of the possible combinations, so that we can count them. Our plan is to represent each combination by a sequence of 5 x's and 3 slashes. For example, the combination

{a, a, b, c, d}

is represented by the sequence

x x / x / x / x

and the combination

{a, a, a, b, d}

is represented by the sequence

x x x / x // x

In general, if we want to represent a given combination of size 5 with unrestricted repetitions as a sequence of x's and slashes, we proceed as follows. Starting from the left, we write down as many x's as there are a's in the particular combination. Then we write a slash, followed by as many x's as there are b's in the combination, then another slash followed by as many x's as there are c's in the combination, and finally another slash followed by as many x's as there are d's in the combination.

Thus, each sequence has as many x's as there are elements in the combination; that is, each sequence has 5 x's. The 3 slashes serve to separate the x's into 4 groups—one for each of the elements of S.

Notice that, in the second example above, the combination does not contain the letter c, which is why there are two slashes next to each other in the corresponding sequence of x's and slashes. Let us do a few more examples.

{a, b, c, d, d}	becomes	x / x / x / x x
{a, b, b, b, c}	becomes	x / x x x / x /
{b, b, b, b, d}	becomes	/ x x x x / / x
{d, d, d, d, d}	becomes	/ / / x x x x x

As you can see, by using this method each combination with unrestricted repetitions, taken from S, corresponds to exactly one sequence of 5 x's and 3 slashes. Also, every such sequence of x's and slashes represents exactly one combination with unrestricted repetitions, taken from S. For example, the sequence

x / / x x / x x

represents the combination

$$\{a, c, c, d, d\}$$

In the language of Section 1.2, we have found a one-to-one correspondence between the elements of the set of all combinations of size 5 with unrestricted repetitions and the elements of the set of all sequences consisting of 5 x's and 3 slashes.

Hence, the number of combinations of size 5 with unrestricted repetitions is the same as the number of sequences of 5 x's and 3 slashes. But, we already know how to compute the number of such sequences. After all, a sequence of 5 x's and 3 slashes is just a *permutation* of size 8 with *fixed* repetitions, taken from the set {x, /}, where x has repetition number 5 and / has repetition number 3. According to Theorem 5.1.2, there are

$$\binom{8}{5,\ 3} = \frac{8!}{5!\ 3!} = 56$$

such permutations. Hence, there are 56 combinations of size 5 with unrestricted repetitions, taken from the set S.

The same technique that we used here will work in general to provide us with a formula for the number of combinations of size k with unrestricted repetitions, taken from any set of size n.

Theorem 5.2.1

Let S be a set of size n. Then the number of combinations of size k with unrestricted repetitions, taken from S, is

$$\binom{n+k-1}{k}$$

PROOF Suppose that $S = \{a_1, a_2, \ldots, a_n\}$. Then just as we did in the example above, we can represent each combination of size k with unrestricted repetitions, taken from the set S of size n, by a sequence of k x's and $n - 1$ slashes. (The $n - 1$ slashes will serve to divide the x's into n groups—one for each element of S.)

In order to represent a given combination by a sequence of x's and slashes, we proceed as follows. Starting from the left, we write down as many x's as there are a_1's in the particular combination. Then we write a slash, followed by as many x's as there are a_2's in the combination, then another slash followed by as many x's as there are a_3's in the combination, and so on. Finally, we write the $(n - 1)$-st slash followed by as many x's as there are a_n's in the combination.

In this way, we see that each combination of size k with unrestricted repetitions, taken from the set S, corresponds to exactly one sequence of k x's and $n - 1$ slashes. Also, every such sequence of x's and slashes corresponds to exactly one combination. Hence, the number of combinations is the same as the number of sequences.

But a sequence of k x's and $n - 1$ slashes is just a permutation of size $n + k - 1$ with fixed repetitions, taken from the set {x, /}, where x has repetition number k and / has repetition number $n - 1$. According to Theorem 5.1.2, there are

$$\binom{n+k-1}{k,\ n-1} = \frac{(n+k-1)!}{k!\ (n-1)!} = \binom{n+k-1}{k}$$

such permutations. Hence, this is also the number of combinations of size k with unrestricted repetitions, taken from the set S. This proves the theorem. ∎

Let us have some examples.

Example 5.2.3

According to Theorem 5.2.1, the number of combinations of size 2 with unrestricted repetitions, taken from a set of size 3 is

$$\binom{3+2-1}{2} = \binom{4}{2} = 6$$

(In this case, $n = 3$ and $k = 2$.) Also, the number of combinations of size 3 with unrestricted repetitions, taken from a set of size 3, is

$$\binom{3+3-1}{3} = \binom{5}{3} = 10$$

(In this case, $n = 3$ and $k = 3$.) Both of these agree with the results of Example 5.2.1.

The number of combinations of size 4 with unrestricted repetitions, taken from a set of size 2, is

$$\binom{2+4-1}{4} = \binom{5}{4} = 5$$

($n = 2$, $k = 4$), which agrees with Example 5.2.2. □

Example 5.2.4

The number of combinations of size 7 with unrestricted repetitions, taken from a set of size 12, is

$$\binom{12+7-1}{7} = \binom{18}{7} = 31{,}824$$

($n = 12$, $k = 7$). On the other hand, the number of combinations of size 12 with unrestricted repetitions, taken from a set of size 7, is

$$\binom{7+12-1}{12} = \binom{18}{12} = 18{,}564$$

($n = 7$, $k = 12$). □

Example 5.2.5

A certain candy store sells 6 different types of chocolates. How many ways are there to fill a box with chocolates if the box holds 10 pieces? (The order of the chocolates in the box is irrelevant.)

Filling a box with chocolates simply amounts to choosing a combination of size 10 with unrestricted repetitions, from a set of size 6. Hence, there are

$$\binom{6+10-1}{10} = \binom{15}{10} = \binom{15}{5} = 3003$$

different ways to fill a box with chocolates. □

Example 5.2.6

Consider the program segment

```
X := 0;
FOR I := 1 TO N DO
    FOR J := 1 TO I DO
        X := X + 1;
PRINT X
```

Let us determine the value of X in two different ways. First, we observe that this program does the same thing as the simpler program

```
X := 0;
FOR I := 1 TO N DO
        X := X + I;
PRINT X
```

and this program prints the value

$$X = 1 + 2 + 3 + \cdots + N$$

On the other hand, looking at the original program, we see that the instruction X := X + 1 is executed once for each choice of integers I and J satisfying the inequality

$$1 \le J \le I \le N$$

But there is one pair of integers (J, I) satisfying this inequality for each combination of size 2, with unrestricted repetitions, taken from the set $\{1, 2, \ldots, N\}$. Hence, according to Theorem 5.2.1, X also has the value

$$X = \binom{N+1}{2}$$

Comparing these two values of X, we get the formula

$$1 + 2 + 3 + \cdots + N = \binom{N+1}{2}$$

Of course, we proved this formula in Chapter 1, but at that time we did not know where the formula came from. (One of the disadvantages of induction is that, although it is very useful for proving formulas such as this one, it is not of much help in *finding* such formulas.) This example shows one method for actually deriving this very useful formula. □

Let us now turn our attention to combinations with restricted repetitions. As we pointed out earlier in this section, in general there is no simple formula for the number of combinations of size k with restricted repetitions, taken from a set of size n. Nevertheless, we can obtain a very nice formula in a special case.

In particular, we want to find a formula for the number of combinations of size k, taken from a set $S = \{a_1, a_2, \ldots, a_n\}$, with the property that the element a_i must be repeated *at least* s_i times in the combination, for all $i = 1, 2, \ldots, n$. In other words, these are combinations with restricted repetitions where the numbers t_i are all equal to infinity, and so there is no restriction on the *maximum* number of times an element of S may appear.

Let us start with a simple example. How many combinations of size 5 are there of the set $S = \{a, b, c\}$ with the property that a must be repeated *at least* 2 times, and b must be repeated *at least* once in each combination?

In order to answer this question, all we have to do is observe that forming such a combination simply amounts to writing down the required 2 a's and 1 b and then adding to that *any* combination of size 2 with *unrestricted* repetitions, taken from S,

$$\{a, a, b, \underbrace{\cdots\cdots\cdots}\}$$

↑

any combination of size 2
with unrestricted repetitions
taken from S

Since, according to Theorem 5.2.1, there are

$$\binom{3+2-1}{2} = \binom{4}{2} = 6$$

different combinations of size 2 with unrestricted repetitions, taken from S, there are also 6 combinations of size 5 with the desired restrictions.

We can easily generalize this line of reasoning to find the formula that we are looking for. Let us put this into a theorem.

Theorem 5.2.2

Let $S = \{a_1, a_2, \ldots, a_n\}$ be a set with n elements. Then the number of combinations of size k, taken from S, with the property that the element a_i must be repeated *at least* s_i times in the combination, for all $i = 1, 2, \ldots, n$, is

$$\binom{n + (k - s_1 - s_2 - \cdots - s_n) - 1}{k - s_1 - s_2 - \cdots - s_n} \tag{5.2.1}$$

This can also be written

$$\binom{n + (k - s_1 - s_2 - \cdots - s_n) - 1}{n - 1} \tag{5.2.2}$$

PROOF We use the same reasoning as we did for the example. In order to form a combination of size k from the set S, with the required restrictions, we simply take the required s_1 a_1's, s_2 a_2's, ..., s_n a_n's and add to them any combination of size $k - s_1 - s_2 - \cdots - s_n$ with *unrestricted* repetitions, taken from S. Since, according to Theorem 5.2.1, there are

$$\binom{n + (k - s_1 - s_2 - \cdots - s_n) - 1}{k - s_1 - s_2 - \cdots - s_n}$$

such combinations with unrestricted repetitions, this is also the number of combinations of size k with the required restrictions. This proves the first part of the theorem. We will leave it as an exercise to show that the binomial coefficients in 5.2.1 and 5.2.2 are equal. ∎

Example 5.2.7

Let us consider again the problem in Example 5.2.5. This time, however, we want to count the number of ways to fill a box of chocolates from the 6 types, call them $T_1, T_2, \ldots, T_6$, if each box must contain at least 2 pieces of type T_1 and at least 1 piece of type T_2.

In this case, filling a box is simply a matter of choosing a combination of size 10 from the set $S = \{T_1, T_2, \ldots, T_6\}$ in such a way that T_1 is repeated at least 2 times and T_2 is repeated at least 1 time. Hence, in this case, the numbers s_i of Theorem 5.2.2 are

$$s_1 = 2 \quad , \quad s_2 = 1 \quad , \quad s_3 = s_4 = s_5 = s_6 = 0$$

and so there are

$$\binom{6 + (10 - 2 - 1 - 0 - 0 - 0 - 0) - 1}{6 - 1} = \binom{12}{5} = 792$$

different ways to fill a box with 10 pieces of chocolate, under these restrictions. (This compares with 3003 ways to fill a box when there are no restrictions.) □

Example 5.2.8

Suppose that a certain exam is being given on 3 consecutive days and that 25 students are required to take the exam. How many ways are there to assign each of the students to 1 of the 3 days, assuming that there must be at least 5 students assigned to each of the first 2 days? (Here we are assuming that it does not matter which students are assigned to which days; only the number of students assigned to each day matters.)

Assigning the 25 students to the 3 days, with these restrictions, amounts to choosing a combination of size 25, taken from the set $S = \{\text{day } 1, \text{day } 2, \text{day } 3\}$, where day 1 must be repeated at least $s_1 = 5$ times, day 2 must be repeated $s_2 = 5$ times, and day 3 need not be repeated at all ($s_3 = 0$). Hence, according to Theorem 5.2.2, there are

$$\binom{3 + (25 - 5 - 5 - 0) - 1}{3 - 1} = \binom{17}{2} = 136$$

ways to assign the students. (How does this compare with the number of ways to assign the students if there are no restrictions?) □

One of the most common uses of Theorem 5.2.2 is to determine the number of combinations of size k, taken from a set S of size n, with the property that each element of S must appear *at least once* in the combination. In this case, each of the numbers $s_1, s_2, \ldots, s_n$ in Theorem 5.2.2 is equal to 1, and we get the following corollary to Theorem 5.2.2. (When a result is a direct consequence of a theorem, as in this case, the result is usually called a *corollary* to the theorem.)

Corollary 5.2.3

Let S be a set of size n. Then the number of combinations of size k, taken from the set S, with the property that each element of S must appear at least once in the combination is

$$\binom{k-1}{n-1} \tag{5.2.3}$$

PROOF This result is a direct consequence of Theorem 5.2.2. We simply set $s_1 = s_2 = \cdots = s_n = 1$ in the expression 5.2.2 to get 5.2.3. ∎

Example 5.2.9

Once again referring to Example 5.2.5, how many ways are there to fill a box of chocolates if the box must contain at least one piece of each type?

This problem is equivalent to determining the number of combinations of size 10, taken from the set S of 6 types of chocolates, with the property that each of the elements of S must be included in the combination. According to Corollary 5.2.3, this number is

$$\binom{10-1}{6-1} = \binom{9}{5} = 126$$

□

Exercises

1. List all of the combinations of size 3, with unrestricted repetitions, from the set $\{a, b, c, d\}$. (It might help to know how many there are before listing them.)
2. List all of the combinations of size 5 from the set $S = \{1, 2, 3\}$ with the property that each element of S must appear at least once. How can you be certain that you didn't miss any in your list?
3. How many ways are there to choose 12 coins from a large supply of pennies, nickels, dimes, and quarters if at least 1 coin of each type must be chosen?
4. At a certain store, candles come in 4 colors. How many combinations are there of 6 candles? How many ways are there to arrange 6 candles on a mantlepiece?
5. Suppose that you have 10 each of 4 differently colored balls. (The balls are indistinguishable except by their color.)
   a) How many choices of 5 balls can you make?
   b) How many choices of 5 balls can you make if you must choose at least 1 ball of each color?
   c) How many choices of 8 balls can you make if you must choose the same number of balls of each color?
6. A certain bakery makes 10 different kinds of donuts.
   a) How many ways are there to choose a dozen donuts?
   b) How many ways are there to choose a dozen donuts if you must choose at least one of each type?

7. Suppose that you roll 12 identical dice at one time. How many possible outcomes are there?
8. A candy store sells 8 flavors of jellybeans. How many ways are there to choose 25 jellybeans if you must choose at least 2 jellybeans of each flavor?
9. Do combinations with fixed repetitions make any sense? If so, what can you say about them?
10. How many ways are there to place n identical balls into k boxes?
11. Show that the binomial coefficients in 5.2.1 and 5.2.2 are equal.
12. A word is said to be **increasing** if its letters are in alphabetical order, read from left to right. For example, the word aabhhkx is increasing, but the word aabclk is not. How many increasing words are there of length 10?

For Exercises 13 through 15, let

$$\left\langle {n \atop k} \right\rangle = \binom{n+k-1}{k}$$

13. Prove that

$$\left\langle {n \atop k} \right\rangle = \left\langle {n \atop k-1} \right\rangle + \left\langle {n-1 \atop k} \right\rangle$$

14. Prove that

$$\left\langle {n \atop k} \right\rangle = \frac{n}{k}\left\langle {n+1 \atop k-1} \right\rangle$$

15. Prove that

$$\left\langle {n \atop k} \right\rangle = \frac{n+k-1}{k}\left\langle {n \atop k-1} \right\rangle$$

16. Let $S = \{a_1, a_2, \ldots, a_n\}$ be a set of size n. How many combinations of size $k = (\frac{1}{2})n^2 + (\frac{1}{2})n + 2$ are there taken from S with the property that the element a_1 must appear at least one time in the combination, a_2 must appear at least two times in the combination, and so on?
17. As we did in Example 5.2.6, use the program segment

```
X := 0;
FOR I1 := 1 TO N DO
    FOR I2 := 1 TO I1 DO
        FOR I3 := 1 TO I2 DO
            X := X + 1;
PRINT X
```

to show that

$$\binom{2}{2} + \binom{3}{2} + \binom{4}{2} + \cdots + \binom{N+1}{2} = \binom{N+2}{3}$$

18. What value of X does the following program segment print?

```
X := 0;
FOR I_1 := 1 TO N DO
    FOR I_2 := 1 TO I_1 DO
        FOR I_3 := 1 TO I_2 DO
            .
              .
                .
                  FOR I_K := 1 TO I_(K-1) DO
                      X := X + 1;
PRINT X
```

5.3 Linear Equations with Unit Coefficients

Let us consider the following problem. How many different solutions, in nonnegative integers, are there of the equation

$$x + y + z = 10 \tag{5.3.1}$$

For example,

$$x = 2 \quad , \quad y = 5 \quad , \quad z = 3$$

$$x = 5 \quad , \quad y = 3 \quad , \quad z = 2$$

and

$$x = 10 \quad , \quad y = 0 \quad , \quad z = 0$$

are all solutions of Equation 5.3.1 in nonnegative integers, but

$$x = 8 \quad , \quad y = 9 \quad , \quad z = -3$$

and

$$x = \tfrac{1}{2} \quad , \quad y = \tfrac{11}{2} \quad , \quad z = 8$$

are not. Notice that the first two solutions just listed are considered to be different, even though they involve the same numbers 2, 3, and 5.

As it turns out, our knowledge of combinations with repetitions makes this problem very easy to solve. Consider the set $S = \{x, y, z\}$. Then any solution to Equation 5.3.1 can be represented as a combination of size 10 with unrestricted repetitions, taken from the set S.

For example, the solution

$$x = 2 \quad , \quad y = 5 \quad , \quad z = 3$$

can be represented as the combination

$$\{x, x, y, y, y, y, y, z, z, z\}$$

which contains 2 x's, 5 y's, and 3 z's.

On the other hand, every such combination corresponds to exactly one solution to Equation 5.3.1 in nonnegative integers. For example, the combination

$$\{x, x, x, x, y, y, y, y, z, z\}$$

corresponds to the solution

$$x = 4 \quad , \quad y = 4 \quad , \quad z = 2$$

In general, the solution $x = i, y = j, z = k$ (where $i + j + k = 10$) corresponds to the combination with i x's, j y's, and k z's.

In this way, we see that every solution to Equation 5.3.1 in nonnegative integers corresponds to exactly one combination of size 10 with unrestricted repetitions, taken from S. Also, every such combination corresponds to exactly one solution to 5.3.1 in nonnegative integers.

In the language of Section 1.2, there is a one-to-one correspondence between the nonnegative integer solutions to Equation 5.3.1 and the combinations of size 10, with unrestricted repetitions.

Therefore, the number of solutions in nonnegative integers of Equation 5.3.1 is the same as the number of combinations of size 10 with unrestricted repetitions, taken from the set S. According to Theorem 5.2.1, this number is

$$\binom{3 + 10 - 1}{10} = \binom{12}{10} = \binom{12}{2} = 66$$

and so there are 66 different solutions, in nonnegative integers, of Equation 5.3.1.

We can easily generalize this discussion to obtain the following result.

Theorem 5.3.1

The number of different solutions, in nonnegative integers, of the equation

$$x_1 + x_2 + \cdots + x_n = k \tag{5.3.2}$$

is

$$\binom{n + k - 1}{k}$$

PROOF As before, every solution to Equation 5.3.2 can be represented as a combination of size k with unrestricted repetitions, taken from the set $S = \{x_1, x_2, \ldots, x_n\}$ of size n. In particular, the solution

$$x_1 = i_1 \quad , \quad x_2 = i_2 \quad , \ldots , \quad x_n = i_n$$

can be represented as the combination

$$\{x_1, \ldots, x_1, x_2, \ldots, x_2, \ldots, x_n, \ldots, x_n\}$$

where there are i_1 x_1's, i_2 x_2's, and so on.

In this way, we see that each solution in nonnegative integers to Equation 5.3.2 corresponds to exactly one combination of size k with unrestricted repetitions, taken from S. On the other hand, each such combination corresponds to exactly one solution to 5.3.2 in nonnegative integers.

Hence, the number of solutions in nonnegative integers to Equation 5.3.2 is equal to the number of combinations of size k with unrestricted repetitions, taken from S, which according to Theorem 5.2.1, is equal to

$$\binom{n + k - 1}{k}$$

This completes the proof. ∎

Example 5.3.1

The number of solutions, in nonnegative integers, of the equation

$$x_1 + x_2 + x_3 + x_4 = 16$$

is

$$\binom{4 + 16 - 1}{16} = \binom{19}{16} = \binom{19}{3} = 969$$

The number of solutions, in nonnegative integers, of the equation

$$x_1 + x_2 + \cdots + x_{10} = 15$$

is

$$\binom{10 + 15 - 1}{15} = \binom{24}{15} = 1{,}307{,}504$$

Example 5.3.2

How many integers are there between 0 and 999,999 (inclusive) with the property that the sum of their digits is equal to 5?

Each integer between 0 and 999,999 can be written as a six-digit integer $x_1x_2x_3x_4x_5x_6$, where each digit satisfies $0 \leq x_i \leq 9$. The condition that the sum of these digits is equal to 5 is

$$x_1 + x_2 + x_3 + x_4 + x_5 + x_6 = 5 \quad \textbf{(5.3.3)}$$

Hence, the answer to our question is the same as the number of solutions to Equation 5.3.3 that satisfy the condition $0 \leq x_i \leq 9$ for all $i = 1, 2, \ldots, 6$.

But any solution to this equation, in nonnegative integers, *automatically* has the property that $x_i \leq 9$ for all $i = 1, 2, \ldots, 6$. After all, how can any of the x_i be larger than 9 if the sum of all of the x_i is only equal to 5, and all of the x_i are nonnegative?

Therefore, we can drop the condition $x_i \leq 9$, and so the answer to our question is the same as the number of solutions to Equation 5.3.3 in nonnegative integers. Thus, using Theorem 5.3.1, we see that there are

$$\binom{6+5-1}{5} = \binom{10}{5} = 252$$

integers between 0 and 999,999 the sum of whose digits is equal to 5. □

Theorem 5.3.1 tells us how to determine the number of solutions, in *nonnegative* integers, to the equation

$$x_1 + x_2 + \cdots + x_n = k \tag{5.3.4}$$

But, sometimes it is important to know how many solutions there are, in *positive* integers, to this equation. We can use Theorem 5.3.1 to answer this question as well.

Suppose that we subtract n from both sides of Equation 5.3.4, to get

$$x_1 + x_2 + \cdots + x_n - n = k - n$$

This can be rewritten in the form

$$(x_1 - 1) + (x_2 - 1) + \cdots + (x_n - 1) = k - n$$

and, if for convenience, we set $y_i = x_i - 1$ for all $i = 1, 2, \ldots, n$, then we get

$$y_1 + y_2 + \cdots + y_n = k - n \tag{5.3.5}$$

Now, each solution to Equation 5.3.4 in *positive* integers corresponds to exactly one solution to Equation 5.3.5 in *nonnegative* integers, and vice versa. In particular, if

$$x_1 = i_1 \quad , \quad x_2 = i_2 \quad , \ldots , \quad x_n = i_n$$

is a solution to 5.3.4 in *positive* integers, then

$$y_1 = i_1 - 1 \quad , \quad y_2 = i_2 - 1 \quad , \ldots , \quad y_n = i_n - 1$$

is a solution to 5.3.5 in *nonnegative* integers. On the other hand, if

$$y_1 = j_1 \quad , \quad y_2 = j_2 \quad , \ldots , \quad y_n = j_n$$

is a solution to 5.3.5 in *nonnegative* integers, then

$$x_1 = j_1 + 1 \quad , \quad x_2 = j_2 + 1 \quad , \ldots , \quad x_n = j_n + 1$$

is a solution to 5.3.4 in *positive* integers.

In the language of Section 1.2, we have found a one-to-one correspondence between the set of all solutions in positive integers to Equation 5.3.4 and the set of all solutions in nonnegative integers to Equation 5.3.5.

Thus, we see that the number of solutions to 5.3.4 in positive integers is the same as the number of solutions to 5.3.5 in nonnegative integers. But, according to Theorem 5.3.1, there are

$$\binom{n + (k - n) - 1}{k - n} = \binom{k - 1}{k - n} = \binom{k - 1}{n - 1}$$

solutions to Equation 5.3.5 in nonnegative integers. This gives us the following theorem.

Theorem 5.3.2

The number of solutions, in *positive* integers, to the equation

$$x_1 + x_2 + \cdots + x_n = k$$

is

$$\binom{k-1}{n-1}$$

Example 5.3.3

The number of solutions, in positive integers, to the equation

$$x_1 + x_2 + x_3 + x_4 = 16$$

is

$$\binom{16-1}{4-1} = \binom{15}{3} = 455$$

This compares with 969 solutions to this equation in *nonnegative* integers (see Example 5.3.1).

The number of solutions, in positive integers, to the equation

$$x_1 + x_2 + \cdots + x_{10} = 15$$

is

$$\binom{15-1}{10-1} = \binom{14}{9} = 2002$$

as compared with 1,307,504 solutions in *nonnegative* integers! (See Example 5.3.1.)

Example 5.3.4

How many ways are there to place 100 people in 3 different rooms, if each room must be occupied and if it does not matter which people go into which room?

If we let x_1 be the number of people that go into room 1, x_2 be the number of people that go into room 2, and x_3 be the number of people that go into room 3, then we must have

$$x_1 + x_2 + x_3 = 100$$

where $x_i > 0$ for all $i = 1, 2,$ and 3.

Thus, the answer to our question is the same as the number of solutions to this equation in positive integers. According to Theorem 5.3.2, this number is

$$\binom{100-1}{3-1} = \binom{99}{2} = 4851$$

and so there are 4851 different ways to place 100 people into 3 rooms under these conditions.

The method we used to establish Theorem 5.3.2 can be used to establish the following theorem.

Theorem 5.3.3

Let $c_1, c_2, \ldots, c_n$ be integers. Then the number of solutions, in integers, to the equation

$$x_1 + x_2 + \cdots + x_n = k \tag{5.3.6}$$

that have the property that

$$x_1 > c_1 \quad , \quad x_2 > c_2 \quad , \ldots, \quad x_n > c_n \tag{5.3.7}$$

is

$$\binom{k - c_1 - c_2 - \cdots - c_n - 1}{n - 1}$$

PROOF If we subtract $c_1 + c_2 + \cdots + c_n$ from both sides of Equation 5.3.6 we get

$$(x_1 - c_1) + (x_2 - c_2) + \cdots + (x_n - c_n) = k - c_1 - c_2 - \cdots - c_n$$

Setting $y_i = x_i - c_i$ for all $i = 1, 2, \ldots, n$, this becomes

$$y_1 + y_2 + \cdots + y_n = k - c_1 - c_2 - \cdots - c_n \tag{5.3.8}$$

Now, each solution to Equation 5.3.6 in integers satisfying condition 5.3.7 corresponds to exactly one solution to Equation 5.3.8 in *positive* integers, and vice versa. In particular, if

$$x_1 = i_1 \quad , \quad x_2 = i_2 \quad , \ldots, \quad x_n = i_n$$

is a solution to 5.3.6 in integers satisfying 5.3.7, then

$$y_1 = i_1 - c_1 \quad , \quad y_2 = i_2 - c_2 \quad , \ldots, \quad y_n = i_n - c_n$$

is a solution to 5.3.6 in positive integers. On the other hand, if

$$y_1 = j_1 \quad , \quad y_2 = j_2 \quad , \ldots, \quad y_n = j_n$$

is a solution to 5.3.6 in positive integers, then

$$x_1 = j_1 + c_1 \quad , \quad x_2 = j_2 + c_2 \quad , \ldots, \quad x_n = j_n + c_n$$

is a solution to 5.3.6 in integers satisfying condition 5.3.7.

Thus, we see that the number of solutions to Equation 5.3.6 that satisfy condition 5.3.7 is the same as the number of solutions to Equation 5.3.8 in positive integers. But, according to Theorem 5.3.2, there are

$$\binom{k - c_1 - c_2 - \cdots - c_n - 1}{n - 1}$$

solutions to Equation 5.3.8 in positive integers, and so there are also this many solutions to Equation 5.3.6 in integers satisfying condition 5.3.7. This completes the proof. ∎

Incidentally, Theorem 5.3.1 is a special case of Theorem 5.3.5, obtained by setting $c_i = 0$ for all $i = 1, 2, \ldots, n$. Also, Theorem 5.3.2 is a special case of Theorem 5.3.3, obtained by setting $c_1 = -1$ for all $i = 1, 2, \ldots, n$.

Example 5.3.5

The number of solutions to the equation

$$x_1 + x_2 + x_3 + x_4 = 16$$

in integers satisfying

$$x_1 > 5, \quad x_2 > 0, \quad x_3 > -2, \quad x_4 > 6$$

is

$$\binom{16 - 5 - 0 - (-2) - 6 - 1}{4 - 1} = \binom{6}{3} = 20$$

In Section 5.6, we will discuss the problem of finding the number of solutions, in integers, to the equation

$$x_1 + x_2 + \cdots + x_n = k$$

where the numbers x_i must satisfy a condition of the form $c_i < x_i < d_i$ for all $i = 1, 2, \ldots, n$.

Exercises

1. a) How many solutions are there, in nonnegative integers, to the equation

$$x + y + z = 10$$

b) How many solutions are there, in positive integers, to this equation?

2. a) How many solutions are there, in nonnegative integers, to the equation

$$x + y + z + w = 14$$

b) How many solutions are there, in positive integers, to this equation?

3. a) How many solutions are there, in nonnegative integers, to the equation

$$x_1 + x_2 + \cdots + x_8 = 16$$

b) How many solutions are there, in positive integers, to this equation?

4. How many solutions, in integers all greater than 2, are there to the equation

$$x + y + z = 10$$

5. How many solutions, in integers all greater than -2, are there to the equation

$$x + y + z = 5$$

6. How many solutions are there, in integers, to the equation

$$x_1 + x_2 + x_3 + x_4 = 20$$

that satisfy the condition $x_1 > 2$, $x_2 > 2$, $x_3 > 7$, and $x_4 > -2$?

7. How many solutions are there, in integers, to the equation

$$x_1 + x_2 + \cdots + x_{10} = 60$$

that satisfy the condition $x_i \geq i$ for all $i = 1, 2, \ldots, 10$?

8. Show that the equations

$$x_1 + x_2 + \cdots + x_{12} = 4 \quad \text{and} \quad x_1 + x_2 + \cdots + x_5 = 11$$

have the same number of solutions in nonnegative integers.

9. How many integers are there between 0 and 100,000 with the property that the sum of their digits is equal to 7?

10. a) How many integers are there between 100 and 1,000,000 with the property that the sum of their digits is equal to 6?

b) How many integers are there between 100 and 1,000,000 with the property that the sum of their digits is less than 6?

11. If you roll 5 distinct dice, how many ways are there to produce a sum equal to 10?

12. How many ways are there to place 50 people in 4 different rooms if it does not matter which people are placed in which rooms?

13. How many ways are there to put 30 identical softballs into 5 different bins if each bin must contain at least 3 softballs?

14. How many ways are there to give 25 identical pieces of candy to 7 children if the first child must get at least 4 pieces, and all of the other children must get at least 1 piece?

15. How many terms are there in the multinomial expansion of $(x_1 + x_2 + \cdots + x_k)^n$ given in Theorem 4.8.2?

5.4 Distributing Balls into Boxes

The same combinatorial problem frequently can be phrased in many different ways, and one of the most common ways to phrase combinatorial problems is in terms of distributing balls into boxes. For this reason, it is important to devote some time to becoming familiar with this terminology.

As a simple example of an important combinatorial problem that can be phrased in terms of distributing balls into boxes, suppose that a certain atom has two electrons, and three electron shells that can contain these electrons. [See Figure 5.4.1(a).] Suppose also that each shell is allowed to contain only one electron. The problem is to count the total number of possible electron configurations.

In this case, it is not hard to see that there are exactly three possible configurations, as pictured in Figure 5.4.1(a).

Now, the same problem can be phrased in terms of distributing balls into boxes as follows. Suppose that we have 3 boxes and 2 balls, and we wish to place the balls into the boxes in such a way that no box receives more than 1 ball. The problem is to count the total number of ways to do this.

The different possibilities are illustrated in Figure 5.4.1(b). As you can see by comparing Figures 5.4.1(a) and (b), these two problems are essentially the same. The only difference is in their context.

In this section, we want to consider the problem of how to count the number of ways of distributing k balls into n boxes, under various conditions. The conditions that are generally imposed are the following:

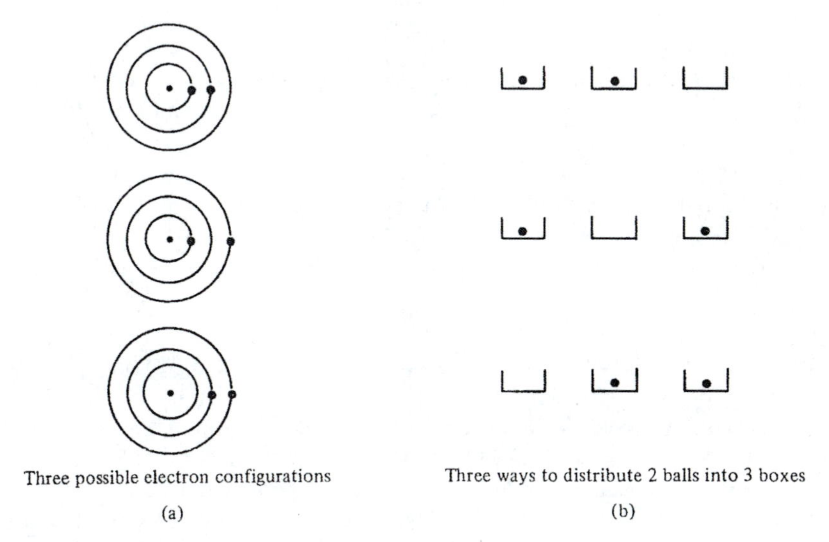

Figure 5.4.1
(a) Three possible electron configurations. (b) Three ways to put two balls into three boxes.

1) The balls can be either *distinguishable* or *indistinguishable*.
2) The boxes can be either *distinguishable* or *indistinguishable*.
3) The distribution can take place either *with exclusion* or *without exclusion*.

Let us discuss these terms briefly. The term "distinguishable" refers to the fact that the balls, or boxes, are marked in some way or have some feature about them that makes each one distinguishable from the others. For example, they may be numbered, each with a different number, they may each be a different color, or they may each be a different size or shape.

For the purposes of our discussion, when we speak of k distinguishable balls, we will assume that they are numbered with the consecutive integers 1 through k, and when we speak of n distinguishable boxes, we will assume that they are numbered with the consecutive integers 1 through n.

The term "indistinguishable" refers to the fact that the balls, or boxes, are so identical that there is no way to tell them apart (not even by their location!). *In particular then, when placing indistinguishable balls into distinguishable boxes, it makes no difference which balls go into which boxes.* In fact, it is not even possible to tell which balls go into which boxes! All that we are able to tell is the number of balls that end up in each box.

In our discussion, we will consider only the case of distinguishable boxes. As you might imagine, the case of indistinguishable boxes turns out to be much more complicated than that of distinguishable boxes and is usually studied in a course in combinatorics.

As to the third condition, the phrase "with exclusion" means that no box can contain more than one ball, and the phrase "without exclusion" means that a box may contain more than one ball.

Fortunately, we can use our knowledge of permutations and combinations to help us with the problems of distributing balls into boxes. Therefore, in each case, we will first try to rephrase the problem in terms of permutations and combinations.

Before considering the various cases, we should clear up one possible point of confusion. Namely, the *order* in which the balls are placed into the boxes is not important. To help keep this in mind, it is a good idea to think of the balls as being placed into the boxes at exactly the same time.

Let us now turn to the various cases. (Incidentally, how many cases are there?)

Case 1

How many ways are there to distribute k distinguishable balls into n distinguishable boxes, with exclusion?

In this case, we consider putting k balls, numbered 1 through k, into n boxes, numbered 1 through n, in such a way that no box receives more than one ball.

Now, we can translate this into the language of permutations and combinations as follows. Putting k distinguishable balls into n boxes, with exclusion, amounts to the same thing as making an ordered selection of k of the n boxes, where the balls do the selecting for us. The ball labeled 1 selects the first box, the ball labeled 2 selects the second box, and so on. In other words, distributing k distinguishable balls into n distinguishable boxes, with exclusion, is the same as forming a permutation of size k, taken from the set of n boxes. This gives us the following theorem.

Theorem 5.4.1

Distributing k distinguishable balls into n distinguishable boxes, with exclusion, corresponds to forming a permutation of size k, taken from a set of size n. Therefore, there are $P(n, k) = (n)_k$ different ways to distribute k distinguishable balls into n distinguishable boxes, with exclusion.

Case 2

How many ways are there to distribute k distinguishable balls into n distinguishable boxes, without exclusion?

In this case, we consider putting k balls, numbered 1 through k, into n boxes, numbered 1 through n, but this time with no restriction on the number of balls that can go into each box.

Again, instead of thinking in terms of putting k balls into n boxes, we can think in terms of selecting k of the n boxes. As before, the balls do the selecting for us, but this time more than one ball may go into the same box, which means that the same box may be chosen more than once.

Therefore, we are still dealing with ordered selections, or permutations, of the boxes, but now with unrestricted repetitions. In particular, we have the following theorem.

Theorem 5.4.2

Distributing k distinguishable balls into n distinguishable boxes, without exclusion, corresponds to forming a permutation of size k, with unrestricted repetitions, taken from a set of size n. Therefore, there are n^k different ways to distribute k distinguishable balls into n distinguishable boxes, without exclusion.

Case 3

How many ways are there to distribute k indistinguishable balls into n distinguishable boxes, with exclusion?

In this case, we have k identical balls, and we wish to place them into n distinguishable boxes in such a way that no box receives more than one ball.

Once such a placement of balls has been made, then, since the balls are identical, all we can say is which boxes have received a ball and which have not. In other words, placing the balls has the same effect as simply choosing k of the n boxes. Those boxes that receive a ball are the ones that are chosen, and those that do not receive a ball are not chosen.

Hence, in this case we are making unordered selections, that is, forming combinations of size k, taken from the set of n boxes. This gives the following theorem.

Theorem 5.4.3

Distributing k indistinguishable balls into n distinguishable boxes, with exclusion, corresponds to forming a combination of size k, taken from a set of size n. Therefore, there are $C(n, k) = \binom{n}{k}$ different ways to distribute k indistinguishable balls into n distinguishable boxes, with exclusion.

Case 4

How many ways are there to distribute k indistinguishable balls into n distinguishable boxes, without exclusion?

In this case, we have k identical balls, to be distributed into n distinguishable boxes, but with no restriction on the number of balls that can occupy a given box.

As in the last case, since the balls are indistinguishable, we can only tell how many balls each box has received. This translates into making a choice of k of the n boxes, but with the possibility that a box may be chosen more than once. Thus, placing k balls into n boxes in this case corresponds to forming an unordered selection, or combination, of size k, taken from the set of n boxes, but with unrestricted repetitions. This gives the following theorem.

Theorem 5.4.4

Distributing k indistinguishable balls into n distinguishable boxes, without exclusion, corresponds to forming a combination of size k with unrestricted repetitions, taken from a set of size n. Therefore, there are $\binom{n + k - 1}{k}$ different ways to distribute k indistinguishable balls into n distinguishable boxes, without exclusion.

We should discuss another condition that is commonly placed on the distribution of balls into boxes, namely, the condition that no box be empty. The next theorem summarizes the possibilities. We will prove part 2 of this theorem in Section 5.7 and leave the other parts for you to prove in the exercises of this section.

Theorem 5.4.5

1) The number of ways to distribute k distinguishable balls into n distinguishable boxes, with exclusion, in such a way that no box is empty, is $n!$ if $k = n$ and 0 if $k \neq n$.

2) The number of ways to distribute k distinguishable balls into n distinguishable boxes, without exclusion, in such a way that no box is empty, is

$$\binom{n}{0}(n-0)^k - \binom{n}{1}(n-1)^k + \binom{n}{2}(n-2)^k - \cdots + (-1)^{n-1}\binom{n}{n-1}(1)^k$$

for $k \geq n$. If $k < n$ then, of course, there is no way. [We have written $(n - 0)^k$ instead of n^k in order to preserve the pattern in the formula. Also, the factor $(-1)^{n-1}$ is in the last term in order to make it positive if n is odd and negative if n is even.]

3) The number of ways to distribute k indistinguishable balls into n distinguishable boxes, with exclusion, in such a way that no box is empty, is 1 if $k = n$ and 0 if $k \neq n$.

4) The number of ways to distribute k indistinguishable balls into n distinguishable boxes, without exclusion, in such a way that no box is empty, is $\binom{k-1}{n-1}$.

For the sake of reference, let us summarize our results in a table.

Table 5.4.1

Balls	*Boxes*	*Exclusion*	*No box empty*	*Number of ways to put k balls into n boxes*	*Reference*
Dist	Dist	with		$(n)_k$	Theorem 5.4.1
Dist	Dist	with	yes	$n!$ if $k = n$ 0 if $k \neq n$	Theorem 5.4.5
Dist	Dist	without		n^k	Theorem 5.4.2
Dist	Dist	without	yes	see reference	Theorem 5.4.5
Indist	Dist	with		$\binom{n}{k}$	Theorem 5.4.3
Indist	Dist	with	yes	1 if $k = n$ 0 if $k \neq n$	Theorem 5.4.5
Indist	Dist	without		$\binom{n+k-1}{k}$	Theorem 5.4.4
Indist	Dist	without	yes	$\binom{k-1}{n-1}$	Theorem 5.4.5

Exercises

Unless mentioned to the contrary, a distribution is assumed to be without exclusion.

1. a) How many ways are there to put 5 distinguishable balls into 7 distinguishable boxes, with exclusion?

b) How many ways are there to put 5 distinguishable balls into 7 distinguishable boxes, without exclusion?

2. a) How many ways are there to put 5 indistinguishable balls into 7 distinguishable boxes, with exclusion?

b) How many ways are there to put 5 indistinguishable balls into 7 distinguishable boxes, without exclusion?

3. a) How many ways are there to put 7 distinguishable balls into 5 distinguishable boxes, with exclusion?

b) How many ways are there to put 7 distinguishable balls into 5 distinguishable boxes, without exclusion?

4. a) How many ways are there to put 7 indistinguishable balls into 5 distinguishable boxes, with exclusion?

b) How many ways are there to put 7 indistinguishable balls into 5 distinguishable boxes, without exclusion?

5. a) How many ways are there to put 6 distinguishable balls into 9 distinguishable boxes, without exclusion, in such a way that no box is empty?

b) How many ways are there to put 6 distinguishable balls into 9 distinguishable boxes, with exclusion, in such a way that no box is empty?

6. a) How many ways are there to put 6 indistinguishable balls into 9 distinguishable boxes, without exclusion, in such a way that no box is empty?

b) How many ways are there to put 6 indistinguishable balls into 9 distinguishable boxes, with exclusion, in such a way that no box is empty?

7. How many ways are there to distribute n indistinguishable balls into m distinguishable boxes in such a way that each box receives at least one ball?

8. How many ways are there to distribute n indistinguishable balls into m distinguishable boxes in such a way that each box receives at least two balls?

9. How many ways are there to distribute n indistinguishable balls into m distinguishable boxes in such a way that the i-th box receives at least j_i balls?

10. Prove part 1 of Theorem 5.4.5.

11. Prove part 3 of Theorem 5.4.5.

12. Prove part 4 of Theorem 5.4.5.

In each of the following exercises, if possible, translate the problem into one involving the distribution of balls into boxes, and then solve the problem. Indicate which theorems you use.

13. A certain computer room has 12 computers and 18 printers. Each printer must be connected to a computer, and each computer must be connected to a printer. In how many ways can the connections be made?

14. Three electrons can each be in one of five different energy levels, say E_1, E_2, E_3, E_4, and E_5.

a) How many possibilities are there if the electrons are considered to be distinguishable?

b) How many possibilities are there if the electrons are considered to be distinguishable, but no two electrons can have the same energy?

c) How many possibilities are there if the electrons are considered to be indistinguishable?

d) How many possibilities are there if the electrons are considered to be indistinguishable, but no two electrons can have the same energy?

15. a) How many ways are there to assign 50 students to 25 desks?

b) How many ways are there to assign 50 students to 25 desks if each desk must be used?

c) How many ways are there to assign 50 students to 25 desks if each desk must be assigned to the same number of students?

16. A certain supermarket has 10 checkout counters. How many ways are there for 7 shoppers to check out their groceries if no 2 shoppers want to use the same checker?
17. How many ways are there to distribute 10 identical red balls and 8 identical blue balls into 5 different boxes?
18. How many solutions are there, in integers, of the equation

$$x + y + z = 9$$

Hint, there are 3 boxes.

19. a) How many ways are there to hand out 10 pieces of chocolate to 5 children?
    b) How many ways are there to hand out 10 pieces of chocolate to 5 children if each child must receive at least 1 piece?
    c) How many ways are there to hand out 10 pieces of chocolate to 5 children if each child must receive at least 2 pieces?
20. How many ways are there to give 12 pieces of chocolate to 8 children if 4 particular children must get at least 1 piece each and 3 particular children must get at least 2 pieces each?
21. In how many ways can 5 people divide 4 apples, 3 oranges, 6 bananas, and 2 pears?
22. In how many ways can 5 people divide 4 apples, 3 oranges, 6 bananas, and 2 pears if each person must get at least 1 banana?

5.5 The Principle of Inclusion-Exclusion—I

Up to now, much of what we have done has depended, either directly or indirectly, on two important principles—the multiplication rule and the pigeonhole principle. In the next few sections, we want to study a third principle every bit as powerful as these two. It is called the *principle of inclusion-exclusion*.

In order to describe this principle, let us begin with a simple situation. Consider a group of 60 students, and suppose that 40 of these students are taking a math class and 30 of them are taking a computer science class. Then, can we tell how many of these students are taking *at least one of* the two classes?

The answer is no. It certainly cannot be $40 + 30 = 70$ since there are only 60 students! The reason that we cannot tell how many students are taking at least one class is that we do not know how many students are taking *both* classes.

So let us suppose that 20 students are taking both classes. Now we can determine the number of students who are taking at least one of the classes as follows. The number $40 + 30 = 70$ is too large, but only because the students who are taking both classes are counted *twice* in this number.

Therefore, since there are 20 students taking both classes, the total number of students taking at least one of the two classes is $40 + 30 - 20 = 50$.

We can summarize this discussion in the following equation:

number of students taking *at least one of* the two classes =

number of students taking the math class + number of students taking the computer science class − number of students taking *both* of the classes (5.5.1)

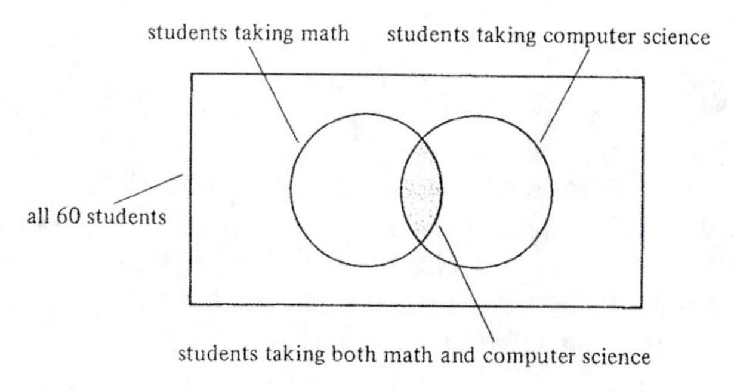

Figure 5.5.1
Students in the shaded region are counted twice.

It might help to remember this equation if we draw a picture, as in Figure 5.5.1. The large box represents the entire group of 60 students. The circle on the left represents the set of students who are taking the math class, and the circle on the right represents the set of students who are taking the computer science class.

The shaded region of intersection of the two circles represents those students who are taking both classes. These are the students that are counted twice when we take the sum of the number of students in each of the two circles, and this is precisely why we must subtract the number of students in this intersection.

Since the same reasoning that we used to solve this problem will work to solve other counting problems, we should try to phrase our result in general terms, so that it will be easier to apply. This is done simply by replacing sets of students by abstract sets.

Let U be a universal set and let A and B be subsets of U, as pictured in Figure 5.5.2. We would like to know how many elements of U are in *at least one of* the two sets A or B. Put another way, we would like to know how many elements there are in the *union* $A \cup B$ of the two sets.

The answer is given in the following equation

number of elements in the union $A \cup B$	=	number of elements in the set A	+	number of elements in the set B	−	number of elements in the intersection $A \cap B$

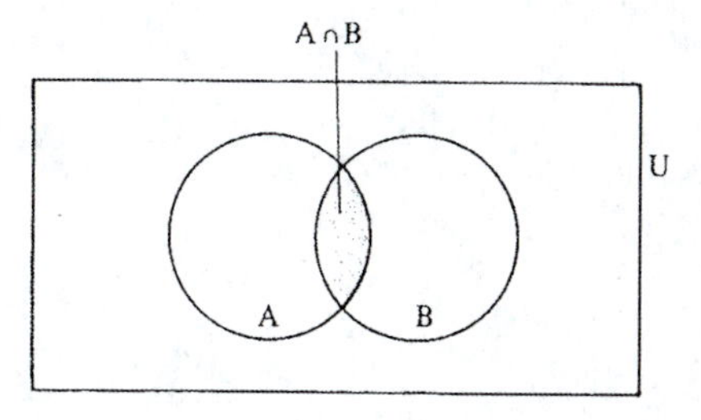

Figure 5.5.2
The shaded region $A \cap B$ is counted twice in $|A| + |B|$. Hence $|A \cup B| = |A| + |B| - |A \cap B|$.

In symbols, this can be written,

$$|A \cup B| = |A| + |B| - |A \cap B| \tag{5.5.2}$$

The reasoning behind Equation 5.5.2 is exactly the same as the reasoning we used to derive Equation 5.5.1. Namely, the sum $|A| + |B|$ counts all of the elements in $A \cup B$, but it happens to count those elements that are in the intersection $A \cap B$ twice. Therefore, in order to have them counted only once, we subtract $|A \cap B|$.

Equation 5.5.2 is the simplest form of the principle of inclusion-exclusion, so let us put it into a theorem.

Theorem 5.5.1 (The Principle of Inclusion-Exclusion for Two Sets)

Let U be a universal set, and let A and B be subsets of U. Then the number of elements of U that are in *at least one of* the sets A or B is

$$|A \cup B| = |A| + |B| - |A \cap B| \tag{5.5.3}$$

Example 5.5.1

In a certain group of 100 kittens, 50 are male, 25 are long haired, and 10 are long-haired males. How many kittens in the group are either male or long haired? (Kittens come in two varieties, long haired and short haired.)

This question can be rephrased, even though a bit awkwardly, as follows. How many kittens in the group are *at least one of* male or long haired? This rewording is a sign that the principle of inclusion-exclusion may apply.

So let

U = the set of all 100 kittens in the group

A = the set of all kittens in the group that are male

B = the set of all kittens in the group that are long haired

Then $A \cup B$ is the set of kittens that are either male or long haired, and so the answer to the question is $|A \cup B|$, which we can compute using the inclusion-exclusion principle.

First, we observe that

$$|A| = 50, \qquad |B| = 25, \qquad |A \cap B| = 10$$

(The last equality follows from the fact that $A \cap B$ is just the set of long-haired males.) Therefore, according to Equation 5.5.3,

$$\begin{aligned} |A \cup B| &= |A| + |B| - |A \cap B| \\ &= 50 + 25 - 10 \\ &= 65 \end{aligned}$$

and so there are 65 kittens that are either male or long haired. ❑

The principle of inclusion-exclusion tells us how to find the number of elements in U that are in *at least one of* the two sets A or B. But we may also want to find the number of elements in U that are in *neither of* the two sets A or B.

This problem can easily be solved by using the principle of inclusion-exclusion. First, we observe that the set of all elements of U that are in neither of the sets A or B is just $(A \cup B)^c$, and we get

$$\begin{aligned} |(A \cup B)^c| &= |U| - |A \cup B| \\ &= |U| - |A| - |B| + |A \cap B| \end{aligned} \tag{5.5.4}$$

This equation is important enough to be made into a theorem. Since it follows almost directly from Theorem 5.5.1, we will call it a corollary to that theorem.

Corollary 5.5.2

Let U be a universal set, and let A and B be subsets of U. Then the number of elements of U that are in *neither of* the sets A or B is

$$|A^c \cap B^c| = |U| - |A| - |B| + |A \cap B| \tag{5.5.5}$$

Example 5.5.2

Referring to Example 5.5.1, we ask the question, How many kittens in the group are short-haired females?

This question can be rephrased as follows. How many kittens in the group are *neither* male nor long haired? In this form, we see that the question is asking for the number of kittens that are in neither of the sets A or B; that is, it is asking for $|A^c \cap B^c|$.

According to Equation 5.5.5, this number is

$$\begin{aligned} |A^c \cap B^c| &= |U| - |A| - |B| + |A \cap B| \\ &= 100 - 50 - 25 + 10 \\ &= 35 \end{aligned}$$

and so there are 35 short-haired female kittens in the group.

Actually, in this case, since we had already computed $|A \cup B|$ in the previous example, it would have been a bit easier to use this information to find $|A^c \cap B^c|$:

$$\begin{aligned} |A^c \cap B^c| &= |U| - |A \cup B| \\ &= 100 - 65 \\ &= 35 \end{aligned}$$

Nevertheless, there will be cases where $|A \cup B|$ has not been computed in advance, and then it is easier to use Equation 5.5.5. □

There is a useful moral in the last two examples. Namely, if a counting problem contains one of the phrases "at least one of" or "neither of," or if it can be reworded so that it contains one of these phrases, then it is a good idea to try the principle of inclusion-exclusion, or its corollary. (Of course, there is no guarantee that this will work, but it is at least worth a try.)

Example 5.5.3

a) How many integers between 1 and 600 have the property that they are divisible by either 3 or 5?

b) How many integers between 1 and 600 have the property that they are divisible by neither 3 nor 5?

If we let P_1 be the property of being divisible by 3, and if we let P_2 be the property of being divisible by 5, then these questions can be rephrased as follows.

a) How many integers are there between 1 and 600 that satisfy *at least one of* the properties P_1 or P_2?

b) How many integers are there between 1 and 600 that satisfy *neither of* the properties P_1 or P_2?

These rewordings are a clue to the fact that we should try the principle of inclusion-exclusion. Let us begin by letting

U = the set of all integers between 1 and 600

A = the set of all integers in U that satisfy property P_1, that is, that are divisible by 3

B = the set of all integers in U that satisfy property P_2, that is, that are divisible by 5

Then since $A \cup B$ is the set of all integers in U that satisfy at least one of the properties, the answer to the first question is $|A \cup B|$. In order to compute $|A \cup B|$, we need to compute $|A|$, $|B|$, and $|A \cap B|$.

As to $|A|$, we must determine the number of integers between 1 and 600 that are divisible by 3. But these integers are just the multiples of 3, that is,

$$3, 6, 9, 12, 15, \ldots$$

and there are $600/3 = 200$ of these. Hence,

$$|A| = 200$$

Similarly, B is the set of multiples of 5, and there are $600/5 = 120$ of these. Hence,

$$|B| = 120$$

Finally, $A \cap B$ is the set of all integers between 1 and 600 that are divisible by both 3 and 5, that is, that are divisible by $3 \cdot 5 = 15$. In other words, $A \cap B$ is the set of multiples of 15. Since there are $600/15 = 40$ of these, we have

$$|A \cap B| = 40$$

Putting the pieces together, we get

$$\begin{aligned} |A \cup B| &= |A| + |B| - |A \cap B| \\ &= 200 + 120 - 40 \\ &= 280 \end{aligned}$$

which tells us that there are 280 integers between 1 and 600 that are divisible by either 3 or 5.

The answer to the second question is $|A^c \cap B^c|$. To compute this, we use Equation 5.5.4,

$$\begin{aligned} |A^c \cap B^c| &= |U| - |A \cup B| \\ &= 600 - 280 \\ &= 320 \end{aligned}$$

(Alternatively, we could use Equation 5.5.5.) Hence, there are 320 integers between 1 and 600 that are divisible by neither 3 nor 5. □

The principle of inclusion-exclusion can be extended to more than just two sets. In the next section, we will extend it to three sets and give some important examples of its use, including an application to counting combinations with restricted repetitions. Then, in the following section, we will extend the principle of inclusion-exclusion to an arbitrary number of sets and give two more important applications.

□ □ Exercises

1. In a group of 200 students, 80 are taking a math class, 60 are taking a chemistry class, and 30 are taking both classes. How many students are taking either a math class or a chemistry class? How many students are not taking either class?
2. In a group of 250 students, 110 are taking a physics class, 80 are taking a biology class, and 20 are taking both classes. How many students are taking at least one of the classes? How many students are taking neither class?
3. In a group of 95 sheep, 70 are shorn, 25 are pregnant, and 10 are both pregnant and shorn. How many of the sheep are pregnant but not shorn? How many of the sheep are neither pregnant nor shorn?
4. How many integers between 1 and 600 are divisible by either 2 or 3? How many integers between 1 and 600 are divisible by neither 2 nor 3?
5. How many integers between 1 and 3500 are divisible by either 5 or 7? How many integers between 1 and 3500 are divisible by neither 5 nor 7?
6. How many integers between 1401 and 14,000 are divisible by either 2 or 7? How many integers between 1401 and 14,000 are not divisible by 2 and not divisible by 7?
7. How many integers between 1 and 1200 are divisible by either 2 or 6? How many integers between 1 and 1200 are not divisible by 2 and not divisible by 6?
8. How many integers between 1 and 4000 are divisible by 4 but not by 100?
9. How many integers between 1 and 30,000 are divisible by 3 but not by 5?
10. Prove that for any two sets, A and B,

$$|A \cup B| + |A \cap B| = |A| + |B|$$

11. How many permutations are there of the numbers 1, 2, . . ., 9, with the property that the permutation must begin with 123 or end with 89? (For example, the permutation 123859467 begins with 123, and the permutation 542317689 ends with 89.) How many permutations are there of 1, 2, . . ., 9, that do not begin with 123 and do not end with 89?
12. How many permutations are there of the numbers 1, 2, . . ., 10, with the property that either 6 immediately precedes 7 or 7 immediately precedes 6? How many permutations are there of the numbers 1, 2, . . ., 10, with the property that 6 and 7 are not adjacent in the permutation?
13. How many permutations are there of the numbers 1, 2, . . , 7, with the property that either 3 immediately precedes 4 or 5 immediately precedes 6?
14. A year is a leap year if it satisfies one of the following two conditions.
    a) It is divisible by 4, but not by 100
    b) It is divisible by 400

For example, the year 1972 was a leap year since it satisfies condition a, and 1600 was a leap year since it satisfies condition b. On the other hand, the year 600 was not a leap year. How many leap years are there between the years 1000 and 4004?

5.6

The Principle of Inclusion-Exclusion—II

In this section, we want to extend the principle of inclusion-exclusion to three sets and give some important examples of its use. In the next section, we will extend the inclusion-exclusion principle to an arbitrary number of sets.

Suppose that A, B, and C are sets, as pictured in Figure 5.6.1. In order to help with the discussion, we have numbered some of the regions in the picture.

Now, we want to find a formula for the number of elements that are in *at least one of* the sets; that is, we want a formula for the number of elements in the *union* $A \cup B \cup C$.

As before, we begin with the sum $|A| + |B| + |C|$. This number counts all of the elements in the union $A \cup B \cup C$, but it happens to count some of these elements more than once. Let us examine the number of times each element of $A \cup B \cup C$ is counted in $|A| + |B| + |C|$.

Those elements of $A \cup B \cup C$ that are in region 1 are counted exactly once in $|A| + |B| + |C|$, since they are counted only in the term $|A|$. Similarly, the elements in regions 2 and 3 are counted exactly once in $|A| + |B| + |C|$.

On the other hand, those elements in region 4 are counted exactly twice in $|A| + |B| + |C|$, since they are counted in both of the terms $|A|$ and $|B|$, but not in the term $|C|$. By similar reasoning, we see that the elements in regions 5 and 6 are also counted exactly twice in $|A| + |B| + |C|$.

Finally, the elements in region 7 are counted exactly 3 times in $|A| + |B| + |C|$ since they are counted in each of the 3 terms.

Now, we cannot fix all of these overcounts at once, but we can fix them one region at a time. Let us first correct the overcount in region 4. Since each element in this region is counted exactly twice, all we have to do is subtract $|A \cap B|$, giving

$$|A| + |B| + |C| - |A \cap B| \tag{5.6.1}$$

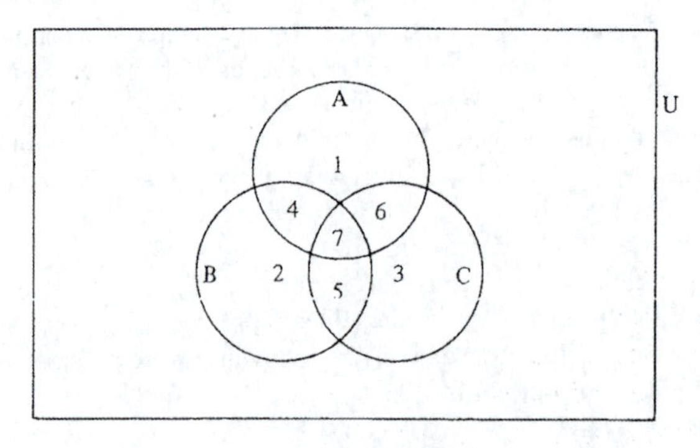

Figure 5.6.1

Now each element in region 4 is counted exactly once in 5.6.1. It is counted with a plus sign in the terms $|A|$ and $|B|$ and with a minus sign in the term $|A \cap B|$, for a net count of one. *Moreover, we have not changed the number of times the elements in regions 1, 2, and 3 are counted.* These elements are still counted exactly once in 5.6.1. (We have changed the number of times the elements in region 7 are counted, but let's not worry about that yet.)

In a similar way, we can correct for the overcount in region 5 by subtracting the term $|A \cap C|$. This will not affect the number of times that the elements in regions 1 through 4 are counted. We can also correct for the overcount in region 6 by subtracting the term $|B \cap C|$, and this does not affect regions 1 through 5. Thus, the expression

$$|A| + |B| + |C| - |A \cap B| - |A \cap C| - |B \cap C| \qquad (5.6.2)$$

counts each element in regions 1 through 6 exactly once. All that is left is to correct the count for the elements in region 7.

So let us see how many times each element of region 7 is counted in 5.6.2. It is counted once with a plus sign for each of the terms $|A|$, $|B|$, and $|C|$ and once with a minus sign for each of the terms $|A \cap B|$, $|A \cap C|$, and $|B \cap C|$, for a net count of 0. In other words, the elements of region 7 are not counted at all in 5.6.2. We can easily fix this, without changing the number of times any other elements are counted, by adding the term $|A \cap B \cap C|$.

Finally then, we arrive at the expression

$$|A| + |B| + |C| - |A \cap B| - |A \cap C| - |B \cap C| + |A \cap B \cap C| \qquad (5.6.3)$$

which counts each of the elements in regions 1 through 7 exactly once. In other words, the expression 5.6.3 equals $|A \cup B \cup C|$. This gives us the following theorem.

Theorem 5.6.1 (The Principle of Inclusion-Exclusion for Three Sets)

Let U be a universal set, and let A, B, and C be subsets of U. Then the number of elements of U that are in *at least one of* the sets A, B, or C is

$$\begin{aligned} |A \cup B \cup C| = {} & |A| + |B| + |C| \\ & - |A \cap B| - |A \cap C| - |B \cap C| \\ & + |A \cap B \cap C| \end{aligned} \qquad (5.6.4)$$

It might help to remember Equation 5.6.4 if we observe that the right side has a definite pattern. First, we add the sizes of all the sets, then we subtract the sizes of the intersections of all pairs of sets, and finally, we add the sizes of the intersections of all triples of sets (there is only one of these).

Example 5.6.1

Suppose that, in a group of 100 students,

a) 50 students are taking math

b) 40 students are taking computer science

c) 35 students are taking chemistry

d) 12 students are taking both math and computer science
e) 10 students are taking both math and chemistry
f) 11 students are taking both computer science and chemistry
g) 5 students are taking all three subjects

How many students are taking *at least one of* the subjects?

Let us take

$$U = \text{the set of all 100 students}$$

$$A = \text{the set of all students in U who are taking math}$$

$$B = \text{the set of all students in U who are taking computer science}$$

$$C = \text{the set of all students in U who are taking chemistry}$$

Then $A \cup B \cup C$ is precisely the set of all students taking at least one of the subjects, and so we want to compute $|A \cup B \cup C|$. Now

$$|A| = 50, \qquad |B| = 40, \qquad |C| = 35$$

and since $A \cap B$ is the set of all students taking both math and computer science,

$$|A \cap B| = 12$$

Similarly, we have

$$|A \cap C| = 10, \qquad |B \cap C| = 11, \qquad |A \cap B \cap C| = 5$$

and so, according to Equation 5.6.4,

$$\begin{aligned} |A \cup B \cup C| &= |A| + |B| + |C| \\ &\quad - |A \cap B| - |A \cap C| - |B \cap C| \\ &\quad + |A \cap B \cap C| \\ &= 50 + 40 + 35 - 12 - 10 - 11 + 5 \\ &= 97 \end{aligned}$$

In other words, exactly 97 students are taking at least one of the subjects. □

Theorem 5.6.1 can be used to determine the number of elements in the universal set U that are in *none of* the sets A, B, and C. First, we write this set in the form $A^c \cap B^c \cap C^c$ and use the fact that

$$A^c \cap B^c \cap C^c = (A \cup B \cup C)^c$$

Then, using Equation 5.6.4, we get

$$\begin{aligned} |A^c \cap B^c \cap C^c| &= |(A \cup B \cup C)^c| \\ &= |U| - |A \cup B \cup C| \\ &= |U| - |A| - |B| - |C| \\ &\quad + |A \cap B| + |A \cap C| + |B \cap C| \\ &\quad - |A \cap B \cap C| \end{aligned}$$

This gives us the following corollary to Theorem 5.6.1.

Corollary 5.6.2

Let U be a universal set, and let A, B, and C be subsets of U. Then the number of elements of U that are in *none of* the sets A, B, or C is

$$
\begin{aligned}
|A^c \cap B^c \cap C^c| = {} & |U| - |A| - |B| - |C| \\
& + |A \cap B| + |A \cap C| + |B \cap C| \\
& - |A \cap B \cap C|
\end{aligned}
\tag{5.6.5}
$$

Example 5.6.2

In a group of 200 people, 100 like Coke, 149 like Pepsi, 83 like Seven-Up, 80 like Coke and Pepsi, 66 like Coke and Seven-Up, 45 like Pepsi and Seven-Up, and 12 like Coke, Pepsi, and Seven-Up. How many of these people like none of the soft drinks?

Let us take

$$
\begin{aligned}
U &= \text{the set of all 200 people} \\
A &= \text{the set of all people in U who like Coke} \\
B &= \text{the set of all people in U who like Pepsi} \\
C &= \text{the set of all people in U who like Seven-Up}
\end{aligned}
$$

Then $A^c \cap B^c \cap C^c$ is the set of all people in U who like none of the soft drinks. In order to compute this number, we use Equation 5.6.5. For this we need

$$|U| = 200$$

$$|A| = 100, \qquad |B| = 149, \qquad |C| = 83$$

$$|A \cap B| = 80, \qquad |A \cap C| = 66, \qquad |B \cap C| = 45$$

$$|A \cap B \cap C| = 12$$

Substituting these values into Equation 5.6.5 gives

$$
\begin{aligned}
|A^c \cap B^c \cap C^c| &= |U| - |A| - |B| - |C| \\
&\quad + |A \cap B| + |A \cap C| + |B \cap C| \\
&\quad - |A \cap B \cap C| \\
&= 200 - 100 - 149 - 83 + \\
&\quad 80 + 66 + 45 - 12 \\
&= 47
\end{aligned}
$$

and so there are exactly 47 people in the group who do not like any of the soft drinks. □

Incidentally, we can see again that the phrases ''at least one of'' and ''none of'' (which is similar to ''neither of'') are clues to the fact that the principle of inclusion-exclusion may apply.

One of the most important applications of the principle of inclusion-exclusion is to counting combinations with restricted repetitions. Let us try some examples.

Example 5.6.3

How many combinations of size 13 are there, taken from the set S = $\{a_1, a_2, a_3\}$, with the property that a_1 can be repeated at most 4 times, a_2 can be repeated at most 5 times, and a_3 can be repeated at most 6 times?

If we let P_1 be the property that a_1 is repeated at least 5 times, P_2 be the property that a_2 is repeated at least 6 times, and P_3 be the property that a_3 is repeated at least 7 times, then this question can be rephrased as follows.

How many combinations of size 13 are there, taken from the set S = $\{a_1, a_2, a_3\}$, that satisfy *none* of the three properties P_1, P_2, or P_3?

This rewording is an indication that we should try using Corollary 5.6.2. So let

U = the set of all combinations of size 13, taken from the set S, with *unrestricted* repetitions

A = the set of all combinations in U that satisfy property P_1

B = the set of all combinations in U that satisfy property P_2

C = the set of all combinations in U that satisfy property P_3

Then $A^c \cap B^c \cap C^c$ is the set of all combinations in U that satisfy none of the properties P_1, P_2, or P_3. In order to compute this number, we use Equation 5.6.5.

According to Theorem 5.2.1,

$$|U| = \binom{3 + 13 - 1}{13} = \binom{15}{13} = 105$$

Now, A is the set of all combinations of size 13, taken from the set S, with the property that a_1 must be repeated at least 5 times. Thus, according to Theorem 5.2.2,

$$|A| = \binom{3 + (13 - 5 - 0 - 0) - 1}{3 - 1} = \binom{10}{2} = 45$$

In a similar way, we get

$$|B| = \binom{9}{2} = 36 \quad \text{and} \quad |C| = \binom{8}{2} = 28$$

The set $A \cap B$ is the set of all combinations of size 13, taken from S, with the property that a_1 must be repeated at least 5 times, *and* a_2 must be repeated at least 6 times. According to Theorem 5.2.2,

$$|A \cap B| = \binom{3 + (13 - 5 - 6 - 0) - 1}{3 - 1} = \binom{4}{2} = 6$$

Similarly, we get

$$|A \cap C| = \binom{3 + (13 - 5 - 0 - 7) - 1}{3 - 1} = \binom{3}{2} = 3$$

and

$$|B \cap C| = \binom{3 + (13 - 0 - 6 - 7) - 1}{3 - 1} = \binom{2}{2} = 1$$

(You should check these equations yourself.)

Finally, $A \cap B \cap C$ is the set of all combinations of size 13, taken from the set S, with the property that a_1 must be repeated at least 5 times, a_2 must be repeated at least 6 times, and a_3 must be repeated at least 7 times. But since $5 + 6 + 7 = 18 > 13$, there are no such combinations. In other words,

$$|A \cap B \cap C| = 0$$

Now we can put all of these pieces together, to get

$$\begin{aligned}|A^c \cap B^c \cap C^c| &= |U| - |A| - |B| - |C| \\ &\quad + |A \cap B| + |A \cap C| + |B \cap C| \\ &\quad - |A \cap B \cap C| \\ &= 105 - 45 - 36 - 28 + 6 + 3 + 1 - 0 \\ &= 6\end{aligned}$$

Thus, there are 6 combinations of the type specified in the problem. ☐

Example 5.6.4

How many combinations of size 25 are there, taken from the set $S = \{a_1, a_2, a_3\}$, with the property that a_1 must be repeated at least 2 times but at most 8 times, a_2 must be repeated at least 3 times but at most 12 times, and a_3 must be repeated at least 7 times but at most 10 times?

The first step in a problem like this is to reduce it to a problem like the one in the last example. This is done simply by observing that each combination of the type described in this problem is formed by writing down the required 2 a_1's, 3 a_2's, and 7 a_3's, and then adding to that any combination of size $25 - 2 - 3 - 7 = 13$, taken from S, with the property that a_1 can be repeated at most $8 - 2 = 6$ times, a_2 can be repeated at most $12 - 3 = 9$ times, and a_3 can be repeated at most $10 - 7 = 3$ times.

Thus, the number of combinations of the type described in this problem is equal to the number of combinations of size 13, taken from the set S, with the property that a_1 can be repeated at most 6 times, a_2 can be repeated at most 9 times, and a_3 can be repeated at most 3 times.

Now, this number can be determined exactly as we did in the last example, by using the principle of inclusion-exclusion.

Let P_1 be the property that a_1 is repeated at least 7 times, and let P_2 be the property that a_2 is repeated at least 10 times, and let P_3 be the property that a_3 is repeated at least 4 times. Also, let

U = the set of all combinations of size 13, taken from the set S, with unrestricted repetitions

A = the set of all combinations in U that satisfy property P_1

B = the set of all combinations in U that satisfy property P_2

C = the set of all combinations in U that satisfy property P_3

Then we want to find $|A^c \cap B^c \cap C^c|$, which we can compute using Equation 5.6.5. We will leave it as an exercise to show that

$$|U| = 105$$

$$|A| = 28 \quad , \quad |B| = 10 \quad , \quad |C| = 55$$

$$|A \cap B| = 0 \quad , \quad |A \cap C| = 6 \quad , \quad |B \cap C| = 0$$

$$|A \cap B \cap C| = 0$$

Hence, according to Equation 5.6.5

$$\begin{aligned} |A^c \cap B^c \cap C^c| &= |U| - |A| - |B| - |C| \\ &\quad + |A \cap B| + |A \cap C| + |B \cap C| \\ &\quad - |A \cap B \cap C| \\ &= 105 - 28 - 10 - 55 + 0 + 6 + 0 - 0 \\ &= 18 \end{aligned}$$

and so there are 18 combinations of the type specified in this problem. □

We can also use the principle of inclusion-exclusion to help us count the number of solutions, in integers, to linear equations with unit coefficients.

Example 5.6.5

How many solutions are there, in integers, to the equation

$$x + y + z = 25 \tag{5.6.6}$$

with the property that

$$2 \le x \le 8, \qquad 3 \le y \le 12, \qquad 7 \le z \le 10 \tag{5.6.7}$$

The proof of Theorem 5.3.1 tells us that the number of solutions to Equation 5.6.6 that satisfy condition 5.6.7 is the same as the number of combinations of size 25, taken from the set $S = \{x, y, z\}$, with the property that x is repeated at least 2 times and at most 8 times, y is repeated at least 3 times and at most 12 times, and z is repeated at least 7 times and at most 10 times.

But, in the last example, we determined that this number is 18, and so there are 18 solutions to Equation 5.6.6 that satisfy condition 5.6.7. □

Exercises

1. In a group of 600 students, 250 like mathematics, 300 like physics, 280 like computer science, 125 like both mathematics and physics, 50 like both mathematics and computer science, 150 like both physics and computer science, and 30 like all three subjects. How many of the students like at least one of the subjects? How many like mathematics but not computer science?
2. Fill in the missing details in Example 5.6.4.
3. How many integers between 1 and 3000 are divisible by either 2, 3, or 5?
4. How many integers between 1 and 1050 are divisible by either 3, 5, or 7?
5. How many combinations of size 7 are there, taken from the set $\{x, y, z\}$, with the property that x cannot be repeated more than once, y cannot be repeated more than 4 times, and z cannot be repeated more than 5 times?

6. How many combinations of size 20 are there, taken from the set $\{a_1, a_2, a_3\}$, with the property that a_1 cannot be repeated more than 5 times, a_2 cannot be repeated more than 5 times, and a_3 can be repeated any number of times?
7. How many combinations of size 10 are there, taken from the set $\{a, b, c\}$, with the property that a cannot be repeated more than 3 times, b cannot be repeated more than 2 times, and c cannot be repeated more than 4 times?
8. How many combinations of size 16 are there, taken from the set $S = \{a_1, a_2, a_3\}$, with the property that a_1 must be repeated at least 1 time and at most 4 times, a_2 must be repeated at least 3 times and at most 7 times, and a_3 must be repeated at least 2 times and at most 6 times?
9. How many combinations of size 22 are there, taken from the set $S = \{1, 2, 3\}$, with the property that 1 must be repeated at least 5 times and at most 10 times, 2 must be repeated at least 3 times and at most 20 times, and 3 can be repeated at most 5 times?
10. How many solutions are there, in integers, to the equation

$$x + y + z = 12$$

with the property that $x > 3$, $y > 5$, and $z > 2$?
11. How many solutions are there, in integers, to the equation

$$x_1 + x_2 + x_3 = 30$$

with the property that $4 \le x_1 < 10$, $7 \le x_2 < 15$, and $10 \le x_3 < 25$?
12. How many combinations of size k are there, taken from the set $\{a_1, a_2, a_3\}$, with the property that a_i cannot be repeated more than r_i times, for $i = 1, 2, 3$?
13. How many ways are there to place three balls, labeled with the integers 1, 2, and 3, into three boxes, labeled 1, 2, and 3, in such a way that none of the balls is put into a box with the same label?
14. How many permutations are there of the numbers 1, 2, . . ., 9, with the property that the permutation either begins with 123, or has a 5 in its fifth place, or ends in 89? (For example, the permutation 123849576 begins with 123, the permutation 478352169 has a 5 in its fifth place, and the permutation 465231789 ends in 89.)
15. How many permutations of size n with unrestricted repetitions are there, taken from the set $\{a, b, c, d\}$, with the property that a, b, and c must appear in the permutation?
16. How many ways are there to distribute k distinguishable balls into n distinguishable boxes, in such a way that the first 3 boxes contain at least 1 ball?
17. Consider the following picture,

S_1	S_2	S_3	S_4

How many ways are there to color each of the squares S_1, S_2, S_3, and S_4 with n different colors, provided that no two adjacent squares can receive the same color? *Hint*, Let U be the set of all possible ways to color the four squares. Let A be the set of all ways to color all four squares in such a way that squares S_1 and S_2 receive the same color. Define B and C similarly.

5.7
The Principle of Inclusion-Exclusion—III

Our goal in this section is to extend the principle of inclusion-exclusion to an arbitrary number of sets.

Let us begin by writing Equations 4.5.3 and 4.6.4 next to each other.

$$|A \cup B| = |A| + |B| - |A \cap B| \tag{5.7.1}$$

$$\begin{aligned} |A \cup B \cup C| = {} & |A| + |B| + |C| \\ & - |A \cap B| - |A \cap C| - |B \cap C| \\ & + |A \cap B \cap C| \end{aligned} \tag{5.7.2}$$

The pattern in these two formulas is rather clear. In Equation 5.7.1, we add the sizes of all of the sets and then subtract the sizes of the intersections of all pairs of the sets. (In this case, there is only one pair.)

In Equation 5.7.2, we add the sizes of all of the sets, then subtract the sizes of the intersections of all pairs of the sets, and finally add the sizes of the intersections of all triples of the sets. (In this case, there is only one triple.)

From these descriptions, you can probably guess what the principle of inclusion-exclusion is in the case of four sets. It is

$$\begin{aligned} |A \cup B \cup C \cup D| = {} & |A| + |B| + |C| + |D| \\ & - |A \cap B| - |A \cap C| - |A \cap D| \\ & - |B \cap C| - |B \cap D| - |C \cap D| \\ & + |A \cap B \cap C| + |A \cap B \cap D| \\ & + |A \cap C \cap D| + |B \cap C \cap D| \\ & - |A \cap B \cap C \cap D| \end{aligned} \tag{5.7.3}$$

Of course, this is only a guess right now, but we will soon prove that our guess is correct. In any case, it does fit the same pattern as Equations 5.7.1 and 5.7.2. Namely, we add the sizes of all the sets, then subtract the sizes of the intersections of all pairs of sets, then add the sizes of the intersections of all triples of sets, and finally subtract the sizes of the intersections of all quadruples of sets.

We would like to generalize this pattern to an arbitrary number of sets. But, as you can see from the last equation, we are going to run into some trouble with length. Just imagine trying to write out the corresponding equation in the case of 10 sets! What we need at this point is some notation that will help keep our formulas under control.

So let $A_1, A_2, \ldots, A_n$ be sets. Then we denote by S_k the sum of the sizes of the intersections of all collections of k of these sets.

For example, in the case of two sets, we have

$$S_1 = |A| + |B|$$

and

$$S_2 = |A \cap B|$$

and with this notation, Equation 5.7.1 can be written

$$|A \cup B| = S_1 - S_2$$

In the case of three sets, we have

$$S_1 = |A| + |B| + |C|$$
$$S_2 = |A \cap B| + |A \cap C| + |B \cap C|$$
$$S_3 = |A \cap B \cap C|$$

and with this notation Equation 5.7.2 can be written

$$|A \cup B \cup C| = S_1 - S_2 + S_3$$

Finally, in the case of four sets, we have

$$S_1 = |A| + |B| + |C| + |D|$$
$$S_2 = |A \cap B| + |A \cap C| + |A \cap D| + |B \cap C| + |B \cap D| + |C \cap D|$$
$$S_3 = |A \cap B \cap C| + |A \cap B \cap D| + |A \cap C \cap D| + |B \cap C \cap D|$$
$$S_4 = |A \cap B \cap C \cap D|$$

and with this notation Equation 5.7.3 can be written

$$|A \cup B \cup C \cup D| = S_1 - S_2 + S_3 - S_4$$

The last example is where the notation really begins to pay off. However, we should point out that notation such as this is a bit deceptive. It certainly makes Equation 5.7.3 look a lot simpler, but we must keep in mind that the complicated nature of Equation 5.7.3 has not disappeared, it has simply been hidden in the notation.

As a matter of fact, some of the sums S_k themselves can get quite long. Nevertheless, it is always easy to tell exactly how many terms there are in any of these sums. For if there are n sets involved, then the sum S_k has exactly $\binom{n}{k}$ terms. (Why?)

With this notation at hand, we can state the principle of inclusion-exclusion for an arbitrary (finite) number of sets.

Theorem 5.7.1 (The Principle of Inclusion-Exclusion)

Let U be a universal set, and let $A_1, A_2, \ldots, A_n$ be subsets of U. Then the number of elements in U that are in *at least one of* the sets $A_1, A_2, \ldots, A_n$ is

$$|A_1 \cup A_2 \cup \cdots \cup A_n| = S_1 - S_2 + S_3 - \cdots + (-1)^{n+1}S_n \qquad (5.7.4)$$

where S_k denotes the sum of the sizes of the intersections of all collections of k of the sets $A_1, A_2, \ldots, A_n$. (Since the signs alternate on the right side of this formula, the last term should be positive when n is odd and negative when n is even. This is the reason that the last term has a factor $(-1)^{n+1}$, which is indeed positive when n is odd and negative when n is even.)

PROOF We will prove this theorem by showing that both sides of Equation 5.7.4 count the same thing. Of course, the left side counts each element of the union

$A_1 \cup A_2 \cup \cdots \cup A_n$ exactly once, and it counts each element of U that is not in this union exactly 0 times. So, we must show that the right side of Equation 5.7.2 also counts this number.

Now, if an element of U is not in the union $A_1 \cup A_2 \cup \cdots \cup A_n$, then it is not in any of the sets $A_1, A_2, \ldots, A_n$, and so it is not counted in any of the sums $S_1, S_2, \ldots, S_n$. Hence, it is counted exactly 0 times on the right side of Equation 5.7.2.

On the other hand, we must show that each element of the union $A_1 \cup A_2 \cup \cdots \cup A_n$ is counted exactly once on the right side of Equation 5.7.2.

So let x be in the union $A_1 \cup A_2 \cup \cdots \cup A_n$. Then, of course, x is in some of the sets $A_1, A_2, \ldots, A_n$. Let us assume that it is in exactly m of these sets. We want to determine how many times x is counted on the right side of Equation 5.7.4.

To make the argument easier to follow, let us consider the case where x is in the first m sets $A_1, A_2, \ldots, A_m$ and not in any of the other sets $A_{m+1}, \ldots, A_n$. As you will see, the same argument will work regardless of which m sets contain x, but this case is easier to describe.

Now, x will be counted exactly $m = \binom{m}{1}$ times in the sum S_1, because it is counted once in each of the terms $|A_1|, |A_2|, \ldots, |A_m|$, but it is not counted in any of the other terms $|A_{m+1}|, \ldots, |A_n|$. Also, it will be counted exactly $\binom{m}{2}$ times in the sum S_2, because it is counted once in each of the terms $|A_i \cap A_j|$ where i and j are both between 1 and m, but it is not counted in any of the terms $|A_i \cap A_j|$ if either of i or j is not between 1 and m. Similarly, x is counted exactly $\binom{m}{3}$ times in the sum S_3, because it is counted once in each term of the form $|A_i \cap A_j \cap A_k|$ where i, j, and k are all between 1 and m, but it is not counted in any of the terms $|A_i \cap A_j \cap A_k|$ if even one of i, j, or k is not between 1 and m.
We can continue in this way until we reach the sum S_m. The element x is counted exactly $\binom{m}{m}$ (= 1) times in the sum S_m, since it is counted only in the term $|A_1 \cap A_2 \cap \cdots \cap A_m|$.

Finally, x is not counted at all in the sums S_i if $i > m$, because it is not in any intersection of more than m of the sets.

Therefore, the total number of times that x is counted on the right side of Equation 5.7.4 is

$$\binom{m}{1} - \binom{m}{2} + \binom{m}{3} - \cdots + (-1)^{m+1}\binom{m}{m} \tag{5.7.5}$$

All we have to do now is evaluate this sum. But we have already done this in Example 4.7.4! According to that example (with n replaced by m) we have

$$\binom{m}{0} - \binom{m}{1} + \binom{m}{2} - \binom{m}{3} + \cdots + (-1)^{m}\binom{m}{m} = 0$$

Moving all but the first term in this sum to the other side of the equal sign gives

$$\binom{m}{0} = \binom{m}{1} - \binom{m}{2} + \binom{m}{3} - \cdots + (-1)^{m+1}\binom{m}{m}$$

But $\binom{m}{0} = 1$, and so the sum in 5.7.5 is equal to 1. Thus x is indeed counted exactly once on the right side of Equation 5.7.4.

This shows that all of the elements in the union $A_1 \cup A_2 \cup \cdots \cup A_n$ are counted exactly once on the right side of Equation 5.7.4 and completes the proof. ∎

As we did in the previous two sections, we can state an important corollary to the principle of inclusion-exclusion. (In fact, some authors even call this corollary the principle of inclusion-exclusion.)

Corollary 5.7.2

Let U be a universal set, and let $A_1, A_2, \ldots, A_n$ be subsets of U. Then the number of elements in U that are in *none of* the sets $A_1, A_2, \ldots, A_n$ is

$$|A_1^c \cap A_2^c \cap \cdots \cap A_n^c| = |U| - S_1 + S_2 - S_3 + \cdots + (-1)^n S_n \qquad (5.7.6)$$

We will leave the proof of this corollary as an exercise. Instead, let us turn now to some applications.

Our first application is to ordinary permutations. Let us agree to say that two permutations of the integers $1, 2, \ldots, n$ are **incompatible** if they do not agree in any of their positions. For example, the permutations 3142 and 4231 are incompatible, but the permutations 3142 and 4312 are not incompatible, since they agree in their fourth positions.

A permutation of the integers $1, 2, \ldots, n$ is called a **derangement** of size n if it is incompatible with the permutation $123 \cdots n$. (In a sense, such permutations completely "derange" the integers $1, 2, \ldots, n$.) Derangements occur in many different situations. For instance, the problem of Example 4.1.6 involves derangements of size 12. (See Exercise 6.)

Corollary 5.7.2 can be used to find a beautiful formula for the number of derangements of size n. We begin by letting

U = the set of all permutations of the integers $1, 2, \ldots, n$

A_1 = the set of all permutations in U with a 1 in the first position

A_2 = the set of all permutations in U with a 2 in the second position

.

.

.

A_n = the set of all permutations in U with an n in the n-th position.

Then $A_1^c \cap A_2^c \cap \cdots \cap A_n^c$ is the set of all derangements of size n. (Why?)

If we expect to use Corollary 5.7.2, then we must first determine the value of each of the sums $S_1, S_2, \ldots, S_n$. Fortunately, it turns out that each term in a given sum S_k has exactly the same value. For instance, the terms in the sum S_1 are $|A_1|$, $|A_2|, \ldots, |A_n|$, and each of these is equal to $(n - 1)!$. (Why?)

In general, for any $1 \le k \le n$, each term in the sum S_k has the form

$$|A_{i_1} \cap A_{i_2} \cap \cdots \cap A_{i_k}|$$

where $i_1, i_2, \ldots, i_k$ are distinct integers between 1 and n. But

$$A_{i_1} \cap A_{i_2} \cap \cdots \cap A_{i_k}$$

is just the set of all permutations of $1, 2, \ldots, n$ in which the k positions $i_1, i_2, \ldots, i_k$ are "fixed."

For example, $A_1 \cap A_2 \cap \cdots \cap A_k$ is the set of all permutations in which the first k positions are fixed to be $123 \cdots k$. But the number of such permutations is $(n - k)!$, since each permutation of this type corresponds to a permutation of the last $n - k$ integers. Thus,

$$|A_1 \cap A_2 \cap \cdots \cap A_k| = (n - k)!$$

But the same reasoning applies to any of the sets

$$A_{i_1} \cap A_{i_2} \cap \cdots \cap A_{i_k}$$

To put it simply, it does not matter which of the positions are fixed; it matters only how many of them are fixed. In other words,

$$|A_{i_1} \cap A_{i_2} \cap \cdots \cap A_{i_k}| = (n - k)!$$

for all distinct integers $i_1, i_2, \ldots, i_k$ between 1 and n.

Thus, since there are $\binom{n}{k}$ terms in the sum S_k, each of which equals $(n - k)!$, we see that

$$S_k = \binom{n}{k}(n - k)! = \frac{n!}{k!}$$

When we substitute this into Equation 5.7.6, we obtain the following formula for the number of derangements of size n,

$$\begin{aligned}|A_1^c \cap A_2^c \cap \cdots \cap A_n^c| &= |U| - S_1 + S_2 - S_3 + \cdots + (-1)^n S_n \\ &= n! - \frac{n!}{1!} + \frac{n!}{2!} - \frac{n!}{3!} + \cdots + (-1)^n \frac{n!}{n!} \\ &= n!\left[\frac{1}{0!} - \frac{1}{1!} + \frac{1}{2!} - \frac{1}{3!} + \cdots + (-1)^n \frac{1}{n!}\right]\end{aligned}$$

(We have written $1/0!$ instead of 1 in order to preserve the pattern.) Let us put this beautiful formula into a theorem. It is customary to denote the number of derangements of size n by D_n.

Theorem 5.7.3

The number D_n of derangements of size n is

$$D_n = n!\left[\frac{1}{0!} - \frac{1}{1!} + \frac{1}{2!} - \frac{1}{3!} + \cdots + (-1)^n \frac{1}{n!}\right]$$

Example 5.7.1

According to Theorem 5.7.3, the number of derangements of size 4 is

$$D_4 = 4!\left[\frac{1}{0!} - \frac{1}{1!} + \frac{1}{2!} - \frac{1}{3!} + \frac{1}{4!}\right]$$
$$= 9$$

Let us make a list of them.

2143	3142	4123
2341	3412	4312
2413	3421	4321

Because we know that there are 9 derangements of size 4, we know that this list is complete.

The number of derangements of size 6 is

$$D_6 = 6!\left[\frac{1}{0!} - \frac{1}{1!} + \frac{1}{2!} - \frac{1}{3!} + \frac{1}{4!} - \frac{1}{5!} + \frac{1}{6!}\right]$$
$$= 265$$

This compares with a total of $6! = 720$ permutations of the integers $1, 2, \ldots, 6$. □

Example 5.7.2

Six married couples come to a dance. In how many ways can the 6 men dance with the 6 women if no husband will dance with his wife?

This is a problem in derangements. Assume that the wives are standing in a line, with their husbands facing them. Number the husbands, from left to right, with the integers 1 through 6. Then, if we let each permutation of the integers $1, 2, \ldots, 6$ correspond to a rearrangement of the men, it is exactly the derangements that correspond to rearrangements in which no man is paired with his wife.

Hence, there are $D_6 = 265$ ways for the men to dance with the women in such a way that no man dances with his wife. □

As a final application of the principle of inclusion-exclusion, we consider a special case of permutations with restricted repetitions.

In particular, we want to find a formula for the number of permutations of size k, taken from the set $S = \{a_1, a_2, \ldots, a_n\}$, with the property that each element of S appears at least once in the permutation.

We begin by letting

$$\begin{aligned}
U &= \text{the set of all permutations of size } k \text{ with } \textit{unrestricted} \text{ repetitions, taken from S} \\
A_1 &= \text{the set of all permutations in U with the property that } a_1 \text{ does } \textit{not} \text{ appear} \\
A_2 &= \text{the set of all permutations in U with the property that } a_2 \text{ does } \textit{not} \text{ appear} \\
&\vdots \\
A_n &= \text{the set of all permutations in U with the property that } a_n \text{ does } \textit{not} \text{ appear}
\end{aligned}$$

Then $A_1^c \cap A_2^c \cap \cdots \cap A_n^c$ is the set of all permutations of S with the property that every element of S appears at least once in each permutation, and so again we want to use Equation 5.7.6.

Now, A_1 is the set of all permutations of size k with unrestricted repetitions, taken from the set S, with the additional property that a_1 does not appear. But this is the same as the set of all permutations of size k with unrestricted repetitions, taken from the set $\{a_2, a_3, \ldots, a_n\}$. Since there are $(n - 1)^k$ of these permutations, we see that $|A_1| = (n - 1)^k$.

By similar reasoning, we have

$$|A_1| = |A_2| = \cdots = |A_n| = (n - 1)^k$$

and so

$$S_1 = n(n - 1)^k$$

As for the other sums $S_2, S_3, \ldots, S_n$, we are again fortunate that each term in a given sum S_m has the same value. A typical term in the sum S_m has the form

$$|A_{i_1} \cap A_{i_2} \cap \cdots \cap A_{i_m}|$$

where $i_1, i_2, \ldots, i_m$ are distinct integers between 1 and n. But

$$A_{i_1} \cap A_{i_2} \cap \cdots \cap A_{i_m}$$

is just the set of all permutations of size k, taken from the set S, with the property that the m elements

$$a_{i_1}, a_{i_2}, \ldots, a_{i_m}$$

do not appear. It is easy to see that there are $(n - m)^k$ such permutations, and so each term in the sum S_m is equal to $(n - m)^k$.

Thus,

$$S_m = \binom{n}{m}(n - m)^k$$

for all $m = 1, 2, \ldots, n$. (This agrees with what we got for S_1.)

Now we can substitute these values of S_m into Equation 5.7.4, along with $|U| = n^k$, to get

$$\begin{aligned}|A_1^c \cap A_2^c \cap \cdots \cap A_n^c| &= |U| - S_1 + S_2 - S_3 + \cdots + (-1)^n S_n \\ &= \binom{n}{0}(n - 0)^k - \binom{n}{1}(n - 1)^k + \\ &\quad \binom{n}{2}(n - 2)^k - \cdots + (-1)^n\binom{n}{n-1}(1)^k\end{aligned}$$

Let us record this formula in a theorem.

Theorem 5.7.4

The number of permutations of size k, taken from the set $S = \{a_1, a_2, \ldots, a_n\}$, with the property that each element of S appears at least once in each permutation, is

$$\binom{n}{0}(n - 0)^k - \binom{n}{1}(n - 1)^k + \binom{n}{2}(n - 2)^k - \cdots + (-1)^n\binom{n}{n-1}(1)^k$$

Incidentally, Theorem 5.7.4 can also be worded in terms of distributing balls into boxes. In particular, the number of permutations of size k, taken from the set $S = \{a_1, a_2, \ldots, a_n\}$, with the property that each element of S appears at least once in each permutation is the same as the number of ways to distribute k distinguishable balls into n distinguishable boxes, without exclusion, in such a way that no box is empty. Thus, Theorem 5.7.4 is just a different version of part 2 of Theorem 5.4.5!

Example 5.7.3

The number of permutations of size 5, taken from the set $S = \{a_1, a_2, a_3\}$, with the property that each element of S appears at least once in each permutation, is

$$\begin{aligned}\binom{3}{0}(3 - 0)^5 - \binom{3}{1}(3 - 1)^5 + \binom{3}{2}(3 - 2)^5 &= 3^5 - 3 \cdot 2^5 + 3 \cdot 1^5 \\ &= 150\end{aligned}$$

(This compares with $3^5 = 243$ permutations of size 5 with unrestricted repetitions, taken from the set S.) □

Exercises

1. How many derangements are there of size 3? List them.
2. How many derangements are there of size 5? List them.

3. How many permutations of size 4 are there, taken from the set $S = \{1, 2, 3\}$, with the property that each element of S appears at least once in each permutation?
4. How many ways are there to put 6 balls into 4 boxes in such a way that no box is empty?
5. List all permutations that are incompatible with the permutation 3412.
6. Solve the counting problem in Example 4.1.6.
7. How many integers are there between 1 and 1,000,000 (inclusive) that are neither perfect squares, perfect cubes, nor perfect fourth powers?
8. Prove Corollary 5.7.2.
9. How many permutations are there that are incompatible with the permutation 52413?
10. How many permutations are there that are incompatible with any given permutation of the integers $1, 2, \ldots, n$?
11. How many solutions, in integers, are there to the equation

$$x + y + z + w = 20$$

with the property that $x \leq 7$, $y \leq 10$, $z \leq 6$, $w \leq 8$?
12. How many solutions are there, in integers, to the equation

$$x_1 + x_2 + x_3 + x_4 = 18$$

with the property that $1 \leq x_i \leq 7$ for all $i = 1, 2, 3, 4$?
13. How many solutions are there, in integers, to the equation

$$x_1 + x_2 + x_3 + x_4 = 0$$

with the property that $-2 \leq x_i \leq 2$ for all $i = 1, 2, 3, 4$?
14. How many solutions are there, in integers, to the equation

$$x_1 + x_2 + \cdots + x_6 = 28$$

with the property that $x_i < 10$ for all $i = 1, 2, \ldots, 6$?
15. How many solutions are there, in integers, to the equation

$$x_1 + x_2 + \cdots + x_{10} = 50$$

with the property that $x_1 > 9$, $x_2 > 20$, $x_3 > 10$, and $x_i > 0$ for all $i = 4, 5, \ldots, 10$?
16. A bag contains 3 pennies, 5 nickels, 2 dimes, and 6 quarters. How many ways can you choose a collection of 5 coins? (Coins of the same denomination are considered to be indistinguishable.)
17. Seven envelopes are opened, and the letters are removed. How many ways can the letters be replaced so that
    a) no letter is put into its original envelope?
    b) exactly one letter is put into its original envelope?
18. How many permutations are there of the integers $1, 2, \ldots, 6$ with the property that 2 is not in the third position, 3 is not in the first position, and 6 is not in the fourth position?
19. How many permutations are there of the integers $1, 2, \ldots, 9$ that agree with the permutation $123 \cdots 9$ in exactly 3 positions?
20. How many permutations are there of the integers $1, 2, \ldots, 9$ with the property that the first 4 positions are occupied by the integers 6, 7, 8, and 9, in any order?
21. Prove that $D_n - nD_{n-1} = (-1)^n$ for $n \geq 2$.
22. (For those readers who are familiar with the concept of a limit.) What is the limit of $D_n/n!$ as n approaches infinity?

5.8
An Introduction to Probability

Probability Theory is a branch of mathematics that deals with methods for describing the likelihood of the outcomes of future experiments, based on certain assumptions about those experiments.

As a simple example, if a coin is perfectly balanced, then we are willing to make the assumption that when the coin is tossed in the air, it is equally likely to land heads up as tails up. In a situation such as this, we would say that the *probability* that the coin will land with heads up (or tails up) is ½.

This type of statement is typical of probability theory. Of course, we can never hope to predict the outcome of a particular experiment with absolute certainty, but we can still obtain very useful information.

Probability theory was first developed, in the late seventeenth and early eighteenth centuries, in an attempt to describe the likelihood of the outcomes of various games of chance, and many of our examples will involve tossing coins, rolling dice, drawing cards, and so on.

Let us begin with a few simple definitions.

Definition

The set of all possible outcomes of an experiment is called the **sample space** of the experiment. Any subset of the sample space, that is, any set of outcomes, is called an **event**. □

In this brief introduction, we will deal exclusively with finite sample spaces, that is, sample spaces with only a finite number of elements.

Example 5.8.1

a) Consider the experiment of tossing a coin in the air and observing which side lands face up. The sample space for this experiment is the set S = {H,T}, where H represents heads and T represents tails.

b) Consider the experiment of rolling a pair of dice. The sample space for this experiment is the set S of all ordered pairs of the form {x,y}, where x is the value on the first die, and y is the value on the second die. Thus,

$$S = \{(1,1), (1,2), (1,3), \ldots, (6,4), (6,5), (6,6)\}$$

Note that the size of S is $6^2 = 36$. The event of getting a sum equal to 7 is the set

$$E = \{(1,6), (2,5), (3,4), (4,3), (5,2), (6,1)\}$$ □

Once the sample space for a given experiment has been determined, the next step is to assign probabilities to each of the possible outcomes. If the sample space has the form

$$S = \{a_1, a_2, \ldots, a_n\}$$

then we denote these probabilities by $p_1, p_2, \ldots, p_n$. That is,

$$\text{Probability of outcome } a_i = P(a_i) = p_i$$

The exact method of assigning probabilities to the outcomes of an experiment depends on the assumptions made about the experiment. For instance, the assumption that a coin is fair is equivalent to the assignments

$$P(H) = \tfrac{1}{2} \quad \text{and} \quad P(T) = \tfrac{1}{2}$$

However the probabilities are assigned, they must satisfy certain simple criteria. In particular, they must be numbers between 0 and 1,

$$0 \le p_i \le 1 \text{ for each } i$$

and their sum must equal 1,

$$\sum_{i=1}^{n} p_i = 1$$

Once probabilities have been assigned to each outcome in a sample space, we can assign a probability to any event.

Definition

Let E be a nonempty event in a sample space S. Then the **probability** of E, denoted by P(E), is the sum of the probabilities of each outcome in the event. We also set $P(\emptyset) = 0$. □

Example 5.8.2

Consider the event of rolling two fair dice. The sample space for this event consists of the 36 ordered pairs described in Example 5.8.1(b). Since we are assuming that the dice are fair, the probabilities of each outcome must be equal, that is $p_i = 1/36$ for all $i = 1, \ldots, 36$. Thus, using the description of E in Example 5.8.1(b), we have

$$\begin{aligned} P(\text{getting a sum of } 7) &= P(E) \\ &= P((1,6)) + P((2,5)) + P((3,4)) + P((4,3)) + P((5,2)) + P((6,1)) \\ &= \frac{1}{36} + \frac{1}{36} + \frac{1}{36} + \frac{1}{36} + \frac{1}{36} + \frac{1}{36} = \frac{1}{6} \end{aligned}$$

□

It is not uncommon for experiments to have the property that each outcome is equally likely. In this case, if the sample space has size n, that is, if $|S| = n$, then the probability of each outcome is $1/n = 1/|S|$. Furthermore, we have the following theorem, whose proof we leave as an exercise.

Theorem 5.8.1

Let S be a finite sample space whose outcomes are equally likely. If E is a nonempty event in S, then

$$P(E) = \frac{|E|}{|S|}$$

We also leave the proof of the following theorem as an exercise.

Theorem 5.8.2

Let S be a finite sample space. Then

1) $P(\emptyset) = 0$
2) $P(S) = 1$
3) $0 \leq P(E) \leq 1$ for all events E
4) $P(E^c) = 1 - P(E)$ for all events E
5) If E and F are events and

$$\text{if } E \cap F = \emptyset \text{ then } P(E \cup F) = P(E) + P(F)$$

6) If $E_1, E_2, \ldots, E_k$ are events and

if $E_1 \cap E_2 \cap \cdots \cap E_k = \emptyset$ then

$$P(E_1 \cup E_2 \cup \cdots \cup E_k) = P(E_1) + P(E_2) + \cdots + P(E_k)$$

(*Hint* for proof: use induction and part 4 to prove part 5.)

It is very important to keep in mind that parts 5 and 6 of the previous theorem hold only when the events are disjoint, that is, only when $E \cap F = \emptyset$ (or $E_1 \cap E_2 \cap \cdots \cap E_k = \emptyset$). In probability (which tends to have its own special vocabulary) disjoint events are said to be **mutually exclusive**.

Example 5.8.3

Five cards are drawn at random (that is, each with equal probability) from a deck of 52 cards. What is the probability of getting exactly 3 aces?

In this case, the sample space is the set of all possible 5-card hands, that is, the set of all combinations of size 5, taken from the set of 52 cards. This space has size

$$|S| = \binom{52}{5}$$

Now we must compute the size of the event E of getting exactly 3 aces. Such a hand can be formed by first choosing 3 of the 4 aces, and this can be done in $\binom{4}{3}$ ways, and then choosing 2 of the remaining 48 (non-ace) cards, which can be done in $\binom{48}{2}$ ways. Thus,

$$|E| = \binom{4}{3}\binom{48}{2}$$

and

$$P(E) = \frac{|E|}{|S|} = \frac{\binom{4}{3}\binom{48}{2}}{\binom{52}{5}} = \frac{4 \cdot 1128}{2598960} \approx 0.0017$$

As you can see, this is a very small probability. For those of you who are familiar with the game of poker, you can see why being dealt 3 aces is a very rare occurrence. □

Example 5.8.4

Five digits $a_1, a_2, \ldots, a_5$ are chosen at random. What is the probability that a_5 is the same as one of the previous 4 digits?

The sample space consists of all permutations of size 5, taken from the set $\{0,1,2,\ldots,9\}$, and so $|S| = 10^5$. Let E be the event that a_5 is the same as one of a_1, a_2, a_3 or a_4. Determining the size of E is a bit awkward. In this case, it turns out to be easier to first determine the size of E^c, which is the event that a_5 is different from a_1, a_2, a_3 and a_4. To determine the size of E^c, we observe that there are 10 possibilities for a_5, but then a_1, a_2, a_3 and a_4 must be taken from the 9 remaining digits. Thus $|E^c| = 9^4 \cdot 10$, and

$$P(E^c) = \frac{|E^c|}{|S|} = \frac{9^4 \cdot 10}{10^5} = 0.6561$$

Finally, we may use part 4 of the previous theorem to get

$$P(E) = 1 - P(E^c) = 0.3439$$

□

The Principle of Inclusion-Exclusion

As you know, if E and F are subsets of a set S, then the Principle of Inclusion-Exclusion says that

$$|E \cup F| = |E| + |F| - |E \cap F|$$

and

$$|E^c \cap F^c| = |S| - |E| + |F| + |E \cap F|$$

If we divide both of these equations by $|S|$, and express the result in terms of probabilities of the events E and F, we get

$$P(E \cup F) = P(E) + P(F) - P(E \cap F)$$

and

$$P(E^c \cap F^c) = 1 - P(E) - P(F) + P(E \cap F) \qquad (5.8.1)$$

These equations can be thought of as the probabalistic versions of the Principle of Inclusion-Exclusion. (Of course, these can be extended to more than 2 sets.)

Example 5.8.5

Five cards are chosen at random from a deck of 52 cards. What is the probability that there is at least one spade and at least one club?

Let E be the probability that no spade has been drawn, and let F be the probability that no club has been drawn. We seek $P(E^c \cap F^c)$. In order to use 5.8.1, we must compute

$$P(E) = P(\text{no spade has been drawn})$$

$$= \frac{|\{\text{5-card hands taken from the 39 non-spades}\}|}{|\{\text{all 5-card hands}\}|} = \frac{\binom{39}{5}}{\binom{52}{5}}$$

and similarly,

$$P(F) = \frac{\binom{39}{5}}{\binom{52}{5}} \quad \text{and} \quad P(E \cap F) = \frac{\binom{26}{5}}{\binom{52}{5}}$$

Thus, according to 5.8.1,

$$P(E^c \cap F^c) = 1 - \frac{\binom{39}{5}}{\binom{52}{5}} - \frac{\binom{39}{5}}{\binom{52}{5}} + \frac{\binom{26}{5}}{\binom{52}{5}} = 0.582$$ □

Independent Events

We say that two events E and F are **independent** if

$$P(E \cap F) = P(E) \cdot P(F)$$

Intuitively speaking, two events are independent if knowing whether one event occurs gives us no information about whether or not the other event will occur.

Example 5.8.6

Suppose we toss a fair coin twice. Let E be the event that the first toss results in heads, and let F be the event that the second toss results in heads. Then $P(E) = P(F) = \frac{1}{2}$ and since $E \cap F$ is the event that both tosses result in heads, we have $P(E \cap F) = \frac{1}{4}$. Thus $P(E \cap F) = P(E) \cdot P(F)$ and so the events E and F are independent. □

Example 5.8.7

A card is chosen at random from a deck of 52 cards. Let E be the event that the card is an ace or a deuce, and let F be the event that the card is an ace, king, queen or jack. Then $P(E) = 8/52 = 2/13$ and $P(F) = 16/52 = 4/13$. But since $E \cap F$ is the event that the card chosen is an ace, we have $P(E \cap F) = 4/52 = 1/13$. Now $P(E) \cdot P(F) = (2/13)(4/13) = 8/169 \neq P(E \cap F)$ and so the events are not independent. □

Let us conclude with some examples taken from information theory. Suppose that we are sending data, in the form of 0's and 1's, over a noisy communications line. Assume that, because of the noise, the probability that a bit (0 or 1) is received correctly is p, where $\frac{1}{2} \leq p \leq 1$. Assume also that the event that one bit is received correctly is independent of the event that another bit is received correctly.

Example 5.8.8

a) What is the probability that a message consisting of n bits will be received correctly?

b) What is the probability that exactly k of the n bits are received correctly?

c) What is the probability that at least k bits are received correctly?

Solutions:

a) If E_i is the event that the i-th bit is received correctly, then $P(E_i) = p$, and since $E_1, \ldots, E_n$ are independent, we have

$$\begin{aligned} P(\text{message received correctly}) &= P(E_1 \cap \cdots \cap E_n) \\ &= P(E_1) \cdots P(E_n) = p^n \end{aligned}$$

b) Consider the case where the first k bits are received correctly, and the rest are not. The probability of this occurring is $p^k(1 - p)^{n-k}$. But this probability would be the same if any set of k bits were received correctly, and since there are $\binom{n}{k}$ possibilities for the k correct bits, we have

$$P(\text{exactly } k \text{ correct bits}) = \binom{n}{k} p^k(1 - p)^{n-k}$$

c) From part b, we have

$$P(\text{at least } k \text{ bits are received correctly}) = \sum_{i=k}^{n} \binom{n}{i} p^i(1 - p)^{n-i}$$

Example 5.8.9

Suppose that, in the hope of increasing the probability of each bit being received correctly, we send it three times (in succession). Then, if exactly one error occurs in the three tries, we can (by taking the majority) still get the correct bit. What is the probability of a single bit being received (that is, interpreted) correctly?

The probability that all three bits are sent correctly is p^3, and the probability of exactly one error is $3p^2(1 - p)$. Hence, the probability of being able to correctly interpret that bit is

$$p^3 + p^2(1 - p) = 3p^2(3 - 2p)$$

Here are some values for this probability

p	$p^2(3 - 2p)$
.5	.500
.6	.648
.7	.748
.8	.896
.9	.972
1.0	1.000

(I am indebted to Professor Ira Gessel for this example.)

Exercises

1. A card is drawn at random from a deck of 52 cards.
   a) Find the probability of drawing a face card.
   b) Find the probability of drawing either a king or a black card.
2. A pair of fair dice are rolled.
   a) Find the probability of obtaining a sum of 5.
   b) Find the probability of obtaining a sum that is even.
3. Three fair dice are rolled. What is the probability of getting exactly one 6?
4. Four fair coins are tossed. Find the probability of getting at least 2 heads.
5. A basket contains 5 red balls, 3 black balls and 4 white balls. Two balls are chosen (at the same time) at random from the basket.
   a) What is the probability that they are both white?
   b) What is the probability that they are of different colors? *Hint:* what is the probability that they have the same color?
6. A certain true-false test has 10 questions. A student guesses randomly at all questions.
   a) What is the probability that the student gets all 10 correct?
   b) What is the probability that the student gets at least 9 correct?
   c) What is the probability that the student gets at least 8 correct?
7. Suppose that 5 cards are drawn at random from a deck of cards.
   a) What is the probability of getting a straight? (A straight consists of 5 cards in a row, such as Ace, 2, 3, 4, 5 or 7, 8, 9, 10, J or 10, J, Q, K, Ace.)
   b) What is the probability of getting a flush? (A flush consists of 5 cards of the same suit.)
8. A fair die is rolled and a card is drawn at random. What is the probability that the number on the die matches the number on the card?
9. If E and F are events in S that are both mutually exclusive and independent, what can you deduce about P(E) and P(F)?
10. Six dice are rolled. What is the probability that there is at least one 3 and at least one 4?
11. Ten bits are sent over a noisy channel, as in Example 5.8.8. What is the probability that at least 8 of the bits are received correctly?
12. Repeat the analysis of Example 5.8.9 where each bit is sent 5 times. If $p = .6$, how much greater is the probability that a bit is received correctly when repeating 5 times instead of 3 times?
13. With regard to Example 5.8.9, show that $p^2(3 - 2p) > p$ for all p satisfying $\frac{1}{2} \leq p \leq 1$. What do you conclude from this?
14. With regard to Example 5.8.9, for what value of p does the method used in that example give the greatest *increase* in the probability of error-free transmission?
15. Prove Theorem 5.8.1.
16. Prove Theorem 5.8.2.

My heart, which by a secret harmony
Still moves with thine, join'd in
connection sweet.
—Milton, *Paradise Lost*

Map me no maps, sir; my head is a map,
a map of the whole world.
—Fielding, *Rape upon Rape*

Geographers crowd into the outer edges
of their maps the parts of the world
which they know nothing about, adding
a note, "What lies beyond is sandy desert
full of wild beasts," or "blind marsh," or
"Scythian cold," or "frozen sea."
—Plutarch, *Lives: Theseus*

Chapter SIX

An Introduction to Graph Theory

6.1 Introduction

In 1736, the citizens of the town of Konigsberg (now called Kaliningrad and located in the U.S.S.R.) had a small problem. A part of their city bordered both sides of the river Pregel and included two islands, as shown below.

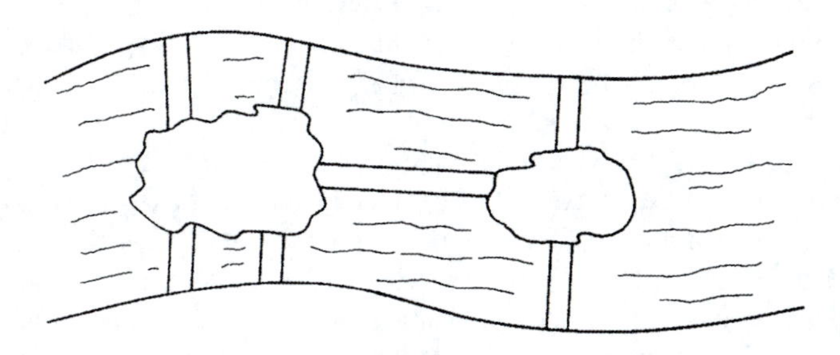

As you can see from this picture, the islands are connected to the mainland and to themselves by seven bridges.

Now, the citizens of Konigsberg liked to take frequent walks, and the question arose as to whether it was possible for a citizen to start at his home and make a journey that traversed each of the seven bridges exactly once.

Apparently, no one in Konigsberg could solve this problem. But the brilliant Swiss mathematician Leonhard Euler (1707–1783), who was at that time a professor of mathematics in St. Petersberg, could solve the problem, and in 1736 he published his solution in a paper.

Euler's idea was very simple. First, he replaced the map of Konigsberg with the following diagram.

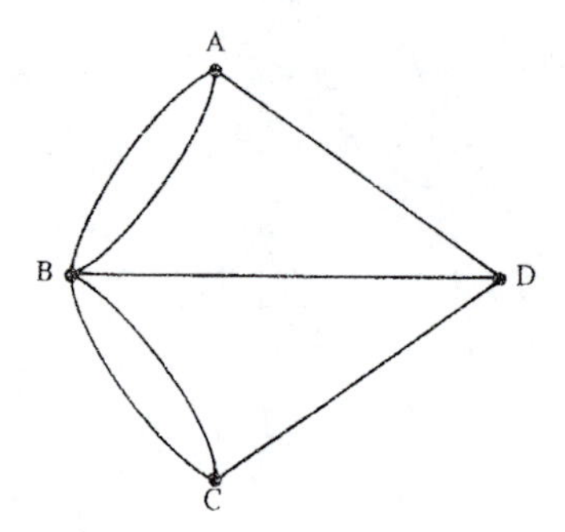

Figure 6.1.1

The points A and C represent the two banks of the river, the points B and D represent the two islands, and the seven arcs connecting these points represent the seven bridges.

Euler could then state the **Konigsberg bridge problem**, as it is now called, in the following way. Is it possible to start at any one of the four points A, B, C, or D, travel along each of the arcs in the diagram exactly once, and return to the starting point?

Euler showed that the answer to this question is "no" by reasoning as follows. Suppose that it is possible to take such a journey. Then any traveler who does so must enter and leave each of the points A, B, C, and D an *even* number of times.

If the trip starts at A, for example, then each time the traveler enters one of the points B, C, or D, he must also leave it. Therefore, the number of times he enters and leaves each of the points B, C, and D must be even. The same is true for the starting point A, but in this case we also must include the original departure from A and the final return to A. Of course, a similar reasoning applies if the traveler starts at any of the four points A, B, C, or D.

Therefore, since the traveler must enter and leave each of these four points along a different arc, and he must use all of the arcs by the end of the journey, there must be an even number of arcs touching each of the points A, B, C, and D. Since we can easily see from Figure 6.1.1 that this is not the case, there can be no such journey.

The diagram in Figure 6.1.1 is called a *graph* (not to be confused with the graph of a function, which is entirely different), and Euler's paper started a new branch of mathematics known as **graph theory**. Graph theory is one of the most widely applicable branches of mathematics, having important applications to the fields of computer science, physics, chemistry, engineering, biology, psychology, the social sciences, economics, operations research, urban planning, cybernetics, linguistics, and others. In this chapter, we want to introduce you to some of the basic principles of this fascinating branch of mathematics. (We discussed some aspects of graph theory in Chapter 3, but we will not assume any knowledge of that material here.)

Let us begin our study of graph theory with the definition of a graph. Intuitively speaking, we want a graph to be a set of points, together with a set of arcs that connect pairs of these points. To be as general as possible, we want to allow more than one arc to connect the same pair of points, as in Figure 6.1.1. Also, we want to allow an arc to connect a point to itself, forming a loop. As an example, Figure 6.1.2 is a graph.

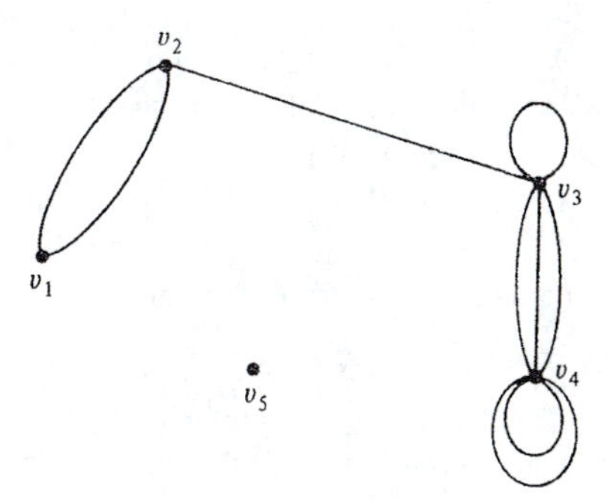

Figure 6.1.2

The precise way in which the graph in Figure 6.1.2 is drawn is not important. What is important are the points of the graph and the number of arcs between each pair of points (including loops). For instance, the diagrams

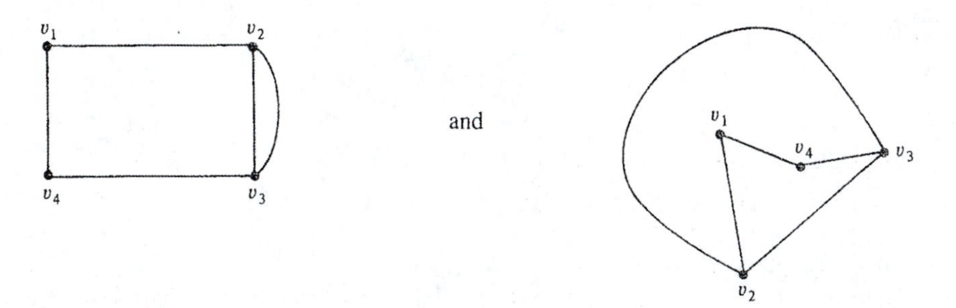

represent exactly the same graph.

Because of this, the important features of the graph in Figure 6.1.2 can be described by giving the set

$$V = \{v_1, v_2, \ldots, v_5\}$$

of its points, and the *multiset*

$$E = \{\{v_1, v_2\}, \{v_1, v_2\}, \{v_2, v_3\}, \{v_3\}, \{v_3, v_4\}, \{v_3, v_4\}, \{v_3, v_4\}, \{v_4\}, \{v_4\}\}$$

of its arcs. (A *multiset* is different from a set in that an element can be repeated several times in a multiset. See Section 5.2 for more on multisets.) Notice that E contains one set of the form $\{u, v\}$ for each arc connecting the points u and v, and one set of the form $\{v\}$ for each arc connecting v to itself. For example, there are two sets $\{v_1, v_2\}$ in E, since the graph contains two arcs connecting v_1 and v_2, and there are two sets $\{v_4\}$ in E, since the graph contains two arcs connecting v_4 to itself.

The elements of the set V are called the *vertices*, or *nodes*, of the graph (the singular of vertices is vertex), and the elements of the multiset E are called the *edges* of the graph. An edge of the form $\{v\}$ is called a *loop* at v. Now we can give a formal definition of the term "graph."

Definition

A **graph** G = (V, E) is an ordered pair, where V is any nonempty finite set, and E is any multiset of subsets of V, each of which has size 1 or size 2. The elements of V are referred to as the **vertices**, or **nodes**, of the graph, and the elements of E are referred to as the **edges** of the graph. An edge that has the form $\{v\}$, where v is a vertex, is called a **loop** at v. We use the word *multiset* for E, instead of set, since we allow the possibility that the same subset may appear more than once in E. □

We now have both a formal definition of a graph as an ordered pair (V, E) and an informal definition of a graph as a set of points and a collection of arcs connecting pairs of these points. As is customary, we will use the informal definition of a graph whenever there is no danger of confusion. But we can be secure in the fact that we can always fall back on the formal definition if it becomes necessary.

If a graph, such as the one in Figure 6.1.2, has more than one edge connecting a pair of vertices, then we say that the graph has *multiple edges*. Graphs that have neither multiple edges nor loops are referred to as **simple graphs**.

The graph in Figure 6.1.3 illustrates an important point about the way graphs are drawn.

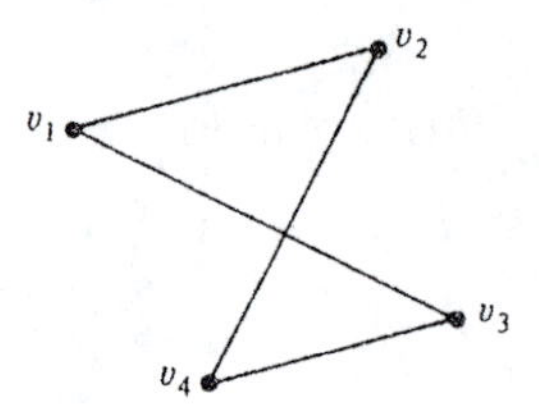

Figure 6.1.3

Namely, the edges of a graph may cross at places other than the vertices of the graph. In this case, the graph has four vertices, which are clearly labeled v_1, v_2, v_3, and v_4. The fact that the edges intersect at other points has no bearing on the graph. In fact, if we redraw this graph as in Figure 6.1.4, then the edges intersect only at the vertices of the graph.

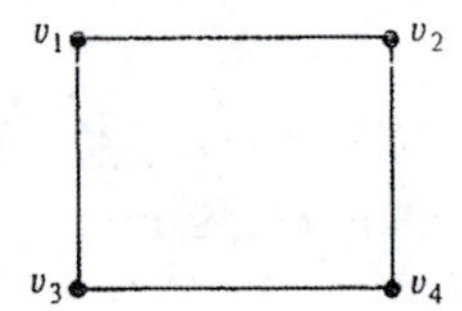

Figure 6.1.4

Another point that we should make is that the edges of a graph do not have to be *straight* line segments.

Incidentally, there are graphs with the property that, no matter how they are drawn in the plane (with unbroken edges), the edges will always intersect at points that are not vertices of the graph. For example, one such graph is

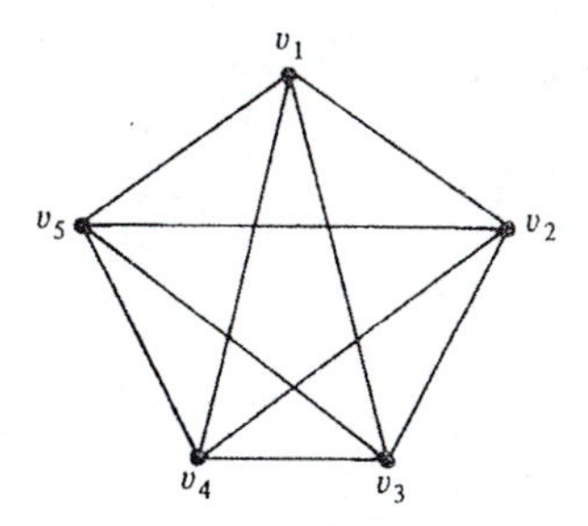

Graphs that can be drawn in the plane without such intersections are called **planar** graphs. The question of whether or not a graph is planar is an interesting one, and there are various criteria that can be used to determine planarity. We will discuss the concept of planarity in Section 6.4.

Before turning to examples of graphs, we should point out that sometimes we will label both the vertices and the edges of a graph, but if the labels are not needed for a particular discussion, then we will not bother to include them.

Let us now consider some examples of how graphs can arise in various contexts.

■ Example 6.1.1

Graphs can be used to represent lines of communication. For example, in the following graph, each vertex represents a country and each edge represents a direct telephone line between the corresponding governments.

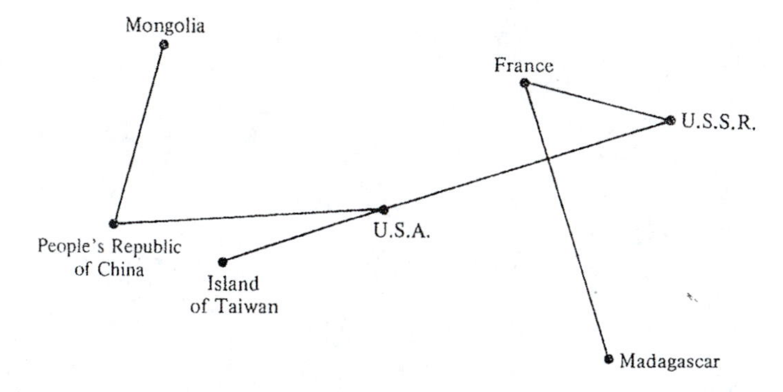

Thus, in this case, there is a direct telephone line between the governments of the United States and the U.S.S.R., but there is no direct line between the governments of the People's Republic of China and the Island of Taiwan. □

Example 6.1.2

Graphs can also be used to represent transportation routes. For example, in the following graph each vertex represents a city and each edge represents a highway between the corresponding cities.

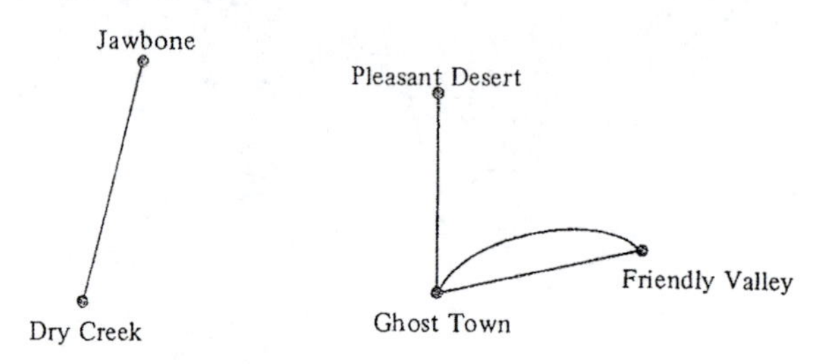

In a similar way, graphs can be used to describe air routes, mail routes, train routes, and so on. □

Example 6.1.3

Graphs can be used to describe relationships of various kinds. One of the most common examples of this is the family tree. In this case, the vertices represent individuals in a family, and there is an edge connecting two members of the family if and only if one member is the son or daughter of the other member. As an example, the following graph is part of the family tree of Princess Anastasia of Russia.

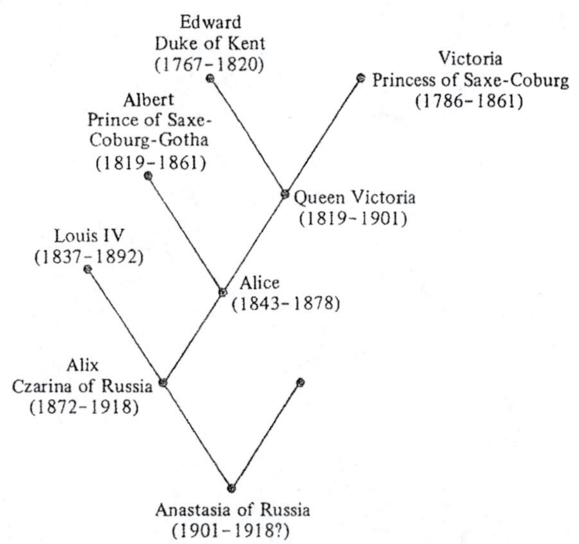

Another type of graph that describes relationships is the so-called *acquaintance graph*. This is a graph in which the vertices represent people, and there is an edge between two vertices if and only if the corresponding people know each other. Acquaintance graphs can be used to study certain types of combinatorial problems. □

Example 6.1.4

Graphs can be used to describe the process of sorting. For example, the following graph describes one way to sort prospective athletes in order to decide in which sport they should compete. A prospective athlete simply starts at the top of the graph and follows the path that describes his or her physical features. The last vertex gives the sport.

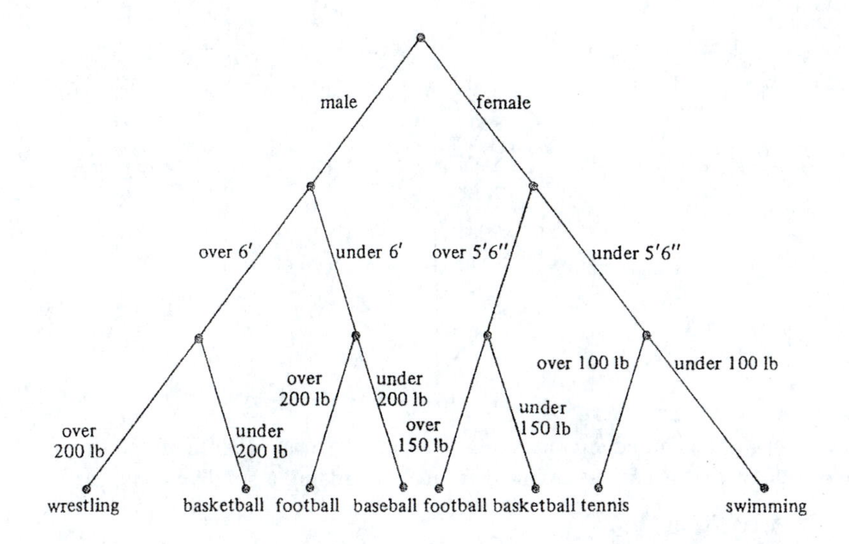

Graphs that sort, or classify, are useful in the biological sciences. For example, the following graph describes a portion of the classification of the animal kingdom.

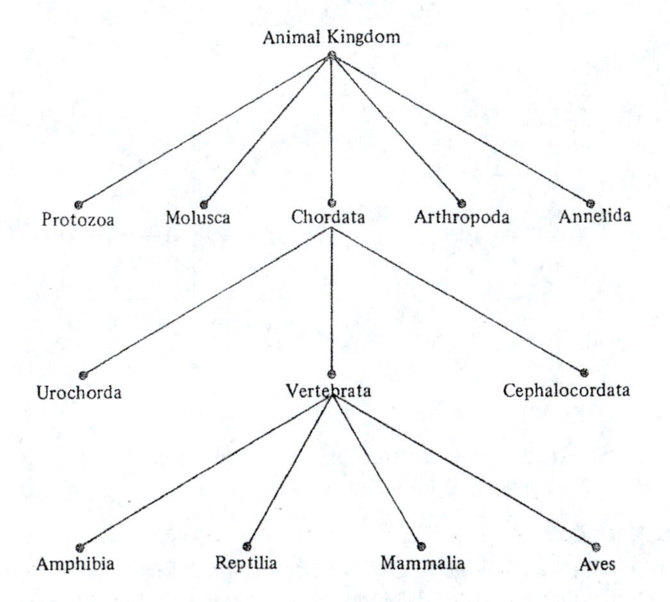

Example 6.1.5

Graphs can be very useful in other branches of mathematics. For instance, graphs can be used to describe, in an orderly manner, the power set (the set of all subsets) of a set S. The following graph describes the power set of the set S = {1, 2, 3}.

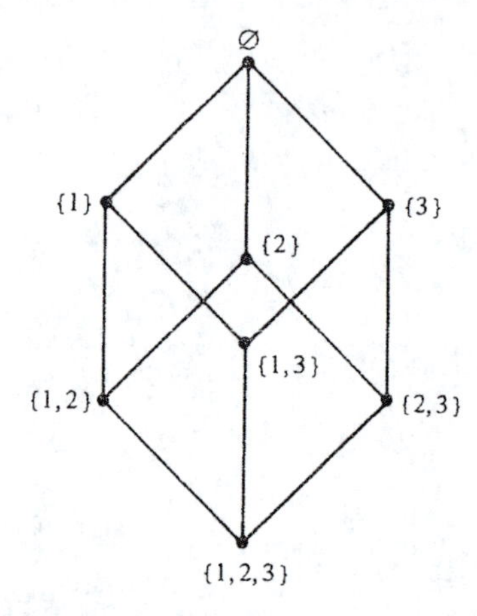

Notice that there is an edge connecting two subsets A and B of S if and only if A is a subset of B, and B can be obtained from A by adding a single element to A. □

Example 6.1.6

Graphs find many uses in the physical sciences. For example, in 1857 the mathematician Arthur Cayley (1821–1895) used graphs to help describe and enumerate the number of isomers of the hydrocarbon molecules C_nH_{2n+2}. As an example, there are two possible isomers of the molecule C_4H_{10}, represented by the following graphs.

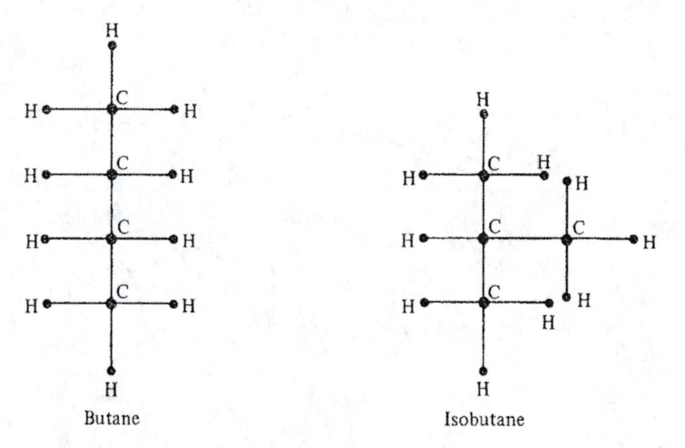

In this case, each vertex represents an atom of either carbon or hydrogen, and each edge represents a chemical bond between two atoms. □

Example 6.1.7

A graph in which each edge has been assigned a number is called an **undirected network**. We will study undirected networks in Section 6.5, but it is easy to see now that they can be very useful. For example, if we label each edge of the transportation graph in Example 6.1.2 with the distance between the corresponding cities, then we obtain a transportation network. Another possibility is to label each of the edges with the maximum number of cars that can pass any given point on the highway in 1 hour.

As another example, if we label each edge of the communications graph in Example 6.1.1 with the cost of the communication per unit of time, then we get a communications network. □

These examples should give you an idea of how useful the concept of a graph can be, and we will see some other important uses for graphs (especially in computer science) during the course of our studies. In the remainder of this section, we want to discuss a few definitions and simple results relating to graphs.

If G is a graph, then we will denote the set of vertices of G by $\mathcal{V}(G)$, and the multiset of edges of G by $\mathcal{E}(G)$. A vertex v is said to be **adjacent** to a vertex w if the set $\{v, w\}$ is an edge of G. A vertex v is adjacent to itself if and only if there is a loop at v. Two edges of G are said to be **adjacent** if they share a common vertex.

If v is a vertex of G and e is an edge of G, then we say that v is **incident** with e if e is either a loop at v or if it has the form $e = \{v, w\}$ for some vertex w of G. In this case, we also say that e is **incident** with v. Loosely speaking, a vertex v is incident with an edge e when v is one of the vertices that make up the edge e.

One way to keep the distinction between the terms *adjacent* and *incident* clear is to remember that adjacent is used when the two objects are of the same type—both vertices or both edges—and incident is used when the two objects are of different types—one vertex and one edge.

The **degree** of a vertex v, denoted by $deg(v)$, is simply the number of times an edge of G meets v. This is not quite the same as the number of edges that are incident with v, since we must count any loops at v *twice* when determining the degree of v. A vertex of degree zero has no edges incident with it, and so it is called an **isolated vertex**.

Let us illustrate these definitions with the following graph.

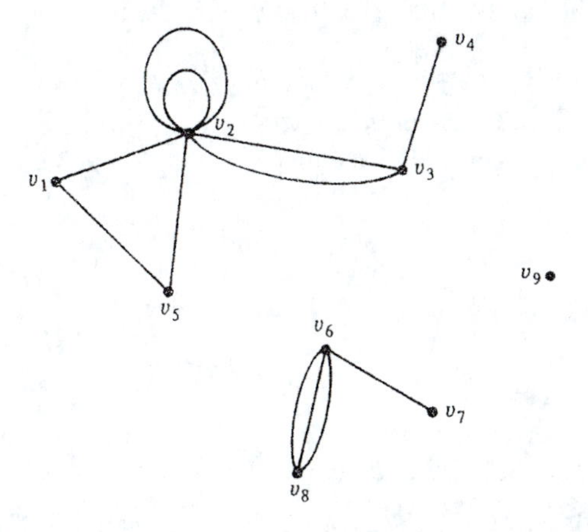

Vertices v_1 and v_2 are adjacent, but vertices v_1 and v_3 are not adjacent. Also, the edges $\{v_1, v_2\}$ and $\{v_2, v_3\}$ are adjacent, but the edges $\{v_1, v_2\}$ and $\{v_3, v_4\}$ are not adjacent.

The vertices v_3 and v_4 are incident with the edge $\{v_3, v_4\}$, and no other vertices are incident with this edge. The degrees of the various vertices are

$$deg(v_1) = 2, \quad deg(v_2) = 8, \quad deg(v_3) = 3$$

$$deg(v_4) = 1, \quad deg(v_5) = 2, \quad deg(v_6) = 4$$

$$deg(v_7) = 1, \quad deg(v_8) = 3, \quad deg(v_9) = 0$$

Since the degree of vertex v_9 is zero, v_9 is an isolated vertex.

There is a simple connection between the degrees of the vertices in a graph and the number of edges in the graph. Let us put it into a theorem.

Theorem 6.1.1

If G is a graph and if $\mathcal{V}(G) = \{v_1, v_2, \ldots, v_n\}$, then

$$deg(v_1) + deg(v_2) + \cdots + deg(v_n) = 2|\mathcal{E}(G)|$$

In words, the sum of the degrees of all of the vertices in a graph G is equal to twice the number of edges of G.

PROOF In order to prove this theorem, let us count the number of incidences of edges and vertices in the graph G. We can count this number in two different ways—first from the point of view of the edges, and then from the point of view of the vertices.

From the point of view of the edges, we observe that each edge that is not a loop is incident with exactly two vertices, and each loop is incident with the same vertex twice. Hence, the number of incidences of edges and vertices is precisely $2|\mathcal{E}(G)|$.

On the other hand, from the point of view of the vertices, we observe that, according to the definition of the term "degree," the number of incidences of the edges of G with the vertex v_i is precisely $deg(v_i)$. Hence, from the point of view of the vertices, the total number of incidences of edges and vertices is the sum

$$deg(v_1) + deg(v_2) + \cdots + deg(v_n)$$

Now, regardless of which point of view we use when counting the number of incidences of edges and vertices, we must always get the same answer. Therefore, we must have

$$deg(v_1) + deg(v_2) + \cdots + deg(v_n) = 2|\mathcal{E}(G)|$$

which is what we wanted to prove. ▮

Example 6.1.8

If a graph G has vertices of degrees 2, 2, 2, 3, 3, 3, 4, 4, 5, and 6 then, according to Theorem 6.1.1, the number of edges in G is

$$|\mathcal{E}(G)| = (\tfrac{1}{2})(2 + 2 + 2 + 3 + 3 + 3 + 4 + 4 + 5 + 6) = 17$$

Since the sum

$$2 + 2 + 2 + 3 + 3 + 3 + 4 + 4 + 5 + 5 + 6$$

is odd, there can be no graph with vertices of degrees 2, 2, 2, 3, 3, 3, 4, 4, 5, 5, and 6 (and no other vertices). □

There is one type of graph that deserves special mention, since it occurs frequently. The **complete graph of order n**, denoted by K_n, is the graph that has n vertices and exactly one edge connecting *each* of the possible pairs of distinct vertices. For example, the following are complete graphs.

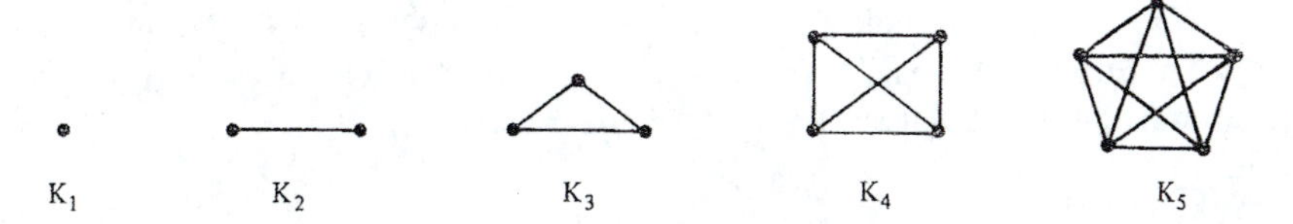

Of course, complete graphs are simple graphs; that is, they have no multiple edges or loops. Also, since there are $\binom{n}{2}$ possible pairs of the n vertices, the complete graph K_n has exactly $\binom{n}{2}$ edges, and each vertex has degree $n - 1$.

A graph H is called a **subgraph** of a graph G if every vertex of H is also a vertex of G and every edge of H is also an edge of G. In symbols, H is a subgraph of G if $\mathcal{V}(H) \subset \mathcal{V}(G)$ and $\mathcal{E}(H) \subset \mathcal{E}(G)$.

As an example, consider the graph G given by

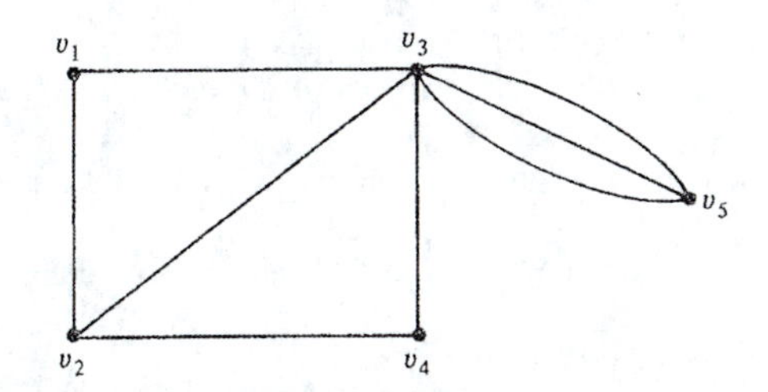

Then each of the following graphs are subgraphs of G.

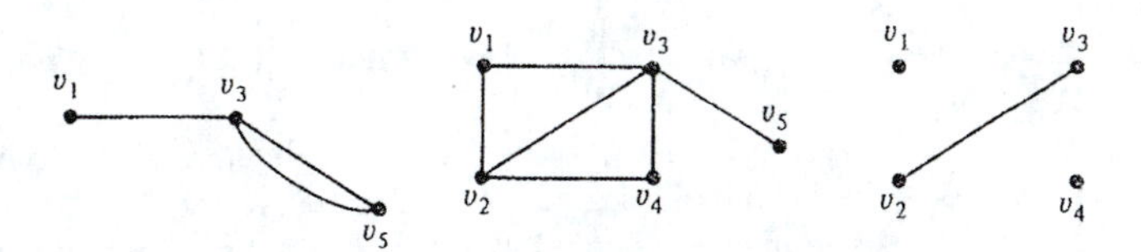

We want to conclude this section by making an important observation. For some reason, graph theory suffers, more than any other branch of mathematics by far, from the problem that different authors define the basic terms used in graph theory in different ways. Therefore, it is very important, whenever you pick up a book that contains a discussion of graphs, to check what definitions the author is using.

As a matter of fact, even the definition of the term "graph" varies from book to book! Some authors define a graph as we have done, but others define a graph to be what we are calling a simple graph. In other words, some authors use the term "graph" to mean a graph with no multiple edges and no loops. Among these authors, some use the term "*multigraph*" for graphs that can have loops and multiple edges. On the other hand, some authors use the term "*multigraph*" for graphs that can have multiple edges but not loops. Then they use the term "*pseudograph*" for graphs that can have both loops and multiple edges. (You should not worry about remembering all of these different terms. Just remember the terms that we are using in this book, and be aware of the problem of varying definitions.)

Exercises

1. Consider the following graph G.

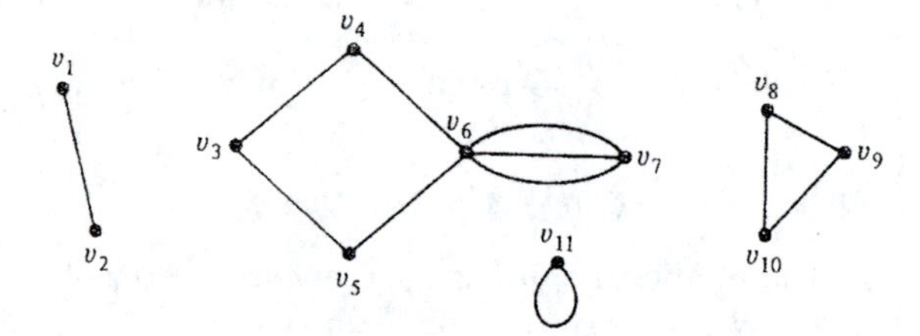

a) Are vertices v_1 and v_2 incident? Explain.
b) Which vertices of G, if any, are adjacent to themselves?
c) Is vertex v_3 adjacent to vertex v_6? Explain.
d) Is G a simple graph? Explain.
e) Find the degrees of each of the vertices of G, and verify Theorem 6.1.1 for this graph.

2. Can a graph have vertices of degrees 2, 2, 3, 4, 5, 5, 6, and 8 and no other vertices? Justify your answer.

3. If a graph has vertices of degrees 1, 2, 3, 3, 4, and 5, how many edges does it have? Justify your answer.

4. Draw a picture of the complete graph K_6.

5. How many edges does the graph K_{10} have?

6. Give an example of a simple graph
a) having no vertices of odd degree.
b) having no vertices of even degree.

7. Show that there is no graph G whose vertices have degrees equal to 2, 3, 3, 4, 4, and 5. *Hint*, use Theorem 6.1.1

8. Show that there is no *simple* graph G whose vertices have degrees 1, 3, 3, and 3. Can there be any type of graph with these degrees?

9. If m and n are any positive integers, find a graph G with the property that every vertex has either degree m or degree n.

10. Find all subgraphs of the graph

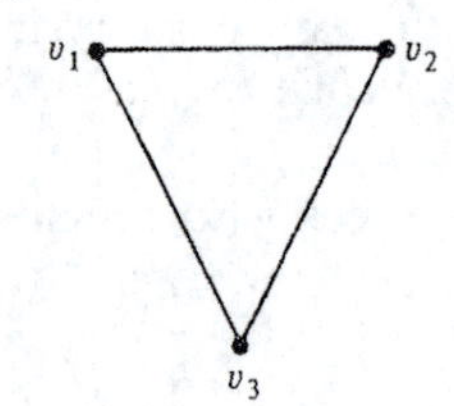

A graph G with the property that all of its vertices have the same degree r is called a ***regular graph****, and the number r is called the* ***degree*** *of the graph. For example, the following graph is regular and has degree 3.*

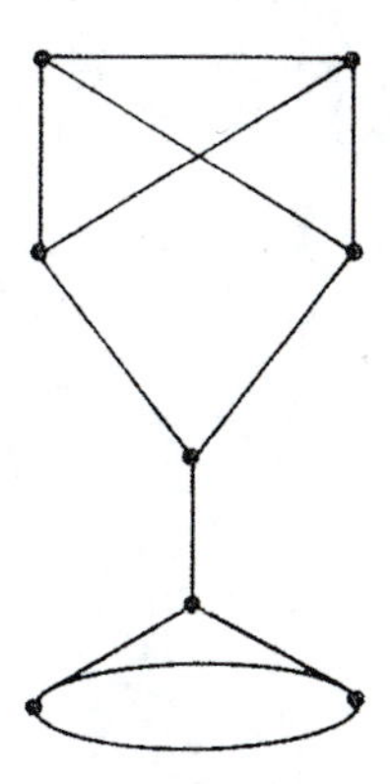

11. a) Give an example of a regular simple graph of degree 1 that is not a complete graph.
b) Give an example of a regular simple graph of degree 2 that is not a complete graph.
c) Give an example of a regular simple graph of degree 3 that is not a complete graph.

12. Is the complete graph K_n regular? If so, what is the degree of K_n? Justify your answer.

13. Prove that if G is a regular graph of degree r, and if G has n vertices, then it must have exactly $nr/2$ edges. *Hint*, use Theorem 6.1.1.

14. Let $S = \{A_1, A_2, \ldots, A_n\}$ be a set of sets. The **intersection graph** of S is defined to be the simple graph that has one vertex for each set, and an edge connecting two vertices if and only if the corresponding sets have nonempty intersection. Find the intersection graph of $S = \{A_1, A_2, A_3, A_4\}$, where

$$A_1 = \{1, 2, 3, 4, 5\} \quad , \quad A_2 = \{2, 4, 6, 8\}$$

$$A_3 = \{3, 5, 12\} \quad , \quad A_4 = \{5, 8, 10\}$$

15. Prove that no simple graph can have the property that all of its vertices have a different degree. *Hint*, use the pigeonhole principle and the fact that the degree of a vertex must be less than the total number of vertices in the graph.

16. Prove that in any graph G, the number of vertices of odd degree is even. *Hint*, divide the vertices of G into two groups, those with even degree and those with odd degree. Then use Theorem 6.1.1 to deduce that the sum of the degrees of those vertices with odd degrees is even. Use this to show that there must be an even number of vertices of odd degree.

Let G be a simple *graph. The* ***complement*** *of G is the simple graph G^c defined as follows. The vertices of G^c are the same as the vertices of G, but G^c has an edge between the vertices u and v if and only if G does* not *have an edge between u and v. For example, if G is the graph*

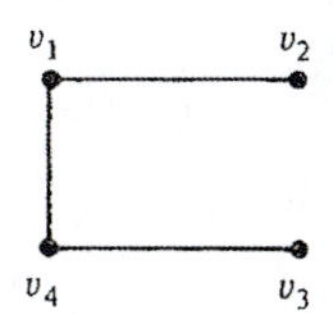

then G^c is the graph

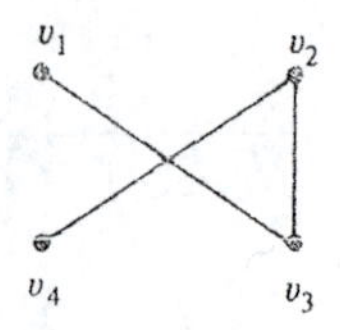

17. Find the complement of the graph

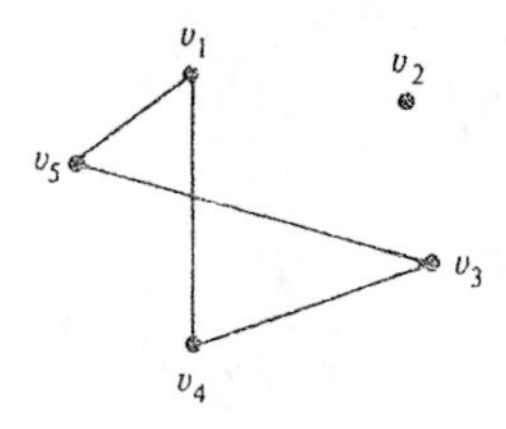

18. Find the complement of the graph

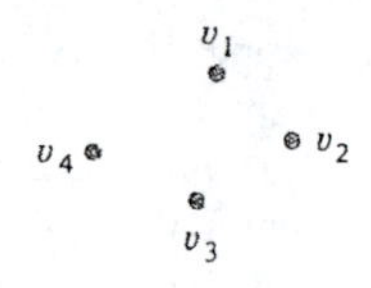

19. What is the complement of the complete graph K_n?

20. Find a relationship between the number of edges of a graph and the number of edges of its complement.

*A simple graph G is said to be **bipartite** if its vertices can be divided into two sets A and B with the property that all of the edges in the graph connect a vertex in A to a vertex in B. Put another way, none of the edges in G connect vertices within the same set A or B. For example, the graph*

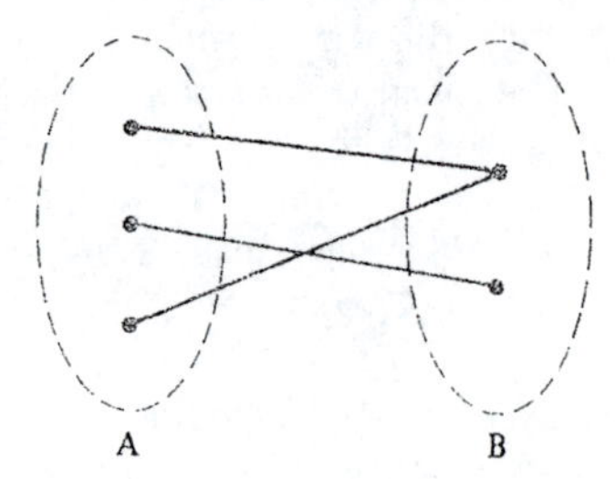

is bipartite, and so is the graph

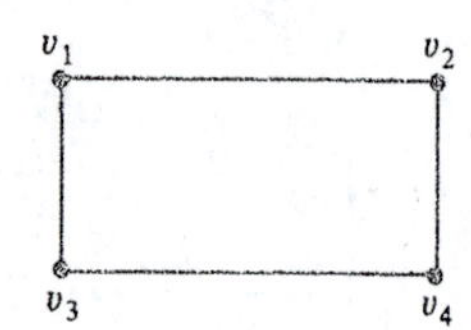

since it can be redrawn in the form

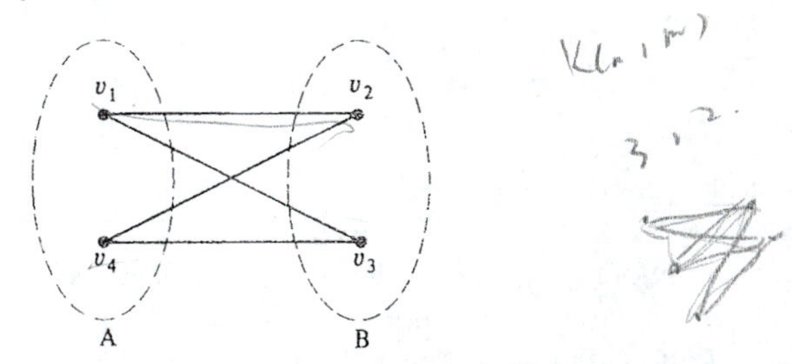

21. Determine which of the following graphs are bipartite and which are not. For those that are bipartite, redraw them so that the sets A and B are evident.

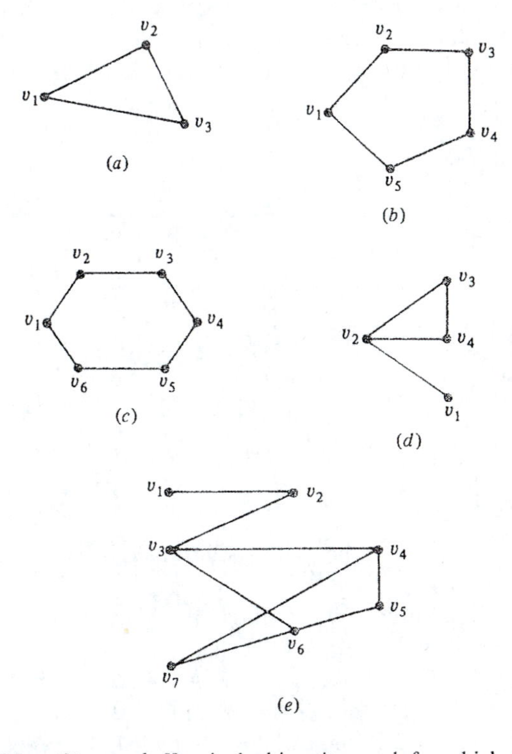

22. The **complete bipartite graph** $K_{n,m}$ is the bipartite graph for which A is a set of n vertices and B is a set of m vertices, and there is an edge between each vertex in A and each vertex in B. Thus, for example, the graph $K_{3,2}$ is

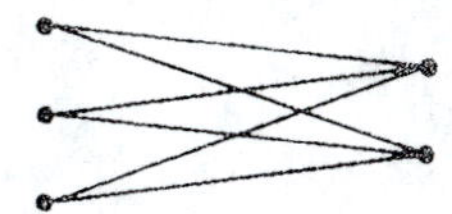

a) Draw the complete bipartite graphs $K_{2,2}$, $K_{3,3}$, and $K_{4,5}$

b) How many edges does the graph $K_{n,m}$ contain?

c) Describe the complement of the graph $K_{n,m}$

6.2

Paths and Connectedness

Many problems in graph theory involve the question of whether or not it is possible to get from one vertex in a graph to another by following the edges of the graph. In order to discuss these questions, we must first give some definitions. Once again, it is important to remember that books on graph theory differ in their terminology, and although the definitions that we are about to give seem to be the most popular, some authors use other definitions.

Let u and v be vertices of a graph G. (We allow the possibility that u and v are the same vertex.) Then a **walk** from u to v is an alternating sequence of vertices and edges of G, beginning with the vertex u and ending with the vertex v, with the property that each edge is incident with the vertex immediately preceding it and the vertex immediately following it in the sequence. For example, in the graph

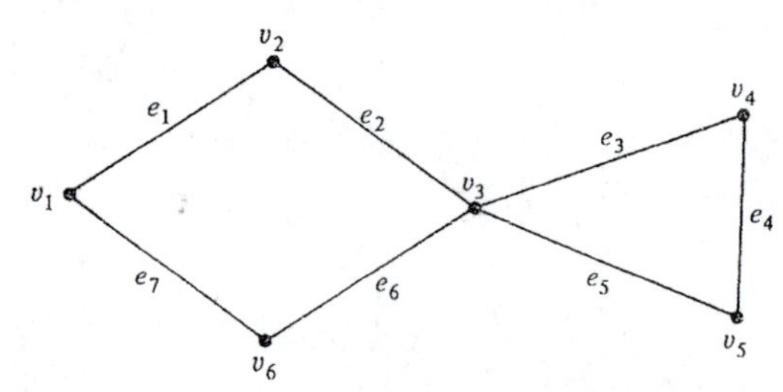

Figure 6.2.1

the sequence

$$v_1, e_7, v_6, e_6, v_3, e_3, v_4, e_3, v_3, e_5, v_5 \tag{6.2.1}$$

is a walk from v_1 to v_5. (To see this, you should trace out this sequence of vertices and edges on the graph itself.)

In loose terms, a walk from u to v is simply a way to get from u to v by following the edges and vertices of the graph. A walk that begins and ends at the same vertex v, that is, a walk from v to v, is said to be a **closed walk**. On the other hand, if u and v are different, then a walk from u to v is said to be an **open walk**.

Since a walk from u to v is completely determined by giving just the sequence of edges, we will describe walks in this way whenever it is convenient. (One reason that it might not be convenient is if we want to have a list of the vertices of the walk.) Thus, for example, the walk in 6.2.1 can be described by the sequence of edges

$$e_7, e_6, e_3, e_3, e_5$$

A **trail** from u to v is a walk from u to v in which no *edge* is used more than once. (A vertex can be used more than once, however.) Thus, for example, the walk in 6.2.1 is not a trail, since the edge e_3 is used twice. On the other hand, the walk

$$v_2, e_2, v_3, e_3, v_4, e_4, v_5, e_5, v_3 \tag{6.2.2}$$

is a trail from v_2 to v_3. (Again, you should trace this trail on the graph itself.) This trail can also be described by giving only its edges e_2, e_3, e_4, e_5.

A trail that begins and ends at the same vertex v, that is, a trail from v to v, is called a **closed trail**, or a **circuit**. For example, the trail

$$e_1, e_2, e_3, e_4, e_5, e_6, e_7 \qquad (6.2.3)$$

is a circuit. A trail that is not closed is called an **open trail**.

A **path** from u to v is a walk from u to v in which no *vertex* is used more than once. However, in a path we do allow one exception, namely, that the first and last vertices in a path can be the same. (But of course, they do not have to be.) The walk in 6.2.2 is not a path, since the vertex v_3 is used twice, but the walk

$$v_1, e_1, v_2, e_2, v_3, e_5, v_5$$

is a path from v_1 to v_5.

A path that begins and ends at the same vertex is called a **closed path**, or a **cycle**. For example, the circuit in 6.2.3 is not a cycle, since it uses the vertex v_3 twice. (To see this, trace the circuit on the graph.) On the other hand, the path

$$e_1, e_2, e_6, e_7$$

is a cycle. A path that is not closed is called an **open path**.

In order to help remember these definitions, let us put them into a table.

Beginning Vertex u **Ending Vertex v**	$u \neq v$	$u = v$
Walk no restriction on the number of times an edge or vertex can appear	open walk	closed walk
Trail no edge can appear more than once	open trail	closed trail or circuit
Path no vertex can appear more than once, with the possible exception that u and v may be the same vertex	open path	closed path or cycle

(Actually, the concept of a walk is a bit too general to be of much use, and so you should concentrate on the concepts of a trail and a path.)

Two vertices u and v in a graph G are said to be **connected** if there is a path in G from u to v. (Of course, if there is a path from u to v, then there is also a path from v to u. Can you describe it?)

A graph G in which every pair of vertices is connected is called a **connected graph**. This is a very important definition. Loosely speaking, it says that a graph is connected if it is possible to go from any vertex to any other by following the edges of the graph. A graph that is not connected is said to be **disconnected**.

As an example, the graph in Figure 6.2.1 is connected, but the following graph is disconnected.

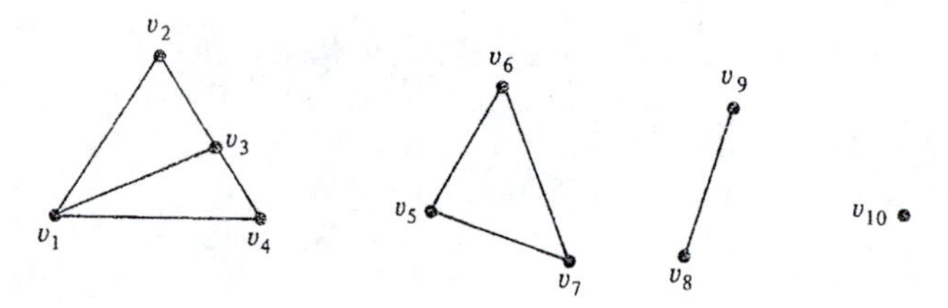

Figure 6.2.2

A disconnected graph is made up of two or more disjoint connected subgraphs. For example, the graph in Figure 6.2.3 is made up of four connected subgraphs,

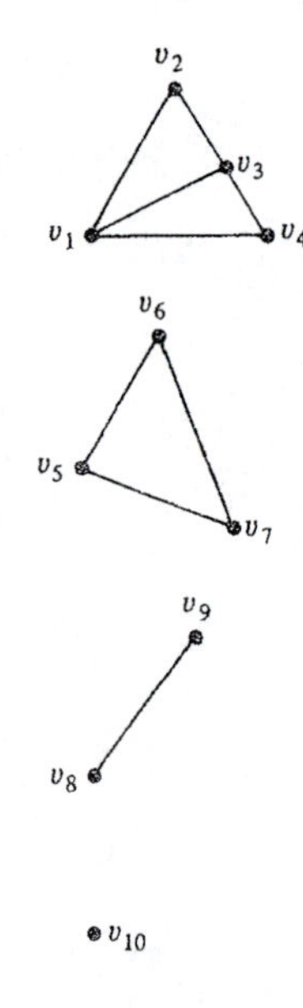

Figure 6.2.3

These subgraphs can be described by saying that they are connected subgraphs that cannot be "enlarged" and still remain connected. We can make this idea more precise in the following definition.

Definition

Let G be a graph. A subgraph H of G is called a **connected component** of G if it has the following two properties.

1) H is connected.
2) There is no *connected* subgraph of G that is "larger than" H, in the sense that it has H as a subgraph, but also contains some vertices or edges that are not in H. □

Thus, the four subgraphs in Figure 6.2.3 are the connected components of the graph in Figure 6.2.2. As another example, the graph

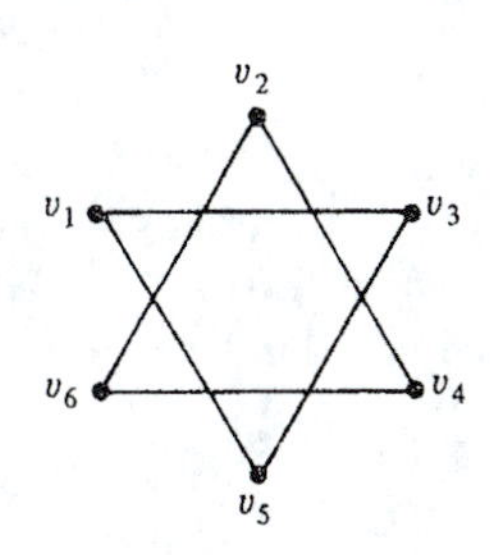

Figure 6.2.4

has two connected components

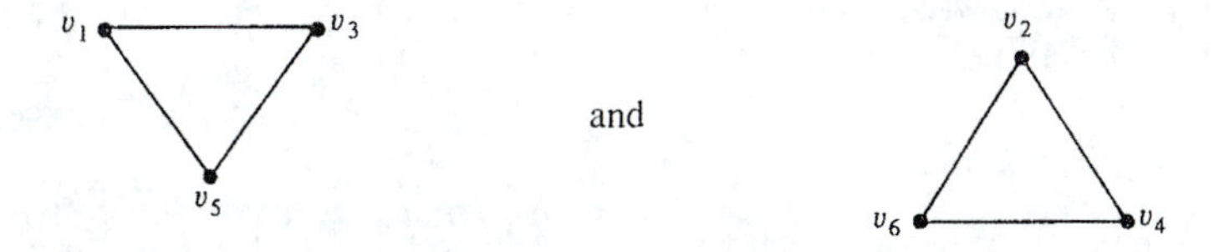

(Remember that the edges of a graph may intersect at points that are not vertices of the graph. This is the case for the graph in Figure 6.2.4.)

Using the definition of connected component, we can state that a graph G is connected if and only if it has precisely one connected component, namely, G itself.

It seems reasonable that if a graph has very few edges, relative to the number of vertices, then it cannot be connected. As a simple example of this, if a graph G has 10 vertices and only 2 edges, it is certainly not connected. We can prove the following very useful result along these lines.

Theorem 6.2.1

Any *connected* graph with n vertices must have at least $n - 1$ edges.

PROOF We prove this theorem by using induction. Let $P(n)$ be the statement that any connected graph with n vertices must have at least $n - 1$ edges. Then $P(1)$ is certainly true. Let us assume that $P(1), P(2), \ldots, P(k)$ are true, and show that $P(k + 1)$ must also be true. Thus, we must show that any connected graph with $k + 1$ vertices has at least k edges.

From among all connected graphs with $k + 1$ vertices, let us pick one, call it G, that has the fewest number of edges. (There may be several such graphs, but we need only one.) If we can show that G must have at least k edges, then we will know that *any* connected graph with $k + 1$ vertices must have at least k edges.

Let us remove any edge from G. The resulting graph, call it H, has fewer edges than G, and so it cannot be connected. In fact, H must consist of exactly two connected components, say H_1 and H_2 (see Exercise 10). If we denote the number of vertices in H_1 by n_1 and the number of vertices in H_2 by n_2, then n_1 and n_2 must satisfy $n_1 \leq k$, $n_2 \leq k$ and

$$n_1 + n_2 = k + 1$$

Now since $n_1 \leq k$ and $n_2 \leq k$, according to our assumption, $P(n_1)$ and $P(n_2)$ must be true. That is, H_1 must have at least $n_1 - 1$ edges, and H_2 must have at least $n_2 - 1$ edges. Hence, G must have at least

$$(n_1 - 1) + (n_2 - 1) + 1 = n_1 + n_2 - 1 = k$$

edges. This shows that $P(k + 1)$ is true, and completes the proof. ∎

It also seems reasonable that, if a simple graph has a very large number of edges, relative to the number of vertices, then it will have to be connected. As a simple example of this, if a simple graph has 10 vertices and 45 edges, then it must be the complete graph K_{10} (why?), and so it must be connected. We can prove the following result along these lines.

Theorem 6.2.2

If a *simple* graph with n vertices has more than $\binom{n-1}{2}$ edges, then it must be connected.

PROOF We can prove this theorem by showing that any graph with n vertices that is *not* connected can have at most $\binom{n-1}{2}$ edges. This is the contrapositive of the statement in the theorem. (See Section 2.3 for the definition of the term "contrapositive.")

From among all *disconnected* graphs with n vertices, let us pick one, call it G, that has the largest number of edges. Then if we can show that G has exactly $\binom{n-1}{2}$ edges, we will know that *all* disconnected graphs with n vertices have *at most* $\binom{n-1}{2}$ edges, and this will prove the theorem.

Now, since G is not connected, it must have more than one connected component. But G cannot have more than two connected components, for if it did, then we could add an edge to G and still have a disconnected graph. (How?) Let us denote the connected components of G by G_1 and G_2.

Since G has the largest number of edges possible for a disconnected graph with n vertices, it is easy to see that each of the components G_1 and G_2 must be a complete graph. For if not, then we could add more edges to G and still have a disconnected graph. Let us suppose that G_1 is the complete graph K_k on k vertices and G_2 is the complete graph K_{n-k} on $n - k$ vertices, where $1 \leq k \leq n - 1$. (The number of vertices in G_1 and G_2 must add up to n.)

All we have to do now is ask ourselves what k must be in order for G to have the largest number of edges. To answer this question, we observe that the number of edges in G is

$$\binom{k}{2} + \binom{n-k}{2} = \binom{n}{2} + k(k-n)$$

But, this number is as large as possible when $k = 1$ or $k = n - 1$ (see Exercise 16), and so G must consist of the two components K_1 and K_{n-1}. Therefore, G has $\binom{n-1}{2}$ edges, and the proof is complete. ∎

Actually, Theorems 6.2.1 and 6.2.2 can be generalized as follows.

Theorem 6.2.3

Let G be a simple graph with n vertices and k components. Then the number m of edges of G must satisfy

$$n - k \leq m \leq \binom{n-k+1}{2}$$

Let us conclude this section with one final definition. An edge e in a *connected* graph G is called a **bridge** if the graph obtained from G by removing the edge e is not connected. For example, the graph

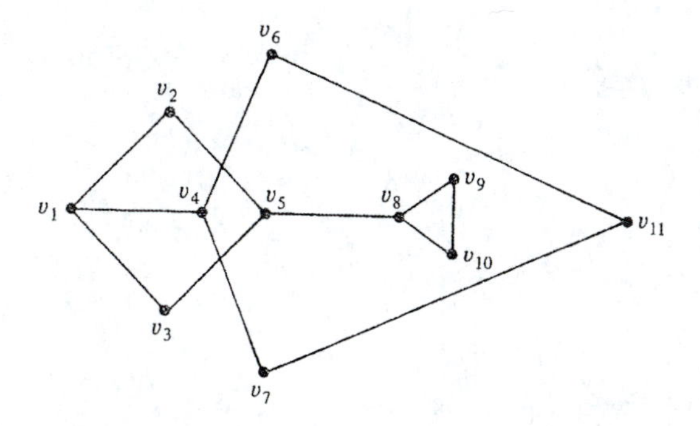

has two bridges, namely, the edges $\{v_1, v_4\}$ and $\{v_5, v_8\}$. None of the other edges are bridges.

Exercises

1. For the graph

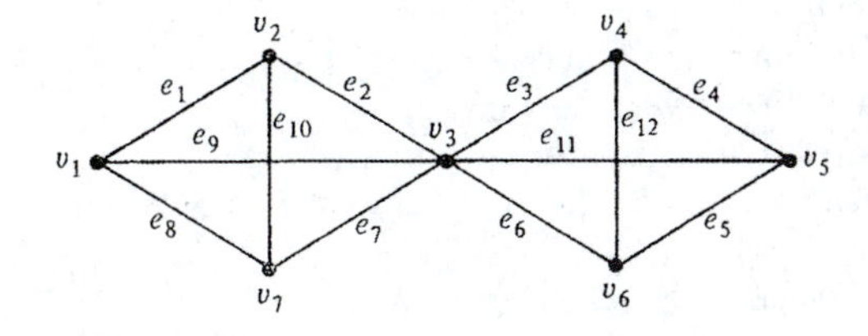

state whether the given sequence is a walk, a trail, a path, or none of these. Also, state whether it is open or closed.

a) $v_1, e_1, v_2, e_2, v_3, e_3, v_4, e_{12}, v_6, e_6, v_3$
b) $e_1, e_2, e_3, e_4, e_5, e_6, e_7, e_8$
c) $v_1, e_8, v_7, e_7, v_3, e_7, v_7, e_8, v_1$
d) e_9, e_{11}, e_4, e_5
e) e_3, e_4, e_5, e_6

2. For the graph

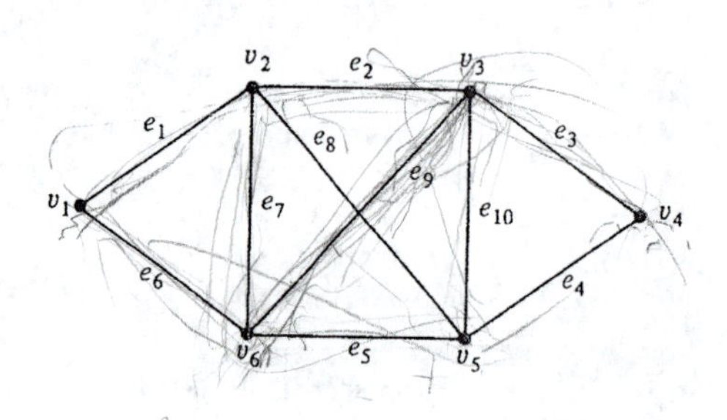

find

a) four different paths from v_1 to v_4
b) four different trails from v_1 to v_4, none of which are paths
c) four different walks from v_1 to v_4, none of which are trails

3. For the graph

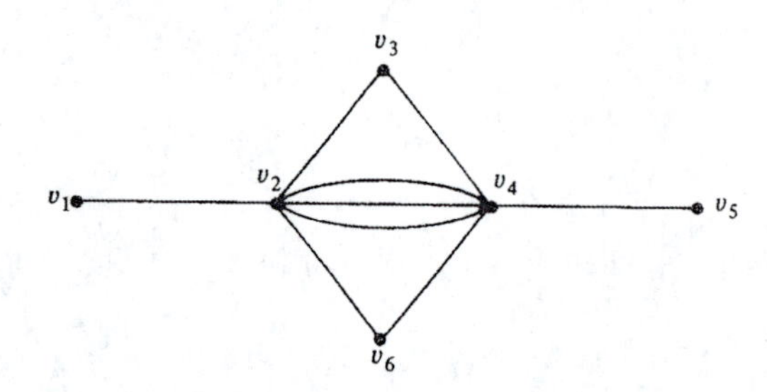

a) How many different paths are there from v_1 to v_5?
b) How many different trails are there from v_1 to v_5?
c) How many different walks are there from v_1 to v_5?

4. Explain why any path in a graph is also a trail.

5. Can a cycle contain only one edge? Can a cycle contain only two edges? Justify your answers.

6. Find all of the bridges in the following graphs.

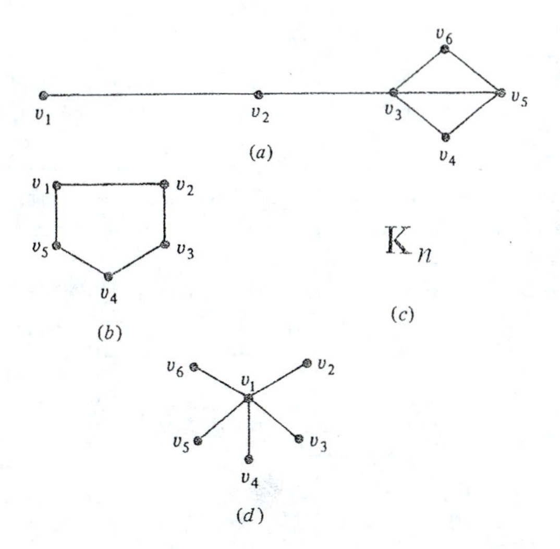

7. Let G be a graph, and let u and v be vertices in G. If there is a walk in G from u to v, show that there must also be a path from u to v. *Hint*, if W is a walk from u to v, and if the vertex w appears twice in W, consider what happens if you remove all of the edges and vertices in W between the two occurrences of w.

8. If a graph G contains a circuit, show that it must also contain a cycle.

9. Among all disconnected graphs with n vertices, let H be one with as many edges as possible. Show that H has exactly two connected components.

10. Among all connected graphs with n vertices, let G be one with as few edges as possible. Prove that if we remove any of the edges of G, the resulting graph must consist of exactly *two* connected components.

11. Let G be a connected graph that contains a circuit C. Show that it is possible to remove some edge from C and still have a connected graph.

12. Prove that any connected graph with the property that every edge is a bridge cannot have any circuits.

13. Prove that any connected graph G that has no circuits must have at least two vertices of degree 1. *Hint*, from among all of the paths in G pick one that has the most edges. Suppose this path goes from u to v. Must u and v be distinct vertices? What about their degrees?

14. Prove that a connected graph with n vertices that has no circuits must have exactly $n - 1$ edges. *Hint*, use the result of Exercise 13 and mathematical induction.

15. Use Theorem 6.2.3 to prove Theorems 6.2.1 and 6.2.2.

16. Show that if k is an integer and $1 \leq k \leq n - 1$, then the expression

$$\binom{n}{2} + k(k - n)$$

is largest when $k = 1$ or $k = n - 1$. *Hint*, rewrite this expression in the form $(\frac{1}{4})n^2 - (\frac{1}{2})n + (k - n/2)^2$.

17. Prove Theorem 6.2.3. *Hint*, let G consist of the connected components $G_1, G_2, \ldots, G_k$, and suppose that G_i has n_i vertices. To prove the left-hand inequality, apply Theorem 6.2.1 to each of the components G_i. For the right-hand inequality, assume that G has the most number of edges possible, and deduce that each component G_i must be a complete graph. Show that the most number of edges occurs when one component is the complete graph K_{n-k+1}, and the other $k - 1$ components are the complete graphs K_1.

*Let G be a graph. The **length** of a path P in G is defined to be the number of edges in P. If u and v are distinct vertices in G, we define the **distance** d(u, v) from u to v to be the length of the shortest path from u to v. If u and v are not connected, then we say that the distance from u to v is infinite. Also, we set $d(v, v) = 0$ for all vertices v in G.*

18. Find the distance between the vertices u and v in the following graphs.

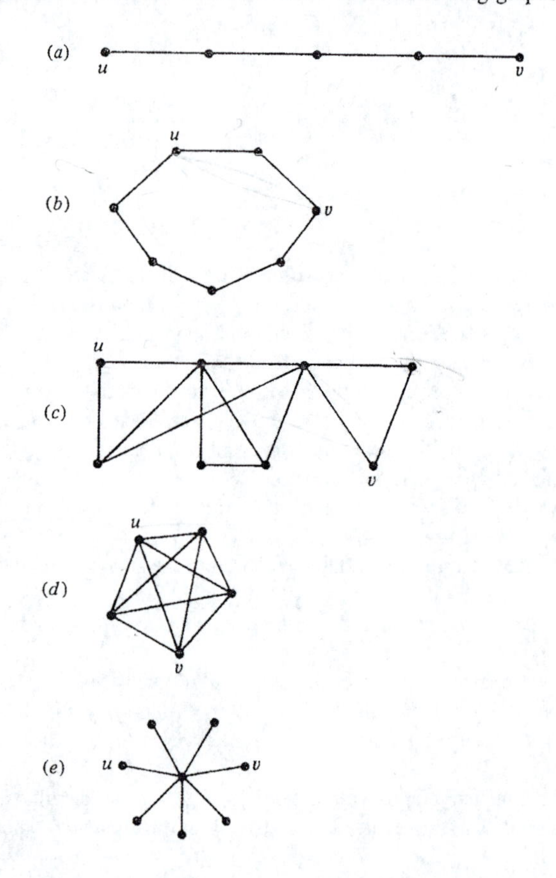

19. For each of the following graphs, determine which vertices have the greatest distance from vertex v_1.

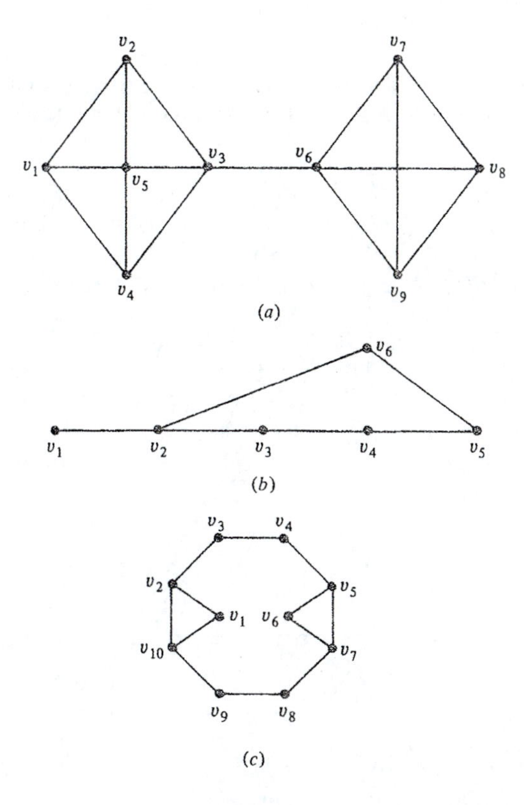

20. The **diameter** of a connected graph G is defined to be the maximum distance between all pairs of vertices in G. Thus, for example, the diameter of the graph

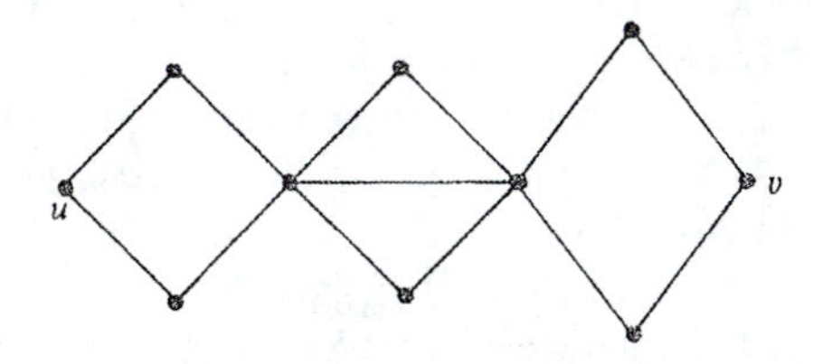

is 5 since the vertices u and v satisfy $d(u, v) = 5$ and no other pair of vertices are at a distance greater than 5 from each other. Find the diameter of each of the graphs in Exercise 18.

6.3 Eulerian and Hamiltonian Graphs

One of the most common problems in graph theory is that of finding a trail or path in a given graph. For example, in the communications graph of Example 6.1.1, a path from one vertex to another represents a way to send a message from one government to another, perhaps through intermediate governments. For instance, the government of Mongolia can send a message to the government of the U.S.S.R. by using either of the two paths

$$\{\text{Mongolia, China}\}, \{\text{China, U.S.A.}\}, \{\text{U.S.A., U.S.S.R.}\}$$

or

$$\{\text{Mongolia, China}\}, \{\text{China, France}\}, \{\text{France, U.S.S.R.}\}$$

As another example, a trail in the transportation graph of Example 6.1.2 represents a route from one city to another that does not use any highway more than once.

Even the famous Konigsberg bridge problem can be stated in terms of trails. The problem amounts to determining whether or not there is a closed trail, or circuit, in the graph

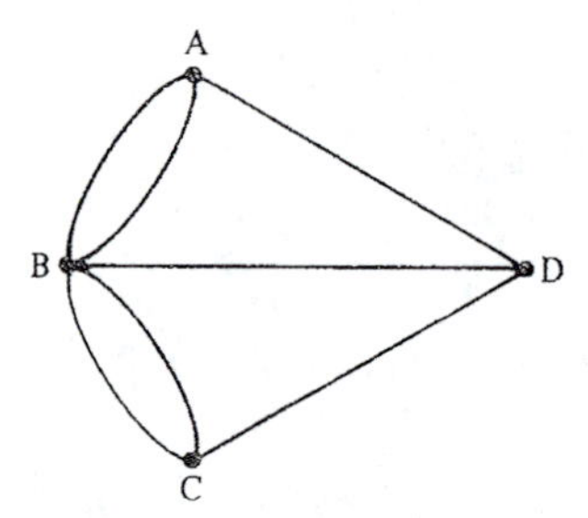

that uses *every* edge of the graph. As we saw at the beginning of Section 6.1, no such circuit exists.

It is easy to see how we can use the concept of a circuit to generalize the Konigsberg bridge problem. In honor of Leonhard Euler, we define an **Eulerian circuit** in a connected graph G to be a circuit that uses every edge of G. If a graph G has an Eulerian circuit, it is known as an **Eulerian graph**. Of course, an Eulerian graph must be connected. Now we may generalize the Konigsberg bridge problem by asking the following questions.

Is there a simple way to tell whether or not a connected graph is Eulerian?

If we know that a graph is Eulerian, is there a simple way to actually find an Eulerian circuit in the graph?

Happily, the answer to both of these questions is yes. In Section 6.1, we showed that there was no Eulerian circuit in the Konigsberg graph by reasoning that, if there were such a circuit, then all of the vertices must have even degree, which does not happen to be the case for this particular graph. As we will see in a moment, the same reasoning applies to any connected graph. That is, in order for a connected graph to have an Eulerian circuit, all of its vertices must have even degree. Also, the converse of this statement is true, namely, every connected graph that has the property that all of its vertices have even degree does have an Eulerian circuit. Let us put this into a theorem.

Theorem 6.3.1

A connected graph is Eulerian if and only if all of its vertices have even degree.

PROOF First let us suppose that a connected graph G is Eulerian and show that all of its vertices must have even degree. Actually, this is quite simple to do if we imagine ourselves traveling along an Eulerian circuit in G, starting in the *middle* of one of the edges of the circuit and returning to the same point. Then each time we pass through a vertex, we contribute two to the degree of that vertex. But since every edge appears exactly once in the circuit, we can conclude that each vertex must have degree equal to the sum of a certain number of twos. Hence, each vertex must have even degree.

Now let us suppose that all of the vertices of a connected graph G have even degree, and show that G must be Eulerian. In fact, let us actually show how to construct an Eulerian circuit in G. (This will show that the answer to our second question is yes.) The details of the construction are a bit involved, but the idea is very simple. We begin by finding any circuit in G, and then we increase the size of this circuit, if necessary, by combining it with other circuits, until we finally have an Eulerian circuit.

To find a circuit in G, we can simply start at any vertex v in G and travel along *different* edges of G until we return to v. We know that this is possible since each vertex of G has even degree, and so whenever we arrive at a vertex other than v, we can always leave it by an edge that has not been used before. Furthermore, since there are only a finite number of edges to traverse, we must eventually return to v.

Let us denote the circuit that we have found by C. (See Figure 6.3.1.)

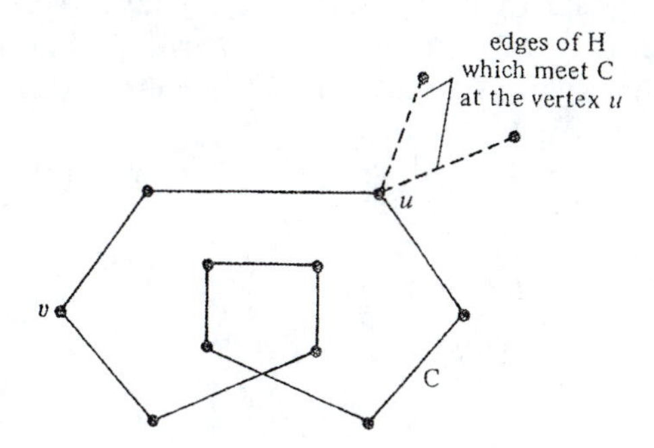

Figure 6.3.1

Now, if it happens that C contains all of the edges of G, then C is an Eulerian circuit and we are done. If not, then let us take a look at the graph obtained from G by deleting the edges of C. We denote this subgraph of G by H. Then H may not be connected, but it still has the property that all of its vertices have even degree. (Why?)

Also, since G is connected, some of the edges of H must meet the circuit C, say at the vertex u (and perhaps other vertices as well). (See Figure 6.3.1.)

Now, starting at u, we can find a circuit *in* H by following the same procedure that we used to find the circuit C. Namely, we start at u and travel along different edges in H until we return to u. This is possible since every vertex of H has even degree. Let us call this circuit D. The two circuits C and D are shown in Figure 6.3.2.

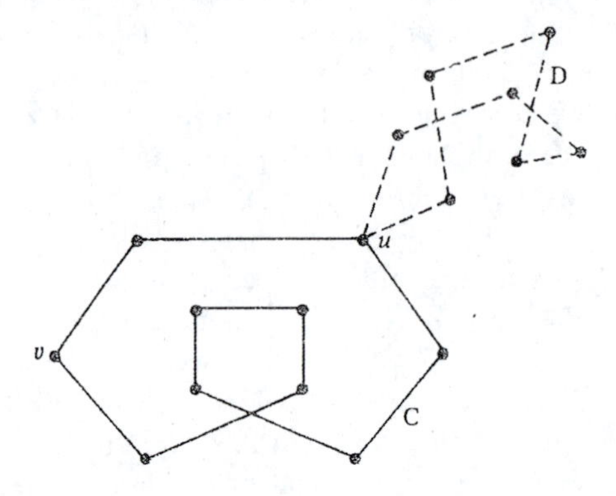

Figure 6.3.2

Now we can put these two circuits together into one large circuit, call it E, by starting at u, traversing all of C, and then traversing all of D. If this new circuit is Eulerian, then we are done. If not, then we can repeat the previous procedure by removing the edges of E from the graph G and finding a circuit in the remaining graph. This circuit can be added to E to create an even larger circuit.

Eventually, by repeating this procedure as many times as is necessary, we must obtain a circuit in G that uses all of the edges of G; that is, we must obtain an Eulerian circuit for G. This shows that G is Eulerian and completes the proof. ∎

Let us consider some examples of the use of Theorem 6.3.1.

Example 6.3.1

Consider the following map.

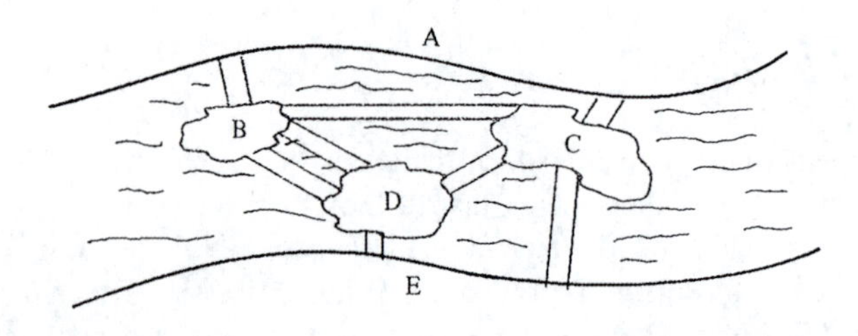

Is there a way to make a round trip through the islands in this map, crossing each bridge exactly once?

In order to answer this question, we first describe the map by a graph G.

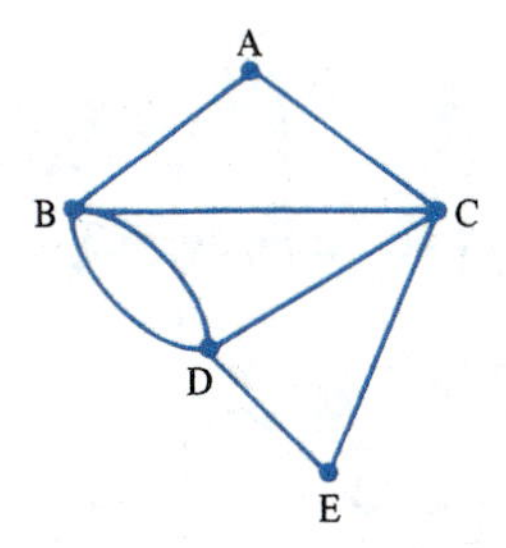

Each vertex in this graph represents a body of land, and each edge represents a bridge. Now, since the degree of each of the vertices in this graph is even, Theorem 6.3.1 tells us that the graph is Eulerian. That is, there is an Eulerian circuit in the graph. Since any such circuit describes a trip through the islands that crosses every bridge exactly once, the answer to the question is yes.

In order to actually find an Eulerian circuit in G, let us follow the steps in the second part of the proof of Theorem 6.3.1. For convenience, we begin by labeling each edge in G.

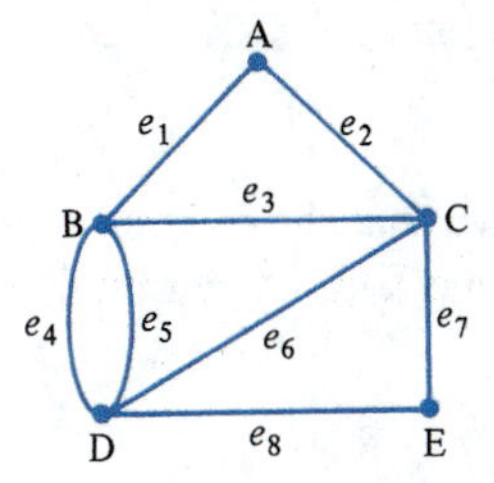

Now we pick any vertex, say vertex A, and find any circuit in G, starting at A, for example,

$$e_1, e_3, e_2$$

Let us draw this circuit in black.

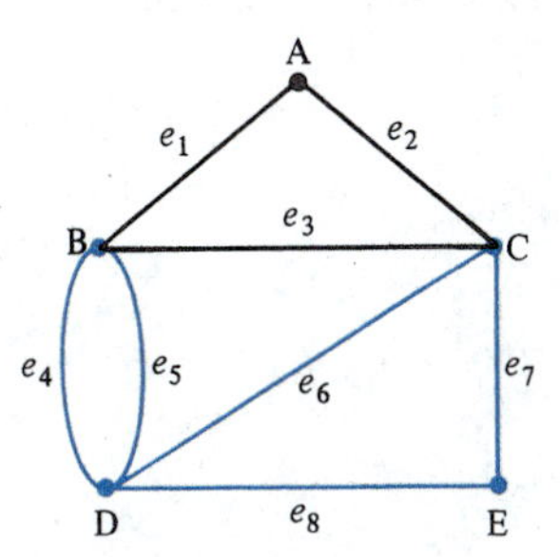

Figure 6.3.3

The next step is to observe that some of the colored edges in Figure 6.3.3 touch the black circuit at vertex B (as well as at C). So, beginning at B, we find a circuit *made up only of colored edges*, say

$$e_4, e_5$$

This circuit is shown in broken lines in Figure 6.3.4.

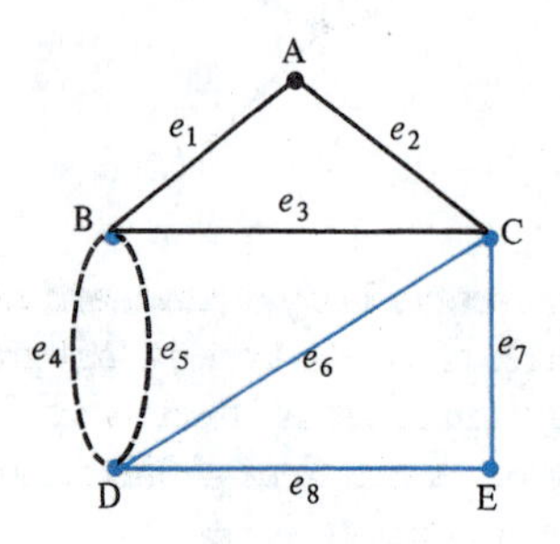

Figure 6.3.4

Now we can put the solid black circuit and the broken black circuit together into one circuit. This can be done by starting at B, traversing the solid black circuit and then the broken black circuit to get

$$e_3, e_2, e_1, e_4, e_5 \tag{6.3.1}$$

The next step is to observe that the colored edges meet the circuit 6.3.1 at the vertex D (as well as at C). So we look for a circuit, made up only of colored edges, starting at D. Such a circuit is

$$e_8, e_7, e_6$$

Finally, by first traversing circuit 6.3.1, starting at vertex D, and then traversing this circuit, we obtain the Eulerian circuit

$$e_5, e_3, e_2, e_1, e_4, e_8, e_7, e_6$$

□

Example 6.3.2

In the following graph, the vertices represent cities and the edges represent highways.

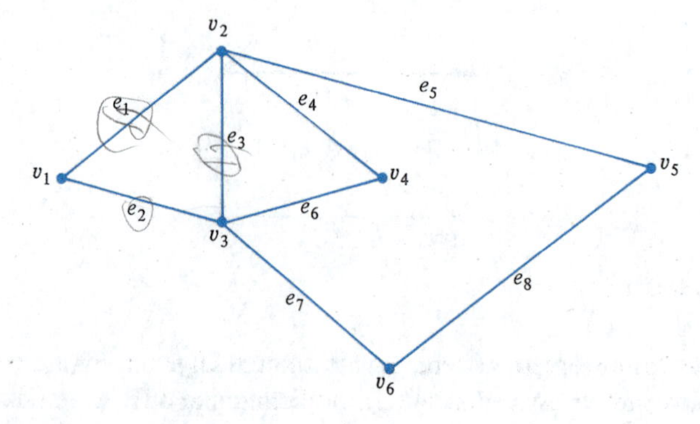

Figure 6.3.5

Is there a way to start at any city and make a round trip that traverses each highway exactly once?

The answer to this question is yes, since the graph in Figure 6.3.5 is Eulerian. To find such a round trip, we must find an Eulerian circuit in this graph. The first step is to pick a vertex, say v_1, and then find a circuit beginning at v_1, say

$$e_1, e_3, e_2$$

This is indicated in black in Figure 6.3.6.

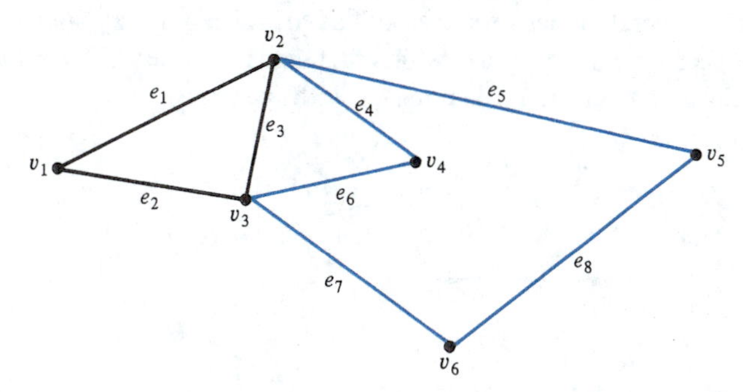

Figure 6.3.6

Then, we observe that the colored edges in Figure 6.3.6 meet the black circuit at vertex v_2. Looking for a circuit made only of colored edges at v_2, we get

$$e_5, e_8, e_7, e_6, e_4$$

Putting this circuit together with the black circuit, we get the Eulerian circuit

$$e_3, e_2, e_1, e_5, e_8, e_7, e_6, e_4$$ □

Finding Eulerian circuits can sometimes be a tedious job, especially in large graphs. There is another method for finding Eulerian circuits, known as **Fleury's algorithm**, which can in some cases be easier to use than the method of the last two examples.

Fleury's Algorithm

Let G be an Eulerian graph. Then an Eulerian circuit can be obtained for G by following the instructions:

1) Start at any vertex v and traverse the edges of G, erasing edges as they are used. Also, erase any isolated vertices as they are created.

2) Never choose an edge that is a bridge unless there is no other choice. That is, never choose an edge that will cause the remaining graph to become disconnected, unless there is no other choice.

Let us try Fleury's algorithm on the graph in Figure 6.3.5. We begin at vertex v_1 and choose first the edge e_1, then the edge e_3. Erasing these edges as we go, we are left at this point with the following graph.

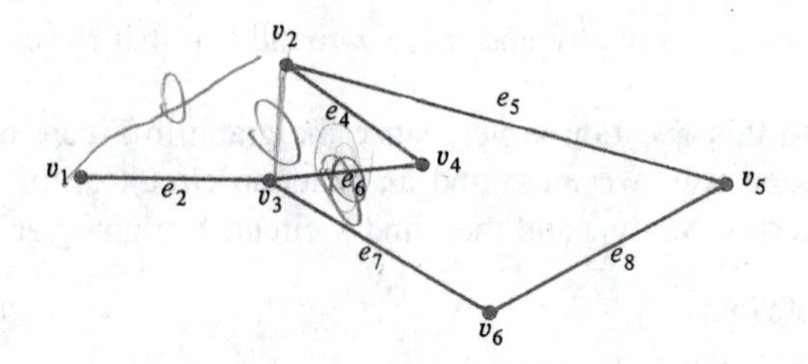

Now we have several choices for the next edge, namely e_2, e_6, and e_7. However, e_2 is a bridge, and so according to the instructions for Fleury's algorithm, we cannot choose e_2. So let us choose e_6, leaving us with the graph

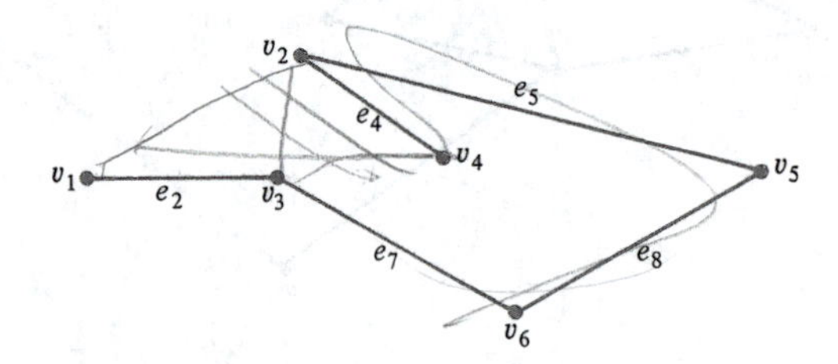

Now we can choose the edges e_4, e_5, e_8, e_7, and e_2, in this order, erasing edges and vertices as we go along. Each of these edges is a bridge, but we are allowed to choose them since there is no other choice.

Finally, putting this sequence of edges together, in the order that we chose them, gives the Eulerian circuit

$$e_1, e_3, e_6, e_4, e_5, e_8, e_7, e_2$$

Example 6.3.3

The following diagram is the floor plan of a certain house.

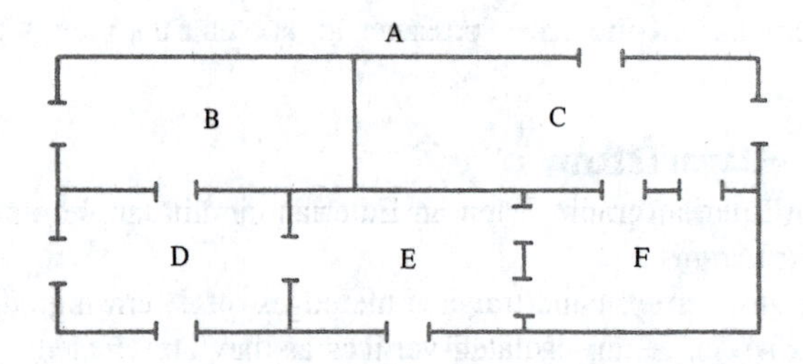

Is there a way to start in any one of the rooms, or outside the house, and make a round trip through the house that goes through each doorway exactly once?

In order to answer this question, we first represent the floor plan by the following graph.

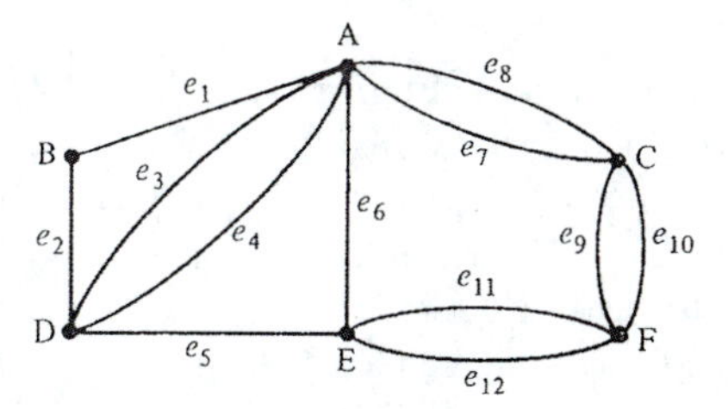

Each vertex of this graph represents a room (vertex A represents the outside of the house), and each edge represents a doorway. Then a round trip through the house that goes through each doorway exactly once corresponds to an Eulerian circuit in this graph.

Since the degree of each vertex in this graph is even, Theorem 6.3.1 tells us that there must be an Eulerian circuit. For example, the circuit

$$e_1, e_2, e_3, e_4, e_5, e_6, e_7, e_9, e_{11}, e_{12}, e_{10}, e_8$$

is an Eulerian circuit. (We found this circuit by using Fleury's algorithm.) ▯

It often happens that we are given a graph G, and two distinct vertices u and v of G, and that we want to know whether or not there is an *open* trail in G from u to v that uses each edge of G exactly once. Such a trail is called an **Eulerian trail** in G. This problem is easily solved with the aid of Theorem 6.3.1. We will leave a proof of the following theorem as an exercise.

■ Theorem 6.3.2

Let G be a connected graph and let u and v be *distinct* vertices in G. Then there is an Eulerian trail in G from u to v if and only if the degrees of u and v are odd and the degrees of all of the other vertices in G are even.

■ Example 6.3.4

Consider the following street map.

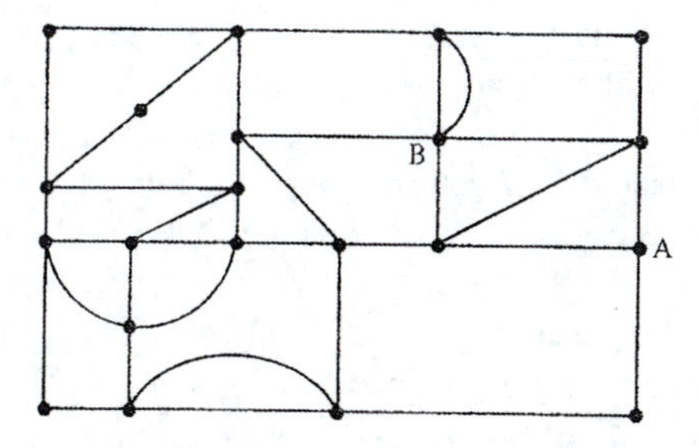

Figure 6.3.7

Each edge represents a street, and each vertex represents a street corner. Suppose that it is your job to deliver the mail to each house along these streets. You want to start at the post office, which is located at corner A and finish at your house, which is at corner B. Naturally, you would like to find a route that covers each street exactly once. Can this be done?

Thinking of the map in Figure 6.3.7 as a graph, we see that such a route is possible if and only if there is an Eulerian trail from vertex A to vertex B. Since the degrees of A and B are odd and the degrees of all the other vertices are even, Theorem 6.3.2 tells us that there is such a trail.

One way to find an Eulerian trail is to temporarily add an edge from A to B and then use either of the methods used earlier to find an Eulerian circuit in the resulting graph. When the new edge is removed from this circuit, the result will be an Eulerian trail from A to B. □

We want to conclude this section by mentioning a problem that is similar to that of determining whether or not a graph is Eulerian. It is the problem of determining whether or not a connected graph has a *cycle* that uses each *vertex* exactly once. Such a cycle is called a **Hamiltonian cycle**, after the Irish mathematician Sir William Rowan Hamilton (1805–1865), and any graph that has a Hamiltonian cycle is called a **Hamiltonian graph**. (In 1857, Hamilton invented a game that involved trying to find a Hamiltonian circuit in the graph

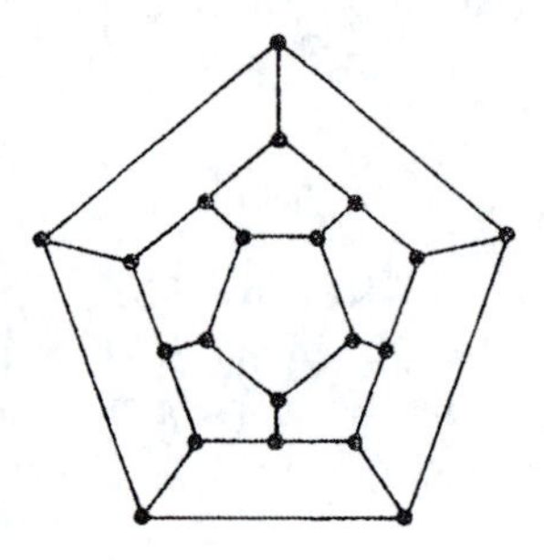

Figure 6.3.8

and this is why his name is associated with the problem. The vertices represented cities, and the object of the game was to find a round trip that covered each city exactly once.)

Even though the problem of determining whether or not a graph is Hamiltonian is similar to that of determining whether or not a graph is Eulerian, for some reason it seems to be much harder. We say this because no one has ever been able to find a reasonable criterion for determining whether or not a graph is Hamiltonian! In fact, this is one of the major unsolved problems in graph theory. (Of course, we could always consider all possible closed walks in a given graph to see if any of them are Hamiltonian cycles, but this is far from being a practical solution to the problem. A solution must consist of a criterion, such as the one given in Theorem 6.3.1 for Eulerian graphs, that is simple enough to be useful.)

Some progress has been made in this problem, however. For example, we have the following theorem, which we will not prove.

Theorem 6.3.3

If G is a simple graph with $n \geq 3$ vertices, and if

$$deg\ u + deg\ v \geq n \tag{6.3.2}$$

for all pairs of *nonadjacent* vertices u and v, then G is Hamiltonian.

Using this theorem, we can prove the following result (see Exercise 18).

Corollary 6.3.4

If G is a simple graph with $n \geq 3$ vertices, and if

$$deg\ v \geq \frac{n}{2} \tag{6.3.3}$$

for all vertices v, then G is Hamiltonian.

Example 6.3.5

The graph

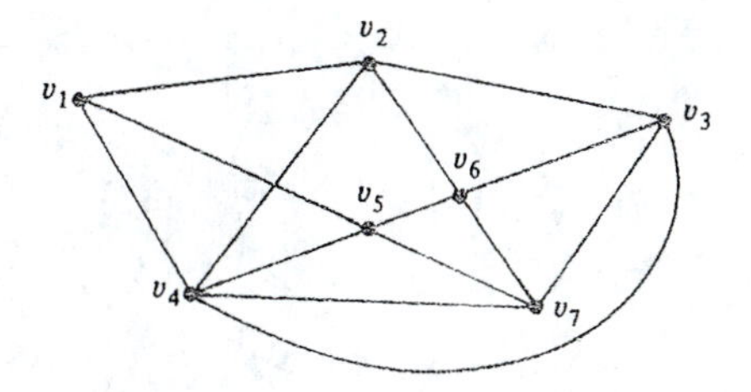

Figure 6.3.9

satisfies condition 6.3.2, namely,

$$deg\ u + deg\ v \geq 7$$

for all pairs of nonadjacent vertices u and v. Therefore, according to Theorem 6.3.3, it is Hamiltonian. The graph

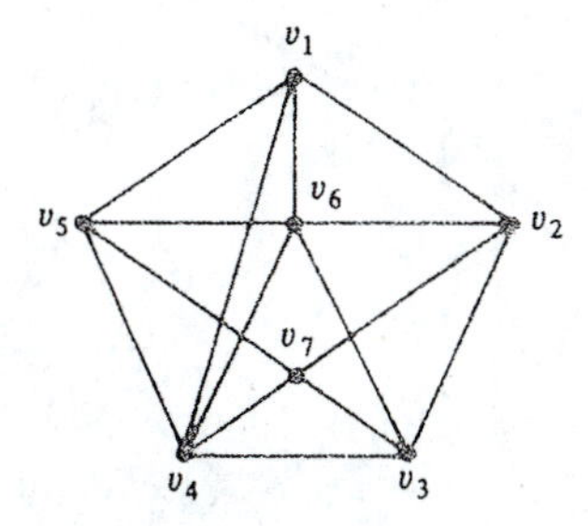

Figure 6.3.10

satisfies condition 6.3.3, namely,

$$deg\ v \geq \frac{7}{2}$$

for all vertices v. Therefore, according to Corollary 6.3.4, it is Hamiltonian. □

We should observe that condition 6.3.3 is easier to check than condition 6.3.2, and so, as a general rule, it is a good idea to see if the corollary applies before trying the theorem. However, as you can see from the first graph in Example 6.3.5, just because the corollary does not apply does not mean that the theorem will not apply. (See Exercise 19.)

□ □ Exercises

In Exercises 1 through 5, decide whether or not the graph is Eulerian, and if it is, then find an Eulerian circuit in the graph.

1.

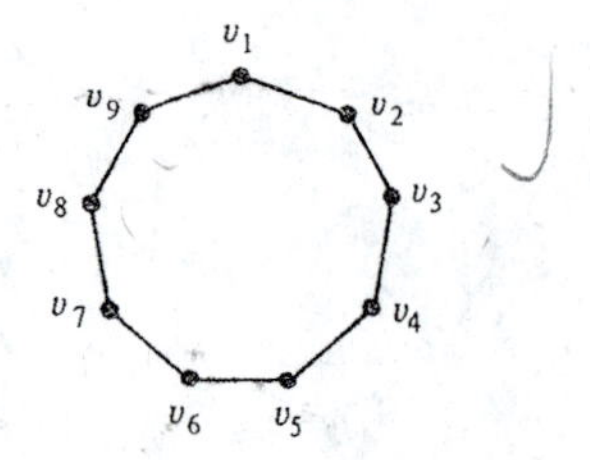

2.

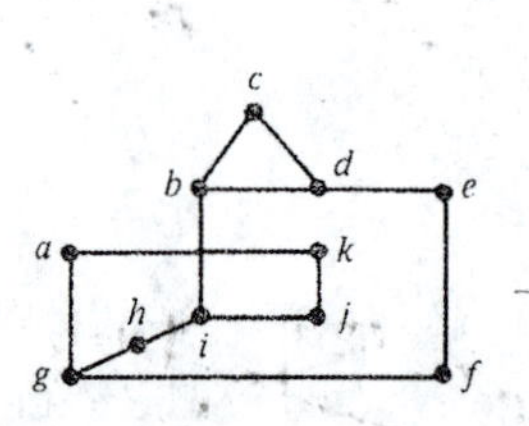

3.

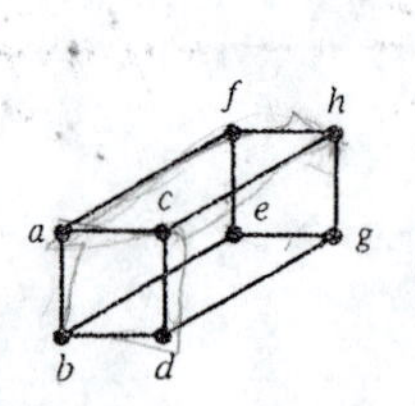

4.

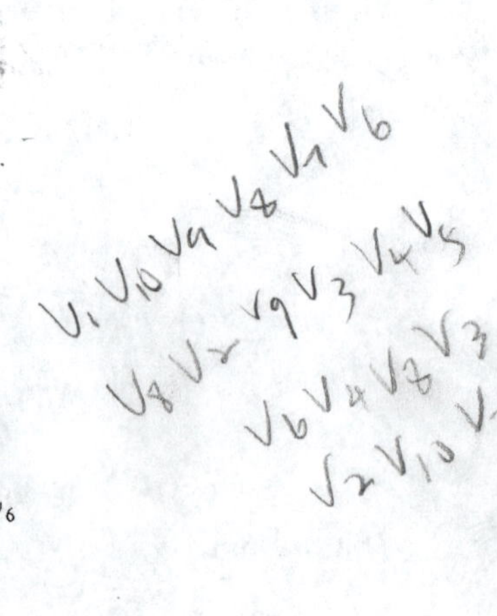

5.

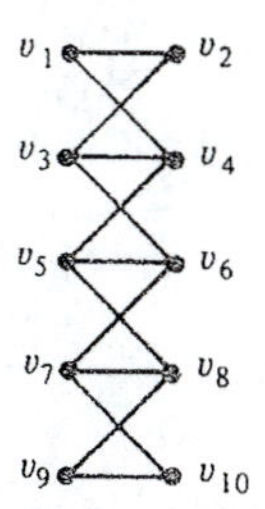

6. Is the complete graph K_n Eulerian? Explain.

7. Is the complete bipartite graph $K_{n,m}$ Eulerian? Explain. (The complete bipartite graph is defined following Exercise 20 of Section 6.1.)

8. Find two distinct Eulerian circuits in the graph

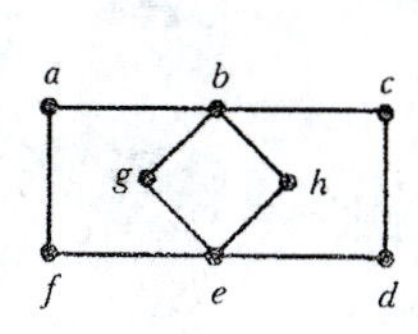

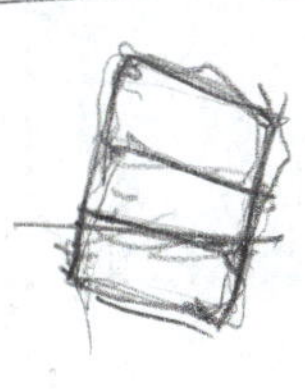

In Exercises 9 through 11, decide whether or not an Eulerian trail exists from u to v. If so, find such a trail.

9.

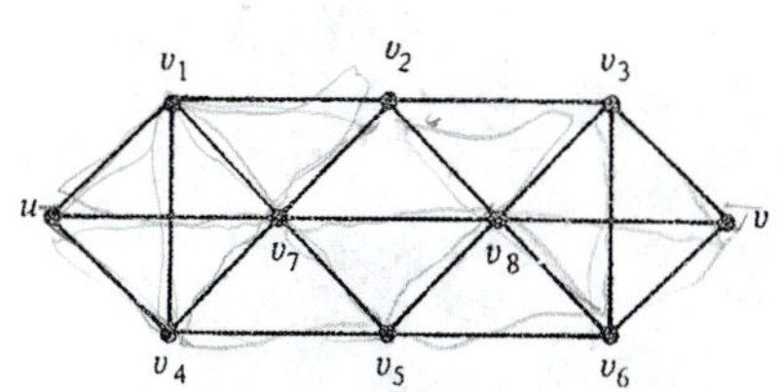

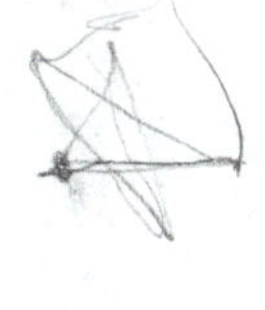

10.

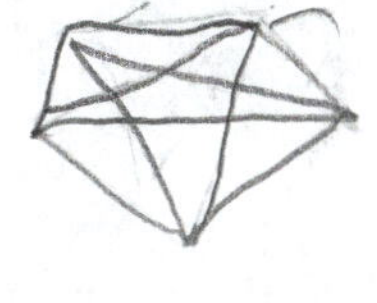

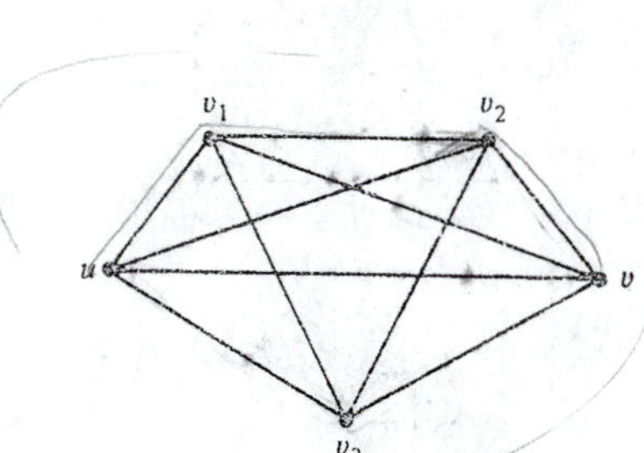

11.

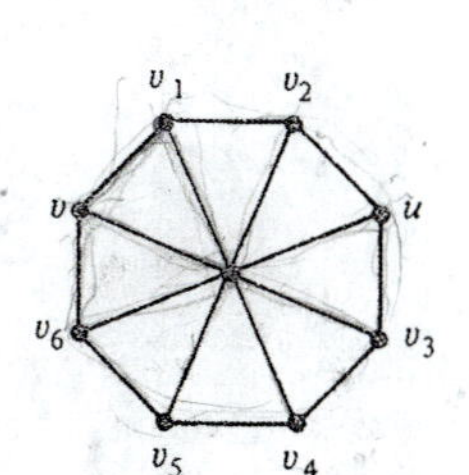

12. Find two distinct Eulerian trails from u to v in the graph

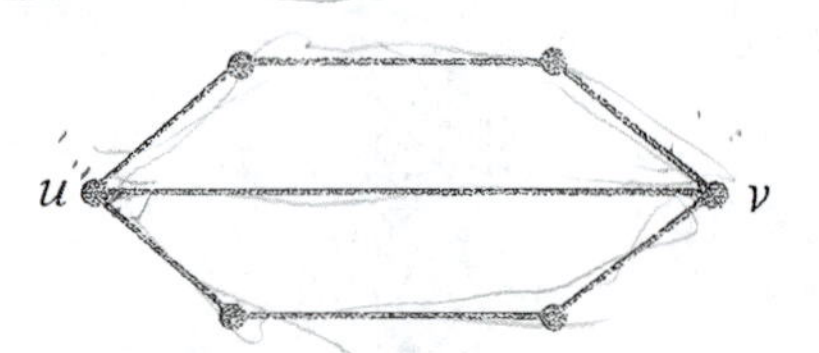

13. Consider the following road map.

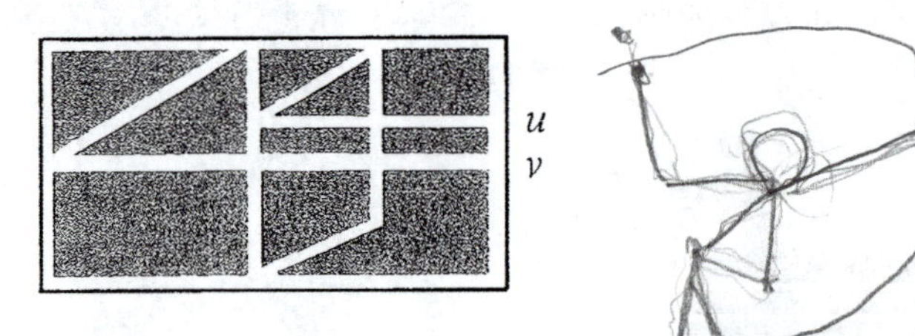

Suppose that it is your job to inspect the roads in this map for potholes and other obstructions to traffic. Can you do this in such a way that you start at corner u, finish at corner v, and cover each street exactly once? If so, find such a way. (Use Fleury's algorithm.)

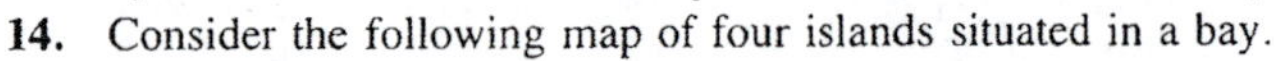

14. Consider the following map of four islands situated in a bay.

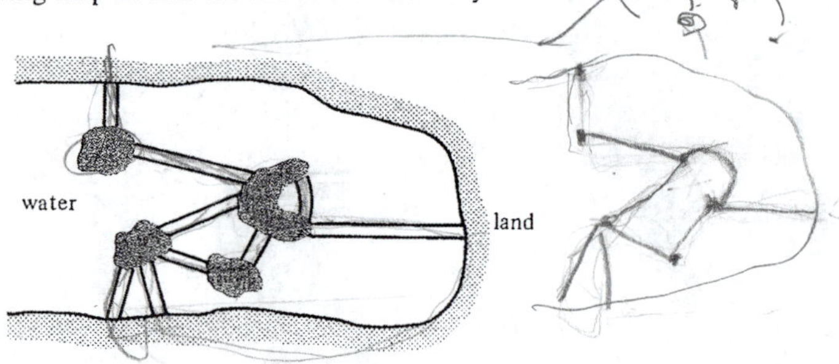

Is there a way to start on the shore and make a round trip through these islands, using each bridge exactly once? If so, find such a trip.

15. Consider the following floor plan of a house.

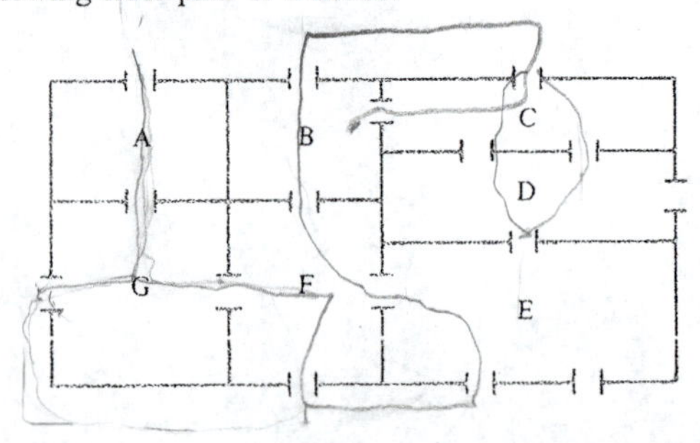

Suppose that we want to start in one of the rooms, or outside, and tour this house in such a way that we go through each doorway exactly once. Is this possible? If so, can we start and end in any room that we choose? Explain.

16. Prove Theorem 6.3.2. *Hint*, add a new edge to the graph G connecting u and v. Then apply Theorem 6.3.1.

17. A graph G is said to be **arbitrarily traversable** from a vertex v if whenever we start at v and construct a trail by continuing to add edges until we cannot go any further, the result is always an Eulerian circuit.

a) Show that the following graph is arbitrarily traversable from the vertex v.

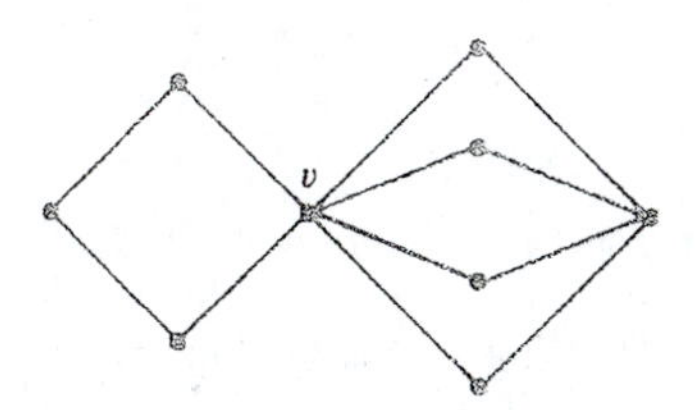

b) Show that the complete graph K_5 is Eulerian, but that it is not arbitrarily traversable from any of its vertices.

c) Can you think of any reason why an arbitrarily traversable graph would make a good model for the floor plan of a museum? What would be the significance of the vertex v?

18. Prove Corollary 6.3.4 by using Theorem 6.3.3.

19. Show that the first graph in Example 6.3.5 satisfies condition 6.3.2, but not condition 6.3.3. Hence, Theorem 6.3.3 applies to this graph even though Corollary 6.3.4 does not.

20. Find a Hamiltonian cycle in the graph of Figure 6.3.8.

21. Find a Hamiltonian cycle in the graph of Figure 6.3.9.

22. Find a Hamiltonian cycle in the graph of Figure 6.3.10.

In Exercises 23 through 26, determine whether or not the graph is Hamiltonian. If it is, find a Hamiltonian cycle in the graph.

23.

v_1, v_2, v_3, v_4, v_5, v_6

24.

a, b, c, d, e

25.

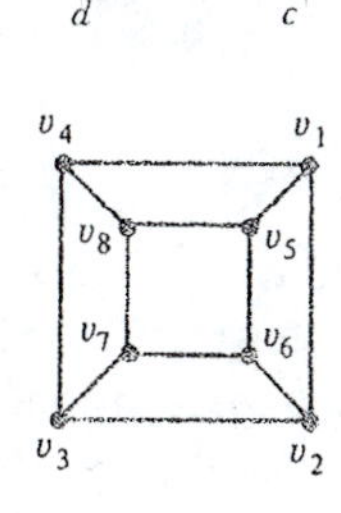

26.

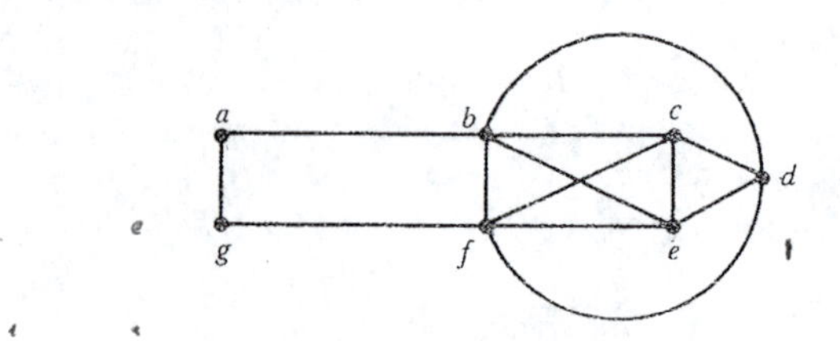

27. Is the complete graph K_n Hamiltonian? Explain.

28. Is the complete bipartite graph $K_{n,n}$ Hamiltonian? Explain.

29. a) Find a graph that is Eulerian but not Hamiltonian.

b) Find a graph that is Hamiltonian but not Eulerian.

6.4
Graph Isomorphisms; Planar Graphs

There is not much question that the graphs in Figure 6.4.1 are the same. One graph is simply drawn a bit differently than the other. Similarly, the graphs in Figure 6.4.2 are the same, although it is a little more difficult to see this than it was for the graphs in Figure 6.4.1.

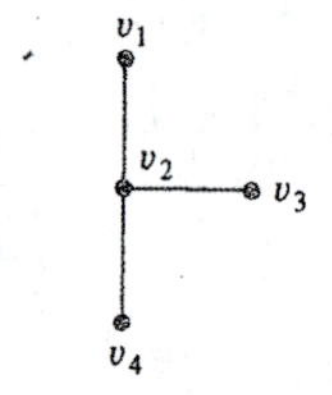

v_1 v_2 v_3 v_4

Figure 6.4.1

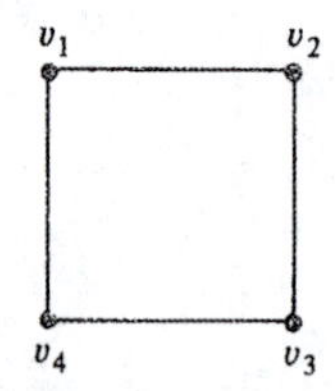

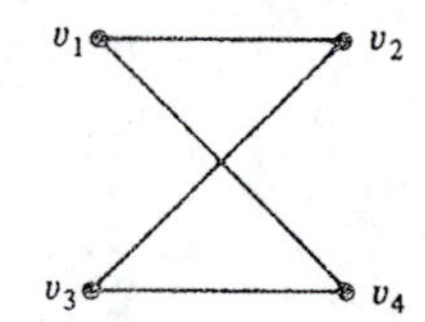

Figure 6.4.2

Now, consider the graphs G_1 and G_2 in Figure 6.4.3. These graphs are not quite the same. For instance, $\{v_2, v_3\}$ is an edge of G_1, but it is not an edge of G_2. Nevertheless, these graphs are "almost" the same. In fact, if we interchange the labels v_3 and v_4 in G_2, then it will be the same as G_1.

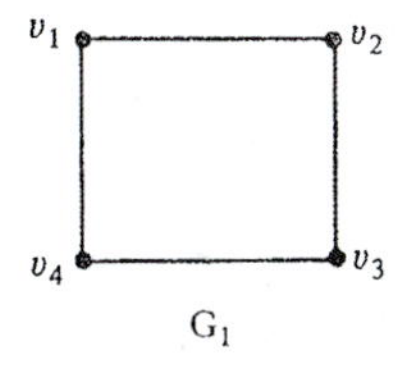

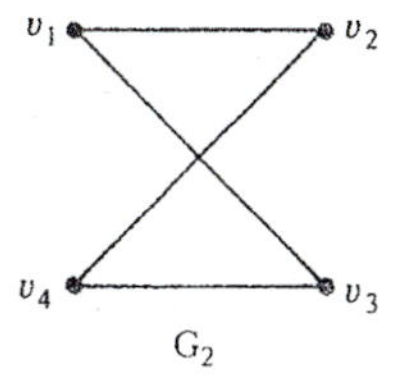

Figure 6.4.3

Let us put this in a more mathematical way. There is a one-to-one correspondence between the vertices of the two graphs, namely

Vertices of G_1		*Vertices of* G_2
v_1	$\longleftrightarrow$	v_1
v_2	$\longleftrightarrow$	v_2
v_3	$\longleftrightarrow$	v_4
v_4	$\longleftrightarrow$	v_3

with the property that if we "match" the vertices of each graph, not by the fact that they have the *same* labels, but rather by using this correspondence, then the two graphs are the same.

The similarity between graphs G_1 and G_2 is so strong that the two graphs have very similar properties, and for many purposes, we do not need to distinguish between them. This leads us to make the following definition. We will restrict our attention to simple graphs.

Definition

Let $G_1 = (V_1, E_1)$ and $G_2 = (V_2, E_2)$ be *simple* graphs. Then a function $f: V_1 \to V_2$ is called an **isomorphism** between G_1 and G_2 if it satisfies the following two conditions.

1) The function f is a bijection (one-to-one and onto). That is, f is a one-to-one correspondence between the vertices of G_1 and the vertices of G_2.
2) $\{v_i, v_j\}$ is an edge of G_1 if and only if $\{f(v_i), f(v_j)\}$ is an edge of G_2. □

If there is an isomorphism between two graphs G_1 and G_2, then we say that the graphs are **isomorphic**. (The term *isomorphic* comes from the Greek—*iso* from the Greek *isos*, meaning *the same as*, and *morphic* comes from the Greek *morphe*, meaning *form*. Hence, *isomorphic* means *the same form*.)

The second condition in this definition says that, once we have "matched" the vertices $v_1, v_2, \ldots, v_n$ in G_1 with the vertices $f(v_1), f(v_2), \ldots, f(v_n)$ in G_2, then the edges of G_1 "match" the edges of G_2.

Let us consider some other examples.

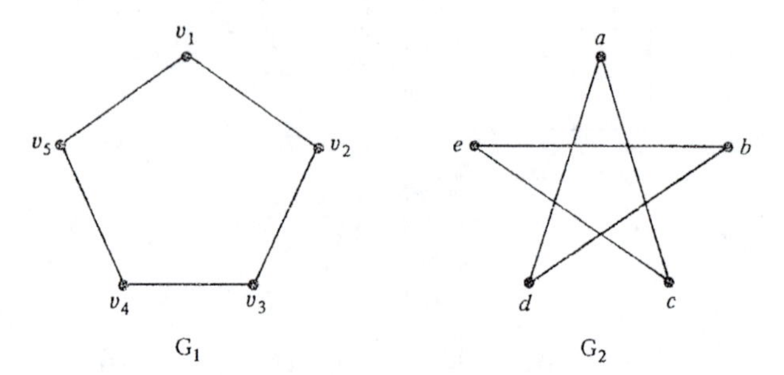

Figure 6.4.4

Example 6.4.1

The graphs in Figure 6.4.4 are isomorphic, since the function f defined by

$$f(v_1) = a$$

$$f(v_2) = c$$

$$f(v_3) = e$$

$$f(v_4) = b$$

$$f(v_5) = d$$

is an isomorphism between the two graphs. We can see this by listing the edges of the two graphs

Edges of G_1	*Edges of* G_2
$\{v_1, v_2\}$	$\{a, c\} = \{f(v_1), f(v_2)\}$
$\{v_2, v_3\}$	$\{c, e\} = \{f(v_2), f(v_3)\}$
$\{v_3, v_4\}$	$\{e, b\} = \{f(v_3), f(v_4)\}$
$\{v_4, v_5\}$	$\{b, d\} = \{f(v_4), f(v_5)\}$
$\{v_5, v_1\}$	$\{d, a\} = \{f(v_5), f(v_1)\}$

As we can see, the edges of these two graphs "match," when we match the vertices using the one-to-one correspondence f. □

Example 6.4.2

The graphs in Figure 6.4.5 are *not* isomorphic. We can see this by reasoning as follows. In order for these two graphs to be isomorphic, there must be a one-to-one correspondence between their vertices that also "matches" the edges of the two graphs. But such a matching means that both graphs must have the same number of edges! Since the graphs in Figure 6.4.5 do not have the same number of edges, they cannot be isomorphic.

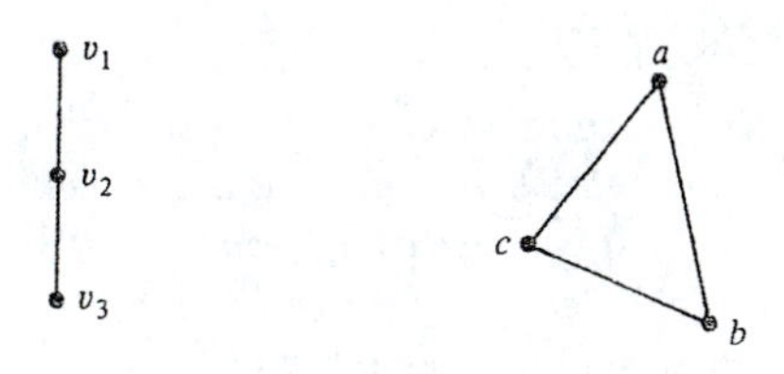

Figure 6.4.5 □

In general, it is very difficult to determine whether or not two graphs are isomorphic. As in the previous example, it is sometimes possible to show that two graphs are *not* isomorphic by an indirect method. The idea rests on the fact that certain properties of graphs are shared by graphs that are isomorphic. For example, isomorphic graphs must have the same number of vertices, as well as the same number of edges. Therefore, if two graphs do not have the same number of vertices, or the same number of edges, then they cannot be isomorphic!

A property that is shared by isomorphic graphs is called an **isomorphic invariant**. For example, the property of having a certain number of vertices is an isomorphic invariant, as is the property of having a certain number of edges. Another important isomorphic invariant is the property of having a certain number of cycles of a certain length. For example, if a graph G has a cycle of length 3, then so must all graphs that are isomorphic to G. We use this invariant in the next example.

Example 6.4.3

The two graphs in Figure 6.4.6 are *not* isomorphic, since the second graph has seven cycles of length 3 but the first graph has only one.

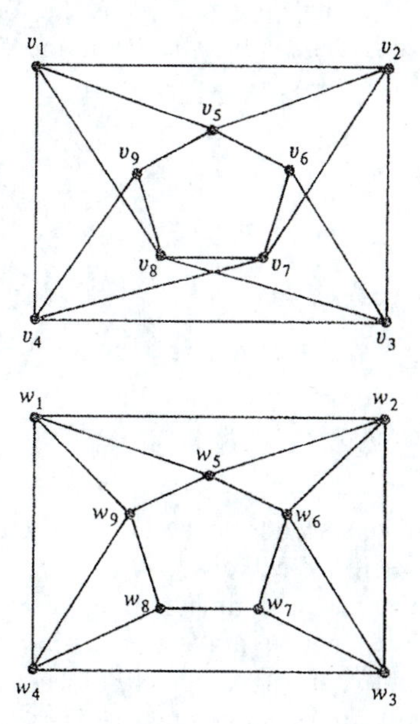

Figure 6.4.6 □

Notice that the graphs in Figure 6.4.6 have the same number of vertices, and the same number of edges. However, as we saw in the previous example, they are *not* isomorphic. This shows that *just because two graphs have the same number of vertices and the same number of edges does not necessarily mean that they are isomorphic.*

Now let us turn our attention to a topic that we mentioned very briefly in Section 6.1. A graph G is said to be **planar** if it can be drawn in the plane in such a way that the edges of the graph intersect only at the vertices of the graph. For example, the graph shown below is planar

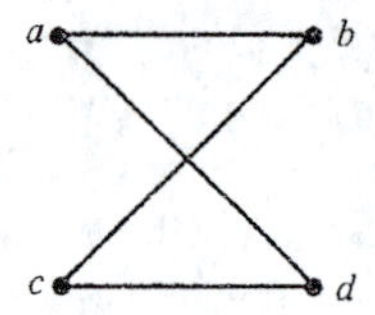

since it can also be drawn as follows

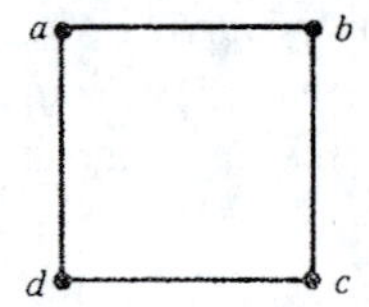

where the edges intersect only at the vertices a, b, c, and d.

As another example, the complete graph K_5 and the complete bipartite graph $K_{3,3}$ shown in Figure 6.4.7 are not planar. (For the definition of complete bipartite graph, see Exercise 22 of Section 6.1.)

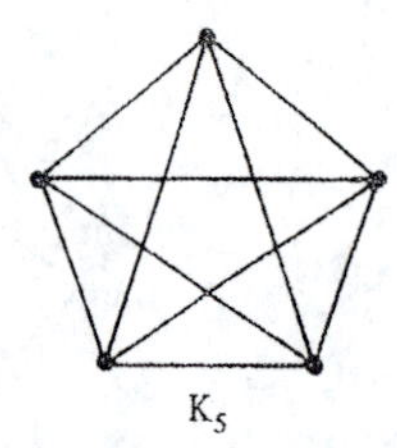

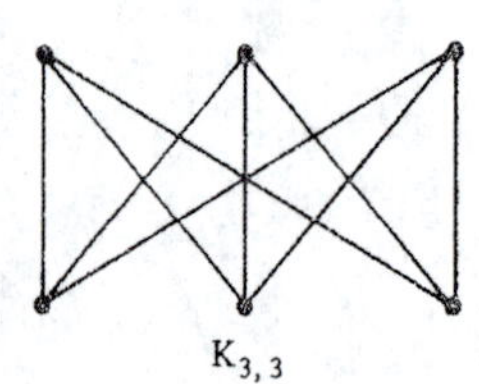

Figure 6.4.7

We will show a bit later in this section that $K_{3,3}$ is not planar, and we will ask you to show that K_5 is not planar in the exercises.

The concept of planarity has many important applications in the real world. For example, when printed circuit boards are designed, it is desirable to have as few unwanted crossings as possible. Of course, the most ideal situation would be to design a circuit that is planar, and therefore has no unwanted crossings. However, in practice this is sometimes difficult to arrange, and so the next best thing is to design the circuit to have as few unwanted crossings as possible. For instance, the circuit

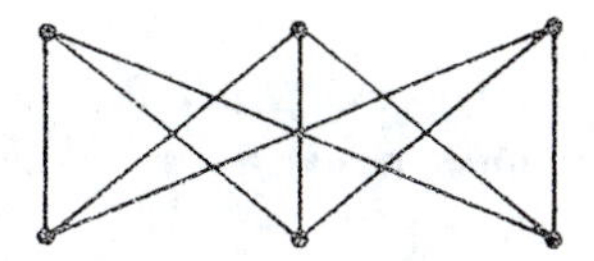

can be redesigned in the form

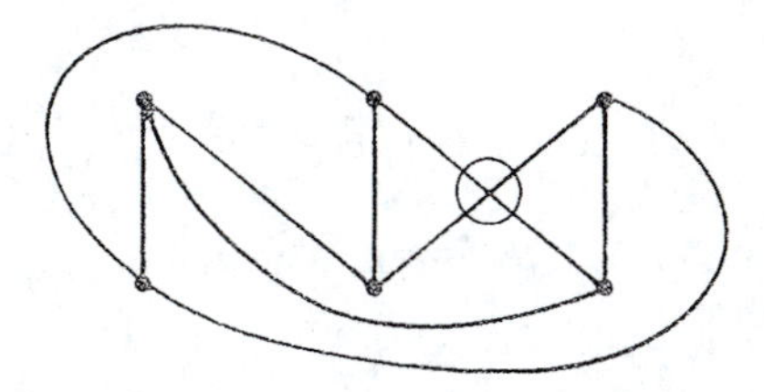

These two graphs are isomorphic, and so the circuits will have the same behavior. However, the second design has only one unwanted crossing (circled), rather than seven unwanted crossings.

Let us agree to use the term **crossing** for a point where the edges of a graph touch that is not a vertex of the graph. When a planar graph is drawn in the plane without crossings, it divides the plane into distinct regions, which are usually called **faces**. For example, the planar graph G shown in Figure 6.4.8 has four faces, labeled f_1, f_2, f_3, and f_4. We have also labeled the vertices and the edges of this graph.

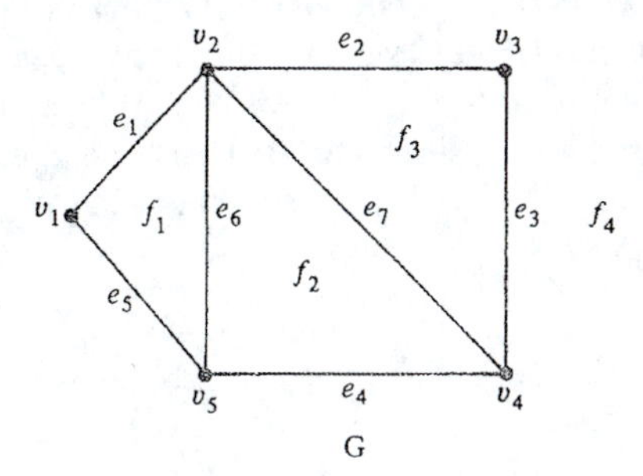

Figure 6.4.8

Now, if we denote the number of vertices of G by v, the number of edges of G by e, and the number of faces of G by f, then we have

$$v = 5, \qquad e = 7, \qquad f = 4$$

Hence, we have $f - e + v = 4 - 7 + 5 = 2$; that is

$$f - e + v = 2 \tag{6.4.1}$$

It is an amazing fact that this formula is true for *all* connected planar graphs! Formula 6.4.1 is known as **Euler's formula**. (Remember Leonhard Euler from the Konigsberg bridge problem?) Let us put it into a theorem.

Theorem 6.4.1 (Euler's Formula)

If a connected planar graph is drawn in the plane with no crossings, then we have

$$f - e + v = 2$$

where f is the number of faces, e is the number of edges, and v is the number of vertices of the graph.

PROOF We will prove Euler's formula by induction. Let P(n) be the proposition that Euler's formula is true for any connected planar graph that has exactly n edges. First, we must show that P(1) is true. But if the graph has only one edge, then it must have one of the two forms

We leave it to you to check that Euler's formula holds in either case.

Now we must show that if P(k) is true, then so is P($k + 1$). That is, we must show that if Euler's formula holds for any connected planar graph with k edges, then it must also hold for any connected planar graph with $k + 1$ edges. So suppose that G is a connected planar graph with $k + 1$ edges.

We will consider two cases, depending on whether or not G has any cycles. If G has no cycles, then it must have a vertex, call it x, of degree 1. This can be seen as follows. We simply pick a vertex and traverse a path from that vertex until we can go no farther. This must happen since the graph has no cycles. The vertex that we get "stuck" at has degree 1. Thus, in the case that G has no cycles, it must have an edge a as shown in Figure 6.4.9.

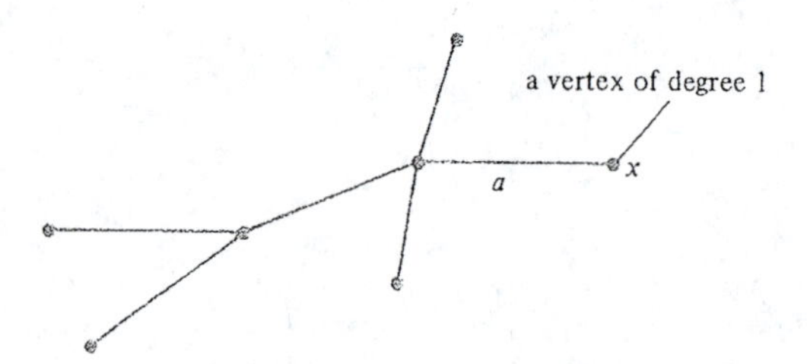

Figure 6.4.9

Now, let us remove the edge a, and the vertex x, and call the resulting graph G'. Certainly G' is a connected planar graph, and it is drawn in such a way that it has no crossings. Furthermore, it has one less edge than G; that is, it has k edges. Therefore, according to the induction hypothesis, Euler's formula holds for the graph G', and so if we let f', e', and v' be the number of faces, edges, and vertices, respectively, of G' then

$$f' - e' + v' = 2 \qquad (6.4.2)$$

But, of course, $f' = f$, $e' = e - 1$, and $v' = v - 1$. Substituting these values into Equation 6.4.2 and simplifying gives

$$f - e + v = 2 \qquad (6.4.3)$$

which is Euler's formula for the graph G. Thus, in the case in which G has no cycles, Euler's formula does hold.

In order to finish the proof, we must show that Euler's formula holds even when G has a cycle. So suppose that G has a cycle, and let a be an edge of that cycle. Then if we remove only the edge a (and no vertices), the resulting graph G' will be connected and planar, and have k edges. Therefore, by the induction hypothesis, Equation 6.4.2 will hold for G'.

But in this case, we have $f' = f - 1$, $e' = e - 1$, and $v' = v$. The reason that $f' = f - 1$ is that the removal of the edge a has the effect of combining two faces into one, as shown in Figure 6.4.10. Substituting these values into Equation

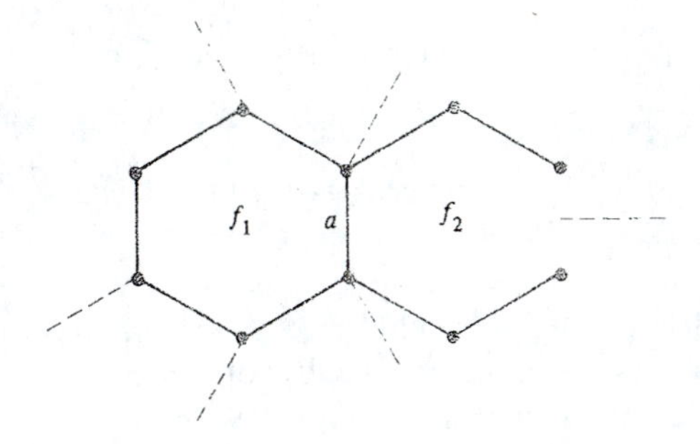

Figure 6.4.10
The removal of edge a has the effect of combining the faces f_1 and f_2.

6.4.2 and simplifying again gives Equation 6.4.3. Thus, we see that in either case, Euler's formula holds for the graph G. This concludes the proof by induction. ∎

Euler's formula can be used to show that the graphs K_5 and $K_{3,3}$ in Figure 6.4.7 are not planar. Let us consider the graph $K_{3,3}$ and leave the complete graph K_5 for the exercises.

Theorem 6.4.2

The graph $K_{3,3}$ is not planar.

PROOF We will prove this theorem by contradiction. That is, we will assume that $K_{3,3}$ is planar and use that assumption, along with Euler's formula, to deduce a contradiction. This will show that $K_{3,3}$ is not planar.

So let us assume that $K_{3,3}$ is planar. Under this assumption, Euler's formula must hold for $K_{3,3}$. ($K_{3,3}$ is certainly connected.) Of course, $K_{3,3}$ has 6 vertices and 9 edges, and so Euler's formula tells us that

$$f = e - v + 2 = 9 - 6 + 2 = 5$$

That is, $K_{3,3}$ must have 5 faces when drawn in the plane with no crossings. In summary, $v = 6$, $e = 9$, and $f = 5$.

Now let us assume that $K_{3,3}$ is drawn in the plane with no crossings. We will use this fact to show that e must be greater than or equal to 10. Of course, this is a contradiction to the fact that $e = 9$.

We begin by counting the number of ordered pairs (x, y), where x is an edge of the graph and y is a face that is bounded by x. Let us denote this number by I. As you can see from the graph in Figure 6.4.7, if $K_{3,3}$ were drawn in the plane with no crossings, each face would be bounded by at least 4 edges. Hence, there are *at least* 4 ordered pairs of the form (x, y) for each face y. Thus, we get

$$I \geq 4f$$

On the other hand, each edge bounds exactly 2 faces, and so there are *exactly* two ordered pairs of the form (x, y) for each edge x in the graph, and so

$$I = 2e$$

Comparing these two expressions, we see that $2e \geq 4f$, or

$$e \geq 2f$$

But since according to Euler's formula $f = 5$, this inequality tells us that $e \geq 10$, which is the promised contradiction. This completes the proof. ∎

To conclude this section, we want to discuss a criterion for deciding whether or not a graph is planar. As we have seen, neither of the graphs K_5 or $K_{3,3}$ is planar. It is not hard to see that any graph that contains either of these graphs as a subgraph cannot be planar. As it turns out, the converse of this is "almost" true. That is, it is almost true that a graph is planar if and only if it does not contain a subgraph of the form K_5 or $K_{3,3}$. Let us explain what we mean by "almost."

Consider the graphs G and G' in Figure 6.4.11. The only difference between these two graphs is that G has an extra vertex v of degree 2. In fact, we can get the graph G' from G by deleting this vertex and connecting the vertices a and b.

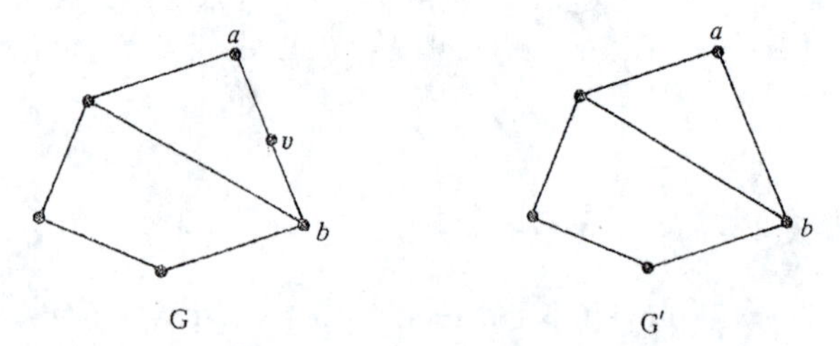

Figure 6.4.11

Similarly, the graph H' shown in Figure 6.4.12 can be obtained from the graph H by inserting a new vertex v on the edge $\{a, b\}$, forming two new edges $\{a, v\}$ and $\{v, b\}$ and discarding the old edge $\{a, b\}$.

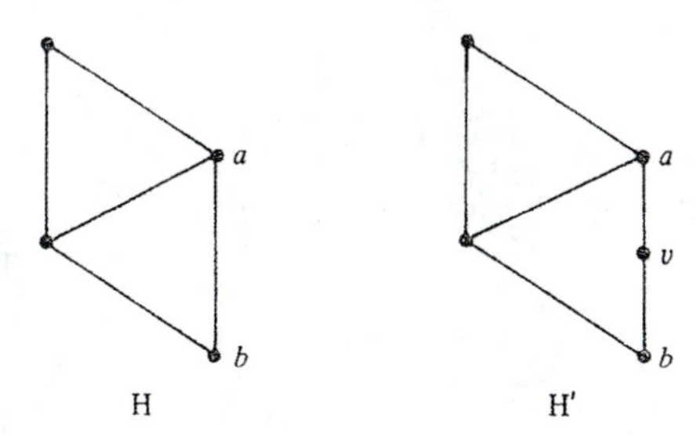

Figure 6.4.12

Let us refer to these two operations of deleting and inserting a vertex of degree 2 as **elementary transformations**. If a graph G can be obtained from a graph H by performing elementary transformations (perhaps of both types), then we say that the graphs G and H are **homeomorphic**. Using these terms, we can state a criterion for a graph to be planar.

Theorem 6.4.3

A graph G is planar if and only if it does not have a subgraph that is homeomorphic to either K_5 or $K_{3,3}$.

Of course, we can also word this theorem in its negative, as follows.

Theorem 6.4.4

A graph G is not planar if and only if it contains a subgraph that is homeomorphic to either K_5 or $K_{3,3}$.

Exercises

The following properties of a graph are isomorphic invariants.

1) has a certain number of vertices
2) has a certain number of edges
3) has a cycle of a certain length
4) has a vertex of a certain degree (see Exercise 13)
5) is connected (see Exercise 14)
6) is Eulerian (see Exercise 15)

In Exercises 1 through 12, either show that the two graphs G_1 and G_2 are isomorphic by finding an isomorphism between them or else prove that they are not isomorphic by exhibiting an invariant from the previous list that is possessed by only one of the two graphs.

1.

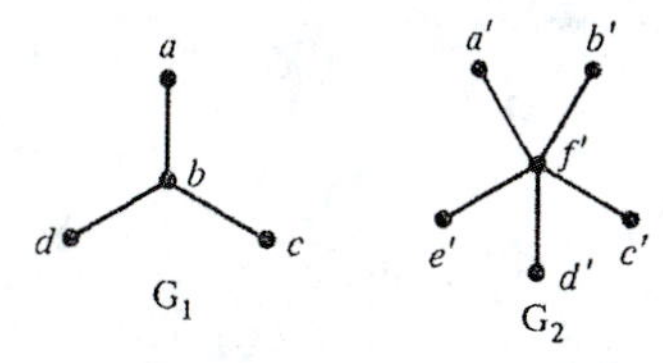

2.

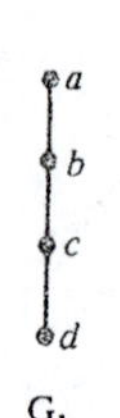

G

y z x w

G_2

3.

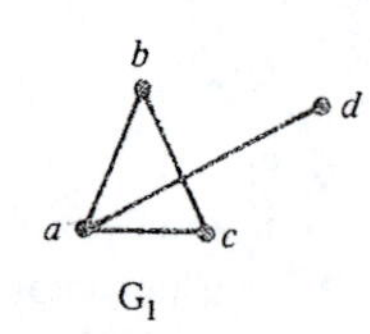

G_1

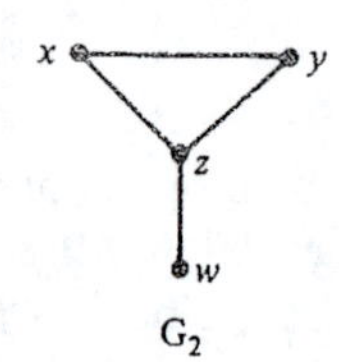

G_2

4.

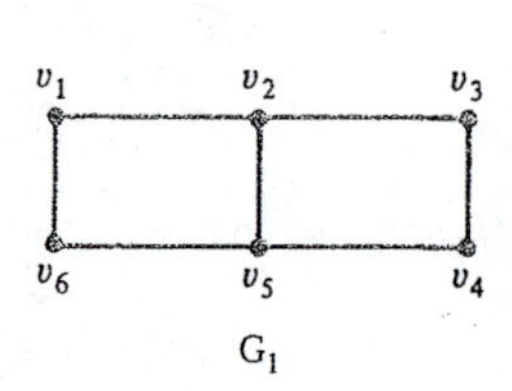

G_1

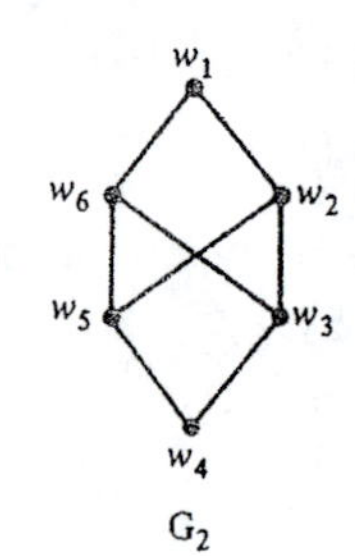

G_2

5.

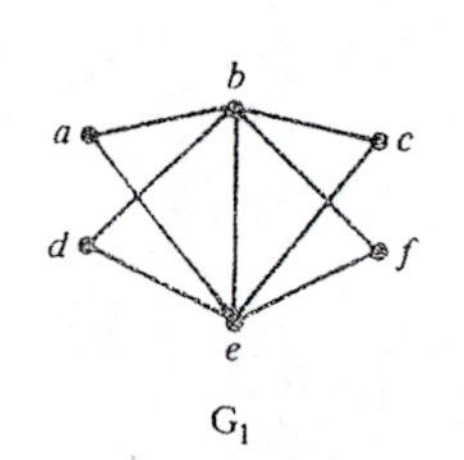

G_1

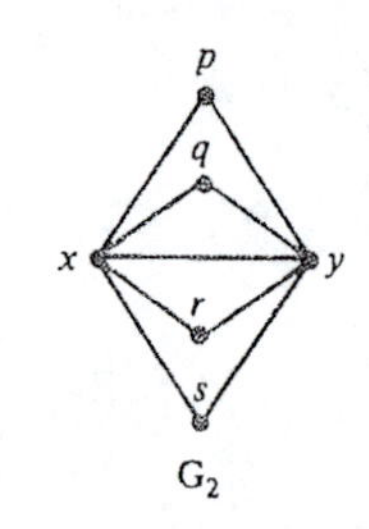

G_2

6.

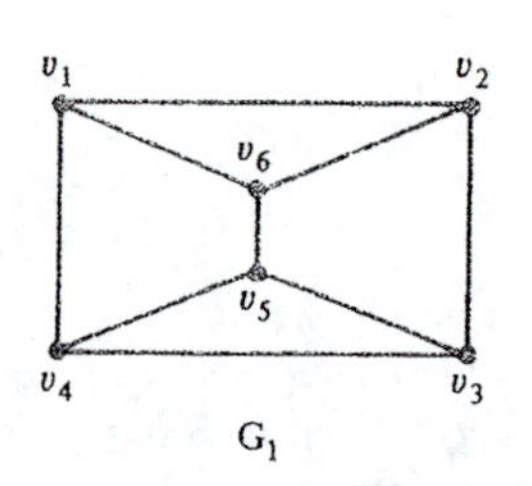

G_1

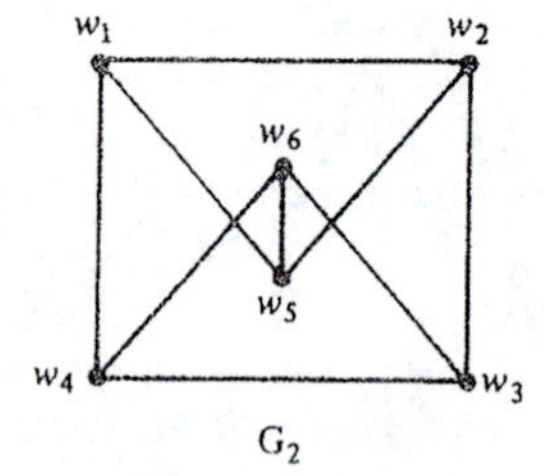

G_2

7.

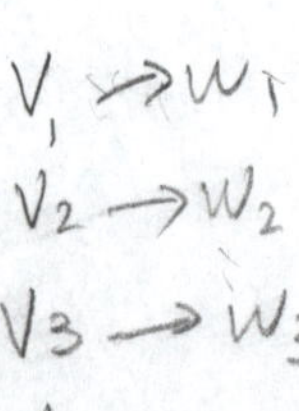

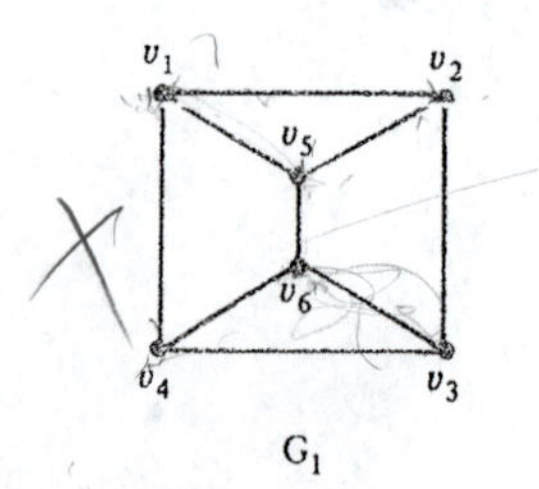

G_1

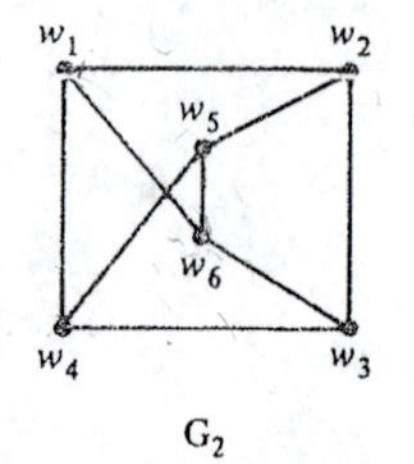

G_2

8.

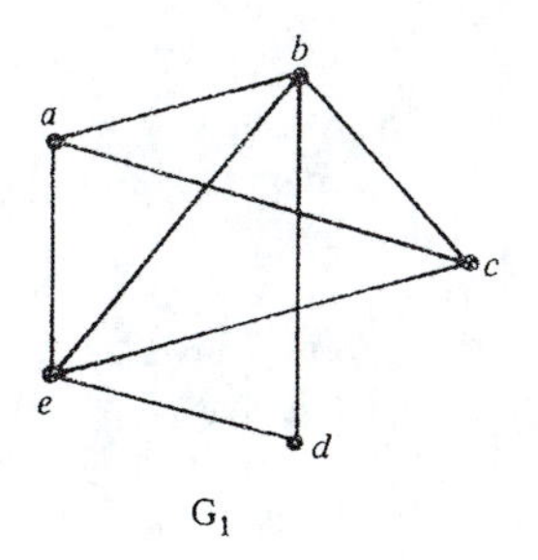

G_1

G_2

9.

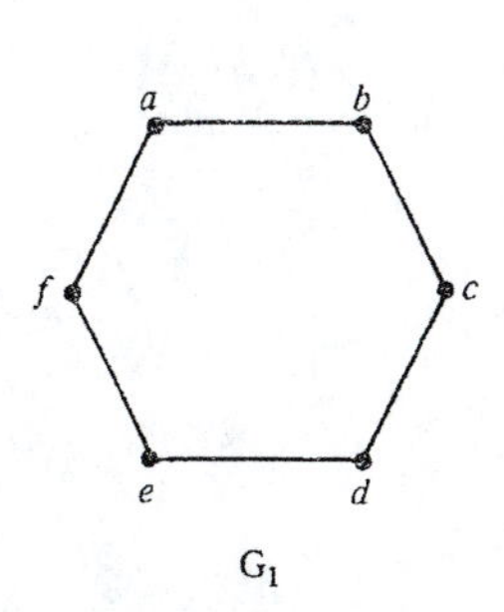

G_1

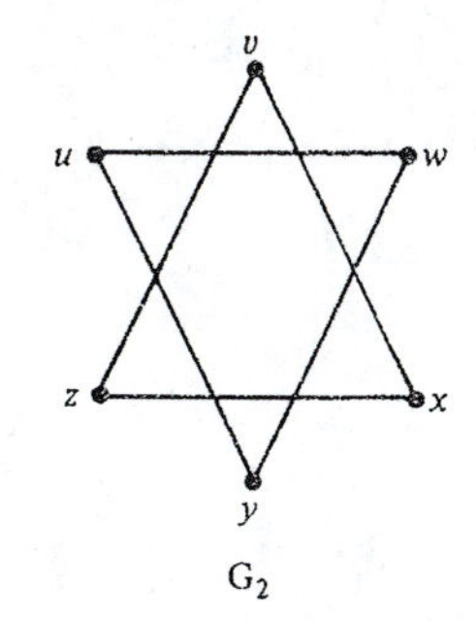

G_2

10.

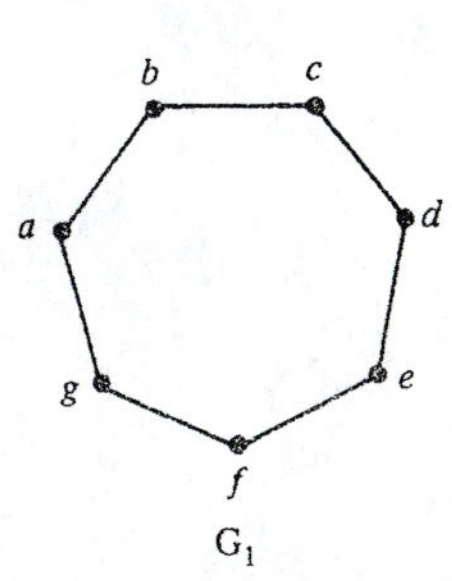

G_1

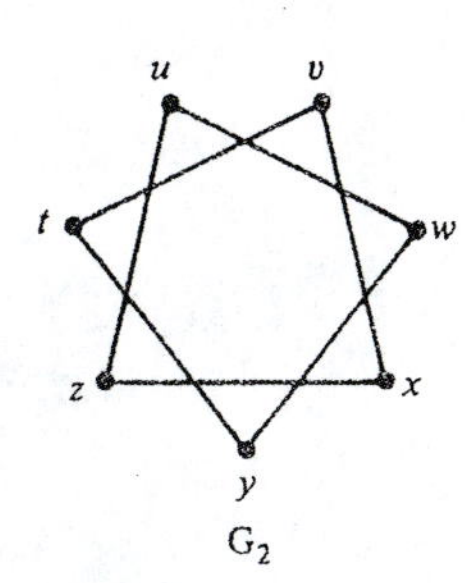

G_2

11.

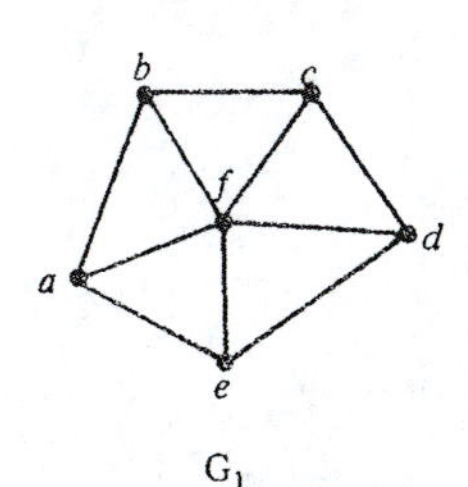

G_1

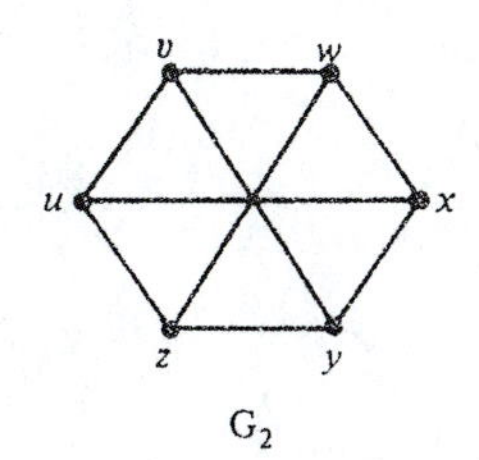

G_2

12.

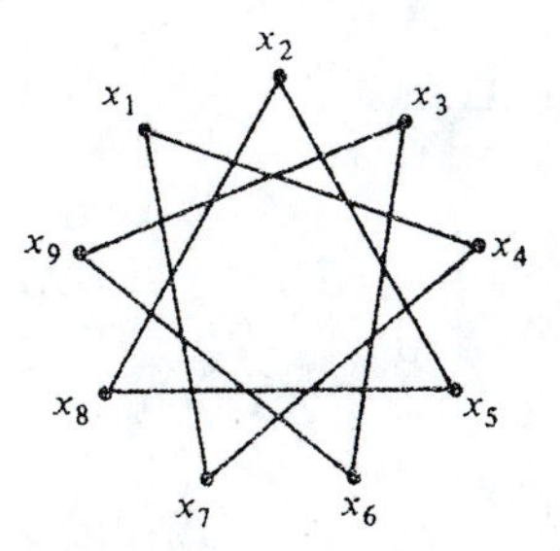

G_1

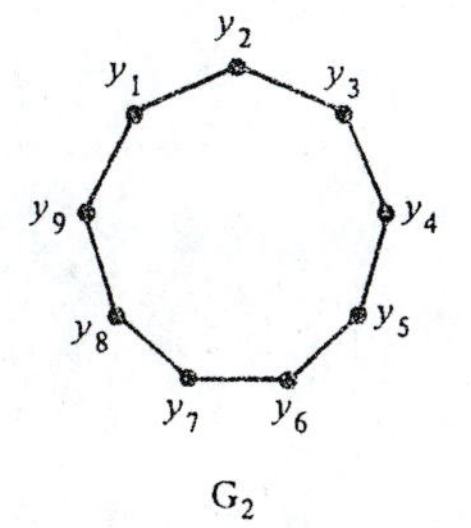

G_2

13. Prove that the property of having a vertex of a certain degree is an isomorphic invariant. That is, prove that if a graph G has a vertex of a certain degree n, then all graphs that are isomorphic to G must also have a vertex of degree n.

14. Prove that the property of being connected is an isomorphic invariant.

15. Prove that the property of being Eulerian is an isomorphic invariant.

16. Is the property of being Hamiltonian an isomorphic invariant? Justify your answer.

17. Is the property of being bipartite an isomorphic invariant? Justify your answer.

18. Draw all nonisomorphic simple graphs with three vertices. That is, make a list of graphs with three vertices so that any simple graph with three vertices must be isomorphic to a graph on your list.

In Exercises 19 through 25, redraw the planar graph with no crossings.

19.

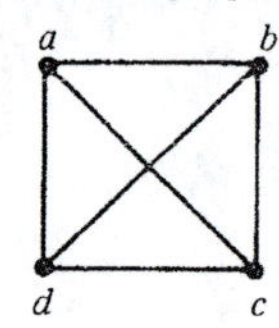

20.

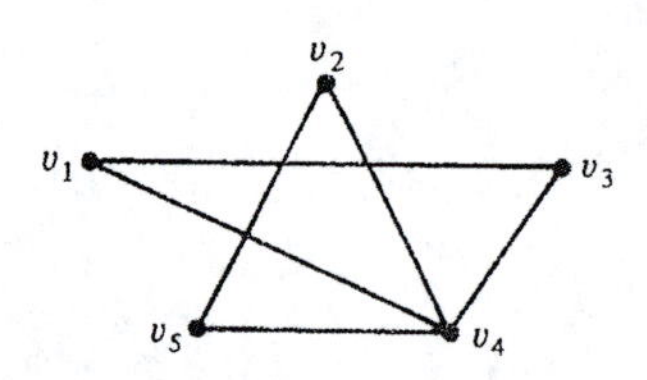

21.

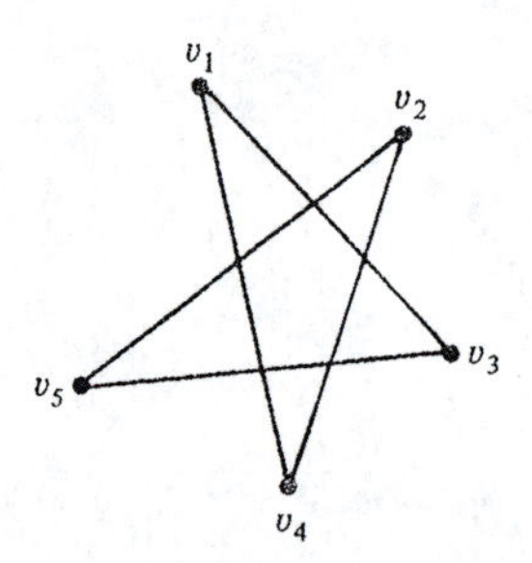

22.

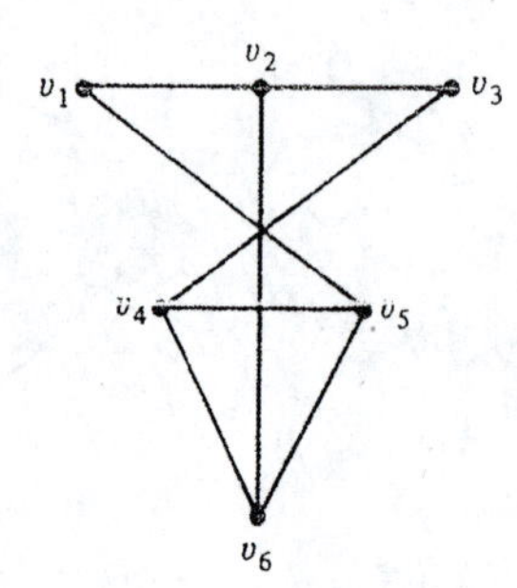

23.

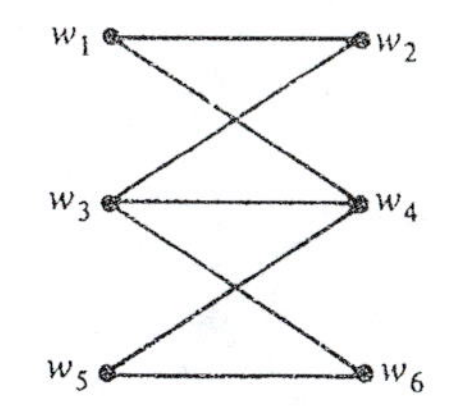

24.

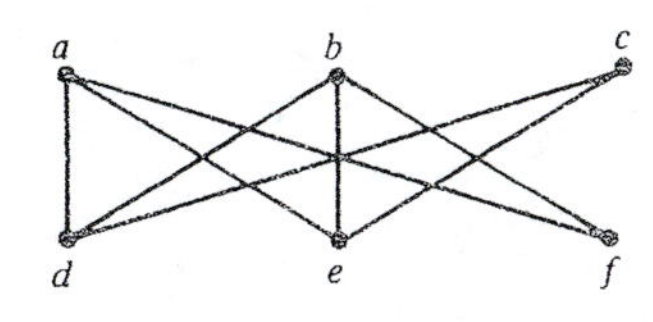

25.

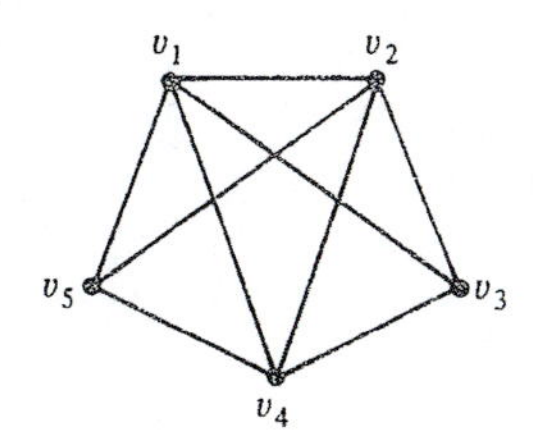

26. Does the addition or deletion of a loop have any effect on the planarity of a graph? Explain your answer. What about the addition or deletion of a multiple edge?

27. Can a connected planar graph have 10 vertices, 16 edges, and 7 faces? Justify your answer.

28. Suppose that a connected planar graph has vertices of the following degrees: 2, 2, 2, 2, 3, 3, 3, 3, 4, 4, 6. How many edges does it have? If it is drawn in the plane with no crossings, how many faces will there be?

29. Prove that any graph with exactly 4 vertices is planar. *Hint*, think about K_4.

30. Prove that any graph that has exactly 5 vertices, one of which has degree 2, must be planar.

31. Prove that if G is a planar graph with at least 3 vertices, then $e \leq 3v - 6$. *Hint*, by a reasoning similar to that in Theorem 6.4.2, show that $3f \leq 2e$. Combine this fact with Euler's formula.

32. Use the result of the previous exercise to show that K_5 is not planar.

33. Derive a formula analogous to Euler's formula for planar graphs that are not necessarily connected. *Hint*, the formula is of the form $f - e + v = ?$, where ? involves the number of connected components of the graph.

6.5

Trees: The Depth First Search

Graphs that do not contain any cycles are especially important in applications. For this reason, we want to study them rather carefully.

A graph that does not contain any cycles is called an **acyclic graph**. For example, the following graphs are acyclic.

Figure 6.5.1

Figure 6.5.2

A connected acyclic graph, such as the one in Figure 6.5.1, is called a **tree**. Also, an acyclic graph, regardless of whether or not it is connected, is called a **forest**. Thus, the graph in Figure 6.5.2 is a forest, but it is not a tree.

If G is a connected graph, then a subgraph T of G is called a **spanning tree** of G if T is a tree and if the vertex set of T is the same as the vertex set of G. (In general, a subgraph H of a graph G is said to **span** G, or to be a **spanning subgraph** of G, if it has the same vertex set as G.)

For example, the graph

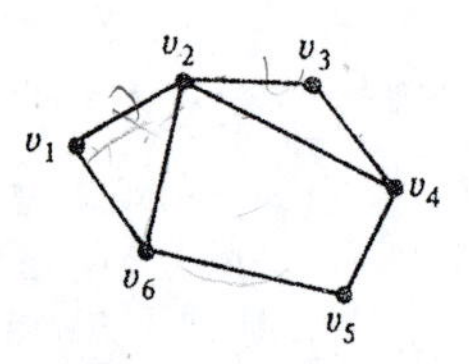

has many spanning trees, including these three:

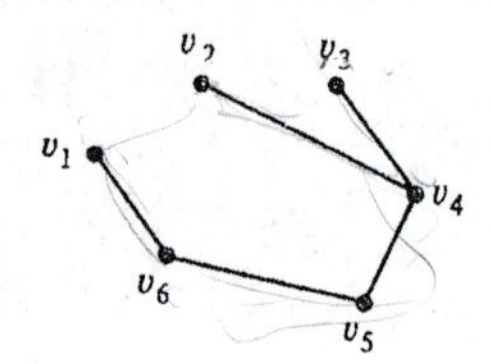

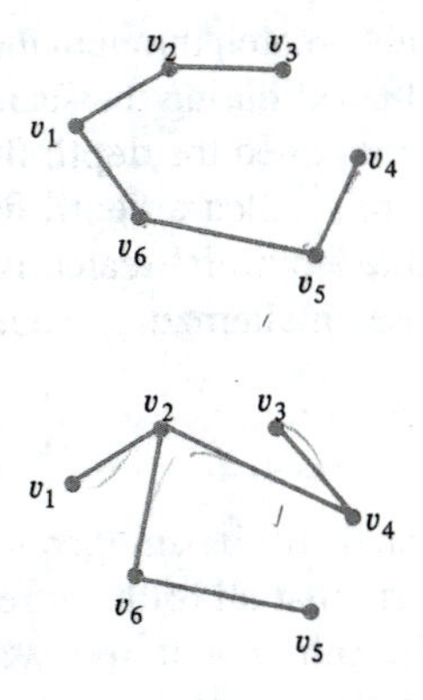

The next example illustrates one use of spanning trees.

Example 6.5.1

In the following graph G, the vertices represent cities and the edges represent highways.

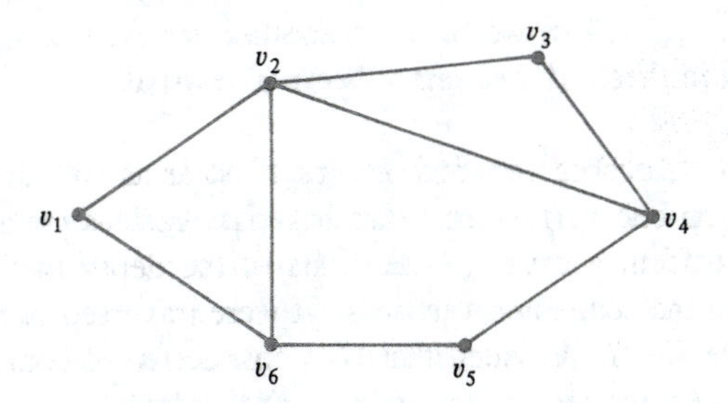

Figure 6.5.3

Suppose that, as the Director of Public Transportation, you decide to close as many of these roads as possible, subject to the condition that it is still possible to get from any city to any other city. How can you accomplish this?

The answer is very simple. All you have to do is find a spanning tree for this graph and close all of the *other* roads. □

It is actually very easy to find a spanning tree in a connected graph G. We simply pick edges in G, in any manner whatsoever, subject only to the condition that we cannot pick an edge if it forms a cycle with any of the previously chosen edges. When we can no longer pick any edges without violating this condition, the edges that we have picked (together with the vertices of G) form a spanning tree for G. We will leave it as an exercise for you to *prove* that this method yields a spanning tree. (Another method for finding a spanning tree is discussed in Exercise 16.)

This method for finding a spanning tree is not very systematic, however, and there are other methods that, although perhaps a bit more complicated, are much more useful. We want to discuss one of these methods now.

The method involves making a search through the edges and vertices of the graph G. Those edges and vertices obtained during the search will form a spanning tree of G. The method we will describe is called the **depth first search**, and a spanning tree obtained from a depth first search is called a **depth first search spanning tree**.

Intuitively, the idea behind a depth first search is the following. We start at any vertex in G, call it u_1, and choose an alternating sequence of vertices and edges in G to form an *open* path

$$u_1, f_1, u_2, f_2, \ldots, u_{n-1}, f_{n-1}, u_n$$

that cannot be further extended and still be an open path. This will happen when we reach a vertex u_n with the property that all of the edges incident with u_n connect it to one of the other vertices in the path. For if we were to add such an edge to our sequence, this would produce a cycle, and we would no longer have an open path.

In a sense, we start at u_1 and *search* as *deeply* as we can into the graph along an open path.

Once we come to the end u_n of this path, we back up one vertex, to u_{n-1} and start over. That is, we search out again, this time starting at u_{n-1}, to form another open path that cannot be extended any further without forming a cycle with *any* of the previous edges, either in the new path or in the first path. If we cannot begin such a path at the vertex u_{n-1}, then we back up another vertex, to u_{n-2}, and try to form a new open path from there. If we cannot begin a new path at u_{n-2}, we back up still further, and so on.

This process of searching out into the graph as far as we can go along an open path, then backing up one vertex and searching out again as far as possible, always being careful not to form a cycle, is the basis of the depth first search. When the search is completed, the edges and vertices that were traversed during the search will form a spanning tree for G (provided that G is connected, of course).

Before giving a more precise formulation of the depth first search method, let us try an example. Consider the graph

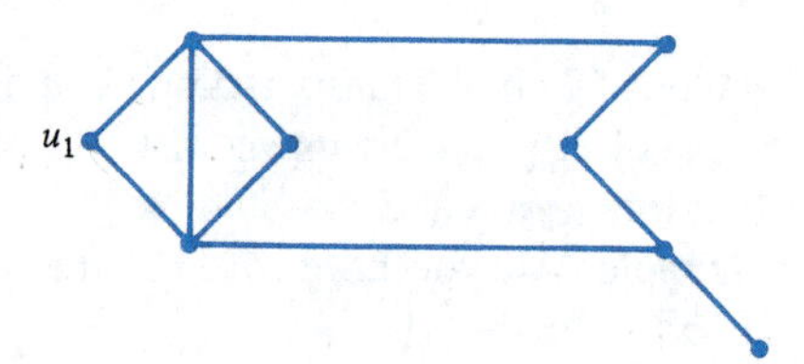

We begin at any vertex, say u_1, and form an open path until we cannot go any further. One such path is indicated below.

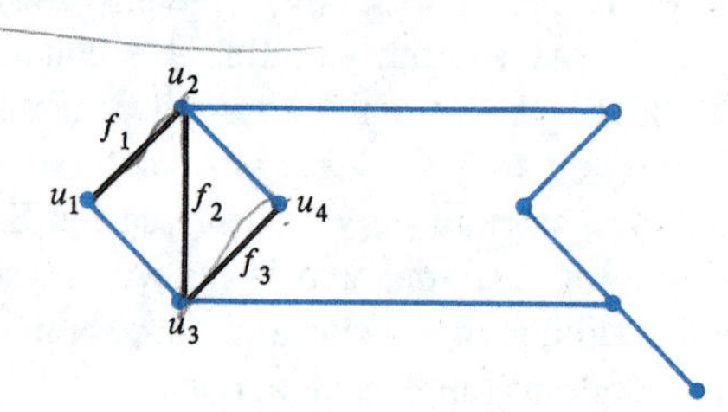

Now, we back up one vertex, to u_3, and form a new open path from there until we cannot go any further, say

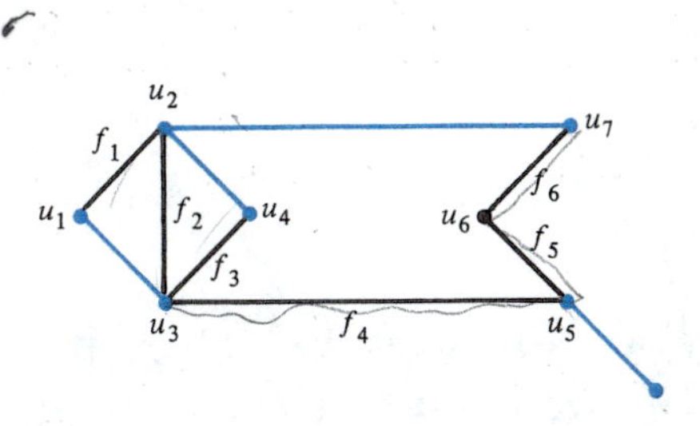

Now we must back up two vertices, to u_5, and form a new open path

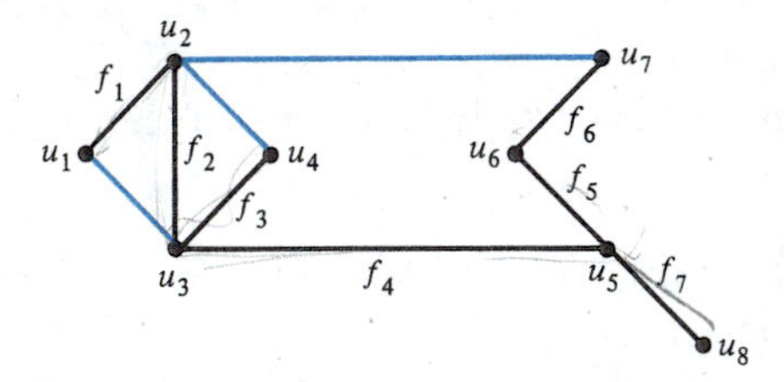

At this point, we cannot continue the search, and the resulting collection of edges and vertices forms a depth first search spanning tree of G.

Now let us describe this search method in a more precise manner. As you can see from the example, the key to the search is that, at each instant, we continue the search by picking that vertex u_i with the largest subscript i from which the search can be continued.

Algorithm for the Depth First Search

Suppose that G has n vertices. Then the algorithm has exactly n steps. At the k-th step, we define two sets V_k and E_k. The set

$$V_k = \{u_1, u_2, \ldots, u_k\}$$

consists of k distinct vertices of G (these are the vertices already encountered in the search), and the set

$$E_k = \{f_1, f_2, \ldots, f_{k-1}\}$$

consists of $k - 1$ distinct edges of G. (These are the edges already encountered in the search.) The set E_1 is the empty set.

The sets V_k and E_k are constructed in such a way that, at each step, the graph (V_k, E_k) is a tree. Thus, when we reach the n-th step, since the set V_n contains n distinct vertices, it must be the set of all vertices of G, and so the graph (V_n, E_n) must actually be a spanning tree of G.

Step 1
Let u_1 be any vertex of G, and let

$$V_1 = \{u_1\}$$
$$E_1 = \varnothing$$

Notice that the graph (V_1, E_1) is a tree. (It consists simply of a single vertex with no edges.)

Once step k has been completed, we have the set $V_k = \{u_1, u_2, \ldots, u_k\}$ of distinct vertices of G and the set $E_k = \{f_1, f_2, \ldots, f_{k-1}\}$ of distinct edges of G. Also, the graph (V_k, E_k) is a tree. Now we are ready for the $(k + 1)$-st step.

Step $k + 1$
Choose the vertex u in V_k *with the largest subscript* having the property that there is an edge of the form $f = \{u, x\}$ that does not form a cycle with any of the edges in E_k. Notice that f cannot be in E_k. (Why?) Then let $u_{k+1} = x$ and $f_k = f$. This gives us the set

$$V_{k+1} = \{u_1, u_2, \ldots, u_{k+1}\}$$

of distinct vertices of G and the set

$$E_{k+1} = \{f_1, f_2, \ldots, f_k\}$$

of distinct edges of G. Also, the graph (V_{k+1}, E_{k+1}) is a tree. (Why?) This completes the $(k + 1)$-st step.

After the n-th step in this algorithm is completed, the set V_n will be the set of all vertices of G and (V_n, E_n) will be a spanning tree of G.

As we mentioned earlier, the depth first search method is very useful. We will see one of its uses in Section 6.8. For now, we can observe that this method provides a convenient way to determine whether or not a graph is connected. For if we apply the depth first search method to a graph G, then the search will reach all of the vertices of G if and only if G is connected. This method for checking connectivity is particularly well suited for use in a computer, for example.

As we know, a tree is defined to be a connected, acyclic graph. But there are many other ways to characterize trees, some of which may at times be more useful than the definition. In the remainder of this section, we want to discuss some of these ways. Let us begin with two simple facts about trees.

Theorem 6.5.1

Any tree T with more than one vertex must have at least two vertices of degree 1.

PROOF Let T be a tree with more than one vertex. Let us pick from among all paths in T, one that has the largest number of edges, call it P. (Of course, there may be more than one such path, but we only need one.) Since T has no cycles, the path P (indeed any path) must be open, and so we may assume that it begins at a vertex u and ends at a *different* vertex v, as pictured below.

But if either u or v had degree greater than 1, then it would be possible to add an edge to the path P, forming a new path in T that has more edges than P. This contradicts the fact that P has the largest number of edges possible for a path in G, and so we can conclude that u and v must both have degree 1. This completes the proof. ∎

Theorem 6.5.2

Any tree with n vertices must have exactly $n - 1$ edges.

PROOF Let us prove this theorem by using induction. Let P(n) be the statement that any tree with n vertices has exactly $n - 1$ edges. Then it is easy to see that P(1) is true. We must show that if P(k) is true, then so is P($k + 1$).

So let T be a tree with $k + 1$ vertices. We want to show that it has k edges. According to Theorem 6.5.1, T must have a vertex, call it v, that has degree equal to 1. (In fact, it must have two such vertices, but we only need one here.) If we remove v, and the edge incident with v, from the tree T, we obtain another tree, call it T', that has exactly k vertices. Hence since P(k) is true, T' must have $k - 1$ edges, and so T has $(k - 1) + 1 = k$ edges. This shows that P($k + 1$) is true and completes the proof. ∎

According to Theorem 6.2.1, any connected graph with n vertices must have at least $n - 1$ edges. Therefore, Theorem 6.5.2 tells us that a tree is a connected graph with the fewest number of edges possible for a graph to have and still be connected. It so happens that the converse of this statement is also true, and this gives us our first characterization of trees.

Theorem 6.5.3

A *connected* graph G with n vertices is a tree if and only if it has exactly $n - 1$ edges.

PROOF If a connected graph G with n vertices happens to be a tree, then according to Theorem 6.5.2, it must have exactly $n - 1$ edges. For the converse, we must show that if a connected graph G has n vertices and $n - 1$ edges, then it must be a tree.

Since we are assuming that G is connected, all we have to do is prove that it is acyclic. We can do this by contradiction. For if G is not acyclic, then it must contain a cycle, call it C. Now if we remove an edge of this cycle from G, the resulting graph H will still have n vertices, and it will still be connected (see Exercise 15), but it will have only $n - 2$ edges. This is a contradiction to Theorem 6.2.1, and so G must be acyclic after all. Thus, G is a tree and the proof is complete. ∎

We can also characterize trees as *acyclic* graphs that have one less edge than vertices.

Theorem 6.5.4

An *acyclic* graph G with n vertices is a tree if and only if it has exactly $n - 1$ edges.

PROOF If an acyclic graph G with n vertices happens to be a tree, then according to Theorem 6.5.2, it must have exactly $n - 1$ edges.

For the converse, let us assume that G is an acyclic graph with n vertices and $n - 1$ edges and show that G must be a tree. By assumption, G is acyclic, and so we only need to show that G must be connected. This we can easily do by contradiction. So let us suppose that G is not connected.

If G is not connected, then it is made up of at least two connected components. Let us denote the connected components of G by $G_1, G_2, \ldots, G_m$, where $m \geq 2$. Now, each component is connected and acyclic, and so it is a tree. Let us denote the number of vertices in the tree G_i by n_i, for $i = 1, 2, \ldots, m$. Then, according to Theorem 6.5.2, the tree G_i must have exactly $n_i - 1$ edges. Hence, the total number of edges in G must be

$$\begin{aligned}(n_1 - 1) + (n_2 - 1) + \cdots + (n_m - 1) &= n_1 + n_2 + \cdots + n_m - m \\ &= n - m\end{aligned}$$

($n_1 + n_2 + \cdots + n_m$ is just the number of vertices in G, which is n.)

But since $m \geq 2$, this is a contradiction to the fact that G has exactly $n - 1$ edges. This contradiction shows that G must be connected, and therefore finishes the proof. ∎

The next theorem characterizes trees in terms of the concept of a bridge. Recall that a bridge in a connected graph is an edge whose removal results in a disconnected graph. We will leave a proof of this theorem as an exercise.

Theorem 6.5.5

A *connected* graph G is a tree if and only if every edge of G is a bridge.

We will also leave the proofs of the next two theorems as exercises.

Theorem 6.5.6

A graph G with no loops is a tree if and only if it has the property that any two distinct vertices in G are connected by exactly one path.

Theorem 6.5.7

A graph G is a tree if and only if it is acyclic and has the property that the addition of *any* new edge to G creates exactly one cycle.

□ □ Exercises

1. Draw all nonisomorphic trees with 3 vertices.
2. Draw all nonisomorphic trees with 4 vertices.

In Exercises 3 through 7, find a spanning tree for the given graph.

3.

4.

5.

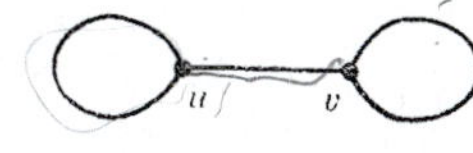

6.

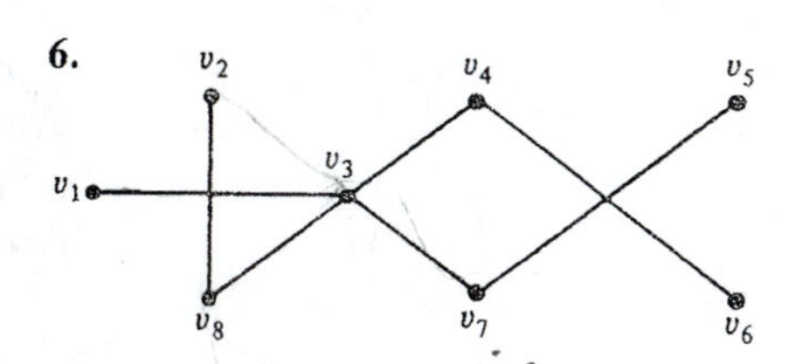

7.

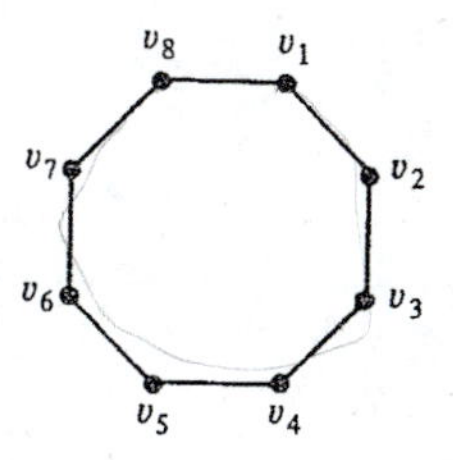

In Exercises 8 through 12, use the algorithm given in this section to find a depth first search spanning tree for the given graph.

8.

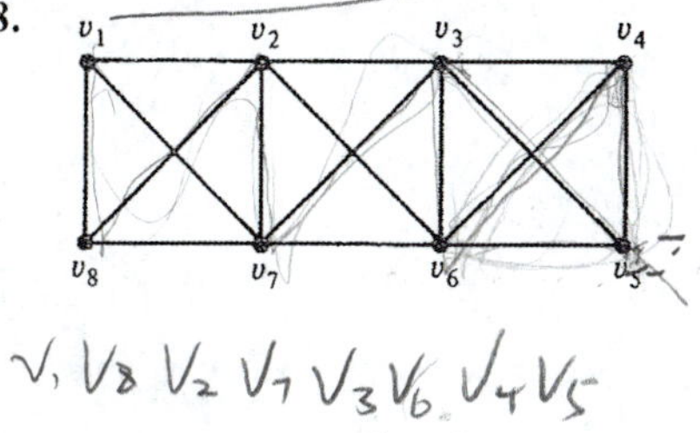

9.

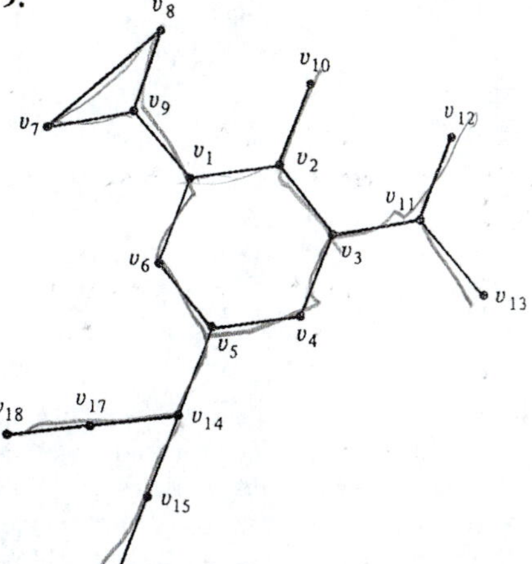

10.

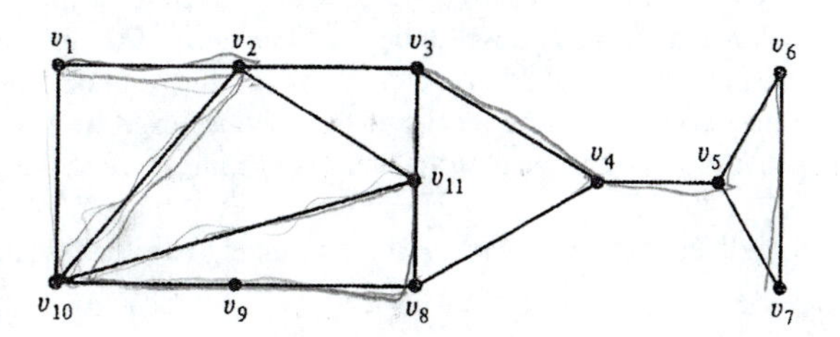

11.

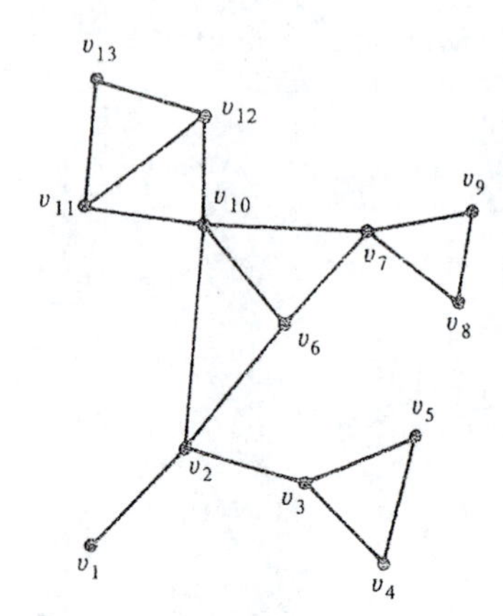

12.

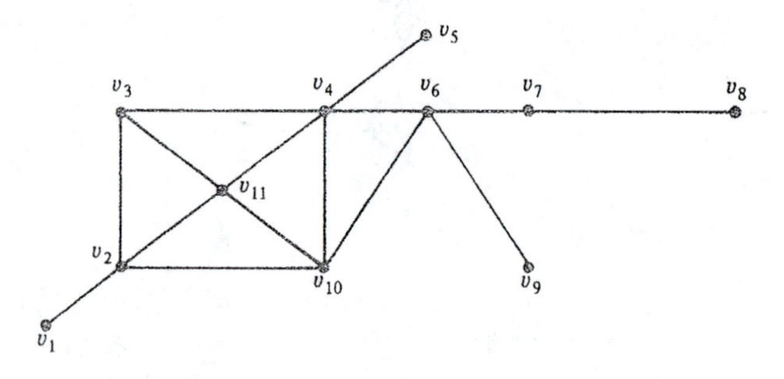

13. Find all the spanning trees of the graph

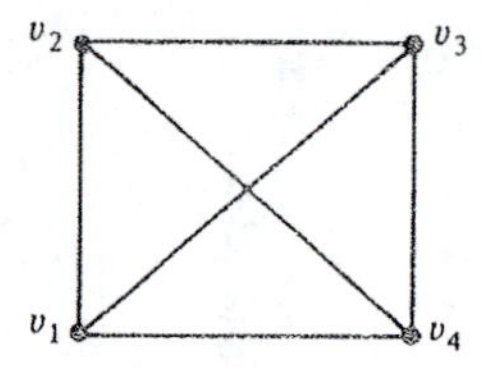

14. Prove that the method of picking edges in a connected graph G, subject only to the condition that an edge cannot be picked if it forms a cycle with the edges already picked, does yield a spanning tree of G.

15. Let G be a connected graph that contains a cycle C. Prove that we can remove any edge of C and the resulting graph will still be connected.

16. Let G be a connected graph. If G has a cycle, remove an edge of this cycle from G. If the resulting graph has a cycle, remove an edge of this cycle. Continue to remove edges of cycles until the resulting graph has no cycles. Prove that the resulting graph is a spanning tree of G. (You may wish to use the results of the previous exercise.)

17. Use the method outlined in Exercise 16 to find a spanning tree for the graph in Figure 6.5.3.

18. Use the method outlined in Exercise 16 to find a spanning tree for the graph

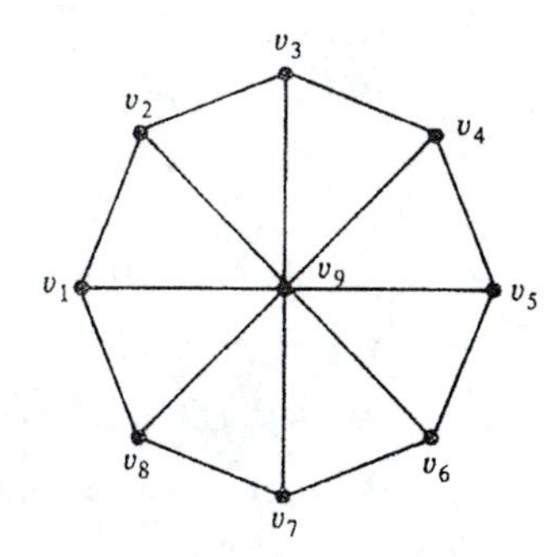

19. Use the method outlined in Exercise 16 to find a spanning tree for the graph

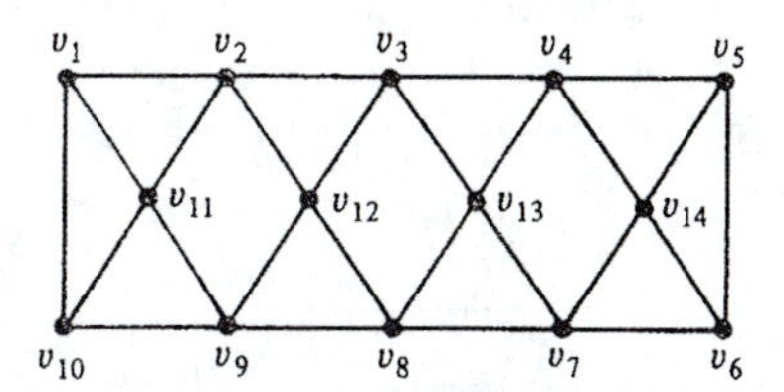

20. Is it true that any two spanning trees of a connected graph must have an edge in common? Either prove that it is true or give a counterexample.

21. Let G be a connected graph with n vertices and k edges. What is the maximum number of edges that can be removed from G such that the remaining graph is still connected? Justify your answer.

22. If F is a forest with n vertices and k connected components, show that F must have exactly $n - k$ edges.

23. Prove Theorem 6.5.5.

24. Prove Theorem 6.5.6.

25. Prove Theorem 6.5.7.

26. At the $(k + 1)$-st step in the depth first search algorithm, we choose the vertex u in V_k with the largest subscript having the property that there is an edge $f = \{u, x\}$ that does not form a cycle with any of the edges in E_k. Suppose we modify the algorithm by choosing instead the vertex u that has the *smallest* subscript, instead of the largest. The resulting search method is called the **breadth first search**.

a) Use the breadth first search method to find a spanning tree for the graph

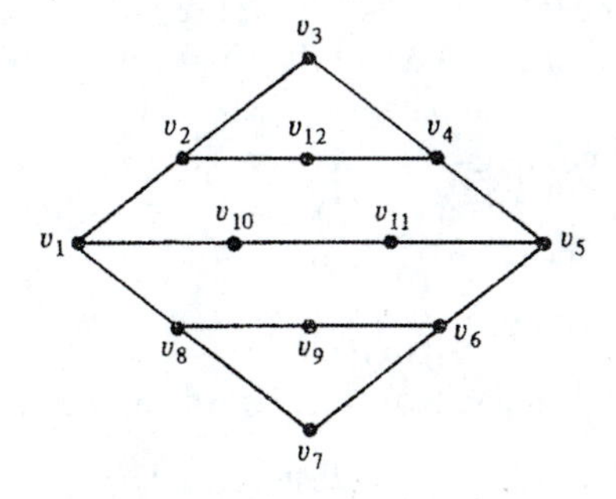

b) Use the depth first search method on the graph in part a.
c) Can you explain why this method is called the breadth first search method?

6.6
Two Applications of Trees: Binary Search Trees and Huffman Codes

In this section, we want to discuss two important applications of trees to computer science. Let us begin with the concept of a binary search tree.

Suppose that we have a large collection of words and that we want to store these words in the memory of a computer in such a way that

1) it is easy to determine whether or not a word is in the collection, and
2) if a word is not in the collection, it can be added easily to the collection.

Perhaps the first thing that comes to mind is to store the words in alphabetical order in successive memory locations,

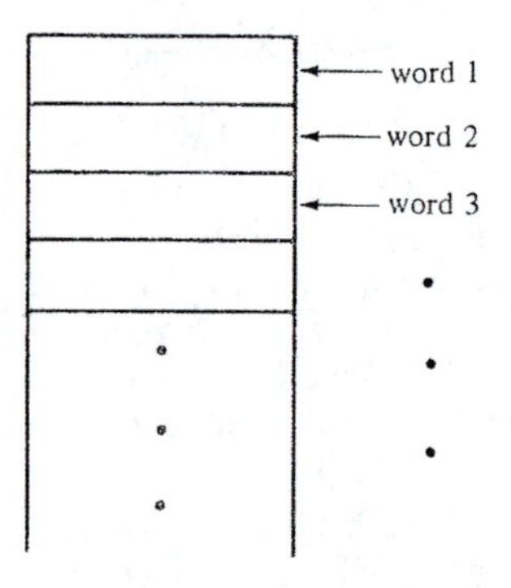

This would certainly make it easy to tell whether or not a word was in the collection. However, it would be difficult to add a new word to the collection, since it would be necessary to move each of the words in the collection that follow the new word down one position in the memory of the computer.

On the other hand, if we simply place the words in successive memory locations in any order whatsoever, then it would be very easy to add a new word to the collection, but difficult to determine whether or not a word was already in the collection, for it might be necessary to compare it with every word in the collection.

It so happens that there is another alternative, which involves the use of trees. The idea is to store the words in the collection at the vertices of a certain tree. In order to describe this more precisely, let us consider a simple example.

Example 6.6.1

Suppose that our collection contains the words in the sentence

graph theory is a very interesting subject

We begin by choosing any word from this collection, say the word "interesting," and labeling a vertex with this word. Then, we connect this vertex to two other vertices, as shown in Figure 6.6.1.

Figure 6.6.1

Then we take another word from the collection, say ''theory,'' and since this word comes *after* the word ''interesting'' in alphabetical order, we label the *right*-hand vertex in Figure 6.6.1 with this word. Again we add two new unlabeled vertices to the graph, giving

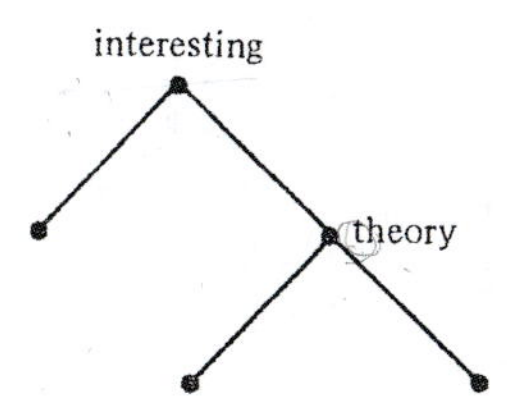

Next, we choose another word, say ''subject.'' In order to determine which unlabeled vertex to label with the word ''subject,'' we make a path in the graph, starting at the top, as follows. Since the word ''subject'' comes *after* the word ''interesting'' in alphabetical order, we move down the graph to the *right*. Then we compare the word ''subject'' with the word ''theory.'' Since ''subject'' comes *before* ''theory'' in alphabetical order, we move down the graph to the *left*. This brings us to an unlabeled vertex, which we label with the word ''subject.'' Then we add two new unlabeled vertices to the graph

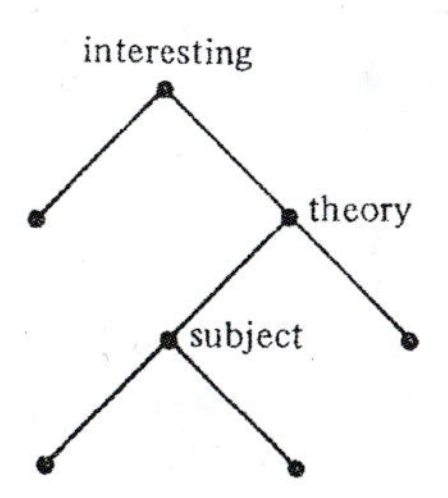

By now the pattern should be clear. In order to add another word from the collection to our graph, we proceed as follows. Starting at the top of the graph, we compare the word with the labels of the vertices. If the word comes before a given label in alphabetical order, then we move down the graph to the left, and if the word comes after a given label, then we move down the graph to the right. Once we arrive at an unlabeled vertex, we label it with the word. Then we add two new unlabeled vertices to the graph.

In our example, if we next add the word "is" to our graph, we obtain the graph

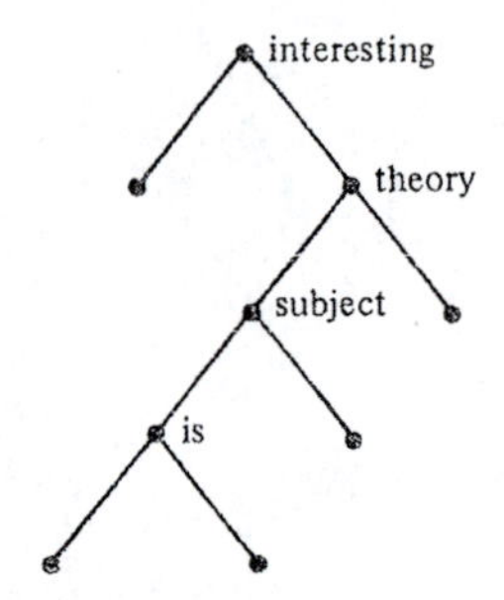

Adding the word "graph" gives

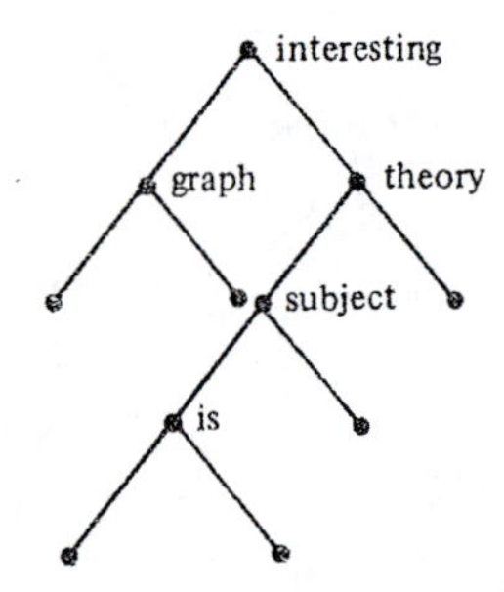

Then adding the word "a," we get

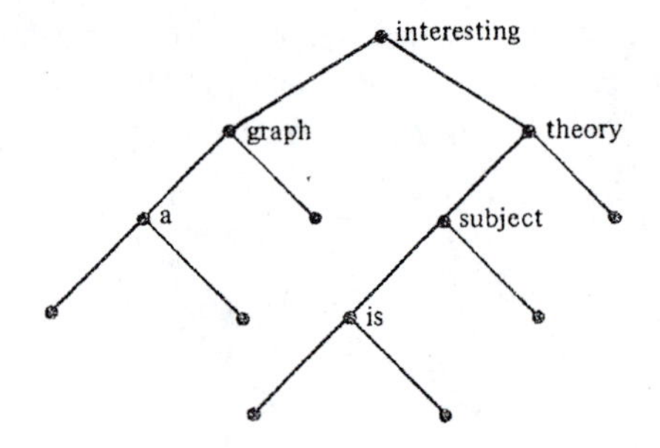

and finally adding the word "very" gives

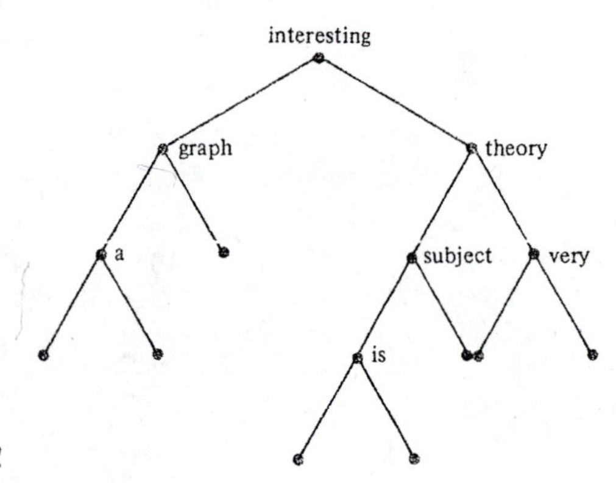

Figure 6.6.2

□

The graph in Figure 6.6.2 is called a **binary search tree**. Storing a collection of words at the vertices of a binary search tree is a very effective way to achieve the goals we want.

Given a word, we can determine whether or not it is in the collection, and if it is not, how it should be added to the collection, by simply trying to add it to the binary search tree, using exactly the same procedure that we used to create the tree in the first place. If, as we proceed, we encounter the word already being used as a label in the tree, then of course the word was already in the collection. On the other hand, if we are successful in reaching an unlabeled vertex, then the new word was not in the collection, and we have, at the same time, found that place in the search tree where it should be added!

There are two reasons why this procedure is effective. First, in general, we can decide whether or not a new word is in the collection without having to compare it to every word in the collection. In fact, we have to compare it only to those words that are in the path that must be followed in attempting to add the new word to the tree. Second, when we add a new word to the tree, none of the other words need to be moved, as would be the case with a simple listing in alphabetical order.

Of course, binary search trees can be used to store types of data other than words, as long as the data is ordered in some way, so that comparisons can be made.

Now let us turn our attention to another application of trees, namely, the coding of information. As we discussed in Chapter 1, when a character such as a letter, a digit, or a punctuation mark is entered into a computer, it is converted into a binary word by means of a *character code*. As you may recall from our discussion in Example 1.2.3, one of the most common character codes is the ASCII code. In this character code, each character is encoded by a binary word of length 7, since at least 7 binary digits are needed in order to store 128 different characters. [In practice, a full byte (8 binary digits) is generally used to encode a character, where the 8-th digit is used either to expand the character set, for such things as Greek letters, other mathematical symbols, and graphics characters, or as a check for errors in the transmission of the character (for example, as a parity bit). However, this will not affect the present discussion.]

The relevant point is that the ASCII code, as well as many other commonly used codes, have the property that each character is given a code of the same length. Such a character code is called a **fixed length character code**.

Fixed length character codes have advantages and disadvantages. One of their advantages is that no special characters are needed to separate the characters in a message. For example, in the ASCII message

$$1010100101001010001011000101$$

we know that the first 7 bits represent the first character, which is T. Similarly, the second set of 7 bits represents the second character, and so on. Hence, we can easily decode this message, to get TREE.

One of the disadvantages of using a fixed length character code, such as the ASCII code, is that characters that are used very frequently, such as the letter e, have codes that are just as long as characters that are seldom used, such as the backward slash $\backslash$. On the other hand, a character code that encodes frequently used characters with fewer bits than seldom used characters would provide a great saving of time and space.

This leads us to the idea of a **variable length character code**, namely, a character code in which frequently used characters have shorter codes than seldom used characters.

However, there is one problem with variable length character codes. It may not be possible to tell when one character stops and another one starts!

In order to illustrate this problem, suppose that our character set C consists of the first six letters of the alphabet, that is C = {a, b, c, d, e, f}, and suppose that a and b are the most frequently used characters. Then one might design a variable length character code as follows

$$\begin{aligned} a &\leftrightarrow 0 \\ b &\leftrightarrow 1 \\ c &\leftrightarrow 00 \\ d &\leftrightarrow 01 \\ e &\leftrightarrow 10 \\ f &\leftrightarrow 11 \end{aligned} \tag{6.6.1}$$

This code is very "tight" in the sense that the individual codes are as short as possible. However, what does the message

010011

mean? Does it mean abaabb, or does it mean dcf, or does it mean abadb, and so on?

Of course, one way to resolve this problem is to use a special character to separate the codes of individual characters. For example, the message

01/00/1/1

is unmistakably dcbb. However, including such a character makes the message longer and helps defeat the purpose for which variable length codes were intended in the first place. (There are other problems as well—how do we encode the separator, and how do we tell when a separator begins, etc?)

As you can see from this discussion, we must be careful when designing a variable length character code to avoid this problem of ambiguity. Fortunately, there are ways to do so, and we want to discuss one method here.

First, we need to make a few definitions. A word w is a **prefix** of a word u if u has the form u = wx, where x is another word. In loose terms, a word w is a prefix for a word u if u "starts with" w. For example, the word 101 is a prefix of the word 1011100.

Now, the problem with the character code 6.6.1 is that the code for one character is a *prefix* of the code for another character. For example, the code for a is a prefix of the code for c. This means that if a message begins with the digit 0, we do not know whether to interpret this as the code for a or as the beginning of the code for c, or d for that matter.

A character code that has the property that no code for a character is the prefix of the code for another character is said to have the **prefix property**. If a character code has this property, then we will not run into the problem that we did for the code 6.6.1. Therefore, any message has a unique meaning. All we have to do is start reading from left to right until we have read the code for a certain character. Then we begin again from that point to read another character. We do not have to worry that we have "stopped too soon," so to speak, since no code is the prefix of another code. Let us consider an example.

Example 6.6.2

Consider the character code shown below for the character set {c, e, o, r, t}

$$c \leftrightarrow 11$$

$$e \leftrightarrow 0$$

$$o \leftrightarrow 100$$

$$r \leftrightarrow 1010$$

$$t \leftrightarrow 1011$$

This character code has the prefix property, since no code is the prefix of another code. Thus, for instance, the message

11100101010100111011

can only be decoded in one way. (Try it for yourself!) □

Naturally, in designing a character code, we will want it to have the prefix property. But there is still one other matter that we must deal with in designing a variable length character code. How do we tell when we have an efficient code? Our goal is to design a code that uses as few bits as possible, on the average. Of course, no code will be the most efficient for all messages, but we can hope to find one that will be the most efficient on the average.

In order to put these matters more precisely, we must have some idea of how frequently each character in the character set occurs. This is done by assigning each character a **frequency**. For example, in the character set C = {a, b, c, d}, we might assign the frequencies

$$f(\mathrm{a}) = 2, \qquad f(\mathrm{b}) = 2, \qquad f(\mathrm{c}) = 8, \qquad f(d) = 5 \tag{6.6.2}$$

This means that, *on the average*, for every 2 occurrences of the character a, there are 2 occurrences of b, 8 occurrences of c, and 5 occurrences of d. Thus, in this hypothetical case, the character c is the most common, and the characters a and b are the least common.

Using the frequency of each character in a character set, we can define the concept of the **weight** of a character code. Intuitively speaking, the weight of a code is a number that describes how efficient the code is, in the sense of using as few bits as possible.

Example 6.6.3

Consider the two character codes shown below

Character code 1	*Character code 2*
a ↔ 11	a ↔ 01010
b ↔ 0	b ↔ 00
c ↔ 100	c ↔ 10
d ↔ 1010	d ↔ 11

Let us compare these two codes using the frequencies in 6.6.2. It might seem at first as though code 1 is more efficient, since it uses a total of only 10 bits, rather than 11 bits for code 2. Also, code 1 has a maximum code length of 4, whereas code 2 has a maximum code length of 5. However, in code 2, the most frequently used characters have the shorter codes, which is not true for code 1. As we will see in a moment, the *weight* of code 2 is less than the *weight* of code 1, and so code 2 is actually more efficient than code 1. □

First let us define the concept of weight.

Definition

Let $C = \{a_1, a_2, \ldots, a_n\}$ be a character set, and suppose that $f(a_1), f(a_2), \ldots, f(a_n)$ are the frequencies of the characters in C. Suppose also that we have a character code for C and that the length of the code for the character a_i is denoted by $l(a_i)$, for $i = 1, 2, \ldots, n$. Then the **weight** of the character code is defined to be the sum

$$\sum_{i=1}^{n} f(a_i)l(a_i) \tag{6.6.3}$$

□

As you can see, each term in this sum is the product of the frequency of occurrence of the character and the length of its code, and so it tells us how many bits occur, on the average, as a result of that particular character. Hence, the entire sum tells us how many bits are required, on the average, by the character code.

Example 6.6.4

Let us compute the weights of the character codes in Example 6.6.3. The first code has weight

$$\begin{aligned} f(a)l(a) + f(b)l(b) + f(c)l(c) + f(d)l(d) &= 2 \cdot 2 + 2 \cdot 1 + 8 \cdot 3 + 5 \cdot 4 \\ &= 50 \end{aligned}$$

and the second code has weight

$$\begin{aligned} f(a)l(a) + f(b)l(b) + f(c)l(c) + f(d)l(d) &= 2 \cdot 5 + 2 \cdot 2 + 8 \cdot 2 + 5 \cdot 2 \\ &= 40 \end{aligned}$$

Thus, the second code has smaller weight and so, on the average, it uses fewer bits to encode messages.

Now it is time that we turned to a method for finding an efficient (in the sense of having small weight) variable length code that has the prefix property. Our plan is to use a tree to help find such a code. The resulting code is called a **Huffman code**. Let us illustrate the procedure with the character set

$$C = \{a, b, c, d, +, -\}$$

and the frequencies given in the following table

Character	*Frequency*
a	32
b	7
c	9
d	25
+	5
−	4

Algorithm for Obtaining a Huffman Code

This algorithm creates a tree whose vertices lie on adjacent levels. The vertices of the tree are labeled with frequencies or with sums of frequencies.

Step 1

Place one vertex for each character on the top level. Label the vertices with the characters and their frequencies, arranged in increasing order of frequency. (We have drawn these vertices as small circles and put the character inside the circle.)

Top level 4 (−) 5 (+) 7 (b) 9 (c) 25 (d) 32 (a)

Step 2

Connect the two leftmost vertices to a new vertex. Label the new vertex with the *sum* of the frequencies of the original vertices and lower this portion of the graph so that the new vertex is at the top level.

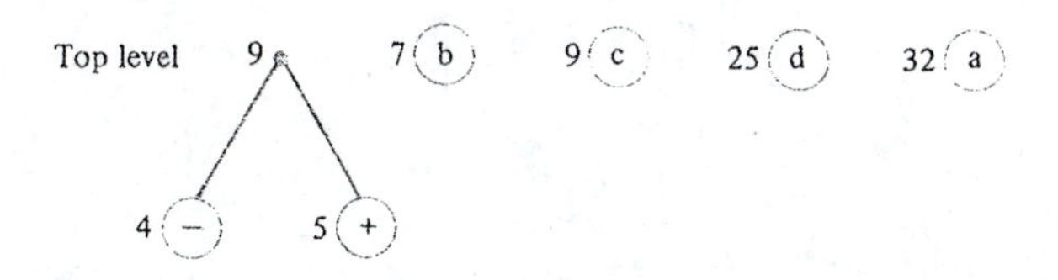

Step 3

Rearrange the graph so that the vertices on the top level are arranged in increasing order of frequency.

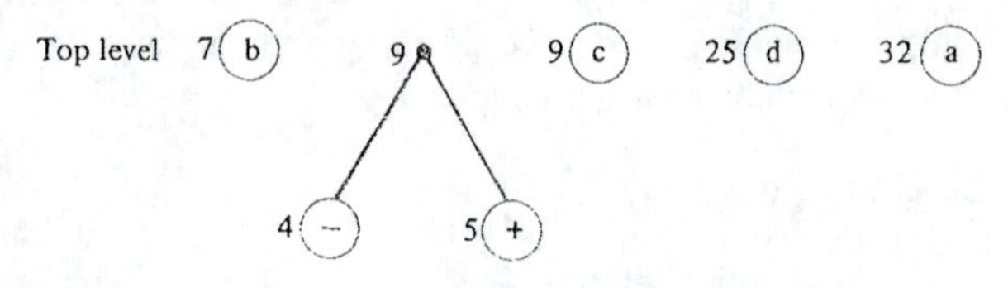

Step 4

Repeat steps 2 and 3 until the top level has only one vertex.

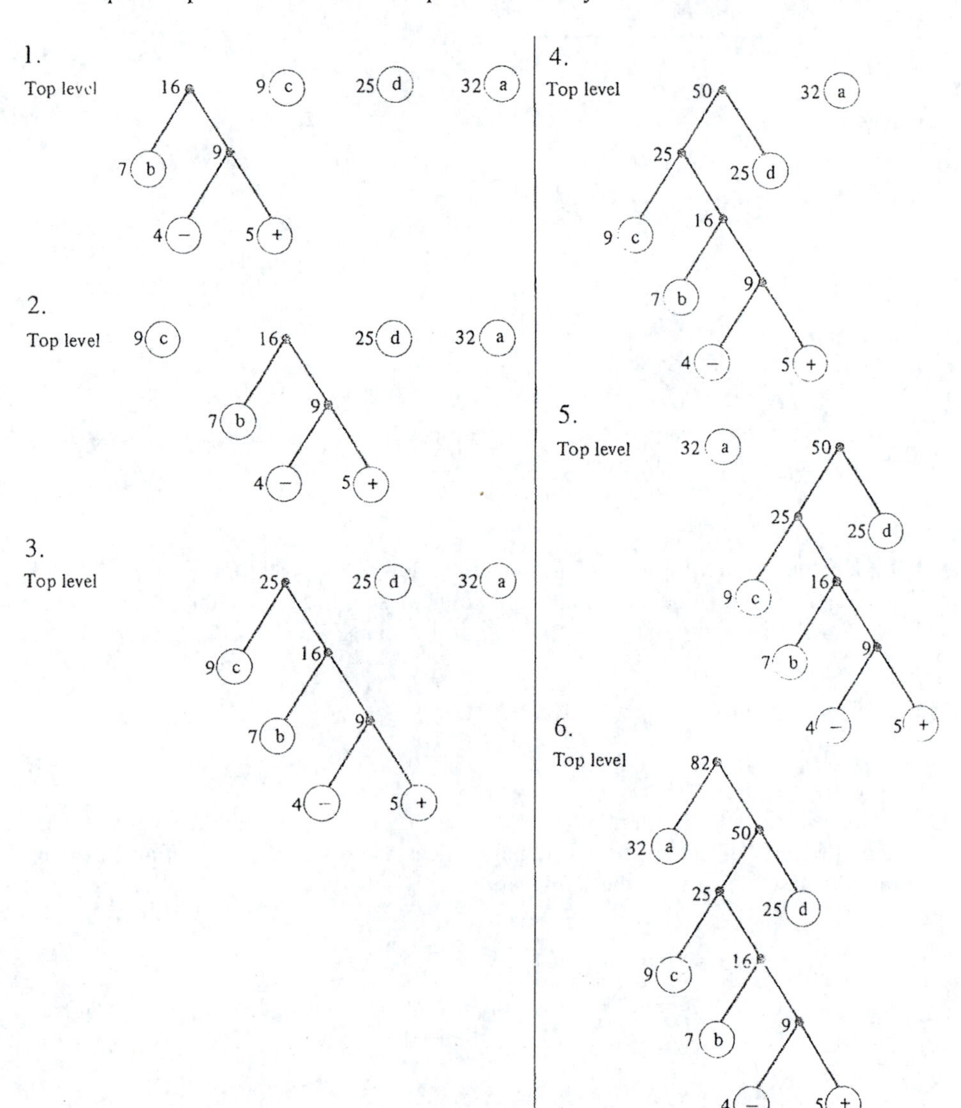

Step 5

Discard all of the frequencies (they are no longer needed) and label each edge of the graph that slants *up* (from left to right) with a 0 and each edge that slants *down* (from left to right) with a 1.

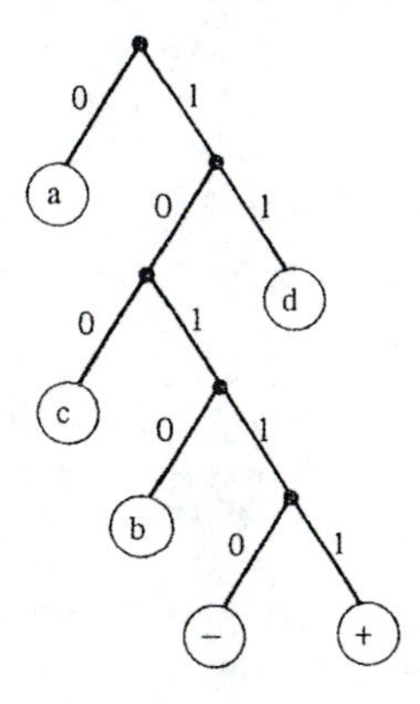

The graph obtained in step 5 is called a **Huffman tree**. In order to find the Huffman code for a particular character, we start at the top of the Huffman tree and follow the edges of the tree until we arrive at the character in question. The sequence of 0's and 1's that we encountered along our path is the Huffman code for that character.

Thus, the Huffman codes for our example are

Character	*Frequency*	*Huffman code*
a	32	0
b	7	1010
c	9	100
d	25	11
+	5	10111
−	4	10110

As you can see, this Huffman code has the prefix property, and characters with higher frequencies have shorter codes than characters with lower frequencies. The weight of this particular Huffman code is

$$32 \cdot 1 + 7 \cdot 4 + 9 \cdot 3 + 25 \cdot 2 + 5 \cdot 5 + 4 \cdot 5 = 182$$

Although we shall not prove it here, it can be proved (by induction in fact) that this code has the smallest weight among all possible codes that have the prefix property. Of course, there is nothing special about this particular example, and we can state the following theorem.

Theorem 6.6.1

A Huffman code for a given character set, with given frequencies, has the prefix property. Furthermore, it has the smallest weight among all character codes (for that character set and set of frequencies) that have the prefix property.

Let us close this section by remarking that for a given set of characters and frequencies, there may be more than one possible Huffman code. In other words, the Huffman code for a given set of characters and frequencies is not necessarily unique. (See Exercise 11.)

Exercises

1. Construct a binary search tree for the collection of words in the sentence "All men are created equal." Add the word "and" to the tree, and then add the word "women."
2. How many different binary search trees are possible for the collection of words {graph, tree, connected}? Find these trees. Add the word "spanning" to each of these trees.
3. Consider the collection of numbers {3245,2219,5555,1919,2351} ordered by size. Find a binary search tree for this collection. Add the numbers 4444 and 7243 to the tree.
4. Construct a binary search tree for the collection of words in the phrase "Being your slave, what should I do but tend." Then add the words, one at a time, from the phrase "upon the hours and times of your desire." (Shakespeare)
5. Construct a binary search tree for the collection of words in the sentence "A fool sees not the same tree as a wise man sees." (William Blake) Add the words "William" and "Blake" to the tree.
6. Construct a binary search tree for the collection of words in the phrase "Poems are made by fools like me." Then add the words, one at a time, from the phrase "But only God can make a tree." (Joyce Kilmer)
7. Construct a binary search tree for the collection of words in the phrase "A lovely Ladie rode him faire beside." Then add the words, one at a time, in the phrase "Upon a lowly Asse more white then snow." [From *The Faire Queene*, by Edmund Spenser (1552–1599).]
8. Construct a binary search tree for the collection of words
   The Sun came up upon the left,
   Out of the sea came he!
   And he shone bright, and on the right
   Went down into the sea.
   (From *The Rime of the Ancient Mariner*, by Samuel Taylor Coleridge.)
9. Can a character code with the prefix property have two characters each having a code of length 1? Explain your answer?
10. Consider the following table

Character	*Frequency*	*Code 1*	*Code 2*
a	2	111	1000
b	12	110	01
c	1	10	11111
d	6	0	10

   a) Compute the weights of the two codes in this table.
   b) Which code is more efficient?
   c) Does code 1 have the prefix property? Justify your answer.

d) Does code 2 have the prefix property? Justify your answer.
e) Which code would you prefer to use? Explain your answer.

11. Show that Huffman codes are not unique by finding a character set, and a set of frequencies, for which there is more than one Huffman code. *Hint*, make the frequencies equal.

In Exercises 12 through 19, we give a character set and the frequency of each character.

12.

Character	*Frequency*
a	3
b	1
c	2

a) Find a Huffman code for the character set using the given frequencies.
b) Compute the weight of the code.
c) Encode the message "aabac."
d) Decode the message 01001001.

13.

Character	*Frequency*
l	2
e	1
h	4
p	3

a) Find a Huffman code for the character set using the given frequencies.
b) Compute the weight of the code.
c) Encode the message "heep."
d) Decode the message 010010111.

14.

Character	*Frequency*
s	1
d	1
y	1
u	1
t	1

a) Find a Huffman code for the character set using the given frequencies.
b) Compute the weight of the code.
c) Encode the message "study."

15.

Character	*Frequency*
x	1
y	1
(	1
)	2
+	2
−	2

a) Find a Huffman code for the character set using the given frequencies.
b) Compute the weight of the code.
c) Encode the message "(x + y)."

16.

Character	*Frequency*
U	11
T	10
S	4
Y	30
D	5

a) Find a Huffman code for the character set using the given frequencies.
b) Compute the weight of the code.
c) Decode the message 01000110001011.

17.

Character	*Frequency*
h	12
p	40
g	15
a	8
r	17

a) Find a Huffman code for the character set using the given frequencies.
b) Compute the weight of the code.
c) Encode the message "harp."
d) Decode the message 1101111000101.

18.

Character	*Frequency*
α	4
β	5
+	7
*	8
(	12
)	25

a) Find a Huffman code for the character set using the given frequencies.
b) Compute the weight of the code.
c) Encode the message "$\alpha(\alpha+\beta)$."
d) Decode the message 11001011101101111110010011010.

19.

Character	*Frequency*
h	13
p	7
y	9
d	22
a	24
*	27

a) Find a Huffman code for the character set using the given frequencies.
b) Compute the weight of the code.
c) Encode the message "(x+y)."

6.7 Undirected Networks; The Minimum Spanning Tree Problem

In Example 6.1.7, we discussed briefly the concept of an undirected network. In this section, we want to study these important graphs in more detail.

Definition

An **undirected network** is a graph N for which each edge has been assigned a real number, called the **weight** of the edge. The sum of the weights of all of the edges of N is called the **weight** of N. □

Incidentally, we will study *directed* networks in Section 6.9. Before considering examples of undirected networks, we should point out that any subgraph H of a network N is also a network, where we assign the same weight to an edge in H as it had as an edge in N.

Example 6.7.1

The graph

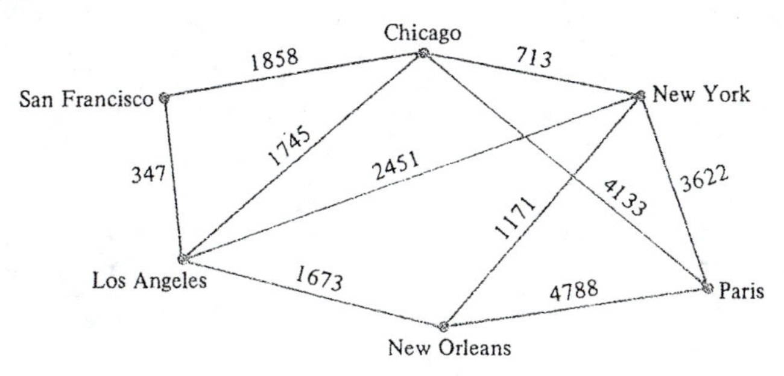

Figure 6.7.1

is an undirected network, where each vertex represents a city, and the weight of each edge is the air distance between the corresponding cities. □

Example 6.7.2

The graph

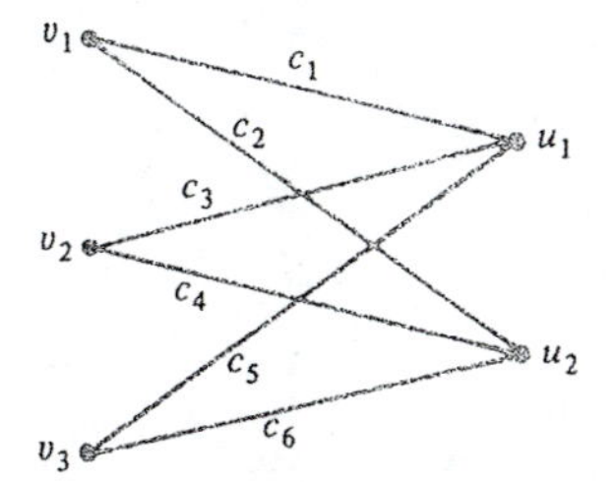

is an undirected network, where each of the vertices v_1, v_2, and v_3 represents a factory that produces a certain product, and each of the vertices u_1 and u_2 represents a retail outlet for this product. The weight of an edge $\{v_i, u_j\}$ is the cost of shipping a fixed quantity of the product from factory v_i to store u_j.

Example 6.7.3

The following road map is an undirected network

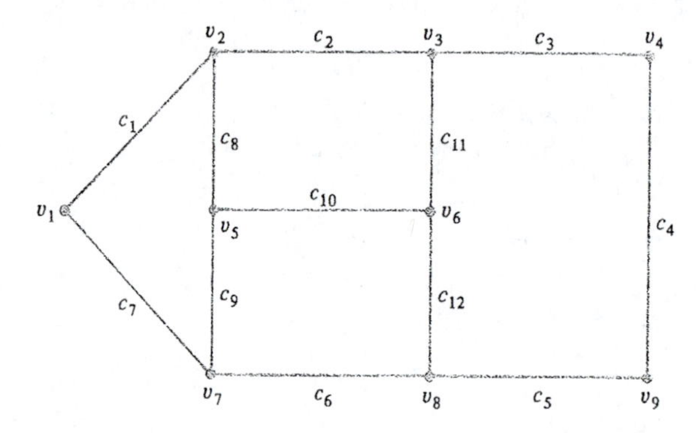

where the weight of an edge is the maximum number of cars that can travel on that road in 1 hour.

Example 6.7.4

Actually, any graph can be represented by an undirected network. For example, the graph

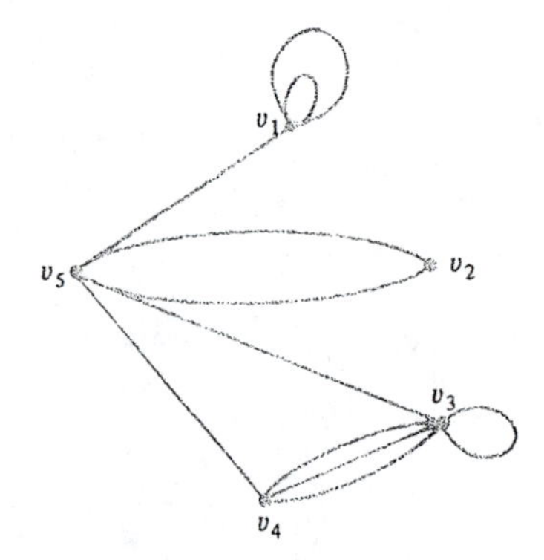

can be represented by the undirected network

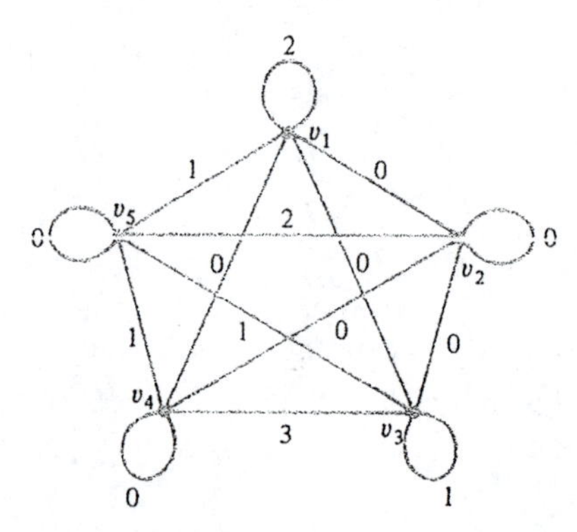

where the weight of each edge $\{v_i, v_j\}$ in the network is equal to the *number* of edges between the vertices v_i and v_j in the original graph. (What is the weight of the entire network in this case?) □

One of the most important problems associated with undirected networks is the so-called *minimum spanning tree problem*. A popular version of this problem is the following. Suppose that a railroad company decides to build a railway system between the cities pictured below.

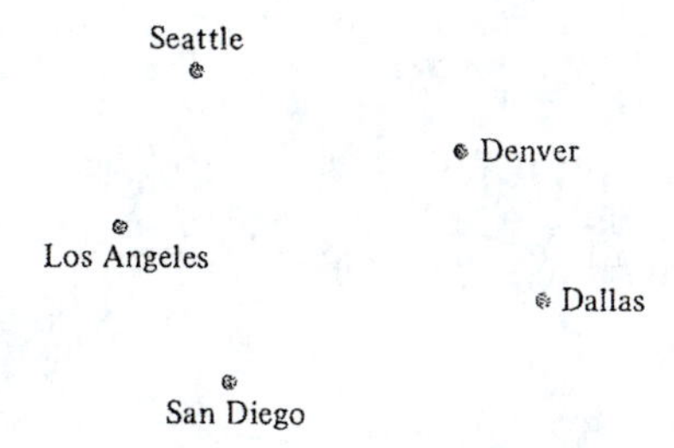

The first step is for the company to determine how much it would cost to build tracks between each pair of cities. Suppose that these costs (in millions of dollars) are given as the weights in the following network, which we denote by N.

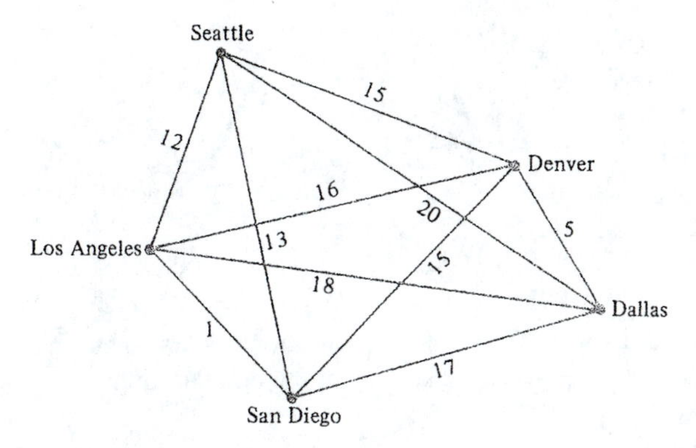

Figure 6.7.2

Thus, for example, it costs 18 million dollars to build a track from Los Angeles to Dallas.

Now, the company wants to save as much money as possible, and so it is certainly not going to build all of the tracks in the network N. However, it does want to be able to take passengers from any one of these cities to any other, perhaps with intermediate stops. The minimum spanning tree problem, in this case, is to determine which tracks the company should build in order to keep the total cost to a minimum.

From a graph theoretical point of view, the problem is to find a subgraph of N of smallest weight that is connected and has the same vertex set as N.

It is not hard to see that such a subgraph must be a *spanning tree* for N. After all, if the subgraph had a cycle, we could simply remove any one of the edges of that cycle, and the resulting subgraph would have smaller weight and still reach all of the vertices of the graph. (This is why the problem is called the minimum spanning tree problem, and not for example, the minimum spanning subgraph problem.)

However, it so happens that not all spanning trees for N will solve the minimum spanning tree problem. It turns out that the tracks in some spanning trees will cost more to build than the tracks in other spanning trees.

For example, the tracks in the spanning tree

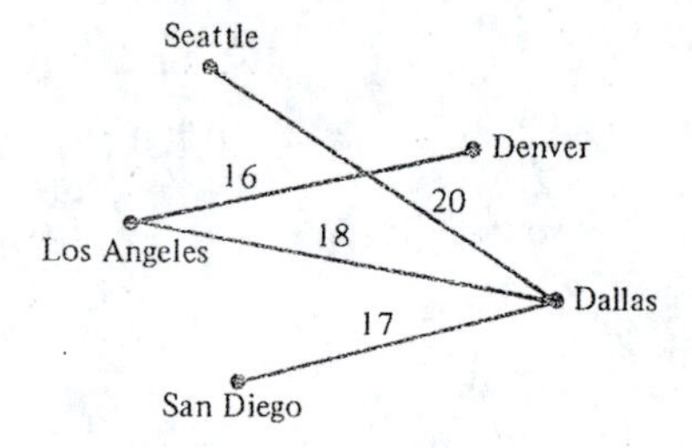

would cost a total of $16 + 17 + 18 + 20 = 71$ million dollars to build, but the tracks in the spanning tree

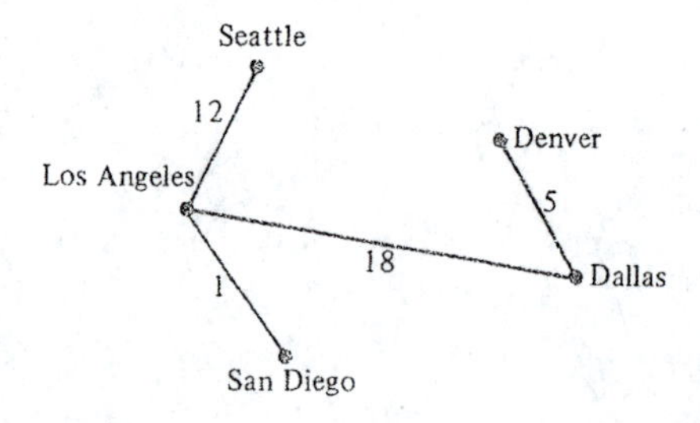

would cost only $1 + 5 + 12 + 18 = 36$ million dollars to build.

This discussion leads us to make the following definitions.

Definition

Let N be a *connected* undirected network with the property that each edge of N has *nonnegative* weight. Then the **minimum spanning tree problem** for N is the problem of finding a subgraph of N of smallest weight from among those subgraphs of N that are connected and that span N (that is, have the same vertex set as N). □

Definition

Let N be a connected undirected network. A spanning tree of N that has smallest weight from among all spanning trees of N (and hence also from among all spanning *subgraphs* of N) is called a **minimum spanning tree** of N. □

At this point, we can say that any minimum spanning tree is a solution to the minimum spanning tree problem. The question is, how do we find a minimum spanning tree for a given network?

Actually, there are several algorithms for finding a minimum spanning tree in a network. We will discuss one here and another in the exercises. The following algorithm is known as **Kruskal's algorithm**.

Kruskal's Algorithm

Let N be a connected undirected network with the property that each edge of N has nonnegative weight. Then the following algorithm defines a minimum spanning tree of N.

1) Choose any edge of N with smallest weight.
2) Continue to choose edges of N, subject to the condition that, at each step, the edge chosen is one of smallest weight from among those remaining edges of N that do not form a cycle with the previously chosen edges.
3) When it is impossible to choose any additional edges without violating this condition, the chosen edges (together with their vertices of course) form a minimum spanning tree of N.

PROOF Let T denote the subgraph of N obtained by using Kruskal's algorithm. We will leave it as an exercise to show that T is a spanning tree of N. Let us prove that it is a minimum spanning tree of N.

First, we observe that any *connected* network must have a minimum spanning tree. All we have to do is consider all possible spanning trees of N and take one, call it S, with the smallest weight.

Now, let us list the edges of T, and below that, the edges of S

$$\text{edges of } T \qquad e_1, e_2, \ldots, e_{n-1}$$

$$\text{edges of } S \qquad f_1, f_2, \ldots, f_{n-1}$$

Let us also suppose that both of these lists are in order of increasing weights.

Of course, it is possible that some of the edges in T are also edges in S. If e_k is the first edge in the list $e_1, e_2, \ldots, e_{n-1}$ that is *not* an edge in S, then we can rewrite these lists as follows:

$$\text{edges of } T \qquad e_1, e_2, \ldots, e_{k-1}, e_k, \ldots, e_{n-1}$$

$$\text{edges of } S \qquad e_1, e_2, \ldots, e_{k-1}, f_k, \ldots, f_{n-1}$$

where the edges e_k and f_k are different.

Our goal now is to find another minimum spanning tree of N that has its first k edges in common with those of T. Then, by repeating this process, we eventually get a minimum spanning tree of N that has *all* of its edges in common with T, and so it must be T. This will show that T is a minimum spanning tree of N.

In order to do this, consider what happens if we add the edge e_k to the graph S. According to Theorem 6.6.7, the resulting graph will have exactly one cycle, call it C, which of course must contain the edge e_k. But the cycle C must also contain an edge f that is in S but not in T. (Why?)

If, after adding the edge e_k to S, we remove the edge f, the resulting graph, call it S', will also be a minimum spanning tree of N. This can be seen as follows. First,

S' is still a spanning tree of N. (Why?) Second, the weight of e_k must be less than or equal to the weight of f. The reason for this is that, after the edges $e_1, e_2, \ldots, e_{k-1}$ were chosen, the edge e_k had the smallest weight from among all remaining edges of N that did not form a cycle with $e_1, e_2, \ldots, e_{k-1}$. Since f is included among these edges, the weight of e_k must be less than or equal to the weight of f. Thus, adding e_k and removing f can only reduce the weight of the resulting graph. Put another way, the weight of S' must be less than or equal to the weight of S. But since S has the smallest weight among spanning trees of N, we conclude that the weight of S' is equal to the weight of S, and so S' is also a minimum spanning tree of N.

Let us compare the edges of S' with the edges of T

$$\begin{array}{ll} \text{edges of } T & e_1, e_2, \ldots, e_{k-1}, e_k, e_{k+1}, \ldots, e_{n-1} \\ \text{edges of } S' & e_1, e_2, \ldots, e_{k-1}, e_k, f'_{k+1}, \ldots, f'_{n-1} \end{array}$$

We can see from this that the minimum spanning tree S' has its first k edges in common with those of T.

By repeating the procedure we just used, with S' in place of S, we will obtain a new minimum spanning tree S'' of N that has its first $k + 1$ edges in common with those of T. Eventually, by repeating the procedure as many times as is necessary, we will obtain a minimum spanning tree of N that has all of its edges in common with those of T. But the only graph with this property is T itself. This shows that T is in fact a minimum spanning tree of N and completes the proof. ∎

Example 6.7.5

Let us test Kruskal's algorithm on the following network N.

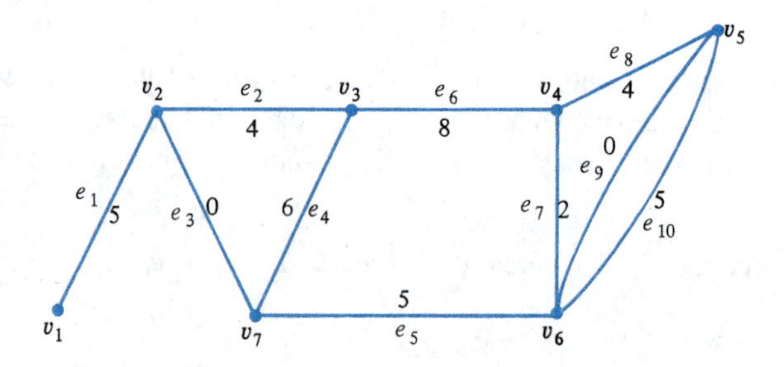

The first step is to pick any edge in N that has smallest weight, say e_3.

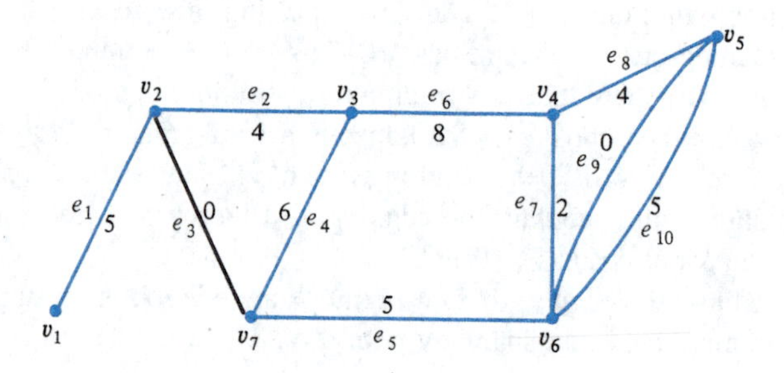

Then, looking among the other edges, we pick one with the smallest weight that does not form a cycle with the edges we have already picked. In this case, we pick e_9.

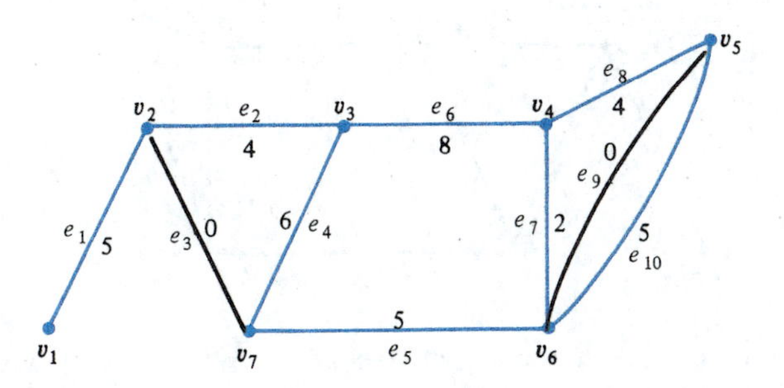

Next, we pick the edge e_7, since it has the smallest weight from among the remaining edges and does not form a cycle with the edges we have already picked.

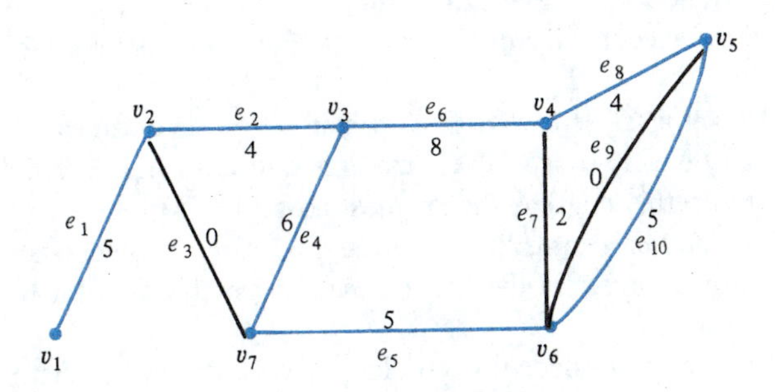

At this stage, there are two edges with smallest weight, namely, e_2 and e_8. However, the edge e_8 forms a cycle with two of the edges that we have already picked, namely, e_7 and e_9, and so we cannot pick this edge. Since e_2 does not form a cycle, we can pick it.

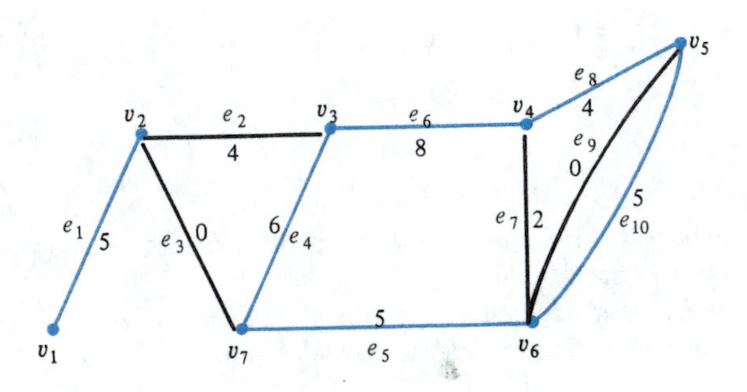

Then we pick e_1, and then e_5. (We cannot pick e_{10} because it forms a cycle with the already chosen edge e_9.)

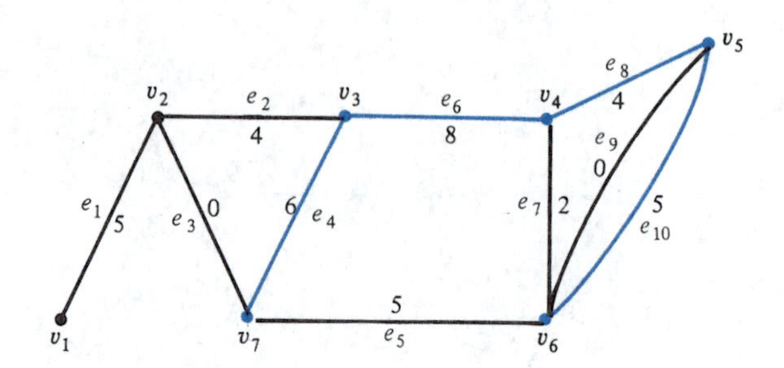

At this point, we cannot choose any more edges from the network N without creating a cycle with the edges that we have already chosen, and so the edges that we have chosen, together with their vertices, form a minimum spanning tree of N. According to Kruskal's algorithm, this spanning tree is a minimum spanning tree of N. □

Let us conclude this section by mentioning another well-known problem involving undirected networks. Let N be a connected undirected network with the property that its edges have nonnegative weights. Suppose that the vertices of N represent cities and that the weight of each edge in N is the cost of traveling between the corresponding cities.

Now, a traveling salesman wishes to start at one city and make a round trip, covering each city exactly once. The **traveling salesman problem** is the problem of finding a way to do this that has the smallest cost.

In graph theoretical terms, the traveling salesman problem is the problem of finding a Hamiltonian cycle in the network N that has the smallest weight among all Hamiltonian cycles in N.

As you might expect, since this problem involves finding a Hamiltonian cycle in a graph, no practical solution has ever been found for the traveling salesman problem. Exercise 19 gives one indication that the problem cannot be solved in a straightforward manner, as we solved the minimum spanning tree problem.

□□ Exercises

1. Represent each of the following graphs as an undirected network.

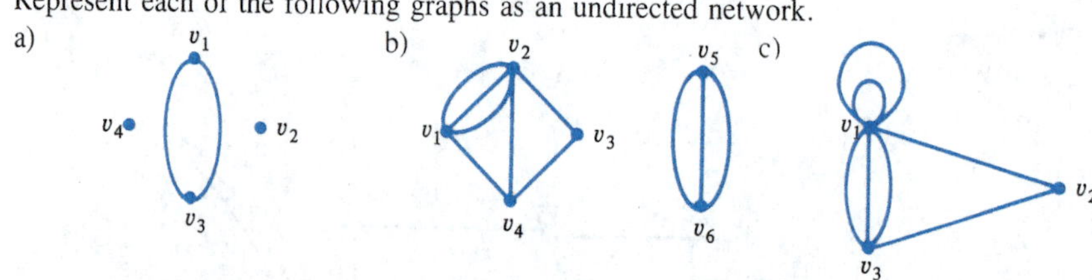

2. If G is a graph and N is the undirected network that represents G, describe the weight of N in terms of the graph G.
3. Find a minimum spanning tree for the network in Figure 6.7.1.
4. Find a minimum spanning tree for the network in Figure 6.7.2.

5. Find a minimum spanning tree for the network

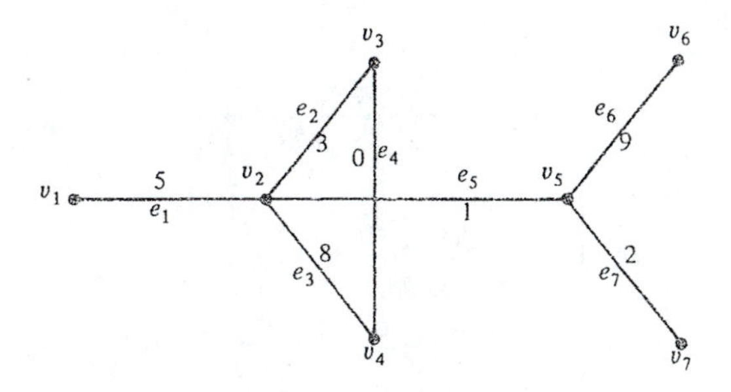

6. Find all minimum spanning trees for the network

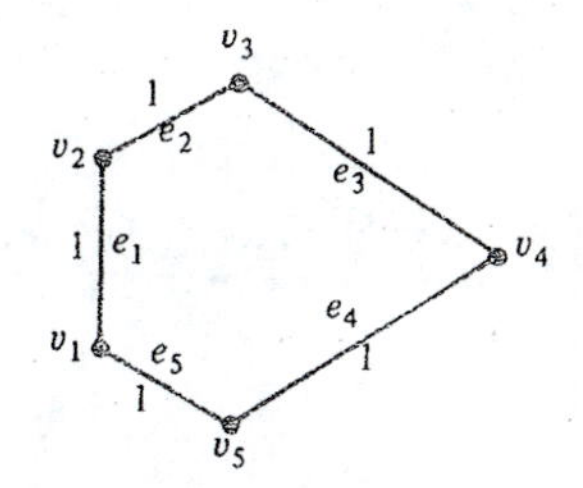

7. Find a minimum spanning tree for the network

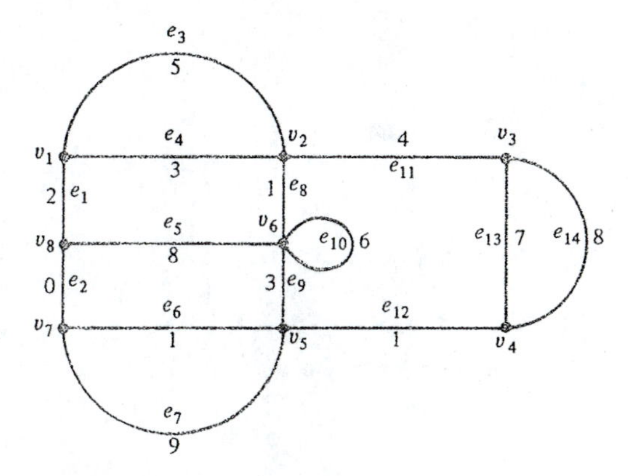

8. Find a minimum spanning tree for the network

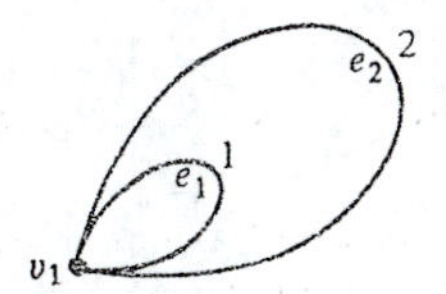

9. Solve the minimum spanning tree problem for the network

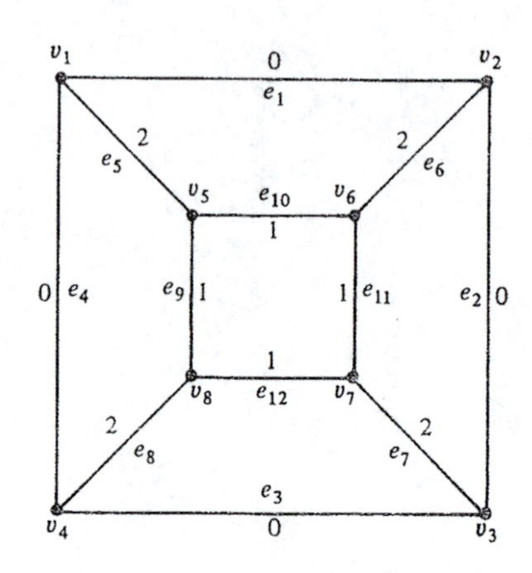

10. Solve the minimum spanning tree problem for the network

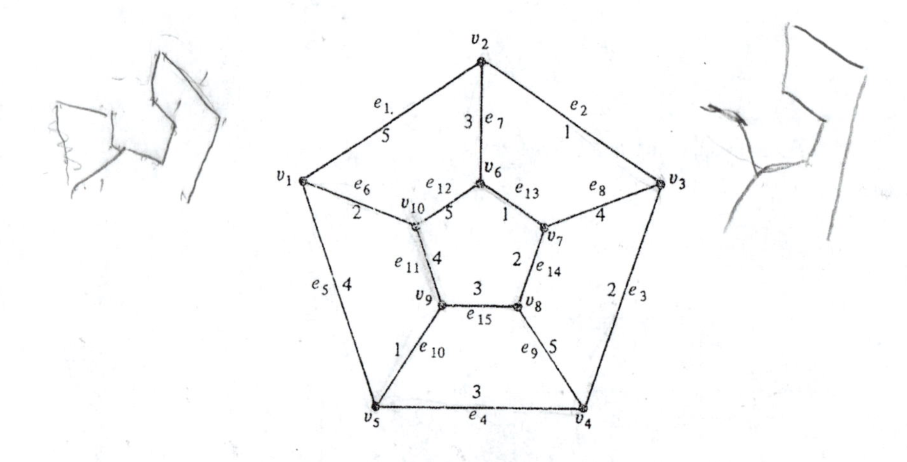

11. Find all minimum spanning trees for the network

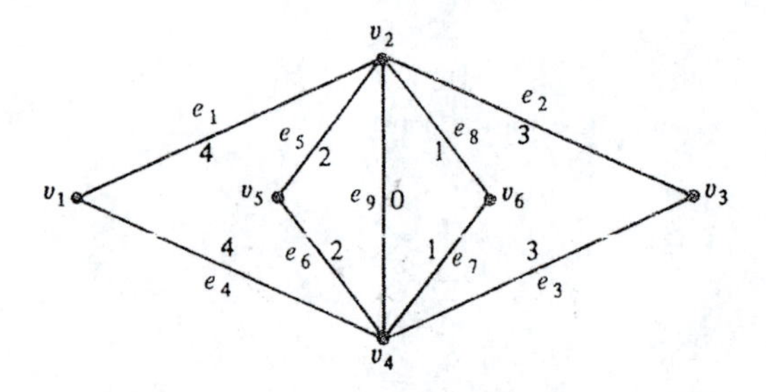

12. Find all minimum spanning trees for the network

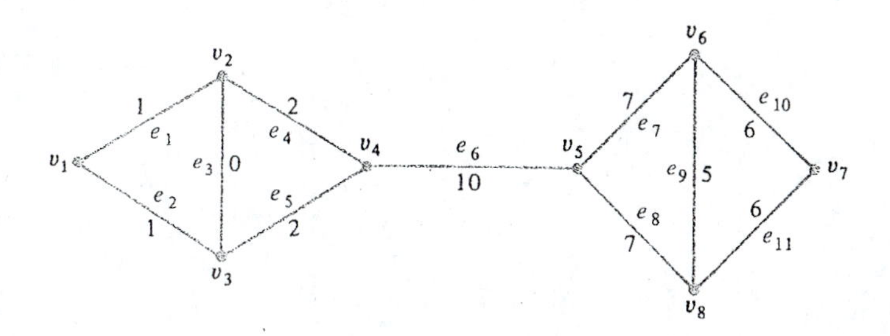

13. Find the spanning subgraph of minimum weight of the following network.

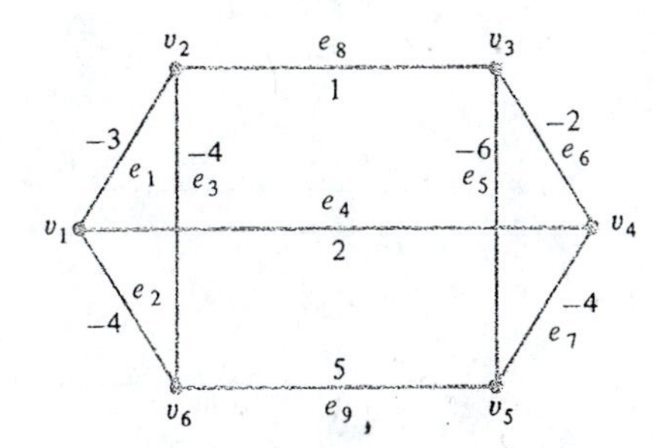

Is your solution a spanning tree for the network? Explain.

14. Let N be a connected, undirected network with the property that each of its edges has nonnegative weight. Prove that a subgraph H of N is a solution to the minimum spanning tree problem for N if and only if H is either

1) a minimum spanning tree of N, or
2) a minimum spanning tree of N, together with some additional edges of N, each of which has zero weight.

15. Referring to the proof of Kruskal's algorithm, show that the graph T obtained by Kruskal's algorithm is a spanning tree for N.

16. State what you think the *maximal* spanning tree problem should be. Can you solve it for any particular class of networks?

17. Solve the traveling salesman problem for the network

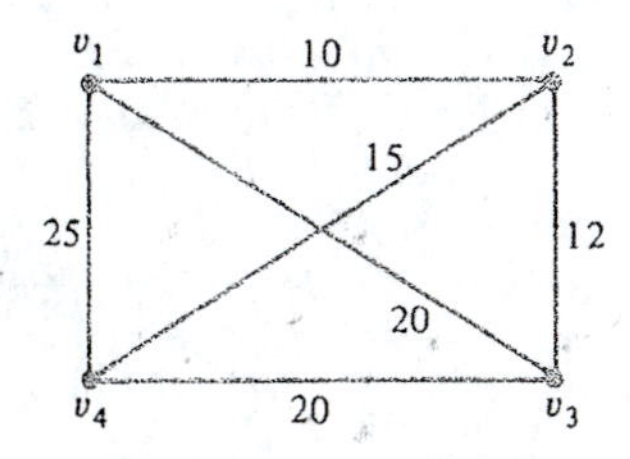

18. Solve the traveling salesman problem for the network

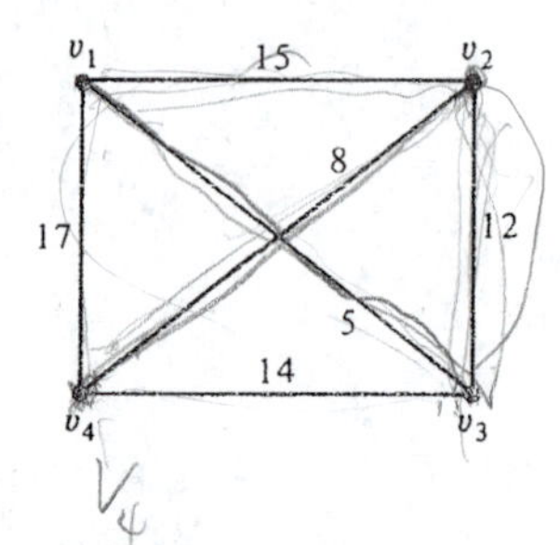

19. Consider the following network.

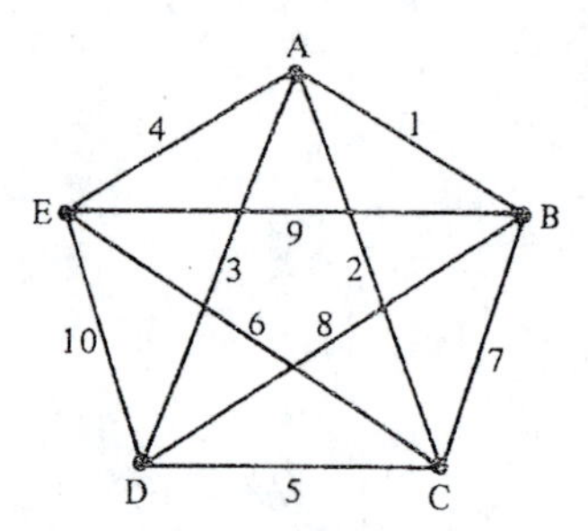

Starting at each of the five vertices, obtain a Hamiltonian cycle by always proceeding to the next available vertex along an edge of smallest weight. Show that none of the five Hamiltonian cycles that you obtain in this way is a solution to the traveling salesman problem.

Another algorithm for finding a minimum spanning tree in a connected network N is ***Prim's algorithm****. Loosely speaking, Prim's algorithm "grows" a tree T that, at each stage, has minimum weight. Once the tree has grown to full size, it is a spanning tree of N. Hence, it is a minimum spanning tree of N.*

Prim's Algorithm

1) *Pick an edge $e = \{u, v\}$ in N with smallest weight among all edges in N. Let T be the graph consisting of the edge e and its vertices.*
2) *Choose an edge of N with the smallest weight from among all edges of N that have one vertex in T and the other vertex not in T. Add this new edge, and its new vertex, to T. (Notice that the new edge cannot form a cycle with any of the other edges in T since one of the vertices of this new edge is not in T. Hence, T is still a tree even after the new edge is added.)*
3) *Repeat step 2 until T is a spanning tree. The resulting graph T is a minimum spanning tree of N.*

20. Use Prim's algorithm to find a minimum spanning tree for the network in Example 6.7.5. Compare the spanning tree with the one found in that example.

21. Use Prim's algorithm to find a minimum spanning tree for the network in Exercise 7.

22. Use Prim's algorithm to find a minimum spanning tree for the network in Exercise 9.
23. Use Prim's algorithm to find a minimum spanning tree for the network in Exercise 10.
24. Use Prim's algorithm to find a minimum spanning tree for the network in Exercise 11.
25. Use Prim's algorithm to find a minimum spanning tree for the network in Exercise 12.

6.8 Directed Graphs; Strong Connectivity

There are many applications of graph theory in which it is useful to be able to assign a direction to each of the edges of a graph. This occurs, for example, in using a graph to represent a communications system, where communication can occur between two points in only one direction. As another example, we might want to use a graph to represent a transportation system whose routes are one way, such as a system of one-way streets. This leads us to make the following definition.

Definition

A **directed graph** is a graph in which each edge has been given a direction. It is customary to refer to the edges of a directed graph as **arcs**. An arc that is directed *from* a vertex u *to* a vertex v is denoted by the *ordered pair* (u, v) (rather than by the set $\{u, v\}$, as is the case with nondirected graphs).

If D is a directed graph, then the graph obtained by ignoring the directions of the arcs in D is called the **underlying graph** of D.

Directed graphs are also known as **digraphs**. Let us have some examples of directed graphs.

Example 6.8.1

The following graph is a directed graph

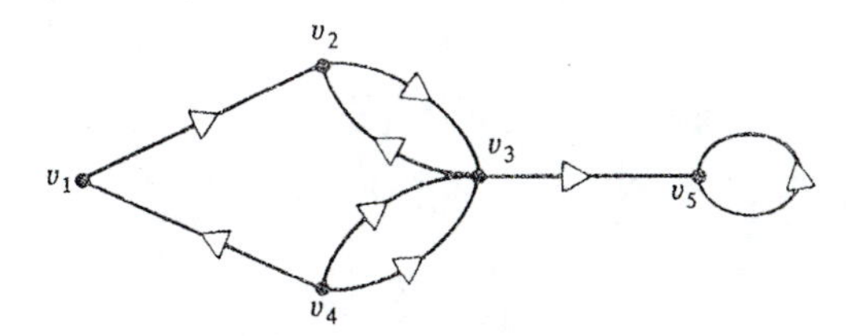

whose underlying graph is

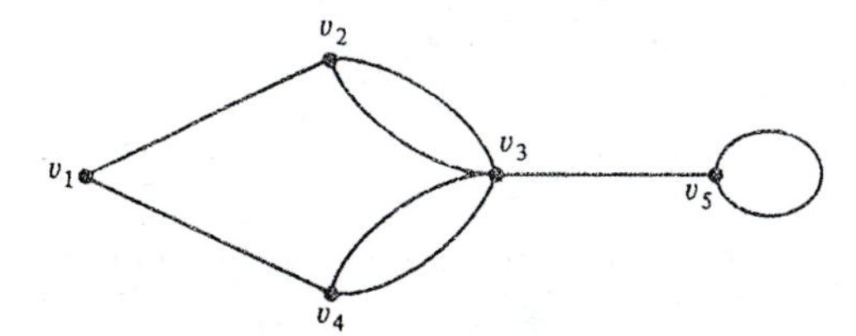

Example 6.8.2

As we have already mentioned, directed graphs can be used to represent systems of one-way streets. For example, in the directed graph

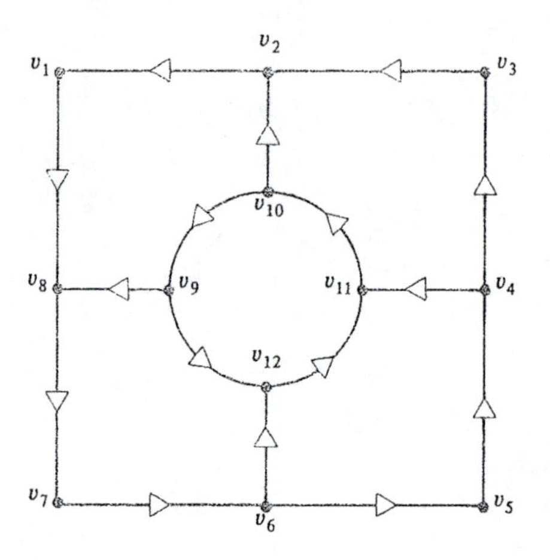

Figure 6.8.1

each vertex represents an intersection and each arc represents a one-way street. (Incidentally, how would you use a directed graph to represent a system of streets, where some of the streets are one-way and some are two-way?) □

Example 6.8.3

The following directed graph represents the outcome of a chess tournament.

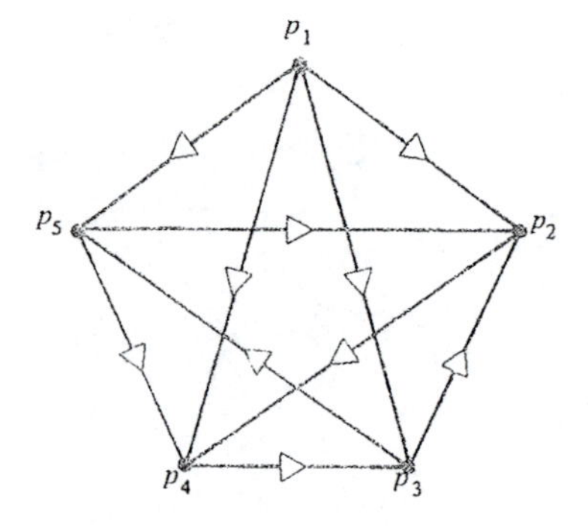

Each vertex represents a player, and there is an arc from vertex u to vertex v if and only if player u defeated player v in the tournament. □

Example 6.8.4

A computer program is often made up of several different procedures. Control can be transferred from certain procedures to certain others, and this leads us to the idea of a directed graph that describes the *flow of control* of the program. The vertices of the graph represent the various procedures, and an arc from vertex v to vertex w means that procedure v can pass control of the program to procedure w.

In a similar way, data (in the form of parameters) can be passed from certain procedures to certain others, and so we can also define a directed graph that describes the *flow of data* from one procedure to another. □

The theory of directed graphs has its own set of definitions, and we need to discuss a few of them now. Let D be a directed graph. If (u, v) is an arc in D, then we say that u is **adjacent to** v, and v is **adjacent from** u. We also call u the **initial point** of the arc (u, v), and v the **terminal point** of (u, v).

The **indegree** of a vertex v, denoted by $indeg(v)$, is the number of arcs of D whose terminal point is v. Similarly, the **outdegree** of a vertex v, denoted by $outdeg(v)$, is the number of arcs in D whose initial point is v. The **degree** of a vertex v is the sum of its indegree and its outdegree. This is the same as the degree of v, thought of as a vertex in the underlying graph of D.

Let us illustrate these concepts with the following directed graph.

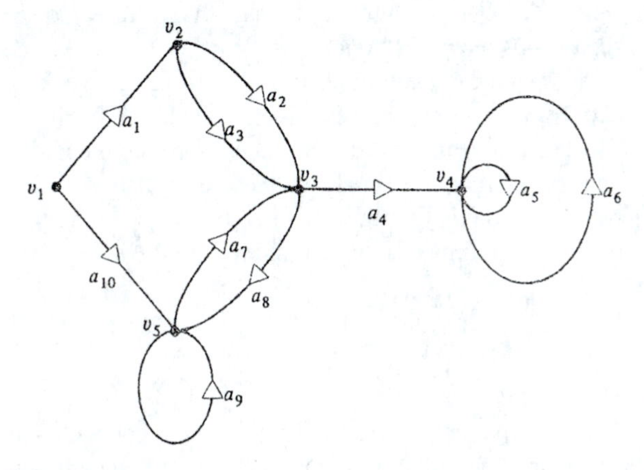

Figure 6.8.2

The arc $a_1 = (v_1, v_2)$ is an arc from v_1 to v_2. Vertex v_1 is adjacent to vertex v_2, and v_2 is adjacent from v_1. Vertex v_3 is both adjacent to and adjacent from vertex v_5, and vertex v_5 is both adjacent to and adjacent from itself (as a result of the loop at v_5). The indegree and outdegree of each vertex is

$$indeg(v_1) = 0, \qquad outdeg(v_1) = 2$$

$$indeg(v_2) = 1, \qquad outdeg(v_2) = 2$$

$$indeg(v_3) = 3, \qquad outdeg(v_3) = 2$$

$$indeg(v_4) = 3, \qquad outdeg(v_4) = 2$$

$$indeg(v_5) = 3, \qquad outdeg(v_5) = 2$$

By a **path** in D, we mean a path in the underlying graph of D. If

$$v_1, a_1, v_2, a_2, \ldots, v_{n-1}, a_{n-1}, v_n$$

is a path in D from v_1 to v_n, and if a_i is an arc *from* v_i *to* v_{i+1}, for all $i = 1, 2, \ldots, n - 1$, then we call this path a **directed path** from v_1 to v_n. Intuitively speaking, a directed path from v_1 to v_n is a path that has the property that we can start at v_1 and reach v_n by following the arcs in the path, *always moving in the direction indicated by the arcs*. For example, in the directed graph of Figure 6.8.2, the path

$$v_1, a_{10}, v_5, a_7, v_3, a_4, v_4$$

is a directed path from v_1 to v_4, but the path

$$v_1, a_{10}, v_5, a_8, v_3, a_4, v_4$$

is not a directed path from v_1 to v_4.

If there is a directed path from u to v, then we say that v is **reachable from** u. A directed graph D is said to be **strongly connected** if any vertex in D is reachable from any other vertex in D. In contrast, we say that a directed graph is **weakly connected** if its *underlying graph* is connected.

Thus, for example, the directed graph in Figure 6.8.2 is weakly connected, but it is *not* strongly connected, since vertex v_1 is not reachable from vertex v_3. On the other hand, the directed graph in Figure 6.8.1 is strongly connected.

Of course, any graph that is strongly connected is also weakly connected, but the graph in Figure 6.8.2 shows that the converse of this statement is false.

The concept of strong connectivity in directed graphs is very important, as the next two examples show.

Example 6.8.5

The fact that the graph in Figure 6.8.1 is strongly connected means that it is possible to drive from any location on the map to any other location by following the directions of the streets. If the graph were not strongly connected, this would mean that some locations could not be reached from other locations.

Example 6.8.6

The fact that a flow of control graph for a particular computer program (see Example 6.8.4) is strongly connected means that control can be passed from any procedure to any other, perhaps through intermediate procedures, however. If such a graph were not strongly connected it would mean that once control was acquired by a certain procedure, it could never be passed on to certain other procedures. Similar statements hold concerning the flow of data in a computer program.

Example 6.8.7

Consider the following directed graph.

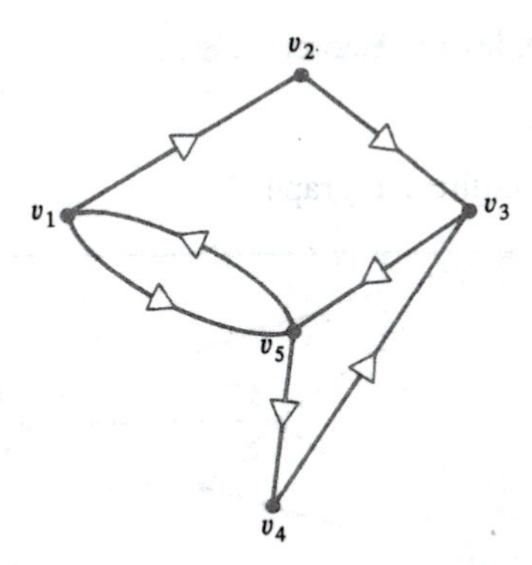

Each vertex of this graph represents a person, and each arc of the form (u, v) represents a line of communication that allows person u to send a message to person v. (But it does not allow person v to send a message to person u.) The fact that this graph is strongly connected means that any person can get a message to any other person, perhaps through the use of intermediaries. □

We will leave as an exercise a proof of the following simple result concerning strongly connected graphs.

Theorem 6.8.1

Let D be a directed graph, and let v be any vertex in D. Then D is strongly connected if and only if every vertex in D is reachable from v and v is reachable from every vertex in D.

It turns out that the method of depth first search, which we discussed in Section 6.5, can be of great help in determining whether or not a graph is strongly connected.

The first step is to modify the depth first search method for directed graphs. This is very easy to do—we just restrict ourselves to searching only in the directions specified by the arcs of the graph. Put another way, instead of finding open paths during the search process, we find only *directed* open paths. [This amounts to modifying the algorithm given in Section 6.5 by replacing the word "edge" by the word "arc" and by replacing $f = \{u, x\}$ by $f = (u, x)$.]

We can now describe a way to determine whether or not a directed graph is strongly connected.

Theorem 6.8.2

Let D be a directed graph, and let v be any vertex of D. Let D' be the directed graph obtained from D by reversing the direction of each of the arcs in D. Then the directed graph D is strongly connected if and only if

1) any depth first search performed on D, starting at v, reaches all of the vertices of D, and

2) any depth first search performed on D', starting at v, reaches all of the vertices of D'.

(Of course, D and D' have exactly the same vertices.)

Let us try some examples of Theorem 6.8.2.

Example 6.8.8

Consider the following directed graph D

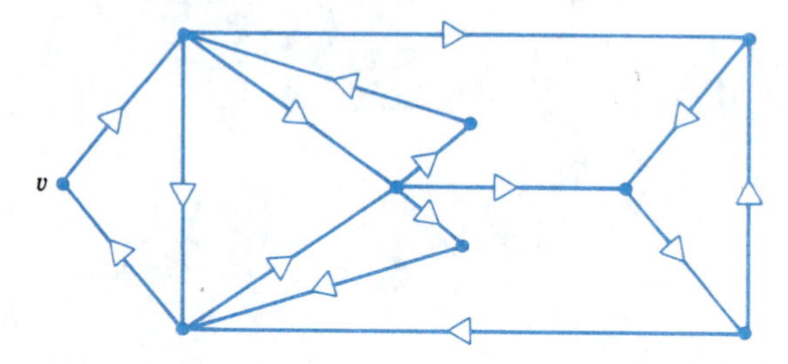

A depth first search in D starting at v, is shown below

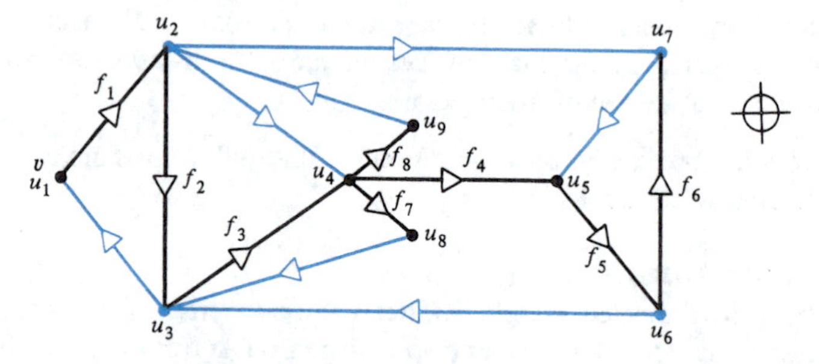

Notice that it does reach every vertex in D. The directed graph D', obtained from D by reversing the directions of its arcs, is

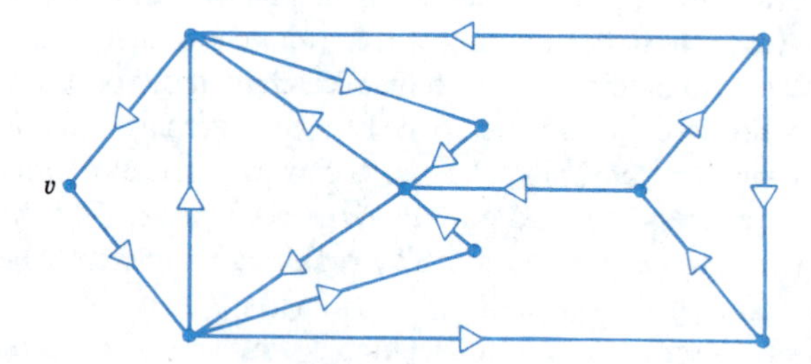

A depth first search in D' is shown below

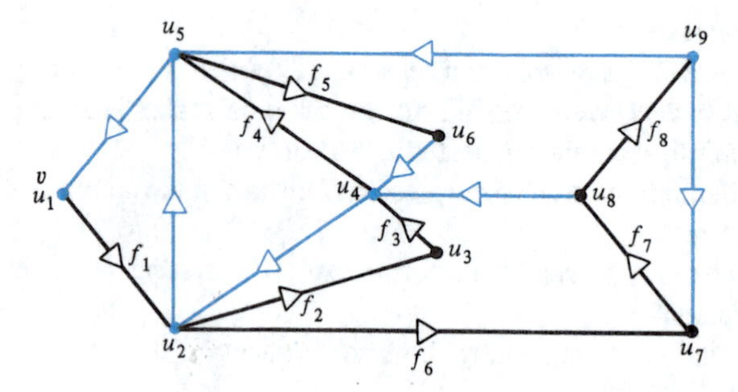

Since this search reaches every vertex in D', Theorem 6.8.2 tells us that D is strongly connected. □

Example 6.8.9

Consider the following directed graph D

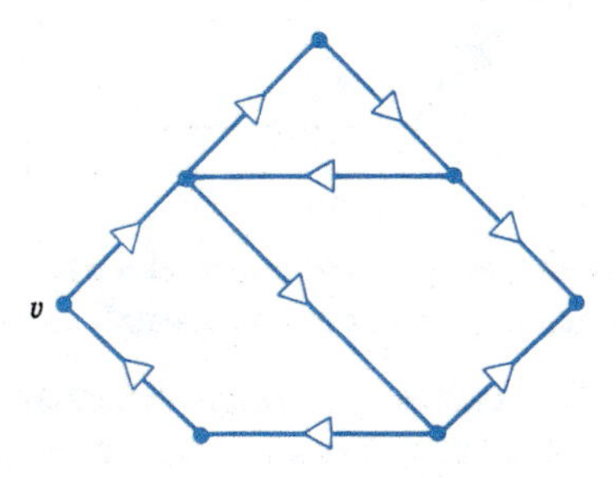

A depth first search in D starting at v is shown below

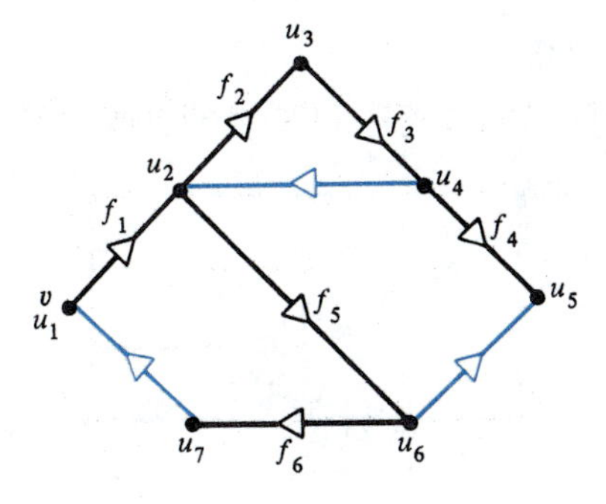

This search does reach every vertex in D. The graph D' is

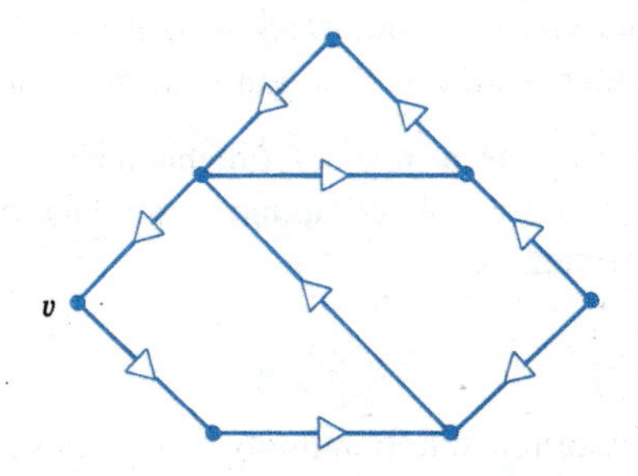

and a depth first search in D' starting at v is

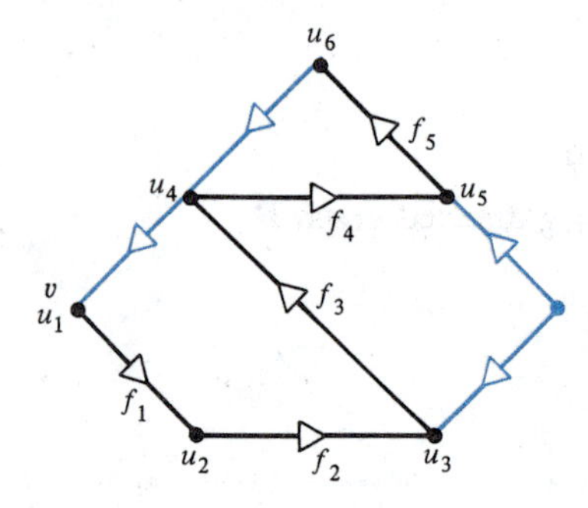

However, in this case, the search does not reach all vertices in D', and so, according to Theorem 6.8.2, the directed graph D is not strongly connected. □

An *undirected* graph G is said to be **strongly orientable** if it is possible to assign a direction to each of the edges of G in such a way that the resulting directed graph is strongly connected. Any such assignment of directions is called a **strong orientation** of the graph G.

The next example indicates one reason we might be interested in whether or not a graph is orientable.

Example 6.8.10

The following (undirected) graph is the street map of a certain town.

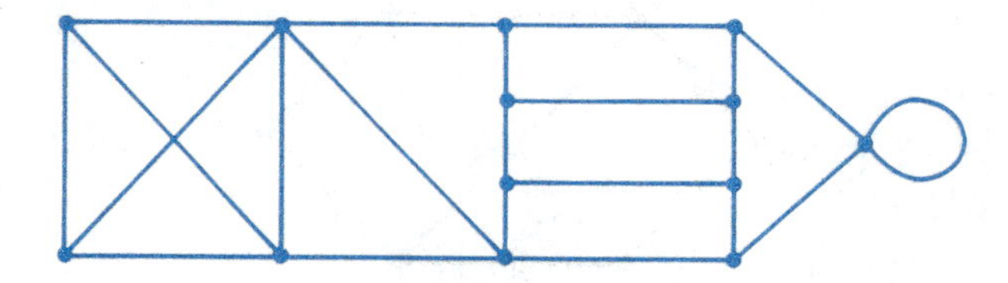

All of the streets are two-way streets, but the mayor of the town wants to make them all into one-way streets. Is there a way he can do this so that it will still be possible to get from any location in the town to any other location?

We can rephrase this question in graph theoretical terms as follows. Is there a way to direct each of the edges of this graph so that the resulting directed graph is strongly connected? In other words, is this graph strongly orientable? □

The following theorem gives a very helpful characterization of orientable graphs. Let us recall that a bridge in a connected graph is an edge whose removal leaves the remaining graph disconnected.

Theorem 6.8.3

A graph G is strongly orientable if and only if it is connected and has no bridges.

Since the proof of this result is a bit involved, we will not give it here. (See Exercise 21, however.) Instead, let us show how the method of depth first search can be used to actually find a strong orientation for any strongly orientable graph.

Algorithm for Finding a Strong Orientation

Let G be a strongly orientable graph. Then the following procedure will produce a strong orientation for G.

1) Pick any vertex in G and perform a depth first search starting at this vertex. Since G is connected, the search results in a depth first search spanning tree $(\mathrm{V}_n, \mathrm{E}_n)$, where $\mathrm{V}_n = \{u_1, u_2, \ldots, u_n\}$ are the vertices of G.

2) If $\{u_i, u_j\}$ is an edge in the depth first spanning tree, orient it by giving it the direction from the vertex with the smaller subscript to the vertex with the larger subscript. That is, if $i < j$, orient the edge from u_i to u_j, and if $j < i$, orient the edge from u_j to u_i.

3) If e is an edge of G that is not in the spanning tree, then it still must have the form $\{u_i, u_j\}$. (Why?) In this case, orient the edge by giving it the direction from the vertex with the larger subscript to the vertex with the smaller subscript. That is, if $i \leq j$, orient the edge from u_i to u_j, and if $j < i$, orient the edge from u_j to u_i.

Let us try this algorithm on the graph

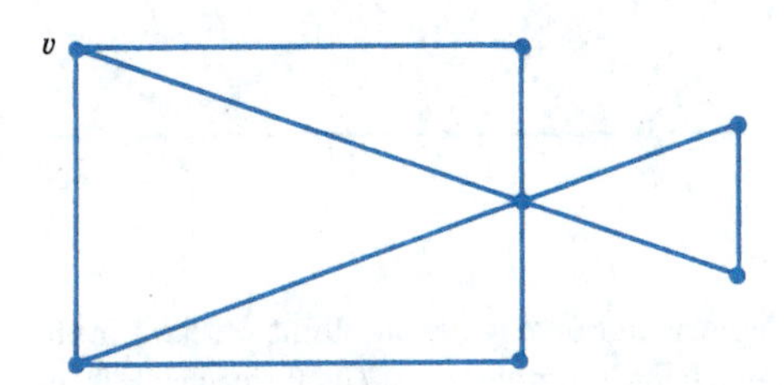

This graph is connected and has no bridges, and so, according to Theorem 6.8.3, it is strongly orientable. We choose the vertex v and perform a depth first search starting at v, to get

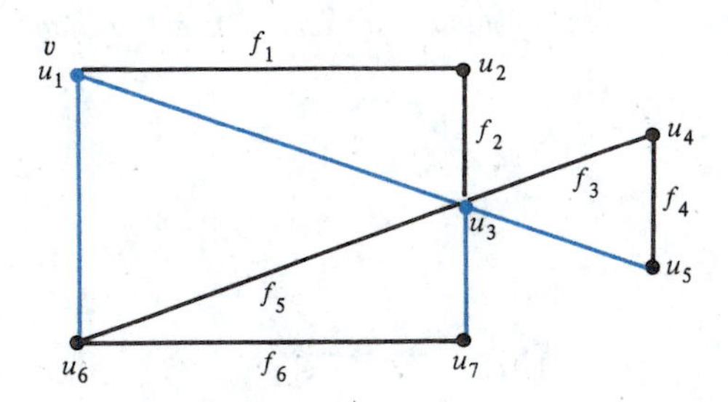

Then we orient the graph according to the rules given in the algorithm.

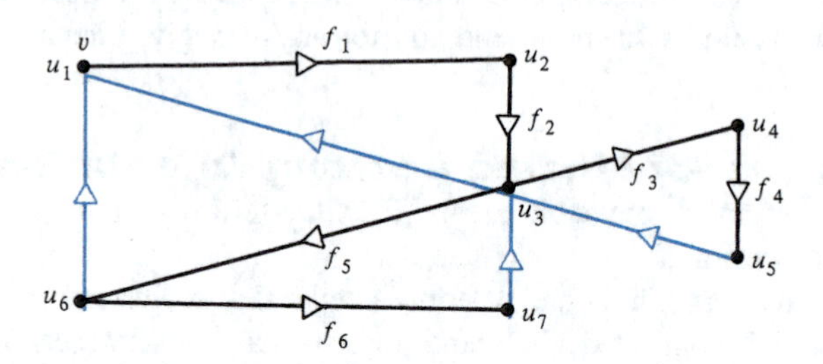

Figure 6.8.3

We will leave it as an exercise to show that this directed graph is strongly connected.

Exercises

1. Consider the following directed graph

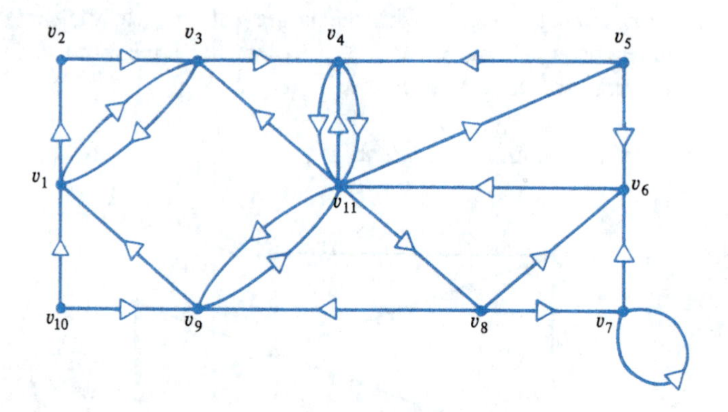

   a) Find the indegrees and outdegrees of all the vertices in this graph.
   b) Is vertex v_1 reachable from vertex v_6? Is v_6 reachable from v_1?
   c) Is vertex v_{10} reachable from vertex v_{11}? Is v_{11} reachable from v_{10}?
2. If D is a directed graph, is it true that the sum of the indegrees of all of the vertices of D is equal to the sum of the outdegrees of all of the vertices? Justify your answer.
3. Let D be a directed graph, and let u, v, and w be vertices of D. If u is reachable from v and v is reachable from w, then is u reachable from w? Justify your answer.

In Exercises 4 through 9, determine whether or not the given directed graph is strongly connected. If it is not, find two vertices u and v such that u is not reachable from v.

4.

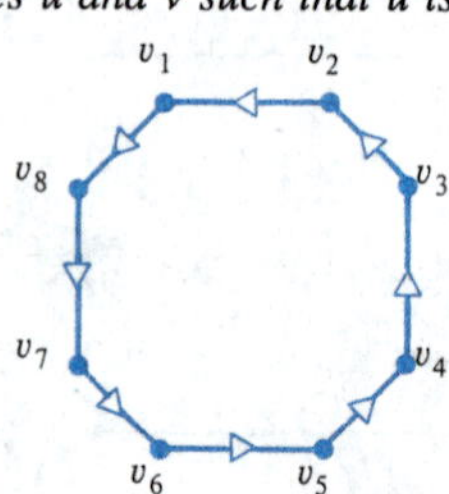

5.

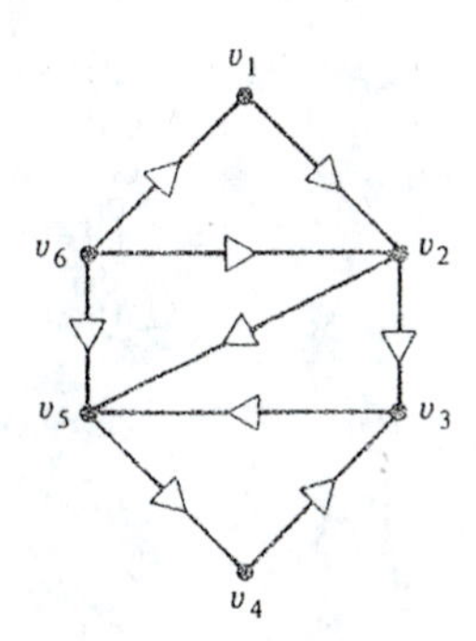
v_1
v_6
v_2
v_5
v_3
v_4

6.

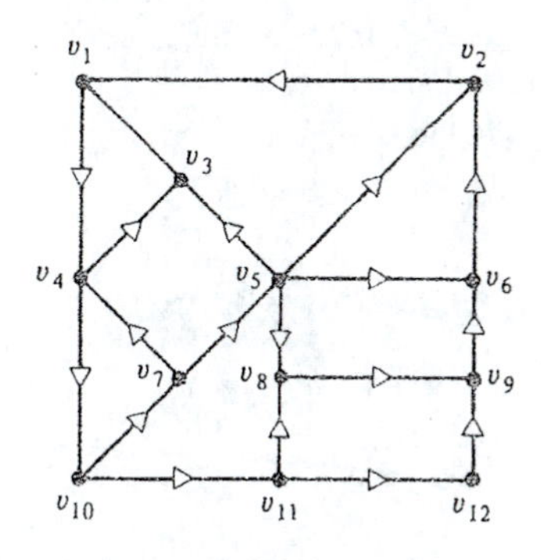
v_1
v_2
v_3
v_4
v_5
v_6
v_7
v_8
v_9
v_{10}
v_{11}
v_{12}

7.

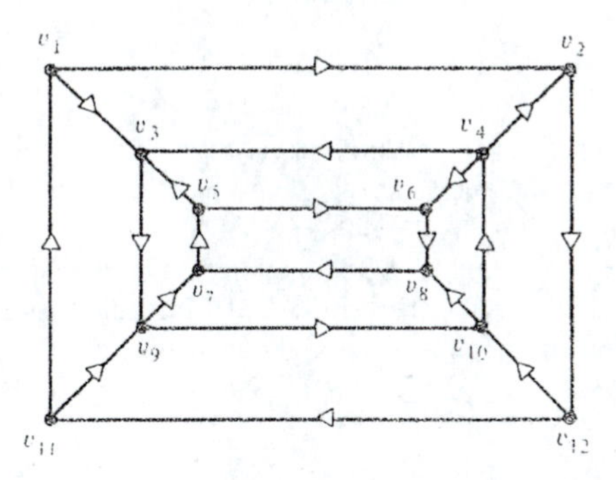
v_1
v_2
v_3
v_4
v_5
v_6
v_7
v_8
v_9
v_{10}
v_{11}
v_{12}

8.

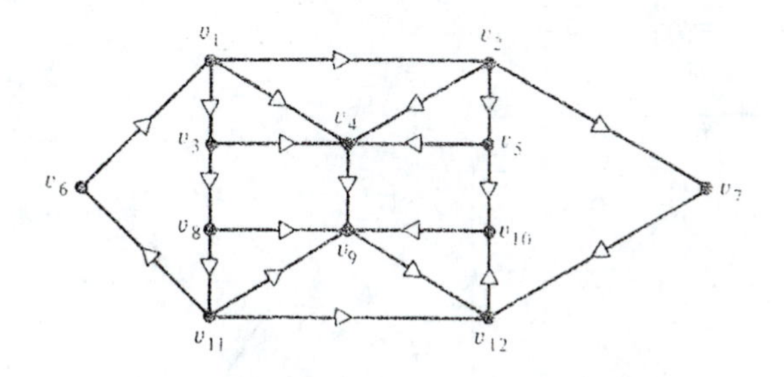
v_1
v_2
v_4
v_3
v_5
v_6
v_7
v_8
v_{10}
v_9
v_{11}
v_{12}

9. Show that the directed graph in Figure 6.8.3 is strongly connected.

10. Consider the graph G that consists of a single cycle with n vertices

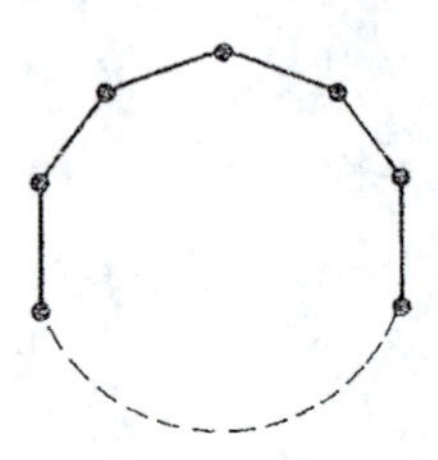

Is this graph strongly orientable? If so, in how many ways can it be oriented to produce a strongly connected directed graph?

11. Consider the following communication system

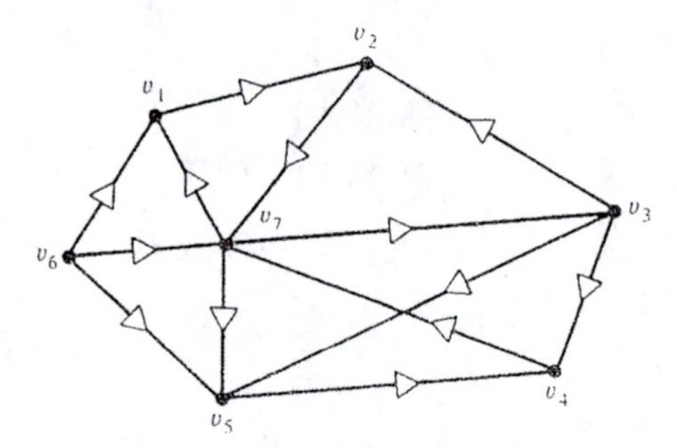

where each vertex represents a person, and each arc represents a one-way communication channel. Can every person get a message to every other person? Justify your answer.

12. Prove Theorem 6.8.1

13. Let D be a directed graph, and let D' be the directed graph obtained from D by changing the direction in each edge of D. The graph D' is sometimes called the **opposite** graph of D. Prove that a directed graph D is strongly connected if and only if its opposite graph is strongly connected.

In Exercises 14 through 17, determine whether or not the given graph is strongly orientable, and if it is, then orient the edges of the graph in such a way that the resulting graph is strongly connected.

14.

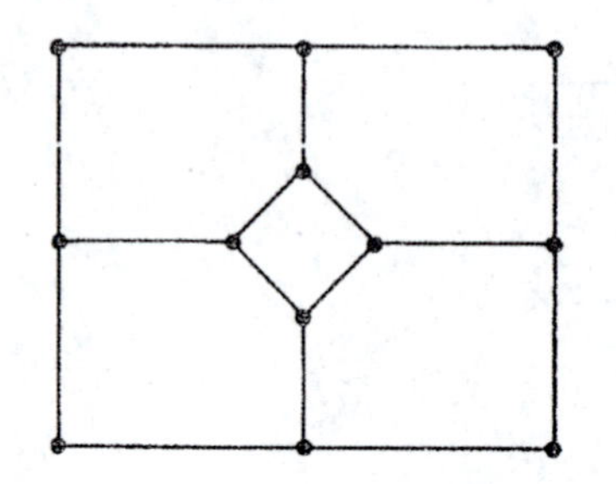

15.

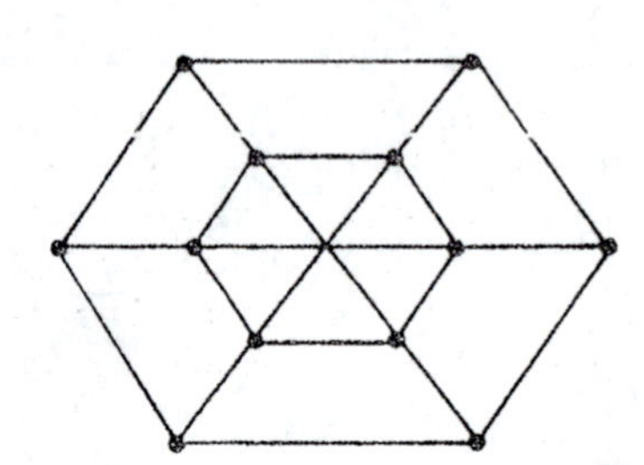

16.

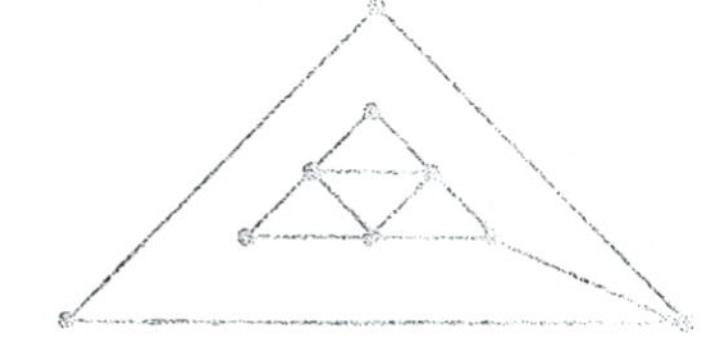

17.

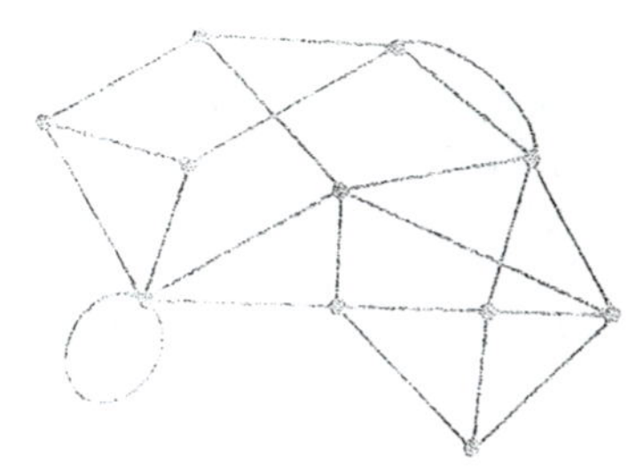

18. Consider the following street map

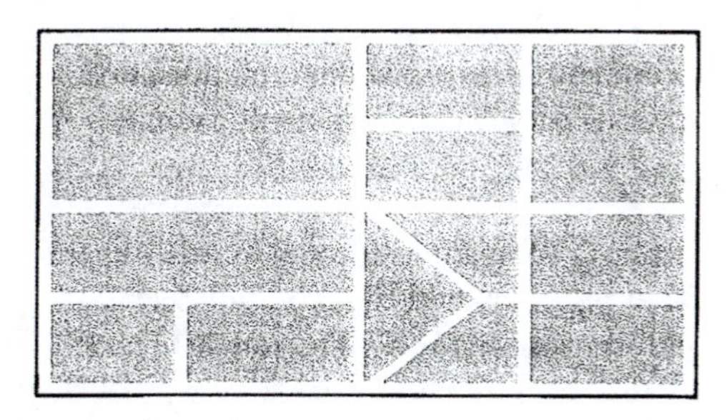

Is it possible to make each of these streets into a one-way street and still be able to get from any location to any other location on the map? If so, find a way.

19. Consider the following street map

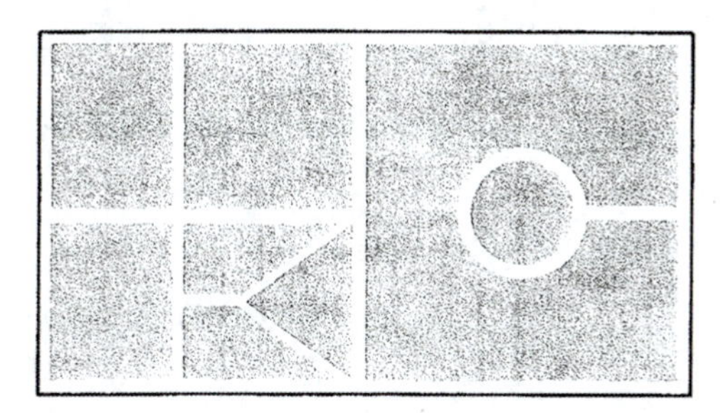

Is it possible to make each of these streets into a one-way street and still be able to get from any location to any other location on the map? If so, find a way.

20. Suppose that a directed graph D has a directed Hamiltonian cycle, that is, a closed, directed path that is a Hamiltonian path for the underlying graph. Prove that G is strongly orientable.

21. Prove that if G is a strongly orientable graph, then G is connected and has no bridges.

22. Show that, given any strongly orientable graph with at least two vertices, there is a way to orient each edge such that the resulting directed graph is *not* strongly connected.

23. Let D be a directed graph. If it is possible to start at any vertex of D and traverse each of the arcs in D exactly once, always traveling in the direction indicated by the arc, and return to the original vertex, then we say that D is an **Eulerian directed graph**. (Such a circuit is called a **directed Eulerian circuit**.) Of course, these definitions are analogous to those given in Section 6.3 for undirected graphs. Prove that a directed graph is an Eulerian directed graph if and only if $indeg(v) = outdeg(v)$ for all vertices in D.

6.9 Directed Networks; The Shortest Path Problem

In Section 6.7, we studied the concept of an undirected network. Let us turn now to a discussion of directed networks. We begin with the definition.

Definition

A **directed network** is a network in which each edge has been given a direction. In other words, a directed network is a network that is also a directed graph. As with any directed graph, we refer to the edges of a directed network as **arcs**.

Whenever we use the term *network* without modification, we will be referring to a directed network. It is not hard to find important examples of networks.

Example 6.9.1

A certain manufacturing company is located in Los Angeles, with a small branch in Kansas City. This company wants to distribute its product to various other cities in the country. In the following network, the weight of each arc gives the cost of shipping the company's product, in dollars per pound, between the corresponding cities.

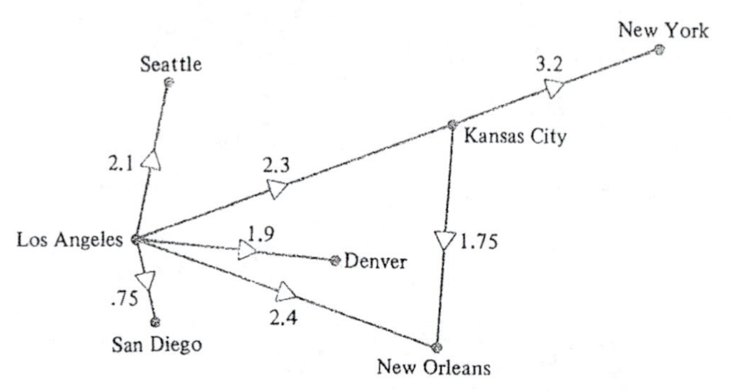

Example 6.9.2

The following network represents a street map, where the streets are all one-way, and the weight of each arc is the maximum capacity of the street, measured by the number of cars that can pass by a certain point on the street in 1 minute.

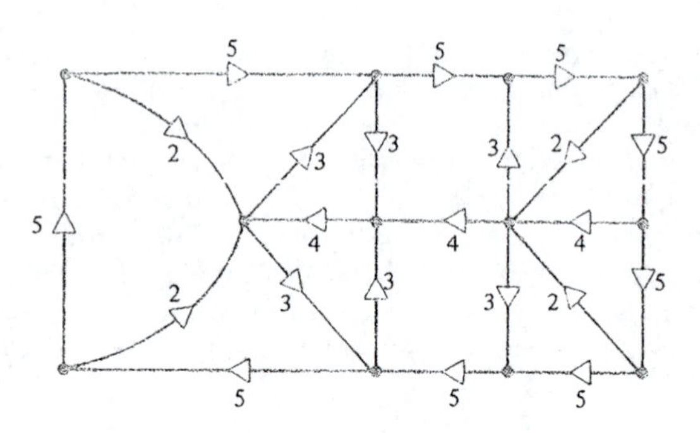

Example 6.9.3

In the following network, each vertex represents a person, and each arc represents a communication channel between the corresponding persons. The direction of each arc gives the direction that the information can flow, and the weight of each arc gives the maximum rate, in words per minute, that the information can be transmitted.

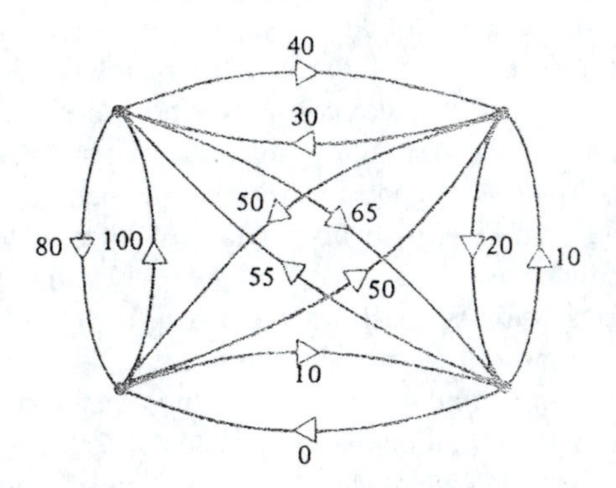

Example 6.9.4

Networks can be of great use in a branch of mathematics known as *probability theory*. As a simple example, suppose that a certain "system" can exist in various "states" $s_1, s_2, \ldots, s_n$. Suppose also that, when the system is in state s_i, there is a certain probability that it will change to state s_j. This is known as the **transition probability** from s_i to s_j. This situation can be described by means of a network, where the vertices of the network represent the states of the system, and the weight of an arc from s_i to s_j gives the transition probability from state s_i to state s_j. For example, the following network represents a system with five possible states. (When there is no arc between two states, it means that the system cannot go from one of these states to the other.)

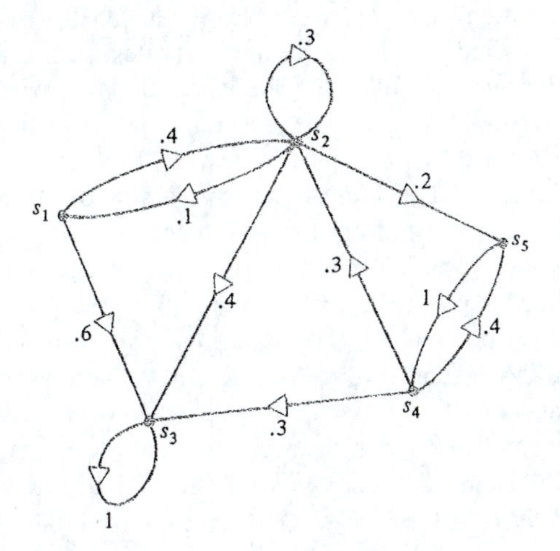

Incidentally, a system such as this is known as a **Markov chain**.

We want to devote the remainder of this section to discussing one particular problem associated with networks. Before stating the problem, let us make a definition.

If N is a network and P is a directed path in N, we define the **weight** of P to be the sum of the weights of all of the arcs in P. If P is a directed path from u to v with the smallest weight, then it is customary to say that P is a **shortest** path from u to v.

Now, given a network N, the **shortest path problem** for N is the problem of finding a shortest path from any vertex u to any other vertex v.

We will approach the shortest path problem from a slightly different angle. The approach that we shall take is to choose a vertex s in N and obtain an algorithm that describes how to find a shortest path from s to *any* other vertex in the network N. The vertex s is sometimes called a **source**, and the problem in this form is sometimes called the **single source shortest path problem** for N.

For convenience in describing the algorithm, let us assume that our networks do not have more than one arc from any vertex u to any other vertex v. (Of course, they may have one arc from u to v and one arc from v to u.) Actually, the case of multiple arcs is not really any harder, and we leave a discussion of it for the exercises. (See Exercise 8.)

The algorithm that we will describe for solving the single source shortest path problem is called **Dijkstra's algorithm**. This algorithm consists of several steps. At the k-th step, we divide the vertex set V of N into two disjoint subsets S_k and $V - S_k$. Let us call the set S_k the *accepted vertex set* and the set $V - S_k$ the *candidate set*.

Also, at the k-th step, we label each vertex v in N with two labels, $D_k(v)$ and $P_k(v)$. If v is in the accepted vertex set S_k, then the label $D_k(v)$ will be the weight of a shortest path from s to v, and the label $P_k(v)$ will be the vertex that immediately precedes v in a shortest path from s to v. [Of course, $P_k(s)$ does not make sense, since no vertex precedes s in a shortest path from s to s. But, for convenience, we will set $P_k(s) = s$.]

Thus, for vertices in the accepted vertex set S_k, the labels $P_k(v)$ give us the answer to the shortest path problem, since we can find a shortest path from s to v by tracing a path *backwards* from v to s, following the labels P_k. (We will demonstrate this a bit later.) Also, the labels $D_k(v)$ give us the weight of a shortest path.

If the vertex v is not in the accepted vertex set, that is, if it is in the candidate set $V - S_k$, then the labels $D_k(v)$ and $P_k(v)$ have a slightly different meaning. The label $D_k(v)$ is the weight of a path from s to v with smallest weight, from among those paths (from s to v) whose vertices (except for v) are in S_k. The label $P_k(v)$ is the vertex that immediately precedes v in such a path from s to v.

As we have already pointed out, the single source shortest path problem is solved for those vertices in the accepted vertex set S_k. Now, the reason that Dijkstra's algorithm works is that, at each step, we increase the size of the accepted vertex set by adding one new vertex to it. Hence, if the network N has n vertices, when we complete the n-th step in the algorithm, the accepted vertex set S_n will contain all of the vertices of N, and so we will have solved the single source shortest path problem for N.

Let us now describe Dijkstra's algorithm in detail.

Dijkstra's Algorithm

Let N be a network with n vertices. We denote the weight of an arc (u, v) in N by $w(u, v)$. Also, if there is no arc of the form (u, v), we set $w(u, v) = \infty$.

Step 1

Let

$$S_1 = \{s\}$$

$$\text{let} \quad D_1(s) = 0 \quad \text{and} \quad P_1(s) = s$$

$$\text{if} \quad v \in V - S_1 \quad \text{let} \quad D_1(v) = w(s, v) \quad \text{and} \quad P_1(v) = s$$

Once the k-th step has been completed, we have the accepted vertex set S_k, the candidate set $V - S_k$, and the labels $D_k(v)$ and $P_k(v)$. We are then ready for the $(k + 1)$-st step.

Step $k + 1$

Find that vertex u in the candidate set $V - S_k$ whose label $D_k(u)$ is as small as possible. Move u to the accepted vertex set, that is, set

$$S_{k+1} = S_k \cup \{u_{k+1}\}$$

As for the labels, if v is in the set S_{k+1}, then we do not change its labels, and so we let

$$D_{k+1}(v) = D_k(v) \quad \text{and} \quad P_{k+1}(v) = P_k(v)$$

On the other hand, if v is in the set $V - S_k$, then we let

$$D_{k+1}(v) = min\{D_k(v), D_k(u) + w(u, v)\}$$

and

$$P_{k+1}(v) = \begin{cases} P_k(v) & \text{if} \quad D_k(v) \leq D_k(u) + w(u, v) \\ u & \text{if} \quad D_k(v) > D_k(u) + w(u, v) \end{cases}$$

The idea behind these equations is the following. If $D_k(v) \leq D_k(u) + w(u, v)$, then we cannot improve upon the number $D_k(v)$ by taking a path from s to v that goes through u (which we can now do since u is in the accepted vertex set S_{k+1}). Hence, we set $D_{k+1}(v) = D_k(v)$ and $P_{k+1}(v) = P_k(v)$. On the other hand, if $D_k(v) > D_k(u) + w(u, v)$, then we can improve upon $D_k(v)$ by taking a path through u. Hence, in this case, we set $D_{k+1}(v) = D_k(u) + w(u, v)$ and $P_{k+1}(v) = u$.

Once the n-th step is complete, the set S_n will be the set of all vertices in N, and the candidate set will be empty. Then the label $D_n(v)$ will be the shortest distance from s to v, and we can find a path of smallest weight from s to v by tracing a path *backwards*, from v to s, by following the labels P_n. (Actually, we can stop the algorithm after the $(n - 1)$-st step, since the labels do not change during the n-th step. Nevertheless, we will take the algorithm through the n-th step.)

We will leave it as an exercise to actually prove that Dijkstra's algorithm solves the single source shortest path problem. Let us try the algorithm on an example.

Example 6.9.5

Let us apply Dijkstra's algorithm to the network

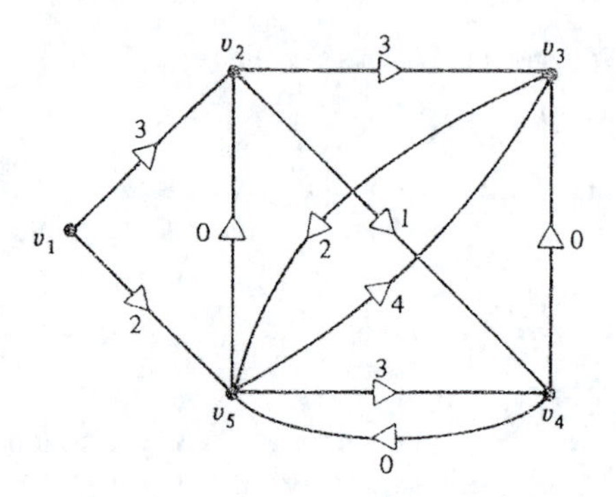

We will take the vertex v_1 as the source s. It is convenient to keep track of the information that we need at each step in a table.

Step 1

The first set of labels is given in the following table.

$$S_1 = \{v_1\} \quad , \quad V - S_1 = \{v_2, v_3, v_4, v_5\}$$

	$D_1(v)$	$P_1(v)$
$v = v_1$	0	v_1
$v = v_2$	3	v_1
$v = v_3$	∞	v_1
$v = v_4$	∞	v_1
$v = v_5$	2	v_1

Step 2

Now, v_5 is the vertex in $V - S_1$ with the smallest D-label, and so we set

$$u = v_5, \qquad S_2 = \{v_1, v_5\}, \qquad \text{and} \qquad V - S_2 = \{v_2, v_3, v_4\}$$

The next set of labels is found in the following table.

	$D_1(v)$	$P_1(v)$	$D_1(v_5) + w(v_5, v)$	$D_2(v)$	$P_2(v)$
$v = v_1$	0	v_1		0	v_1
$v = v_2$	3	v_1	$2 + 0$	2	v_5
$v = v_3$	∞	v_1	$2 + 4$	6	v_5
$v = v_4$	∞	v_1	$2 + 3$	5	v_5
$v = v_5$	2	v_1		2	v_1

Notice that we did not complete the third column in this table. In fact, we filled in only those entries corresponding to the vertices that are not in S_2, since these are the only vertices whose labels may change. In order to determine the value of $D_2(v)$ for one of these vertices v, we simply take the smaller of the entries in the first and third columns of the table.

Step 3

Now v_2 is the vertex in $V - S_2$ with smallest D label, and so we set

$$u = v_2, \qquad S_3 = \{v_1, v_2, v_5\}, \qquad \text{and} \qquad V - S_3 = \{v_3, v_4\}$$

The third set of labels is found in the following table.

	$D_2(v)$	$P_2(v)$	$D_2(v_2) + w(v_2, v)$	$D_3(v)$	$P_3(v)$
$v = v_1$	0	v_1		0	v_1
$v = v_2$	2	v_5		2	v_5
$v = v_3$	6	v_5	$3 + 2$	5	v_2
$v = v_4$	5	v_5	$2 + 1$	3	v_2
$v = v_5$	2	v_1		2	v_1

Step 4

The vertex v_4 is the vertex in $V - S_3$ with the smallest D label, and so we set

$$u_4 = v_4, \qquad S_4 = \{v_1, v_2, v_4, v_5\}, \qquad \text{and} \qquad V - S_4 = \{v_3\}$$

The fourth set of labels is given in the following table.

	$D_3(v)$	$P_3(v)$	$D_3(v_4) + w(v_4, v)$	$D_4(v)$	$P_4(v)$
$v = v_1$	0	v_1		0	v_1
$v = v_2$	2	v_5		2	v_5
$v = v_3$	5	v_2	3 + 0	3	v_4
$v = v_4$	3	v_2		3	v_2
$v = v_5$	2	v_1		2	v_1

Step 5

The final step is to move the last vertex v_3 into the accepted vertex set, and so $S_5 = \{v_1, v_2, v_3, v_4, v_5\}$ and $V - S_5 = \varnothing$. Also, we do not change any of the labels.

	$D_5(v)$	$P_5(v)$
$v = v_1$	0	v_1
$v = v_2$	2	v_5
$v = v_3$	3	v_4
$v = v_4$	3	v_2
$v = v_5$	2	v_1

This table gives us all the information that we need to construct shortest paths from the source v_1 to any other vertex. For example, to find a shortest path from v_1 to v_3, we look first at the third row of the table. This tells us that a shortest path from v_1 to v_3 has weight $D_5(v_3) = 3$ and that the immediate predecessor of v_3 in such a path is $P_5(v_3) = v_4$. In other words, a shortest path from v_1 to v_3 ends in v_4, v_3.

Now, since a shortest path from v_1 to v_3 ends in v_4, v_3, we know that it must consist of a shortest path from v_1 to v_4, followed by the arc (v_4, v_3). Therefore, the vertex that immediately precedes v_4 in a shortest path from v_1 to v_3 must be $P_5(v_4) = v_2$. Now we know that a shortest path from v_1 to v_3 ends in v_2, v_4, v_3.

By a similar reasoning, we see that since $P_5(v_2) = v_5$ and $P_5(v_5) = v_1$, a complete shortest path from v_1 to v_3 is v_1, v_5, v_2, v_4, v_3. This path does indeed have weight equal to 3.

We will leave it as an exercise to find a shortest path from v_1 to the other vertices of this network.

This example, and others like it, can be shortened considerably, and much of the duplication avoided, by building one large table as the steps in the algorithm are performed, rather than several small tables. Thus, in this case, we would construct the table shown in Figure 6.9.1.

	Step 1		Step 2			Step 3			Step 4			Step 5	
	$S_1 = \{v_1\}$		$u = v_5, S_2 = \{v_1, v_5\}$			$u = v_2, S_3 = \{v_1, v_2, v_5\}$			$u = v_4, S_4 = \{v_1, v_2, v_4, v_5\}$			$u = v_3, S_5 = \{v_1, v_2, v_3, v_4, v_5\}$	
	$D_1(v)$	$P_1(v)$	$D_1(v_5) + w(v_5, v)$	$D_2(v)$	$P_2(v)$	$D_2(v_2) + w(v_2, v)$	$D_3(v)$	$P_3(v)$	$D_3(v_4) + w(v_4, v)$	$D_4(v)$	$P_4(v)$	$D_5(v)$	$P_5(v)$
$v = v_1$	0	v_1		0	v_1		0	v_1		0	v_1	0	v_1
$v = v_2$	3	v_1	$2 + \infty$	2	v_5		2	v_5		2	v_5	2	v_5
$v = v_3$	∞	v_1	$2 + 4$	6	v_5	$3 + 2$	5	v_2	$3 + 0$	5	v_4	5	v_4
$v = v_4$	∞	v_1	$2 + 3$	5	v_5	$2 + 1$	5	v_2		3	v_2	3	v_2
$v = v_5$	2	v_1		2	v_1		2	v_1		2	v_1	2	v_1

Figure 6.9.1

□ □ Exercises

1. Find the shortest path from the source to each of the vertices v_1, v_2, v_3, v_4, v_5 in the network of Example 6.9.5.

2. Use Dijkstra's algorithm to find a shortest path from the vertex v_1 to each of the other vertices in the network

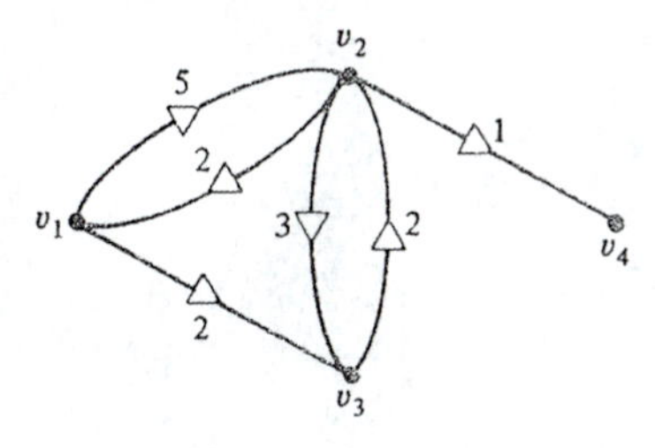

3. Use Dijkstra's algorithm to find a shortest path from vertex v_1 to each of the other vertices in the network

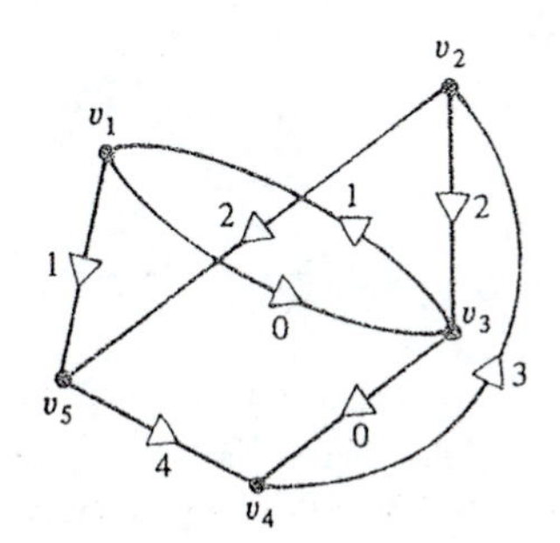

4. Use Dijkstra's algorithm to find a shortest path from vertex v_1 to each of the other vertices in the network

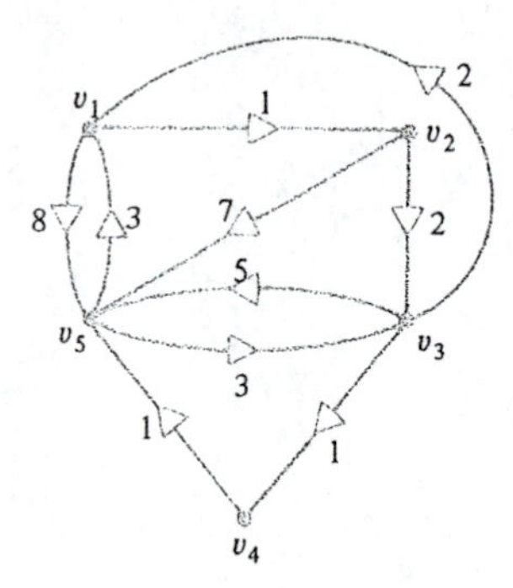

5. Use Dijkstra's algorithm to find a shortest path from vertex v_1 to each of the other vertices in the network

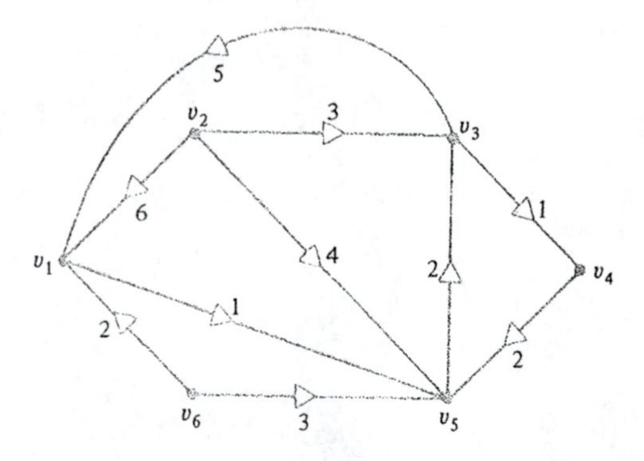

6. In the following undirected network, the weight of an edge $\{u, v\}$ represents the cost, in dollars per pound, of transporting a certain product between the vertices u and v, in either direction.

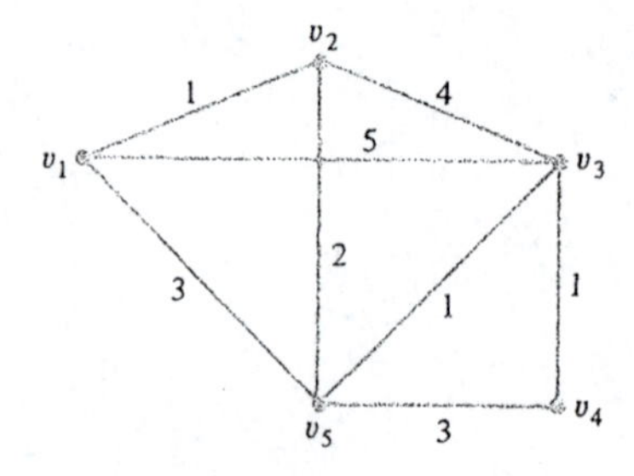

Use Dijkstra's algorithm to find the cheapest way to transport the product from v_1 to v_4. (First, you will have to decide how Dijkstra's algorithm can be applied in this case.)

7. Can you find a clever way to solve the shortest path problem for the following network that uses neither Dijkstra's algorithm nor trial and error?

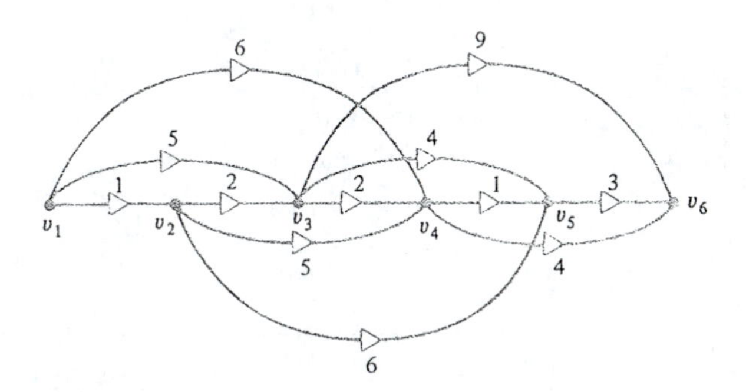

8. If a network N has more than one arc from a vertex u to a vertex v, how could you "prepare" the network so that you could still use Dijkstra's algorithm in the form we have described it?
9. Prove that Dijkstra's algorithm actually does solve the single source shortest path problem. *Hint*, you need to verify the claims made in the text for the labels $D_k(v)$ and $P_k(v)$. First, show that these claims are true for the labels $D_1(v)$ and $P_1(v)$. Then, assuming that they are true for $D_k(v)$ and $P_k(v)$, show that they are also true for $D_{k+1}(v)$ and $P_{k+1}(v)$. Once you have done this, you will know that they are true for $D_n(v)$ and $P_n(v)$.

6.10 Finite State Machines

In this section, we want to discuss an important application of graph theory to computer science. The internal memory of a computer is an example of a type of *system* that can exist in only a finite number of different *states*. It turns out that there are many other important examples of systems with this property. Typically, these systems accept *inputs* of some type and, as a result of the input, enter into a new state (or perhaps stay in their present state). The new state usually depends not only on the input but also on the particular state of the system at the time when the input was received.

With these thoughts in mind, we are ready for a definition.

Definition

A **finite state machine** (also called an **automaton**) consists of the following objects.

1) A finite set S, consisting of the **states** of the machine.
2) A particular state s_0 in S, which is called the **initial state** of the machine.
3) A subset F of S, consisting of the **final states** of the machine.
4) A finite set Σ consisting of the **inputs** for the machine. The set Σ is called the **alphabet** of the machine.
5) A function $\delta: S \times \Sigma \to S$ that maps ordered pairs in $S \times \Sigma$ to elements of S, known as the **transition function** of the machine.

The transition function δ describes how the machine changes from one state to another. If the machine is in state s, and it receives the input i, then the machine enters into state $\delta(s, i)$.

Now, there are two convenient ways of describing a finite state machine. One way is to use a table, known as the **state table** of the machine. If $S = \{s_1, s_2, \ldots, s_n\}$ and $\Sigma = \{i_1, i_2, \ldots, i_m\}$, then the state table is

		Inputs			
		i_1	i_2	$\cdots$	i_m
States	s_1	$\delta(s_1, i_1)$	$\delta(s_1, i_2)$	$\cdots$	$\delta(s_1, i_m)$
	s_2	$\delta(s_2, i_1)$	$\delta(s_2, i_2)$	$\cdots$	$\delta(s_2, i_m)$
	$\vdots$	$\vdots$	$\vdots$		$\vdots$
	s_n	$\delta(s_n, i_1)$	$\delta(s_n, i_2)$	$\cdots$	$\delta(s_n, i_m)$

As you can see, the state table is nothing more than a description of the transition function. Of course, it also gives the states and the inputs of the machine. The state table, together with the initial state s_0 and the set F of final states, completely describes a finite state machine.

Another way to describe a finite state machine is by using a graph, known as the **state diagram** of the machine. Each vertex of the graph represents a state. The arcs of the graph are labeled with the input symbols, and there is an arc from state s to state s', labeled with the input i, if and only if $\delta(s, i) = s'$.

s —— i ——> s'

$\delta(s, i) = s'$

In words, there is an arc from s to s' labeled i if and only if when the machine is in state s and is given the input i, it enters state s'.

It is customary to use little circles to denote the vertices in the graph of a finite state machine. Also, the initial state s_0 is designated by a small arrow at the vertex and the final states are designated by drawing two concentric circles, as shown below.

the initial state　　　　a final state

Let us have an example of a finite state machine.

Example 6.10.1

The following graph is the state diagram of a finite state machine.

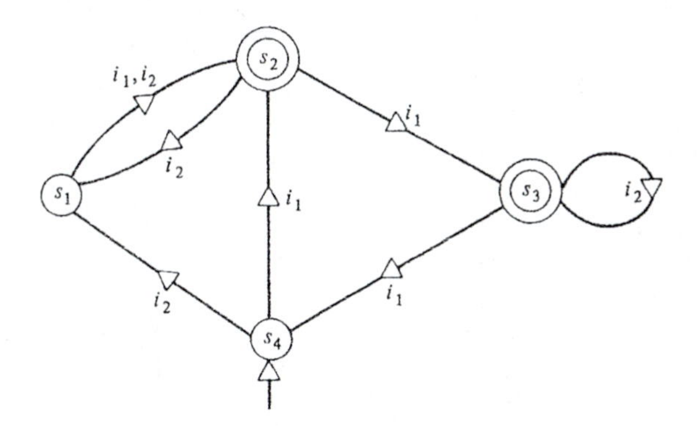

As can be seen from this graph, there are four states, s_1, s_2, s_3, and s_4, and there are two input symbols i_1 and i_2. The initial state is state s_4 and the final states are states s_2 and s_3. (It is possible for the initial state of a finite state machine to also be one of the final states, although that does not happen to be the case in this example.)

If the machine is in state s_4, for example, and it receives the input i_2, then in order to determine what state the machine enters, we simply follow the arc that starts at vertex s_4 and has label i_2. In this case, the machine proceeds to state s_1.

Notice that, because of the loop at state s_3, if the machine is in that state and is given the input i_2, then it remains in that state. Notice also that the arc from s_1 to s_2 has two labels. This simply means that if the machine is in state s_1 and receives either input i_1 or i_2, it then proceeds to state s_2. (This is just a shorthand notation to avoid having to draw two separate arcs from s_1 to s_2, one labeled i_1 and the other labeled i_2.)

The state table for this finite state machine is

	i_1	i_2
s_1	s_2	s_2
s_2	s_3	s_1
s_3	s_4	s_3
s_4	s_2	s_1

Generally, we think of a finite state machine as a machine that "reads" an input and "acts" on that input—entering into a new state. In a sense, we think of it as a "living" object that is capable of "action." In fact, it is customary to think of a finite state machine as being able to read, not just a single input but also a *sequence* of inputs $i_1, i_2, i_3, \ldots, i_k$. The machine processes each input in the sequence in the order in which it is received and enters into a new state (if necessary) each time an input is processed. We always assume that the machine reads the symbols in a sequence *from left to right,* just as we would read a sentence in the English language.

The idea of a machine reading sequences of inputs is so important that it deserves a bit more attention. Recall that we called the set Σ of inputs the alphabet of the machine. A finite sequence of elements of Σ is called a **word**, or **string**, over Σ. (We discussed words over an alphabet many times before, but to refresh your memory, we will discuss them again here.) For example, if $\Sigma = \{a, b, c, 1, *\}$ is an alphabet, then the following are words over Σ,

1, aaa, *bb1c, babbab*1*

We also define the **empty word** over Σ, denoted by θ, to be the word that has no symbols. The set of all words over Σ is denoted by Σ^* and any subset of Σ^* is called a **language** over Σ.

With this terminology, we can say that a finite state machine reads words (or strings), and after a word is read by the machine, it "comes to rest," so to speak, in a certain state. This state depends not only on the word being read, but also on the state of the machine just before the word is given to the machine.

In fact, it is customary to extend the definition of the transition function so that it is defined for ordered pairs of the form (s, w), where s is a state and w is a *word* over Σ. If we use the same symbol δ for this extension, then $\delta(s, w)$ is defined to be the state that the machine is in after it has read the entire word w.

More formally, the extension of the transition function $\delta: S \times \Sigma \to S$ is the function $\delta: S \times \Sigma^* \to S$ that is defined as follows. We first set

$$\delta(s, \theta) = s$$

where θ is the empty word. This says that if the machine reads the empty word θ, then it does not change its state. Now suppose that w is a word that ends with the symbol i. Thus w has the form $w = w'i$ where w' is also a word (perhaps the empty word). Then we define $\delta(s, w)$ by

$$\delta(s, w) = \delta(s, w'i) = \delta(\delta(s, w'), i)$$

This says that, in order to compute $\delta(s, w'\, i)$, we first find the state of the machine after it has read all of the word w except for the last symbol, that is, after it has read the word w'. This state is $\delta(s, w')$. Then we determine what state the machine will enter into if it is in this state $\delta(s, w')$ when it encounters the last symbol i. This is state $\delta(\delta(s, w'), i)$. □

Example 6.10.2

For the finite state machine in Example 6.10.1, let us compute $\delta(s_3, i_2 i_1 i_1 i_2 i_1 i_2)$. As the machine reads the input symbols in the word $w = i_2 i_1 i_1 i_2 i_1 i_2$, it goes through a sequence of states, which is best described in a table

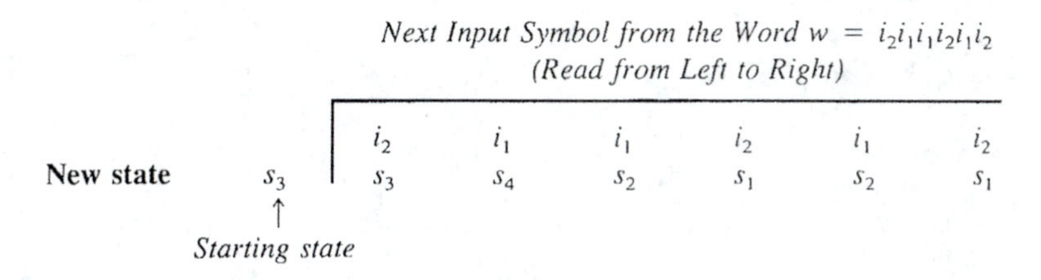

Next Input Symbol from the Word $w = i_2i_1i_1i_2i_1i_2$
(Read from Left to Right)

		i_2	i_1	i_1	i_2	i_1	i_2
New state	s_3	s_3	s_4	s_2	s_1	s_2	s_1
	↑ *Starting state*						

Thus, the machine comes to rest in state s_1 after the entire word w is read, and so $\delta(s_3, i_2i_1i_1i_2i_1i_2) = s_1$. □

Suppose that a finite state machine is in its initial state when it reads a word w. Then if the machine comes to rest in one of its *final* states, we say that the machine has **accepted** the word w. In symbols, the word w is accepted by a finite state machine if $\delta(s_0, w) \in F$. On the other hand, if the machine comes to rest in a state that is not one of the final states, then we say that it has **rejected** the word w. In symbols, w is rejected if $\delta(s_0, w) \notin F$. The set of all words that are accepted by a finite state machine is called the **language accepted by the machine**.

Let us consider some other examples of finite state machines. Recall that a word over the alphabet $\Sigma = \{0, 1\}$ is called a **binary word**.

Example 6.10.3

Let us build a finite state machine that is capable of determining whether or not a binary word contains an even number of 1's. Thus, we want the machine to accept a binary word if it contains an even number of 1's and reject it if it contains an odd number of 1's.

We say that a binary word has **even parity** if it contains an even number of 1's, and it has **odd parity** if it contains an odd number of 1's. Thus, the machine we want to construct is called a **parity check machine**.

Since a parity check machine must be able to read strings of 0's and 1's, its input set must be $\Sigma = \{0, 1\}$. Also, its state set should be $S = \{\text{even}, \text{odd}\}$. Since a binary word is accepted if and only if it has an even number of 1's, we want the final state set to be $F = \{\text{even}\}$. Let us also take the initial state to be even.

Now, imagine a binary word being read by our machine. The initial state of the machine is even. Therefore, if the first digit read is a 0, the machine should not change its state. On the other hand, if the first digit is a 1, then the machine should change to state odd.

As a matter of fact, our parity check machine should have the property that it changes state whenever it reads a 1 but does not change state when it reads a 0. The state diagram that describes such a machine is

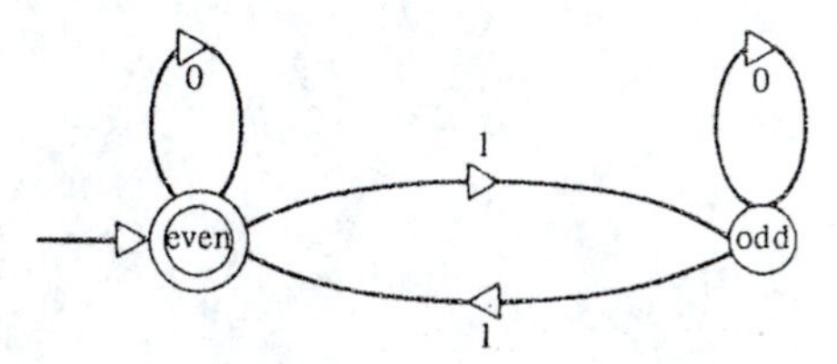

and the state table is

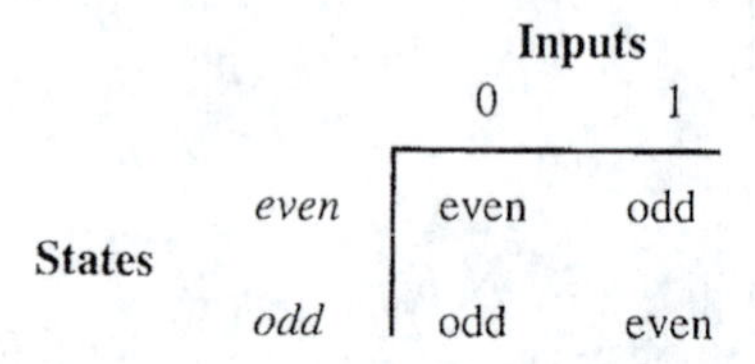

		Inputs	
		0	1
States	*even*	even	odd
	odd	odd	even

As an example of how the parity check machine works, let us feed it the binary word $w = 1001011$. Thus, we want to compute $\delta(\text{even}, 1001011)$. Starting at the initial state even and reading the digits of w from left to right, the machine goes through the sequence of states given in the following table.

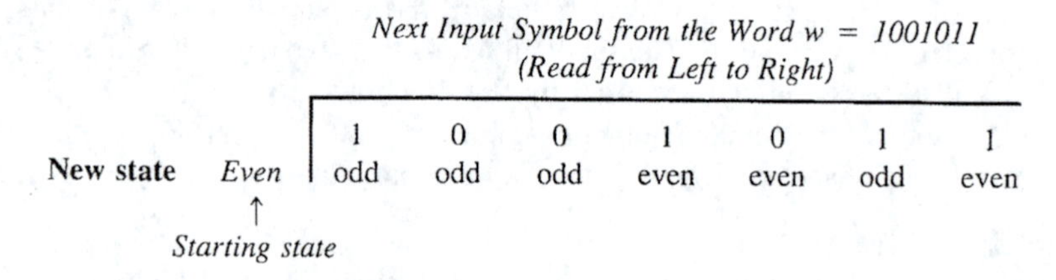

		Next Input Symbol from the Word w = 1001011 (Read from Left to Right)						
		1	0	0	1	0	1	1
New state	*Even* ↑ *Starting state*	odd	odd	odd	even	even	odd	even

Now, the machine is in state even when it has finished reading the word w, and so $\delta(\text{even}, 1001011) = \text{even}$. This tells us that w has even parity, and since even is a final state, w is accepted by the machine. □

It may seem as though we went to a great deal of trouble just to tell whether or not a binary word has an even number of 1's in it. After all, we could just simply count the number of 1's and see if that number is even or not! However, we should remember that this example was chosen simply to illustrate the idea of a finite state machine. We will encounter more complicated machines in the next two examples, and in the exercises.

■ Example 6.10.4

The following graph is the state diagram of a machine that reads binary words and gives the remainder obtained after dividing the number of 1's in the word by 3. The machine accepts a binary word if and only if this remainder is either 0 or 1.

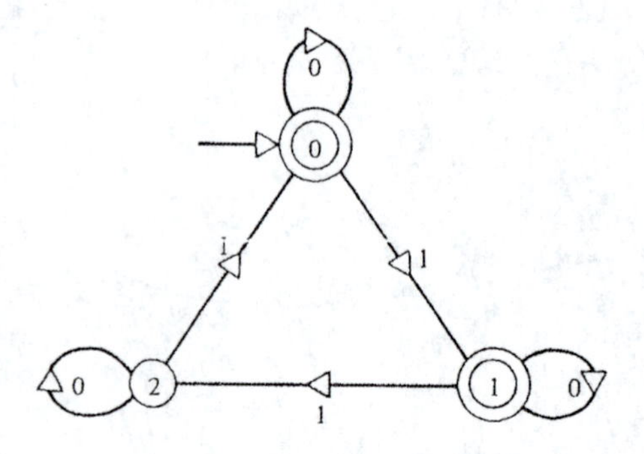

The table below describes how the machine reads the word $w = 11011001$

Next Input Symbol from the Word w = 11011001 (Read from Left to Right)

		1	1	0	1	1	0	0	1
New state	0	1	2	2	0	1	1	1	2
	↑ *Starting state*								

From this table we see that $\delta(0, 11011001) = 2$. Hence, the remainder after dividing the number of 1's in the word $w = 11011001$ by 3 is 2. Since the state 2 is not a final state, the word w is rejected by the machine. □

Example 6.10.5

Let us build a finite state machine that accepts a binary word if and only if it ends in 101. In other words, we want to build a machine whose language is the set of all binary words ending in 101.

We begin with the graph

that accepts the word $w = 101$. Now we must fill in the other arcs. Let us first observe that, if we input 10, the machine goes into state s_3. Therefore, if we *follow* this input by the input 101, the machine goes first to state s_3, and from there it must somehow get to its final state s_4 using the rest of the input, namely 101. This suggests that we add a new arc from s_4 to s_3 labeled 0, shown as below

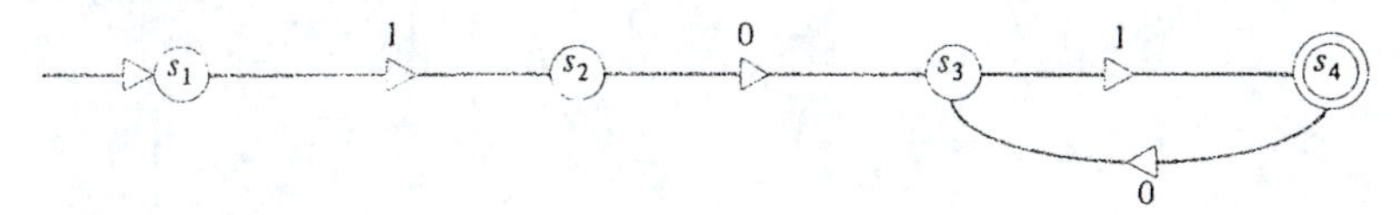

Figure 6.10.1

Incidentally, it also suggests the possibility of simply adding a loop at s_4 labeled 0, as shown below

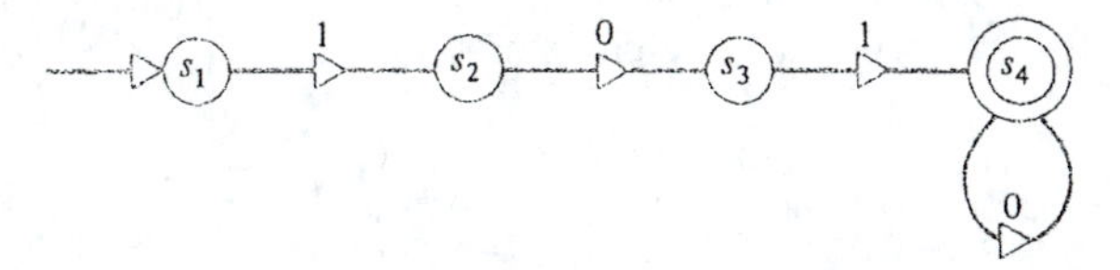

However, as you can see from the diagram, if we were to add such a loop, then the word 1010 would be accepted by the machine, and this we do not want.

So, we now have the graph in Figure 6.10.1. Let us again put our machine into state s_3 by inputting the word 10, and this time follow that input by 0101. This must put the machine again into its final state s_4. One way to accomplish this is to add an arc from s_3 to s_1 labeled 0.

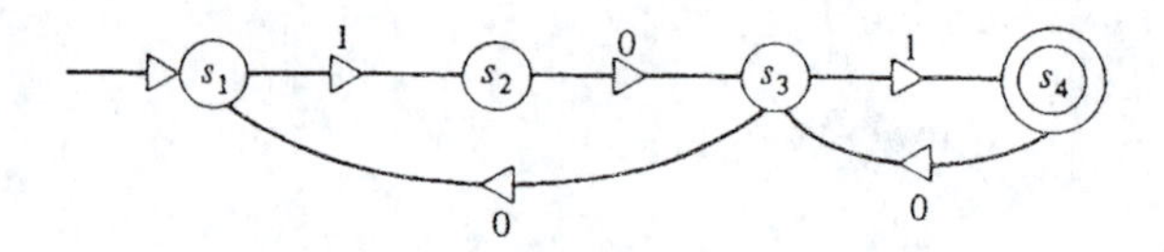

(Another way to accomplish this is to add a loop at s_3 labeled 0, but if we did that, the word 1001 would be accepted by the machine, which again we do not want.)

Now, let us input the word 1, which puts the machine into state s_2. If we follow this input by 101, the machine must again go into state s_4. This suggests that we add a loop at s_2

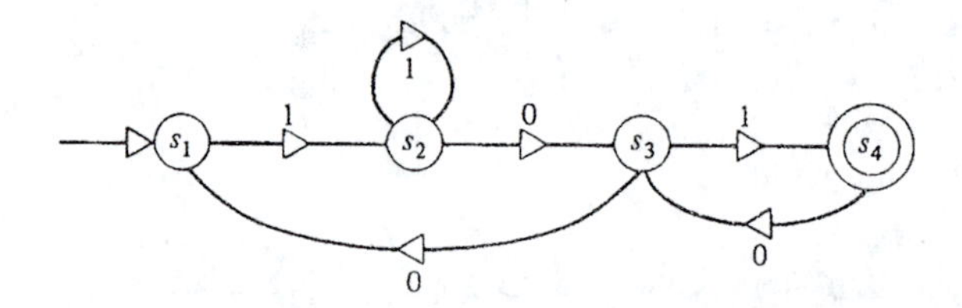

Next, we input a 0, followed by 101. After the 0 is read by the machine, it must be in a state with the property that the further input 101 will take it to the final state s_4. This suggests a loop at s_1 with label 0

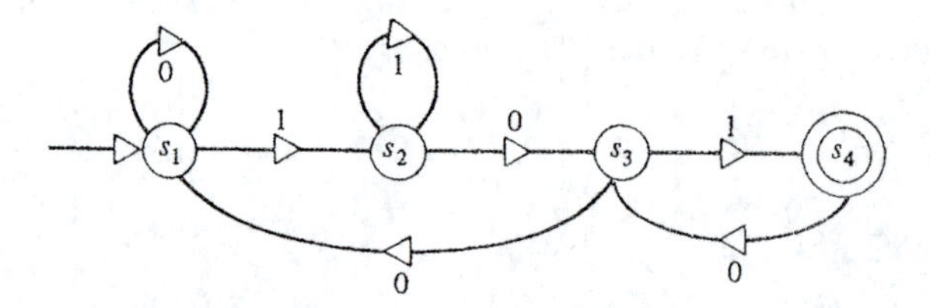

Finally, we need to add an arc at state s_4 with label 1. For this, let us begin with the input 101, which puts the machine into state s_4, and follow that by the input 101. This suggests adding an arc from s_4 to s_2 labeled 1. (It also suggests adding a loop at s_4 labeled 1, but this allows the machine to accept words ending in 11, which we do not want.) Thus, we arrive at the completed state diagram

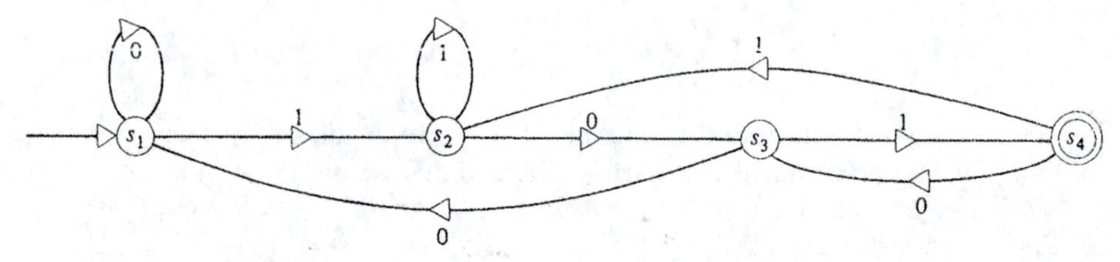

Figure 6.10.2

In order to be certain that we have a machine that does indeed accept words ending in 101, we can reason as follows. Suppose that $w = w'101$ is any word ending in 101. Then, after the first part w' of the word is input, the machine is in one of its four states s_1, s_2, s_3, or s_4. But, as you can easily see from the state diagram, *regardless of which state the machine is in*, the further input 101 will take the machine into the final state s_4. For example, from state s_2, the further input 101 sends the machine through the sequence of states s_2, s_3, s_4. Thus, after reading the last part of the input, the machine will always be in its final state, and so it will accept any word of the form $w = w'101$.

We must still show that any binary word that does not end in 101 is not accepted by the machine. This can be done by reasoning as follows. Suppose that the word $w = w'0$ ends in the symbol 0. Then, after w' is read, the machine will be in one of its four states. But in no case is there an arc from any of these four states into the final state s_4 that is labeled 0. Thus, when the machine reads the last symbol 0 in w, it cannot go into its final state. In other words, it must reject the word $w = w'0$.

Hence, the only possible words that the machine can accept are those ending in a 1. Now suppose that a word $w = w'11$ ends in two 1's. Such a word must be rejected by the machine since, after it has read the word w', there is no way that it can go into its final state by following two arcs, both labeled 1. Hence, the only possible words that the machine can accept are those ending in 01.

Finally, suppose that a word $w = w'001$ ends in 001. Such a word must also be rejected by the machine, since after it has read the word w', there is no way to follow three arcs labeled 0, 0, and 1 and end up in the final state s_4. Thus, we see that the only possible words that this machine can accept are those ending in 101.

This argument proves that the language accepted by the machine whose state diagram is given in Figure 6.10.2 is precisely the set of all binary words ending in 101. ☐

Exercises

1. Consider the finite state machine

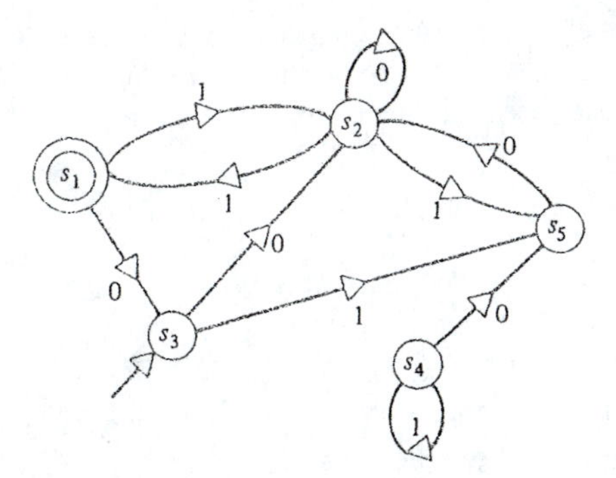

a) Find $\delta(s_1, 10100)$, find $\delta(s_4, 110010)$.

b) Which of the following words are accepted by this machine, and which are rejected? $w_1 = 101$, $w_2 = 1011010$, $w_3 = 11001001$

c) Find the state table for this finite state machine.

2. Consider the finite state machine

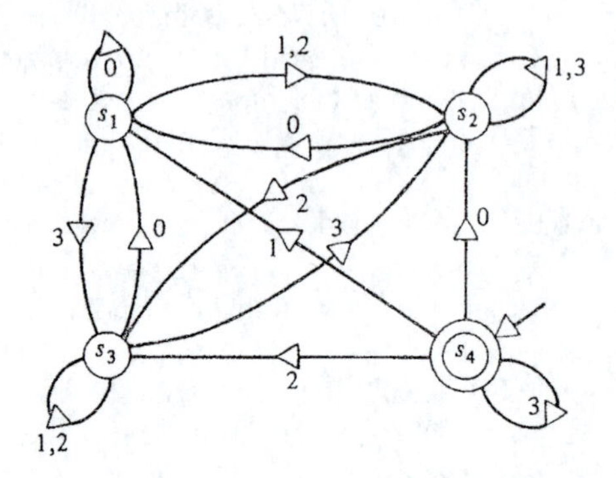

a) Find $\delta(s_1, 1230)$, find $\delta(s_3, 33100231)$.
b) Which of the following words are accepted by this machine, and which are rejected? $w_1 = 3333$, $w_2 = 0000$, $w_3 = 23003201$
c) Describe the language accepted by this machine.
d) Find the state table for this finite state machine.

3. Consider the finite state machine given by

	0	1	2
s_1	s_1	s_2	s_3
s_2	s_2	s_3	s_1
s_3	s_3	s_1	s_2

Initial state s_1
$F = \{s_2\}$

a) Find $\delta(s_2, 12121)$.
b) Find $\delta(s_3, 012012012)$.
c) Is the word $w = 120222$ accepted or rejected by this machine?
d) Find the state diagram for this machine.

4. Find the state table for the machine in Example 6.10.4.
5. Find the state table for the machine in Example 6.10.5.
6. Describe the language accepted by the machine in Example 6.10.3.
7. Describe the language accepted by the machine in Example 6.10.4.
8. Show that the language accepted by the finite state machine

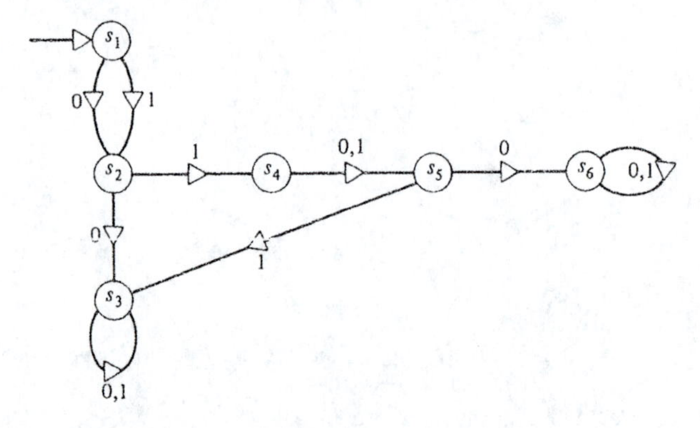

is the set of all binary words in which the second digit is a 0 or the fourth digit is a 1.

9. Show that the language accepted by the machine

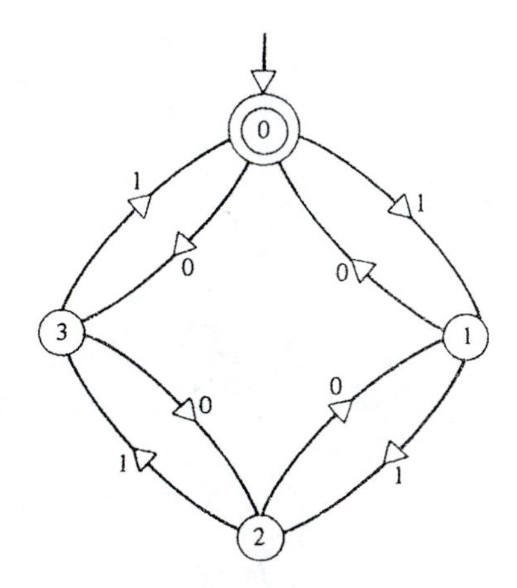

is the set of all binary words with the property that the number of 1's in the word, minus the number of 0's in the word, is divisible by 4.

10. What is the language accepted by the finite state machine

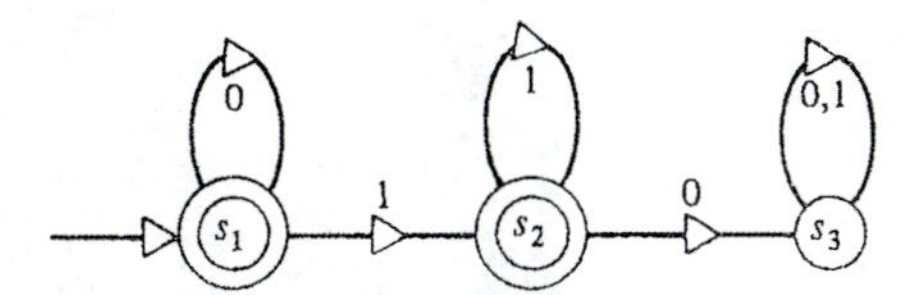

11. Build a finite state machine that accepts a binary word if and only if its second digit is a 1 and its fourth digit is a 0.

12. Build a finite state machine that accepts a binary word if and only if the remainder, after dividing the number of 1's in the word by 4, is equal to 2.

13. Build a finite state machine that accepts a binary word if and only if it ends in 110.

14. Build a finite state machine that accepts a binary word of length 4 (that is, with four digits) if and only if the number of 1's in the word is greater than the number of 0's.

15. Build a finite state machine, with input set equal to the set of letters in the alphabet, that accepts the word ALERT but rejects all other words.

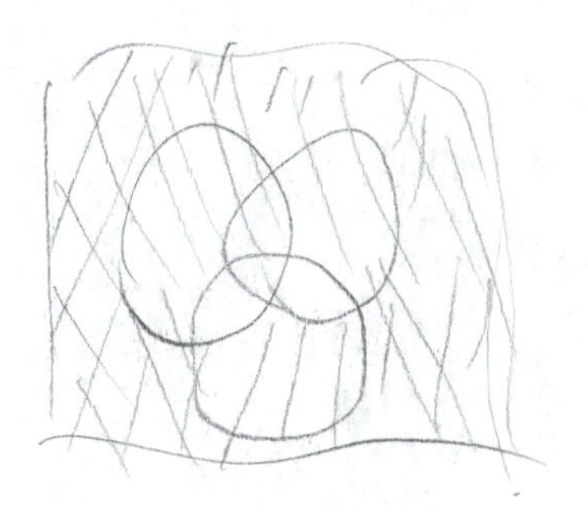

Answers to Selected Odd-Numbered Exercises

Chapter 1

Section 1.1 (page 1)

1. a) {1, 3, 4, 5, 6, 7} b) {3, 5, 7} c) {3, 5, 7}
d) {6} e) {1} f) ∅

3. (S − T) − V = {*b*}; S − (T − V) = {*b*, *c*}. They are not equal.

5. A × B = {(1, *x*), (1, *y*), (1, *z*), (2, *x*), (2, *y*), (2, *z*), (3, *x*), (3, *y*), (3, *z*)}; B × A = {(*x*, 1), (*x*, 2), (*x*, 3), (*y*, 1), (*y*, 2), (*y*, 3), (*z*, 1), (*z*, 2), (*z*, 3)}. They are not the same.

7. A × B × C
= {(1, 1, 2), (1, 1, 4), (1, 4, 2), (1, 4, 4), (2, 1, 2), (2, 1, 4), (2, 4, 2), (2, 4, 4)}

9. a) Σ_3 = {*aaa*, *aab*, *aa*1, *aa*2, *aba*, *abb*, *ab*1, *ab*2,
*a*1*a*, *a*1*b*, *a*11, *a*12, *a*2*a*, *a*2*b*, *a*21, *a*22,
baa, *bab*, *ba*1, *ba*2, *bba*, *bbb*, *bb*1, *bb*2,
*b*1*a*, *b*1*b*, *b*11, *b*12, *b*2*a*, *b*2*b*, *b*21, *b*22,
1*aa*, 1*ab*, 1*a*1, 1*a*2, 1*ba*, 1*bb*, 1*b*1, 1*b*2,
11*a*, 11*b*, 111, 112, 12*a*, 12*b*, 121, 122,
2*aa*, 2*ab*, 2*a*1, 2*a*2, 2*ba*, 2*bb*, 2*b*1, 2*b*2,
21a, 21*b*, 211, 212, 22*a*, 22*b*, 221, 222}

b) $\Gamma_3 = \{\theta, a, b, 1, 2, aa, ab, a1, a2, ba, bb, b1, b2, 1a, 1b, 11, 12, 2a, 2b, 21, 22\} \cup \Sigma_3$, where Σ_3 is given in part a.

11. $\Gamma_n - \Gamma_{n-1} = \Sigma_n$

13. a) $\{1, 2, 5\}$

b) $\{0, \pm 1, \pm 2, \pm 3, \pm 4, \pm 5, \pm 6, \pm 7, \pm 8\}$

c) $\varnothing$

15. $\mathcal{P}(\{1, a, z\}) = \{\varnothing, \{1\}, \{a\}, \{z\}, \{1, a\}, \{1, z\}, \{a, z\}, \{1, a, z\}\}$

17. $\mathcal{P}(\{\varnothing, \{1\}\}) = \{\varnothing, \{\varnothing\}, \{\{1\}\}, \{\varnothing, \{1\}\}\}$

21. No, it is ambiguous. For example, if $A = \{a\}$, $B = \varnothing$, and $C = \varnothing$ then $(A \cup B) \cap C = \varnothing$ but $A \cup (B \cap C) = \{a\}$.

23. Yes. If E = {even integers} and O = {odd integers} then $E \cup O = Z$ and $E \cap O = \varnothing$.

25. If $n = 1$, then the sets $\Sigma_a, \Sigma_b, \ldots, \Sigma_z$ do form a partition of $\Sigma_n = \Sigma_1$. However, if $n > 1$, then these sets are not disjoint, and so they cannot form a partition of Σ_n.

27. a) $B - A$ b) A c) D

29. $0 + 0 = 0$; $1 \cdot 1 = 1$

31. $(x + y) + z = x + (y + z)$; $(x \cdot y) \cdot z = x \cdot (y \cdot z)$

33. See the answer to Exercise 3. The operation of set difference is *not* associative.

35. Let $A_i = \{x \mid x \text{ is an integer and } x \geq i\}$.

39. a) Since $A \subset B \cup A$ and $B \subset B \cup A$, we have $A \cup B \subset B \cup A$. Also, since $B \subset A \cup B$ and $A \subset A \cup B$, we have $B \cup A \subset A \cup B$. Hence $A \cup B = B \cup A$.

43. a) If $x \in (A \cup B)^c$ then $x \notin A$ and $x \notin B$. Hence $x \in A^c$ and $x \in B^c$. In other words, $x \in A^c \cap B^c$ and so $(A \cup B)^c \subset A^c \cap B^c$. On the other hand, if $x \in A^c \cap B^c$ then $x \in A^c$ and $x \in B^c$. In other words, $x \notin A$ and $x \notin B$. Hence, $x \notin A \cup B$; that is, $x \in (A \cup B)^c$ and so $A^c \cap B^c \subset (A \cup B)^c$. Therefore, $(A \cup B)^c = A^c \cap B^c$.

45. a) $\{1, 2, 3, 7, 8\}$ c) $\varnothing$ e) $\{0, 3, 7\}$

47. $A \Delta B = (A - B) \cup (B - A) = (B - A) \cup (A - B) = B \Delta A$. The operation of symmetric difference is commutative.

Section 1.2 (page 17)

1.

A		B		A		B
1	$\leftrightarrow$	1		1	$\leftrightarrow$	y
5	$\leftrightarrow$	5		5	$\leftrightarrow$	x
z	$\leftrightarrow$	x		z	$\leftrightarrow$	5
y	$\leftrightarrow$	y		y	$\leftrightarrow$	1

3. a) 70 | 79 | 82

c) 73 | 110 | 112 | 117 | 116

e) 66 | 65 | 83 | 73 | 67

5. MATH IS FUN

7. $n \leftrightarrow 2n$ for $n \in A$

9. $n \leftrightarrow 7 \cdot 10^n$ for $n \in A$

11.

$\mathcal{P}(\{a, b\})$	Σ_2
$\varnothing$	$\leftrightarrow 00$
$\{a\}$	$\leftrightarrow 01$
$\{b\}$	$\leftrightarrow 10$
$\{a, b\}$	$\leftrightarrow 11$

13. Let 1 correspond to a and 2 correspond to b. This extends to a one-to-one correspondence between Σ^* and Ω^*. For example, 122212 corresponds to *abbbab*.

17. If $S \in \mathcal{P}(\{1, 2, \ldots, n\})$ let $S' = \{2n \mid n \in S\}$. (If $S = \varnothing$ then $S' = \varnothing$.) Then $S' \in \mathcal{P}(\{2, 4, 6, \ldots, 2n\})$ and the correspondence $S \leftrightarrow S'$ is one-to-one.

Section 1.3 (page 26)

1. $n \leftrightarrow n + 1$ for $n \in A$
3. $n \leftrightarrow n - 5$ for $n \in A$
5. $n \leftrightarrow n - 1$ for $n \in A$
7. If n is even then $n = 2m$ for some integer m. Hence, $(-1)^n[(n + 1)/2] = (-1)^{2m}[(2m + 1)/2] = [m + \frac{1}{2}] = m = n/2$. Also, if n is odd then $n = 2m + 1$ for some integer m. Hence, $(-1)^n[(n + 1)/2] = (-1)^{2m+1}[(2m + 2)/2] = -[m + 1] = -(m + 1) = -(n + 1)/2$.
13. The pairing circle $\leftrightarrow$ (radius, center) is a one-to-one correspondence.
15. $(a, r_1, r_2) \leftrightarrow a(x - r_1)(x - r_2) = 0$ is a one-to-one correspondence.

Section 1.4 (page 36)

1. This is a function. However, it is neither injective nor surjective. $\text{Im}(f) = \{1\}$.
3. This is a function. It is injective but not surjective. $\text{Im}(f) = \{\text{positive integers}\}$.
5. This is a function. It is surjective but not injective. $\text{Im}(f) = \Sigma$.
7. This is a function. It is injective but not surjective. $\text{Im}(f) = \{w \in \Sigma^* \mid w \text{ starts with the letter } a\}$.
9. This is not a function since, for example, $f(1) = f(2/2) = 2$ and $f(1) = f(3/3) = 3$.
11. This is a function. It is injective but not surjective. $\text{Im}(f) = \{T \in \mathcal{P}(S \cup \{x\}) \mid x \in T\}$.
13. This is not a function from $\mathbf{N}$ to $\mathbf{Q}$ since $f(0)$ is not defined.
15. It is not injective since, for example, $f(1) = f(-1)$. However, it is surjective since if r is any nonnegative real number, then $f(\sqrt{r}) = (\sqrt{r})^2 = r$, and so $r \in \text{Im}(f)$. Hence, $\text{Im}(f) = \mathbf{R}'$.
17. Both of these functions are bijective.
19. The Hamming distance function is neither injective nor surjective. It is not injective since, for example, $H(u, w) = H(w, u)$ for $u \neq w$. It is not surjective since $\text{Im}(H) = \{0, 1, \ldots, n\}$, which is not equal to $\mathbf{N}$.
21. a) 5 c) 4
23. No.
27. $(g \circ f)(1) = z$, $(g \circ f)(2) = y$, $(g \circ f)(3) = y$. The composition $f \circ g$ is not defined since $\text{Im}(g)$ is not in the domain of f.
29. $(f \circ f)(n) = n^4 + 2n^2 + 2$
31. Yes. The square root function $g:\mathbf{R}^+ \to \mathbf{R}^+$, $g(x) = \sqrt{x}$ is the inverse of f.

Section 1.5 (page 46)

1. $0^5 + 1^5 + 2^5 + 3^5 + 4^5 + 5^5$
3. $(-1)^{-6} + (-1)^{-5} + (-1)^{-4} + (-1)^{-3} + (-1)^{-2} + (-1)^{-1} + (-1)^0 + (-1)^1 + (-1)^2 + (-1)^3 + (-1)^4 + (-1)^5 + (-1)^6$
5. $(0 + 1 + 2 + 3)(0 + 1 + 2 + 3)$
7. $(1 + 1 + 1) + (1 + 1 + 2) + (1 + 2 + 1) + (1 + 2 + 2) + (2 + 1 + 1) + (2 + 1 + 2) + (2 + 2 + 1) + (2 + 2 + 2)$
9. $\sum_{i=0}^{5} 2i$
11. $\sum_{i=0}^{99} 2i$
13. $\sum_{i=0}^{n-1} 2^i$
15. $\sum_{i=2}^{m} \frac{1}{i^2(i-1)^2}$

21. Let P(n) be the proposition that the equation in Exercise 21 holds. Since $1^3 = 1^2$, we see that P(1) is true. Now assume that P(k) is true. Then

$$\sum_{j=1}^{k+1} j^3 = \sum_{j=1}^{k} j^3 + (k+1)^3 = \left(\sum_{j=1}^{k} j\right)^2 + (k+1)^3$$

$$= \left(\frac{k(k+1)}{2}\right)^2 + (k+1)^3 = \left(\frac{(k+1)(k+2)}{2}\right)^2 = \left(\sum_{j=1}^{k+1} j\right)^2$$

Hence, P(k + 1) is true, and so P(n) is true for all $n \geq 1$.

29. $1000 + 999 + \cdots + 1 = 1000 \cdot 1001/2 = 500{,}500$

31. MN

33. N(N + 1)(N + 2)/3

37. 1 if N is odd and 0 if N is even.

42. The set $\mathbf{O}^+$ of all odd positive integers is defined by the following conditions:
1) $1 \in \mathbf{O}^+$
2) If $n \in \mathbf{O}^+$ then $n + 2 \in \mathbf{O}^+$
3) Nothing is in $\mathbf{O}^+$ unless it can be obtained by a finite number of applications of statements 1 and 2.

44. The set **E** of all even integers is defined by the following conditions:
1) $0 \in \mathbf{E}$
2) If $n \in \mathbf{E}$ then $n + 2 \in \mathbf{E}$ and $n - 2 \in \mathbf{E}$
3) Nothing is in **E** unless it can be obtained by a finite number of applications of statements 1 and 2.

46. The set Σ^* of all words over Σ is defined by the following conditions:
1) $\theta \in \Sigma^*$
2) If $w \in \Sigma^*$ then $aw \in \Sigma^*$ for all $a \in \Sigma$
3) Nothing is in Σ^* unless it can be obtained by a finite number of applications of statements 1 and 2.

Section 1.6 (page 57)

1. Assume that $y^2 \leq x^2$. Then $\sqrt{y^2} \leq \sqrt{x^2}$, that is $|y| \leq |x|$. (Recall that $\sqrt{a^2} = |a|$ for any real number a.) But the fact that $|y| \leq |x|$ is a contradiction to the fact that $0 < x < y$. This contradiction implies that y^2 cannot be less than or equal to x^2. That is, $x^2 < y^2$, and the theorem is proved.

3. Assume that $x \neq 0$. Then we may divide the equation $x^2 = x$ by x, to obtain $x = 1$, which is a contradiction to the fact that $x \neq 1$. This contradiction implies that $x = 0$ and proves the theorem.

7. Assume that no such number n exists; in other words, assume that $ne \leq M$ for *all* natural numbers n. Dividing both sides by e gives $n \leq M/e$ for all natural numbers n. In words, this says that *all* natural numbers are less than or equal to the real number M/e. But this is a contradiction to the stated property of the real numbers ($r = M/e$). Hence, the theorem must be true.

Chapter 2

Section 2.1 (page 63)

1. a) Simple statement. b) Not a statement.
c) Not a statement. d) Compound statement.
e) Not a statement. f) Not a statement.
g) Compound statement. h) Simple statement.

i) Simple statement. j) Compound statement.
k) Not a statement. l) Compound statement.

3. a) $p \wedge q$ b) $p \wedge q$ c) $\sim p \wedge q$
d) $\sim(p \wedge \sim q)$ e) $p \leftrightarrow \sim q$ f) $q \rightarrow \sim p$

5. a) It is raining and the sky is not falling.
b) It is not the case that it is not raining.
c) It is not the case that it is raining or the sky is falling.
d) It is not raining or it is raining and the sky is not falling.
e) If the sky is not falling, then it is raining or the sky is not falling.
f) The sky is not falling if and only if it is not raining.
g) It is not raining implies the sky is falling, if and only if the sky is not falling implies it is raining.

7. "John is in love with Mary" = A
"Today is Sunday" = B
"I must go to school" = C
"Frank is interested in Judy" = D
"Judy is interested in Frank" = E
"Frank loves Judy" = F
"Judy loves Frank" = G
"It will be raining next Sunday" = H
"We can go to the park" = I
"The grass will get watered" = J
"It rains" = K
a) $\sim$A b) B $\wedge$ C c) D $\wedge$ E
d) F $\vee$ G $\vee$ (F $\wedge$ G) e) H $\rightarrow$ $\sim$I f) J $\leftrightarrow$ K

9. Let p be "$x = 0$" and let q be "$x^2 = 0$."
a) $(p \rightarrow q) \wedge (q \rightarrow p)$ b) $(p \leftrightarrow q)$

Section 2.2 (page 70)

1.

p	$\sim(\sim p)$
T	T
F	F

$f(\text{T}) = \text{T}$, $f(\text{F}) = \text{F}$. Neither

3.

p	$p \vee \sim p$
T	T
F	T

$f(\text{T}) = \text{T}$, $f(\text{F}) = \text{T}$. Tautology

5.

p	q	$\sim p \leftrightarrow (p \wedge q)$
T	T	F
T	F	T
F	T	F
F	F	F

$f(\text{T, T}) = \text{F}$, $f(\text{T, F}) = \text{T}$,
$f(\text{F, T}) = \text{F}$, $f(\text{F, F}) = \text{F}$.
Neither.

7.

p	$\sim(\sim p \rightarrow \sim(\sim p \rightarrow p))$
T	F
F	F

$f(T) = \text{F}$, $f(\text{F}) = \text{F}$. Contradiction

9.

p	q	$[(p \to q) \land (q \to p)] \leftrightarrow (p \leftrightarrow q)$
T	T	T
T	F	T
F	T	T
F	F	T

$f(\mathrm{T}, \mathrm{T}) = f(\mathrm{T}, \mathrm{F}) = f(\mathrm{F}, \mathrm{T}) = f(\mathrm{F}, \mathrm{F}) = \mathrm{T}$. Tautology.

11.

p	q	$(p \to \sim q) \land (q \leftrightarrow \sim p)$	$[(p \to \sim q) \land (q \leftrightarrow \sim p)] \to \sim(\sim p \lor q)$
T	T	F	T
T	F	T	T
F	T	T	F
F	F	F	T

$f(\mathrm{T}, \mathrm{T}) = \mathrm{T}$, $f(\mathrm{T}, \mathrm{F}) = \mathrm{T}$, $f(\mathrm{F}, \mathrm{T}) = \mathrm{F}$, $f(\mathrm{F}, \mathrm{F}) = \mathrm{T}$. Neither.

13.

p	q	r	$p \to q$	$r \to q$	$(p \to q) \to (r \to q)$
T	T	T	T	T	T
T	T	F	T	T	T
T	F	T	F	F	T
T	F	F	F	T	T
F	T	T	T	T	T
F	T	F	T	T	T
F	F	T	T	F	F
F	F	F	T	T	T

$$f(x, y, z) = \begin{cases} \mathrm{F} & \text{if } x = \mathrm{F}, y = \mathrm{F}, z = \mathrm{T} \\ \mathrm{T} & \text{otherwise} \end{cases}.$$ Neither.

15.

p	q	r	s	$p \land q$	$r \lor s$	$(p \land q) \leftrightarrow (r \lor s)$
T	T	T	T	T	T	T
T	T	T	F	T	T	T
T	T	F	T	T	T	T
T	T	F	F	T	F	F
T	F	T	T	F	T	F
T	F	T	F	F	T	F
T	F	F	T	F	T	F
T	F	F	F	F	F	T
F	T	T	T	F	T	F
F	T	T	F	F	T	F
F	T	F	T	F	T	F
F	T	F	F	F	F	T
F	F	T	T	F	T	F
F	F	T	F	F	T	F
F	F	F	T	F	T	F
F	F	F	F	F	F	T

19. Let $\mathrm{P} = p \land q \land r \land s \land t$, and apply Theorem 2.2.2 to the tautology $\mathrm{P} \to \mathrm{P}$. There are $2^5 = 32$ rows in the truth table of the statement $(p \land q \land r \land s \land t) \to (p \land q \land r \land s \land t)$.

25.

p	q	p exclusive or q
T	T	F
T	F	T
F	T	T
F	F	F

27. Proof by contradiction. Assume that the result is not true; that is, assume that A $\rightarrow$ B and B $\rightarrow$ C are tautologies, but that A $\rightarrow$ C is not a tautology. Then there is an assignment of truth values such that A $\rightarrow$ C is not true; that is, there is an assignment of truth values such that A is true but C is false. (This is the only way that A $\rightarrow$ C can be false.) Now if A is true, then since A $\rightarrow$ B is a tautology, B must also be true. But since B is true and B $\rightarrow$ C is a tautology, C must also be true. This is a contradiction to the fact that C is false, and so the result must be true.
29. If p is false and q is false, then $p \rightarrow (p \vee q)$ is false under column 2. If p is false and q is true, then $p \rightarrow (p \vee q)$ is false under either 3 or 4.

Section 2.3 (page 80)

1. Since c is always false, $p \vee c$ has the same truth value as p. Hence, $p \vee c \equiv p$. Also, $p \wedge c$ is always false, and so $p \wedge c \equiv c$. The properties of t are proved similarly.
3. Show that the truth tables of $p \vee q$ and $q \vee p$ are the same, and similarly for $p \wedge q$ and $q \wedge p$.
5. Use truth tables, as in the answer to Exercise 3.
7. Use truth tables, as in the answer to Exercise 3.
9. Show that the truth tables of $p \leftrightarrow q$ and $(p \rightarrow q) \wedge (q \rightarrow p)$ are the same.
13. a) Letting P $= p \leftrightarrow q$ and Q $= p \wedge \sim q$, the statement becomes $\sim$[P $\vee$ Q] $\equiv$ $\sim$P $\wedge$ $\sim$Q, which is just De Morgan's law, and hence is true. Therefore, according to Theorem 2.3.5, the statement is true.
17. a) Converse: If $x^2 \neq 0$ then $x \neq 0$.
    Contrapositive: If $x^2 = 0$ then $x = 0$.
    Inverse: If $x = 0$ then $x^2 = 0$.
    c) Converse: If x is a real number, then x is a rational number.
    Contrapositive: If x is not a real number then x is not a rational number.
    Inverse: If x is not a rational number, then x is not a real number.
    e) Converse: If $x = 1$ then $x \neq 2$ and $x^2 + 3x + 1 = 0$.
    Contrapositive: If $x \neq 1$ then $x = 2$ or $x^2 + 3x + 1 \neq 0$.
    Inverse: If $x = 2$ or $x^2 + 3x + 1 \neq 0$ then $x \neq 1$.
23. The contrapositive is: If $ax = bx$ then $a = b$ or $x = 0$. *Proof*: If x was not equal to 0, then we could divide both sides of the equation $ax = bx$ by x, to get $a = b$. Hence, either $x = 0$ or $a = b$. This proves the contrapositive, and hence also the original statement.

Section 2.4 (page 88)

1. Valid.
3. Invalid. (p true, q false)
5. Invalid. (p true, q true)
7. Invalid. (p false, q and r true)
9. Valid.
11. Invalid. (p false, q and r true)
13. Invalid. (p true, q, r, and s false)

15. The argument $p \to s, p \therefore s$ is valid. Replacing s by $(q \to r)$ gives the argument in Exercise 15. Hence it too is valid.
17. The argument $t \to u, u \to v \therefore t \to v$ is valid. Replacing t by $(p \wedge q)$, u by $(r \to s)$, and v by $\sim s$ gives the argument in Exercise 17. Hence, it too is valid.
19. Let p be the statement "it is raining" and let q be the statement "it is snowing." Then the argument is $p \vee q, \sim q \therefore p$, which is valid.
21. Let p and q be as in the answer to Exercise 19. Then the argument is $p \vee \sim q, q \therefore \sim p$, which is not valid. (Let p be true and q be true.)
23. Let b be the statement "we have bigger bombs" and let d be the statement "democracy is safe." Then the argument is $b \to d, \sim b \therefore \sim d$, which is invalid. (Let b be false and d be true.)
25. Let p be the statement "it rains today," let q be the statement "the newspaper is right," and let r be the statement "the radio is wrong." Then the argument is $p \to (q \vee r)$, $\sim p \vee \sim q \therefore \sim r$, which is invalid. (Let p and q be false and r be true.)
27. Let p = "is an ostrich"
q = "bird is at least 9 feet tall"
r = "bird is in the aviary"
s = "lives on mince-pies"
t = "bird is mine"

Then the argument is

$$\begin{array}{l} q \to p \\ r \to t \\ p \to \sim s \\ t \to q \\ \hline r \to \sim s \end{array}$$

which is valid.

Section 2.5 (page 99)

1. a, b, d, and f
3. a) 1 b) 0 c) 0
5. No. $p(1, 0) = 1 \wedge 0 = 0$ but $q(1, 0) = 1 \vee 0 = 1$.

7.

x	x'
1	0
0	1

9.

x	y	$(x \wedge y)' \wedge x'$
1	1	0
1	0	0
0	1	1
0	0	1

13. $p(x, y) = (x \wedge y) \vee (x \wedge y') \vee (x' \wedge y)$
15. $p(x, y, z) = (x \wedge y \wedge z) \vee (x' \wedge y \wedge z)$
17. $p(x, y, z) = (x \wedge y \wedge z) \vee (x \wedge y \wedge z') \vee (x \wedge y' \wedge z) \vee (x' \wedge y' \wedge z) \vee (x' \wedge y' \wedge z')$
19. $q(x, y) = (x \wedge y) \vee (x \wedge y') \vee (x' \wedge y)$

21. $q(x, y, z) = (x \wedge y' \wedge z) \vee (x' \wedge y \wedge z) \vee (x \wedge y \wedge z)$

23. $p(x, y, z) = (x' \wedge y \wedge z') \vee (x' \wedge y \wedge z) \vee (x \wedge y' \wedge z') \vee (x \wedge y' \wedge z) \vee (x \wedge y \wedge z') \vee (x \wedge y \wedge z)$

25. $q(x, y, z) = (x' \wedge y' \wedge z') \vee (x' \wedge y' \wedge z) \vee (x' \wedge y \wedge z') \vee (x' \wedge y \wedge z) \vee (x \wedge y \wedge z')$

27. $q(x, y, z) = (x \wedge y' \wedge z) \vee (x \wedge y \wedge z)$

29. $q(x, y, z, w) = \vee$ of all terms of the form $e_1 \wedge e_2 \wedge e_3 \wedge e_4$, *except* for $x' \wedge y' \wedge z' \wedge w'$.

Section 2.6 (page 106)

1.

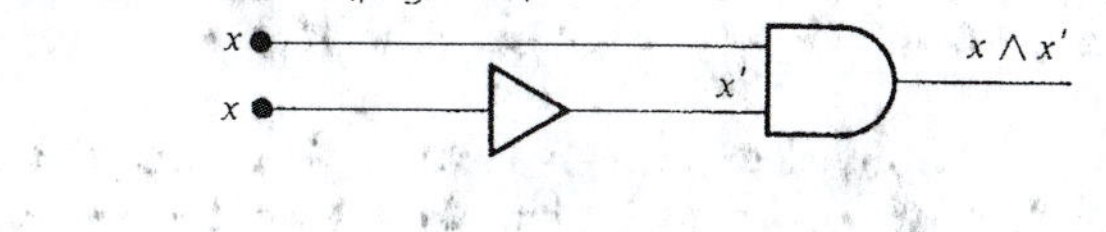

$p(x) = x \wedge x'$

x	$p(x)$
T	F
F	F

3.

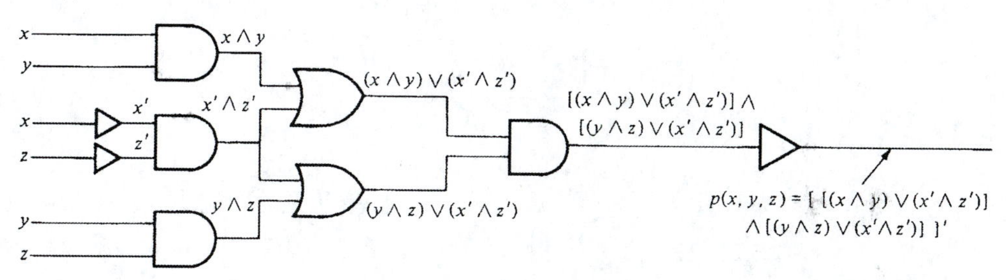

x	y	z	$p(x, y, z)$
0	0	0	0
0	0	1	1
0	1	0	0
0	1	1	1
1	0	0	1
1	0	1	1
1	1	0	1
1	1	1	0

5.

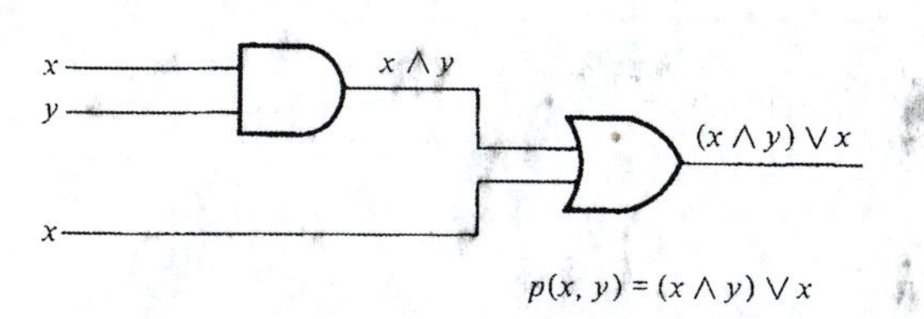

$p(x, y) = (x \wedge y) \vee x$

x	y	$p(x, y)$
0	0	0
0	1	0
1	0	1
1	1	1

7.

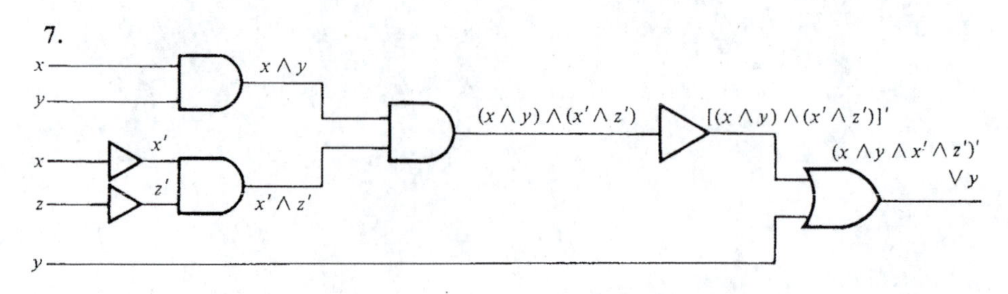

$p(x, y, z) = (x \wedge y \wedge x' \wedge z')' \vee y$

9. $p(x, y) = (x \wedge y') \vee (y \wedge x')$

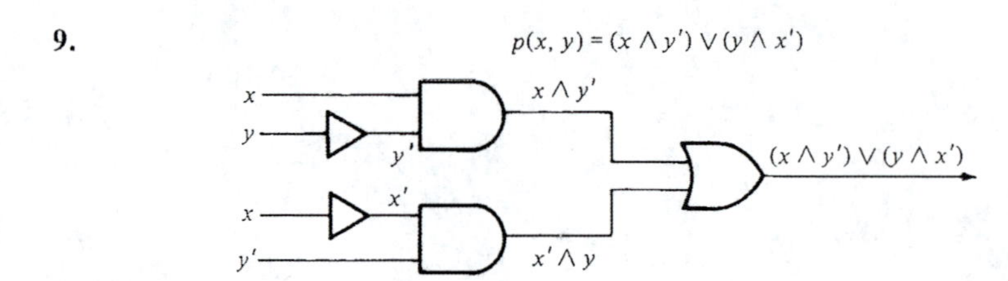

11. $p(x, y) = x' \wedge y$

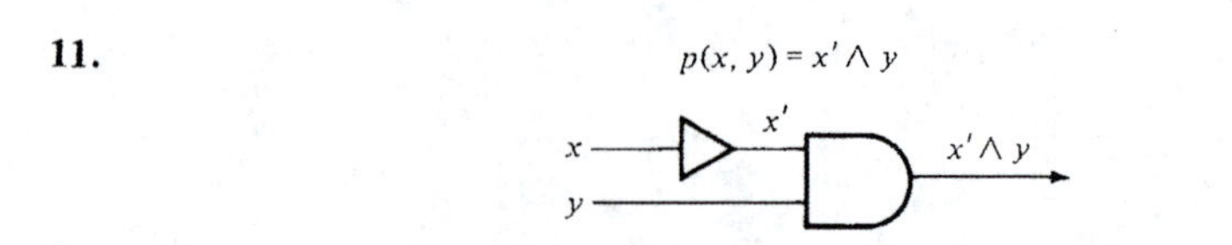

13.

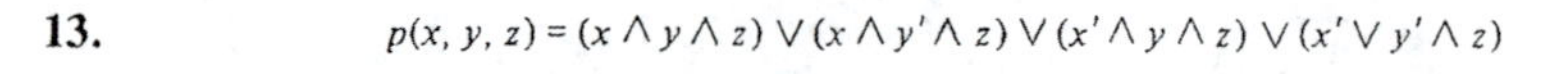

$p(x, y, z) = (x \wedge y \wedge z) \vee (x \wedge y' \wedge z) \vee (x' \wedge y \wedge z) \vee (x' \vee y' \wedge z)$

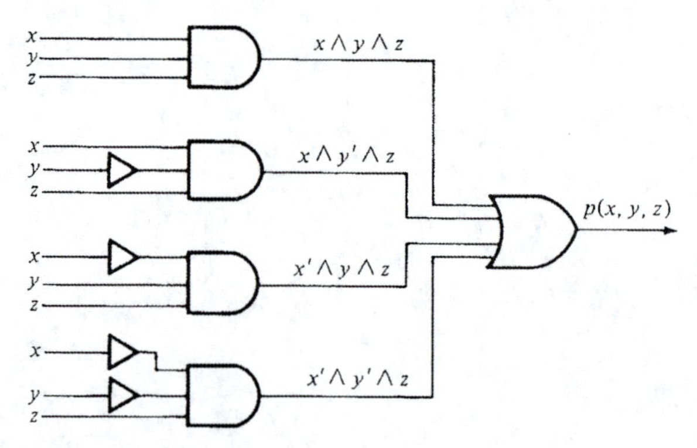

15. $p(x, y) = (x \wedge y) \vee (x \wedge y') \equiv x$

17. $p(x, y, z) = (x \wedge y \wedge z) \vee (x \wedge y \wedge z') \vee (x' \wedge y \wedge z) \vee (x' \wedge y' \wedge z)$
$\equiv (x \wedge y) \vee (x' \wedge z)$

19. $p(x, y, z) = (x \wedge y \wedge z') \vee (x' \wedge y' \wedge z)$

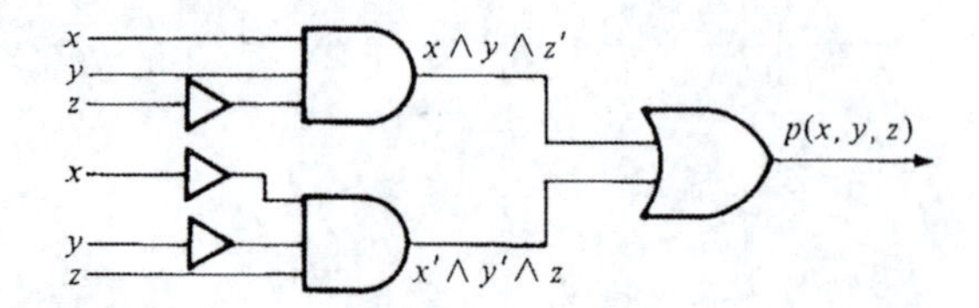

Section 2.7 (page 118)

1. $p(1, 1) = p(1, 0) = p(0, 1) = 1, p(0, 0) = 0;$
   $q(1, 1) = q(1, 0) = q(0, 1) = 1, q(0, 0) = 0.$
   Hence, p and q have the same truth table, and so they are logically equivalent.
3. Show that they all have the same truth table.
5.

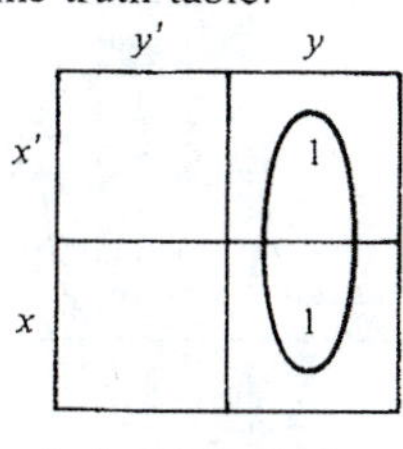

$(x \wedge y) \vee (x' \wedge y) \equiv y$

7.

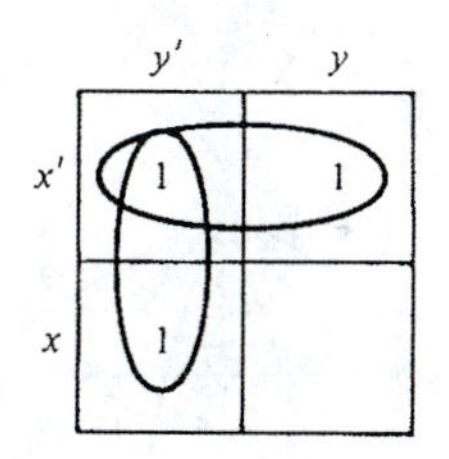

$(x \wedge y') \vee (x' \wedge y) \vee (x' \wedge y') \equiv x' \vee y'$

9. $(x \wedge y' \wedge z) \vee (x' \wedge y' \wedge z') \vee (x' \wedge y' \wedge z) \vee (x \wedge y' \wedge z') \equiv y'$

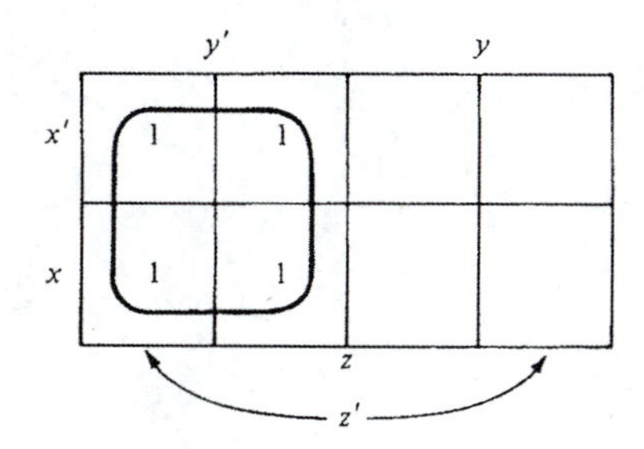

11. $(x' \wedge y' \wedge z') \vee (x' \wedge y \wedge z') \vee (x \wedge y' \wedge z) \vee (x \wedge y \wedge z) \equiv (x \wedge z) \vee (x' \wedge z')$

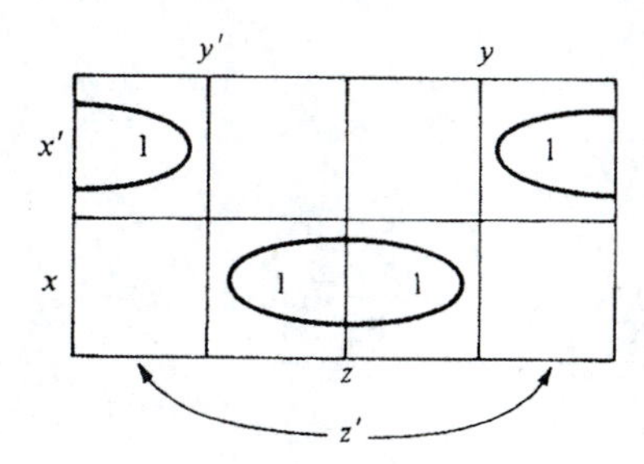

15. $(x' \wedge y \wedge z) \vee (x' \wedge y \wedge z') \vee (x \wedge y' \wedge z') \vee (x \wedge y' \wedge z) \vee (x \wedge y \wedge z)$
$\equiv (x \wedge y') \vee (y \wedge z) \vee (x' \wedge y)$

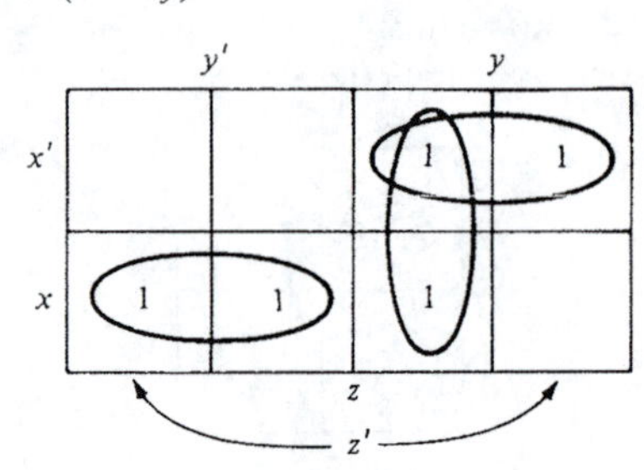

Chapter 3

Section 3.1 (page 127)

1.

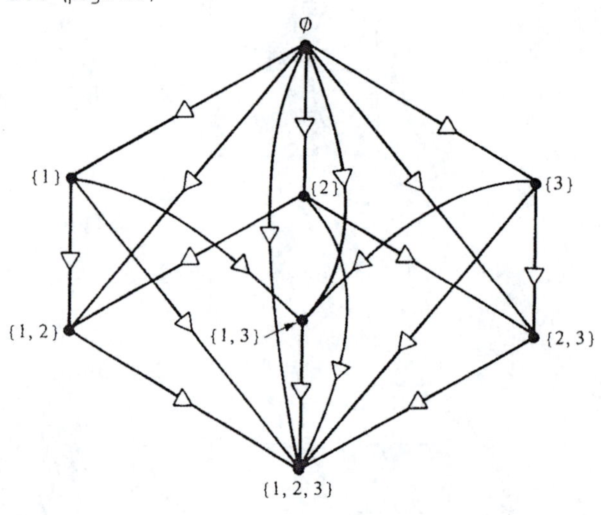

3.

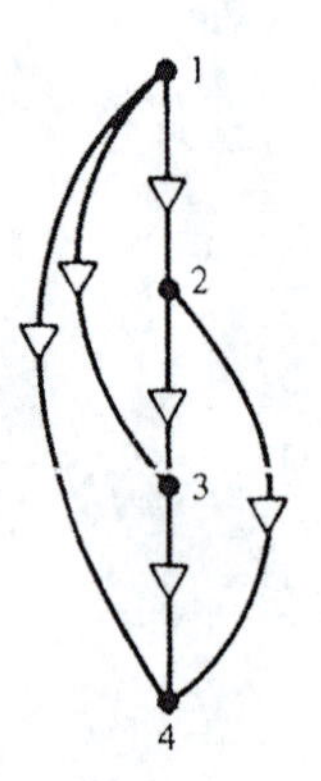

5.

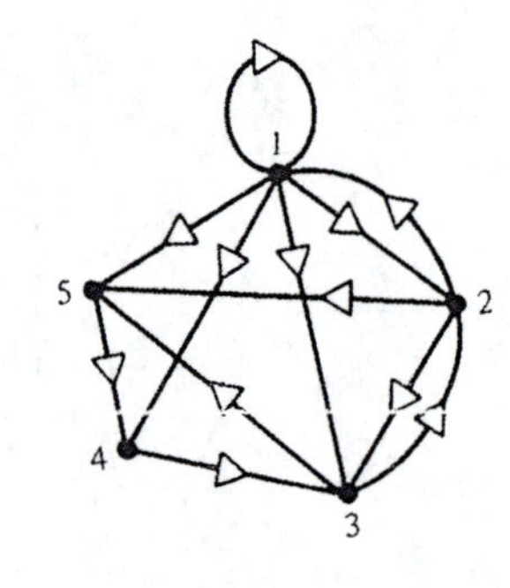

7.

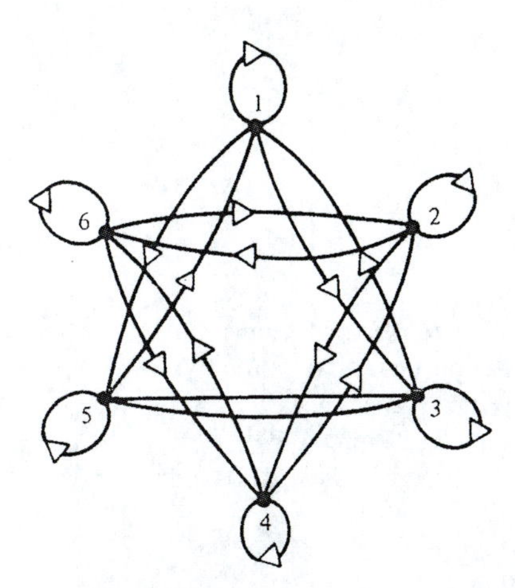

9.

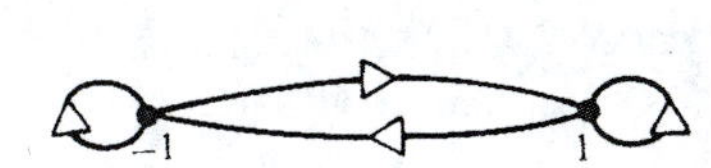

15.

17. a) True c) True

19. The inverse is the relation defined by the equation $2y + 3x = 0$.

21. The inverse is $AR^{-1}B$ if B logically implies A.

23. a) R^{-1} is also the relation of equality.

c) If arc (a, b) is in the graph of R then so is arc (b, a).

Section 3.2 (page 135)

1. Transitive and antisymmetric
3. Reflexive and transitive
5. Transitive and antisymmetric
7. Reflexive, symmetric, and transitive
9. Transitive and antisymmetric
11. Reflexive, symmetric, and transitive

13.

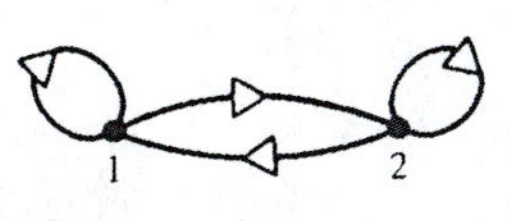

4 • • 3

15.

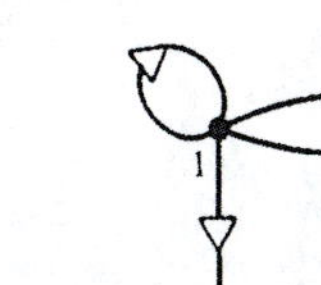

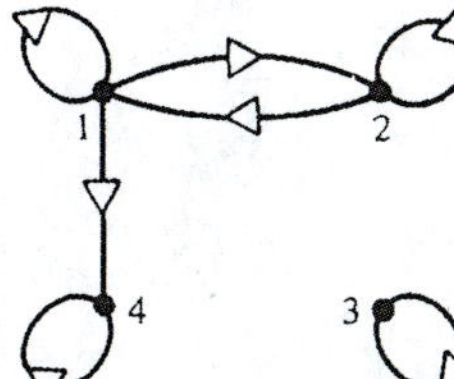

17.

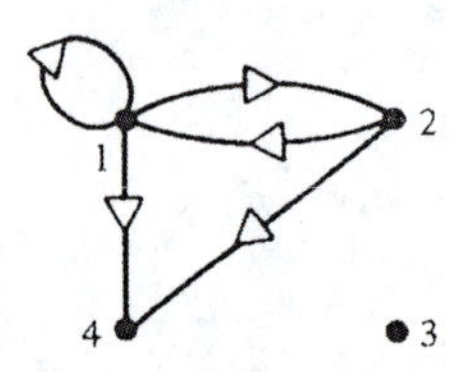

19. $R = \{(a, a) \mid a \in A\}$. In words, each element of A is related to itself and *only* itself.

21. Assume that $m \equiv n \pmod{k}$ and $n \equiv p \pmod{k}$. We want to show that $m \equiv p \pmod{k}$. From the definition of congruence modulo k, we see that $m - n = ak$ and $n - p = bk$, for some integers a and b. But then $m - p = (m - n) + (n - p) = ak + bk = (a + b)k$, and so $m \equiv p \pmod{k}$. Hence, the relation is transitive.

23. a) True b) True

25. True

27. The union of the relations "less than" and "greater than" is the relation "not equal to."

Section 3.3 (page 141)

1. Equivalence relation. The equivalence classes are $\{1, 2\}$ and $\{3\}$.

3. Equivalence relation. The equivalence classes are $\{1\}$, $\{2\}$, and $\{3\}$.

5. Not an equivalence relation. (Not symmetric.)

7. Equivalence relation. The equivalence classes can be described as follows. For each element $b \in \text{Im}(f)$, we get an equivalence class consisting of all $a \in A$ for which $f(a) = b$.

9. Equivalence relation. If $|S| = n$, then there are $n + 1$ equivalence classes. Namely, for each k satisfying $0 \leq k \leq n$, we get the equivalence class consisting of all subsets of S with size k.

11. Not an equivalence relation. (Not reflexive and not symmetric.)

13. Not an equivalence relation. (Not symmetric.)

15. Not an equivalence relation. (Not reflexive and not transitive.)

Section 3.4 (page 148)

1. Not a partial order. (Not reflexive.)

3. Not a partial order. (Not antisymmetric.)

5. This is a partial order.

7.

• 2 • 3 • 5 • 7 • 11 • 13 • 17

9.

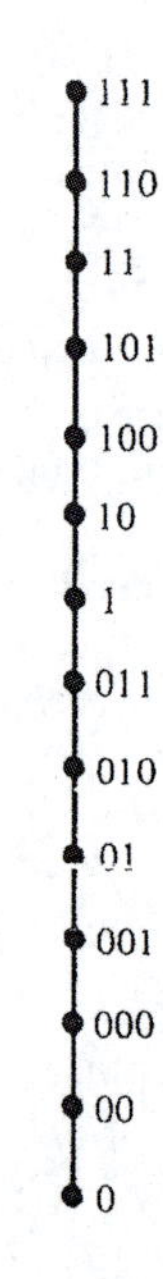

13. Suppose that the relation $\preceq$ is both a partial order and an equivalence relation. Then it is both symmetric and antisymmetric. If $x \preceq y$, then by symmetry, we have $y \preceq x$, and so by antisymmetry we have $x = y$. Hence, $x \preceq y$ if and only if $x = y$. (We use the fact that $\preceq$ is reflexive for the "if" part of this statement.) In other words, the relation $\preceq$ must be equality.

Section 3.5 (page 155)

1. The only maximal element is $\{a, b, c, d\}$, which is also the maximum element. The only minimal element is $\emptyset$, which is also the minimum element.
3. There are no maximal elements, and hence no maximum element. The only minimal element is 1, which is also the minimum element.
5. There are no maximal elements, and hence no maximum element. Each prime number in P is a minimal element. There is no minimum element.
7. Every element is both a maximal and a minimal element. There are no maximum or minimum elements.
9. There are no maximal or minimal elements. Hence, there are no maximum or minimum elements.
13. $\emptyset, \{1\}, \{2\}, \{3\}, \{1, 2\}, \{1, 3\}, \{2, 3\}, \{1, 2, 3\}$
15. $d, 3, c, 4, 2, b, 1, a$

Section 3.6 (page 162)

1. The function f is a bijection. Also, since for a and b in P we have $a \leq b$ if and only if $a^2 \leq b^2$, we conclude that f is an order isomorphism.
3. From the Hasse diagrams

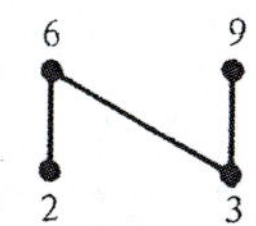

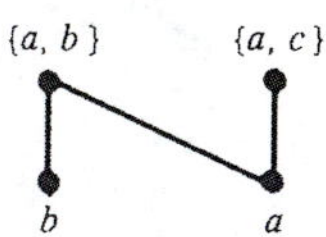

we see that the function defined by $f(2) = \{b\}$, $f(3) = \{a\}$, $f(6) = \{a, b\}$, and $f(9) = \{a, c\}$ is an order isomorphism.
5. a) The function $f(x) = 2x$ is a bijection and $x \leq y$ if and only if $2x \leq 2y$, that is, if and only if $f(x) \leq f(y)$. Hence, f is an order isomorphism.
b) One such function is defined by $g(0) = 2$, $g(1) = 0$, $g(x) = 2x$ for $x \geq 2$.
9. The composition of two bijections is a bijection. Also, since f and g are order isomorphisms, we have $x \preceq_1 y$ if and only if $f(x) \preceq_2 f(y)$ if and only if $g(f(x)) \preceq_3 g(f(y))$, that is, $(g \circ f)(x) \preceq_3 (g \circ f)(y)$.

Chapter 4
Section 4.2 (page 172)

1. $3 \cdot 6 \cdot 4 \cdot 6 = 432$
3. $26 \cdot 2 = 52$
7. $2^4 + 2^5 + 2^6 = 112$
9. a) 4^8
b) $4^0 + 4^1 + \cdots + 4^8 = (1 - 4^9)/(1 - 4) = 87{,}381$
11. 4^{10}
13. $96/3 = 32$
15. $2^5 = 32$. Each sequence of heads and tails can be thought of as a word of length 5 over the alphabet $\Sigma = \{H, T\}$.

17. a) $10^3 = 1000$ b) $10 \cdot 9 \cdot 8 = 720$

19. $9 \cdot 8 \cdot 7 \cdots 2 \cdot 1 = 362{,}880$

21. a) $9 \cdot 9 \cdot 8 = 648$

b) 320 (*Hint:* first choose the units place, then choose the hundreds place, and finally the tens place.)

23. $\frac{n}{2}\left(\frac{1}{k} + \frac{1}{n}\right)$

Section 4.3 (page 177)

1. 11

3. Given any three integers, at least two of the integers must have the same parity (either two even integers or two odd integers). But the difference of two integers that have the same parity is even.

5. $4 \cdot 4 + 1 = 17$

7. Label each person with the number k of friends he has in the group. Then $0 \leq k \leq n - 1$. However, it is not possible for both of the labels 0 and $n - 1$ to be used. Hence, there are at most $n - 1$ possible labels for the n people, and so at least 2 people must have the same label; that is, at least 2 people must have the same number of friends.

9. 3

11. If not, then the i-th box has at most $a_i - 1$ balls, and so the total number of balls is at most $(a_1 - 1) + (a_2 - 1) + \cdots + (a_n - 1) = a_1 + a_2 + \cdots + a_n - n$. However, by assumption there is one additional ball, and so this is a contradiction.

15. 4

Section 4.4 (page 183)

1. a) $7/6$ c) 18 e) $1/9$

7. $6! = 720$

9. $2^{10} - 11 = 1013$

11. $3^6 - 2^6 - 6 \cdot 2^5 = 473$

13. $2^{10} - 11 = 1013$

15. $10^3 - 4 = 996$

17. a) $(7!)^3$ b) $3!(7!)^3$

21. $10!$

23. $10!5!$

25. a) $5! = 120$ c) $5! - 4! = 96$

Section 4.5 (page 191)

1. $(10)_5 = 30{,}240$

3. a) $(26)_4 \cdot 10^2 = 35{,}880{,}000$ c) $(26)_4 \cdot 9^2 = 29{,}062{,}800$

5. $26^3 = 17{,}576$

7. a) $4^3 = 64$ b) $(4)_3 = 24$

9. $(n)_k$ if $n \geq k$, 0 if $n < k$.

11. a) $(20)_{10}$ c) $6 \cdot (10)_5 \cdot (10)_5$

13. a) $P(n, n - 1) = n(n - 1) \cdots 2 = n(n - 1) \cdots 2 \cdot 1 = n! = P(n, n)$

c) $$\frac{P(n, n - k)}{(n - k)!} = \frac{n(n - 1) \cdots (k + 1)}{(n - k)!}$$

$$= \frac{n(n-1)\cdots(k+1)\cdot k!}{(n-k)!\cdot k!}$$

$$= \frac{n!}{k!(n-k)!}$$

Also,

$$\frac{P(n,k)}{k!} = \frac{n(n-1)\cdots(n-k+1)}{k!}$$

$$= \frac{n(n-1)\cdots(n-k+1)\cdot(n-k)!}{k!\cdot(n-k)!}$$

$$= \frac{n!}{k!(n-k)!}$$

15. This equation is the same as the equation

$$5n(n-1)(n-2) = 2(n-1)(n-2)(n-3)(n-4)$$

whose solutions are $n = 1$, 2, and 8.

17. $N(N-1)\cdots(N-K+1) = (N)_K$

Section 4.6 (page 198)

1. a) 1

c) (1,000,000)(999,999)/2

e) 4200

3. a) $\frac{n}{k}\binom{n-1}{k-1} = \frac{n}{k}\frac{(n-1)!}{(n-k)!(k-1)!} = \frac{n!}{(n-k)!k!} = \binom{n}{k}$

c) $\frac{n-k+1}{k}\binom{n}{k-1} = \frac{n-k+1}{k}\frac{n!}{(n-k+1)!(k-1)!}$

$$= \frac{n!}{(n-k)!k!} = \binom{n}{k}$$

7. $\binom{2n}{n} = 2\binom{2n-1}{n-1}$ is even

9. a) $\binom{11}{3} = 165$

b) $\binom{5}{3}\binom{6}{4} = 150$

c) $\binom{11}{6} - \binom{9}{4} = 336$

d) $\binom{5}{4}\binom{6}{3} - \binom{3}{2}\binom{6}{3} = 40$

11. a) $\binom{10}{6} = 210$

b) $2^{10} - \binom{10}{0} - \binom{10}{1} - \binom{10}{2} = 968$

13. $\binom{15}{6}\binom{9}{4} = 630{,}630$

15. $\binom{6}{1}\binom{8}{5}\binom{10}{10} + \binom{6}{2}\binom{8}{6}\binom{10}{9} + \binom{6}{3}\binom{8}{7}\binom{10}{8} + \binom{6}{4}\binom{8}{8}\binom{10}{7} = 13{,}536$ if the order of the students in each row does not matter

$\binom{6}{1}(8)_5(10)_{10} + \binom{6}{2}(8)_6(10)_9 + \binom{6}{3}(8)_7(10)_8 + \binom{6}{4}(8)_8(10)_7$ if the order of the students in each row does matter

17. a) $\binom{12}{10} = 66$

b) $\binom{8}{7}\binom{4}{3} = 32$

c) $\binom{8}{6}\binom{4}{4} + \binom{8}{7}\binom{4}{3} = 60$

19. 15

21. $\binom{4}{1} 9^4 + 2\binom{4}{2} 9^3 + 3\binom{4}{3} 9^2 + 4\binom{4}{4} 9^1 = 36{,}000$

23. $\binom{7}{3} = 35$

25. $\binom{N}{3}$

27. $\sum_{k=0}^{n} \binom{n}{k}$

Section 4.7 (page 207)

5. a) $\binom{13}{7} = 1716$ c) $2^6 \binom{9}{6} = 5376$

7. Take $x = 1$, $y = 2$ in the binomial formula.

17. a) $\binom{2n}{2} = \frac{2n(2n-1)}{2} = n(2n-1) = n(n-1) + n^2 = 2\binom{n}{2} + n^2$

b) The left-hand side counts the number of ways to choose two objects from $2n$ objects. But the right-hand side also counts this number, for we may choose two objects from $2n$ objects by choosing both objects from the first n objects ($\binom{n}{2}$ ways), or both objects from the second n objects ($\binom{n}{2}$ ways), or 1 object from the first n objects and 1 object from the second n objects (n^2 ways).

Section 4.8 (page 216)

1. $x^4 + 4x^3y + 4x^3z + 6x^2y^2 + 6x^2z^2 + 12x^2yz + 4xy^3 + 12xy^2z + 12xyz^2 + 4xz^3 + y^4 + 4y^3z + 6y^2z^2 + 4yz^3 + z^4$

5. $\binom{8}{2,\ 2,\ 2,\ 2} = 2520$

7. $\binom{9}{5,\ 4,\ 0,\ 0} = 126$

9. $\binom{10}{2,\ 2,\ 2,\ 2,\ 2} = 113{,}400$

11. $\binom{2n}{2,\ 2,\ \ldots,\ 2} = \dfrac{(2n)!}{2^n}$

13. $\binom{8}{2,\ 3,\ 3} = 560$

15. $\binom{10}{3,\ 3,\ 2,\ 2} = 25{,}200$

Section 4.9 (page 224)

1. 46, 190, 804, 3556, 16300

3. 3, 3, $6^{1/2}$, $(6^{1/2} + 3)^{1/2}$, $((6^{1/2} + 3)^{1/2} + 6^{1/2})^{1/2}$

5. $s_n = a^n b$

7. $p_n = 2^n n!$

9. $p_0 = 1,\ p_n = 1 \cdot 3 \cdot 5 \cdots (2n - 1),\ n \geq 1$

11. $r_n = (1 + 3^n)/2$

13. $q_n = (2 \cdot 4^n + 1)/3$

15. $s_n = n^2/2 - n/2 + 1$

17. $a_n = 1 + n(n + 1)(2n + 1)/6$

19. $s_n = 1 + n(n + 1)/2 + n(n + 1)(2n + 1)/6$

21. $s_8 = 88$

23. $s_n = s_{n-1} + s_{n-2} + s_{n-3} + s_{n-4};\ s_0 = 1,\ s_1 = 2,\ s_2 = 4,\ s_3 = 8;\ s_7 = 108$

Section 4.10 (page 235)

1. $s_n = 2^n - 1$

3. $s_n = a + (b - a)n$

5. $s_n = \begin{cases} (-3)^{n/2} & \text{if } n \text{ is even} \\ (-3)^{(n-1)/2} & \text{if } n \text{ is odd} \end{cases}$

7. $s_n = [(5 + (-1)^n \cdot 3)/8]4^n = \begin{cases} 4^n & \text{if } n \text{ is even} \\ 4^{n-1} & \text{if } n \text{ is odd} \end{cases}$

9. $s_n = 0$

11. $s_n = [(1 + (-1)^n)/2]\pi^n a + [(1 + (-1)^{n-1})/2]\pi^{n-1} b$

$= \begin{cases} \pi^n a & \text{if } n \text{ is even} \\ \pi^{n-1} b & \text{if } n \text{ is odd} \end{cases}$

Section 4.11 (page 241)

1. $s_n = c4^n - 1/3$

3. $s_n = c4^n - 2n/3 + 7/9$

5. $s_n = c2^n + n + 2$

7. $s_n = c_1(-3)^n + c_2 2^n - 3n/4 - 33/16$

9. $s_n = c(-1)^n - n^2/2 - n - 3/4$

11. $s_n = c_1 2^n + c_2(-2)^n - 2n^n/3 - 41n/9 - 241/27$

13. $s_n = c2^n + (-1)^n/3$

15. $s_n = c6^n - 2^{n-1} - 3^n$

17. $s_n = c2^n - (1/15)2^{-n}$

Section 4.12 (page 241)

1. $g(x) = \dfrac{1}{1 - 3x},\ s_n = 3^n$

3. $g(x) = \dfrac{1}{1 + 2x},\ s_n = (-2)^n$

5. $g(x) = \dfrac{1 + 2x}{(1 - x)^2},\ s_n = 3n + 1$

7. $g(x) = \dfrac{1 - 2x}{(x - 1)^2},\ s_n = 1 - n$

9. $g(x) = \dfrac{x}{(1-x)^2}$, $s_n = n$

11. $g(x) = \dfrac{1}{(1+x)^2}$, $s_n = (-1)^n(n+1)$

13. $g(x) = \dfrac{x^2}{(1-x)^3}$, $s_n = \binom{n}{2}$

15. $g(x) = \dfrac{1-3x}{(1-2x)(1-4x)}$, $s_n = (2^n + 4^n)/2$

Chapter 5

Section 5.1 (page 251)

3. $\binom{10}{4, 4, 2} = 3150$

5. $3^4 = 81$

7. 1

9. $\binom{12}{3, 2, 2, 1, 1, 1, 1, 1} = 19{,}958{,}400$

11. $4^3 = 64$

13. $3^4 + 3^5 = 324$

15. $\binom{13}{3, 4, 5, 1} = 360{,}360$

17. $\binom{12}{4, 3, 5} = 27{,}720$

19. $\binom{5}{2, 3} + \binom{5}{3, 2} = 20$

21. n^k

Section 5.2 (page 256)

3. $\binom{11}{8} = 165$

5. a) $\binom{8}{5} = 56$ b) $\binom{4}{1} = 4$ c) 1

7. $\binom{17}{12} = 6188$

Section 5.3 (page 265)

1. a) $\binom{12}{10} = 66$ b) $\binom{9}{2} = 36$

3. a) $\binom{23}{16} = 245{,}157$ b) $\binom{15}{7} = 6435$

5. $\binom{10}{2} = 45$

7. $\binom{14}{5} = 2002$

9. $\binom{11}{7} = 330$

11. $\binom{9}{4} = 126$

13. $\binom{19}{15} = 3876$

15. $\binom{n + k - 1}{n}$

Section 5.4 (page 272)

1. a) $(7)_5 = 2520$ b) $7^5 = 16{,}807$

3. a) 0 b) $5^7 = 78{,}125$

5. a) 0 b) 0

7. $\binom{n - 1}{m - 1}$

9. $\binom{n - j_1 - j_2 - \cdots - j_m + m - 1}{m - 1}$

13. The printers are distinguishable balls, the computers are distinguishable boxes, and the distribution is so that no box is empty. According to Theorem 5.4.5, the answer is

$$\binom{12}{0}(12 - 0)^{18} - \binom{12}{1}(12 - 1)^{18} + \cdots - \binom{12}{11}(1)^{18}$$

(Using a computer to evaluate this sum gives the number 601,783,536,940,185,600.)

15. a) 25^{50}

b) $\binom{25}{0}(25 - 0)^{50} - \binom{25}{1}(25 - 1)^{50} + \cdots - \binom{25}{24}(1)^{50}$

c) $\binom{50}{2, 2, \ldots, 2} = \dfrac{50!}{2^{25}}$

17. $\binom{14}{10}\binom{12}{8} = 495{,}495$

19. a) $\binom{14}{10} = 1001$ b) $\binom{9}{5} = 126$ c) 1

21. $\binom{10}{6}\binom{8}{4}\binom{7}{3}\binom{6}{2}$

Section 5.5 (page 278)

1. 110, 90

3. 15, 10

5. 1100, 2400

7. 600, 600

9. 8000

11. $6! + 7! - 4! = 5736$; $9! - 5736 = 357{,}144$

13. $6! + 6! - 5! = 1320$

Section 5.6 (page 284)

1. 535, 200

3. 2200

5. $\binom{9}{7} - \binom{7}{2} - \binom{4}{2} - \binom{3}{2} + \binom{2}{2} = 7$
7. 0
9. $\binom{16}{2} - \binom{10}{2} - \binom{10}{2} + \binom{4}{2} = 36$
11. $\binom{11}{9} - \binom{5}{3} - \binom{3}{1} = 42$
13. 8
15. $4^n - 3 \cdot 3^n + 3 \cdot 2^n - 1, n \geq 3$
17. $n^4 - 3 \cdot n^3 + 3 \cdot n^2 - n$

Section 5.7 (page 291)

1. There are two derangements of size 3, namely, 312 and 231.
3. $3^4 - 3 \cdot 2^4 + 3 = 36$
5. 4321, 4123, 4231, 1324, 1234, 1243, 2341, 2134, 2143
7. $1{,}000{,}000 - 1000 - 100 - 31 + 10 + 31 + 3 - 3$
9. 44
11. $\binom{23}{3} - \binom{15}{3} - \binom{12}{3} - \binom{16}{3} - \binom{14}{3} + \binom{4}{3} + \binom{8}{3} + \binom{6}{3} + \binom{5}{3} + \binom{3}{3} + \binom{7}{3} = 298$
13. $\binom{11}{8} - 4\binom{6}{3} = 85$
15. 10
17. a) D_7 b) $7D_6$
19. $\binom{9}{3} D_6$

Section 5.8 (page 301)

1. a) 3/13 b) 4/13
3. 25/72
5. a) 1/11 b) 47/66
7. a) $\dfrac{10 \cdot 4^5}{\binom{52}{5}}$ b) $\dfrac{4 \cdot \binom{13}{5}}{\binom{52}{5}}$

Chapter 6

Section 6.1 (page 309)

1. a) No, v_1 and v_2 are adjacent.
   b) v_{11} is adjacent to itself.
   c) No. There is no edge between v_3 and v_6.
3. The graph has $(\frac{1}{2}) \Sigma_{i=1}^{6} deg(v_i) = (\frac{1}{2}) \cdot 18 = 9$ edges.
5. $\binom{10}{2} = 45$ edges.

7. Since $2 + 3 + 3 + 4 + 4 + 5 = 21$ is not even, there is no such graph.

9.

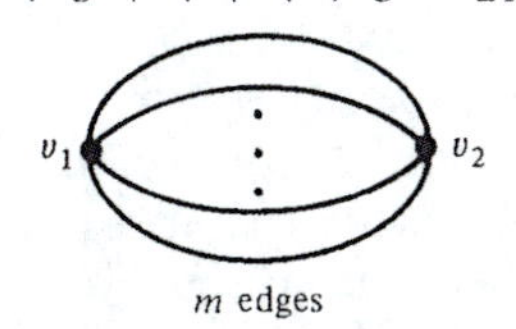

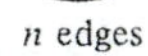

11. a)

b)

c)

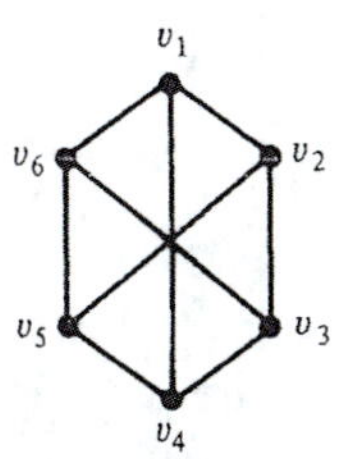

13. According to Theorem 6.1.1, $nr = 2|\mathcal{E}(G)|$ and so $|\mathcal{E}(G)| = nr/2$.

15. If a simple graph has n vertices, then the only possibilities for the degree of its vertices are $0, 1, 2, \ldots, n - 1$. However, a simple graph cannot have a vertex of degree 0 *and* a vertex of degree $n - 1$. Hence, there are only $n - 1$ possibilities for the degrees of the vertices, and since there are n vertices, the pigeonhole principle tells us that at least two vertices must have the same degree.

17.

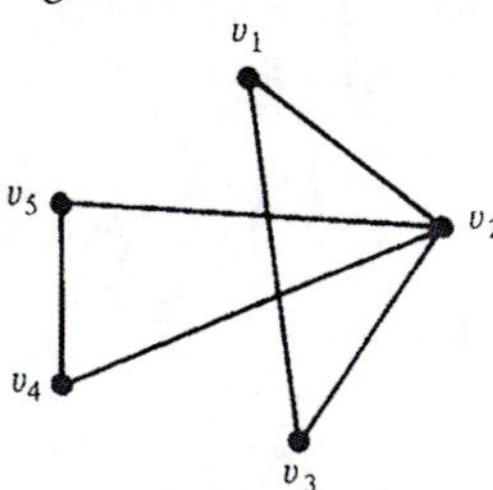

19. The graph consisting of n isolated vertices (and no edges).

21. Graphs c and e are bipartite.

c)

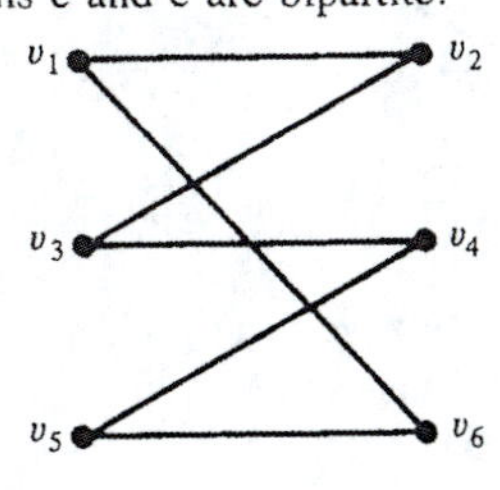

e)

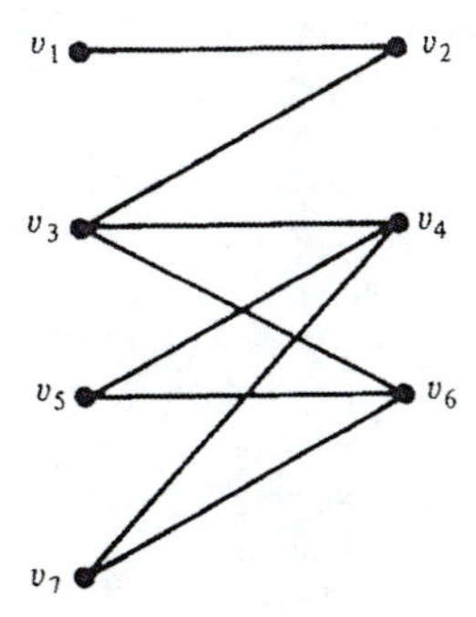

Section 6.2 (page 324)

1. a) Open trail b) Closed trail (circuit) c) Closed walk
d) None of these e) Closed path (cycle)

3. a) 5
b) $5 + (5)_3 + (5)_5 = 185$
c) Infinite

5. A loop is a cycle that contains exactly one edge. Two distinct edges connecting the same two distinct vertices form a cycle with exactly two edges.
7. If the vertex w appears more than once in W, then delete the closed walk in W from the first occurrence of w to the last occurrence of w. Repeat this process until no vertex appears more than once in the walk, except possibly for the first vertex u and the last vertex v. The result will be a path from u to v contained in W.
9. If H had $k > 2$ components, we could add an edge to H connecting two of these components. The result would be a disconnected graph (with $k - 1 > 1$ components) with more edges than H. This contradiction shows that H must have exactly two components.
19. a) $d(v_1, v_8) = d(v_1, v_7) = d(v_1, v_9) = 4$
    b) $d(v_1, v_4) = d(v_1, v_5) = 3$
    c) $d(v_1, v_6) = 5$

Section 6.3 (page 333)

1. Eulerian. The entire graph is an Eulerian circuit.
3. Not Eulerian.
5. Not Eulerian.
7. If n and m are both even, then $K_{n,m}$ is Eulerian. Otherwise it is not.
9. According to Theorem 6.3.2, there is an Eulerian trail from u to v.
11. No Eulerian trail.
13. It is possible since the graph has an Eulerian trail from u to v.
15. The graph associated with this floor plan is

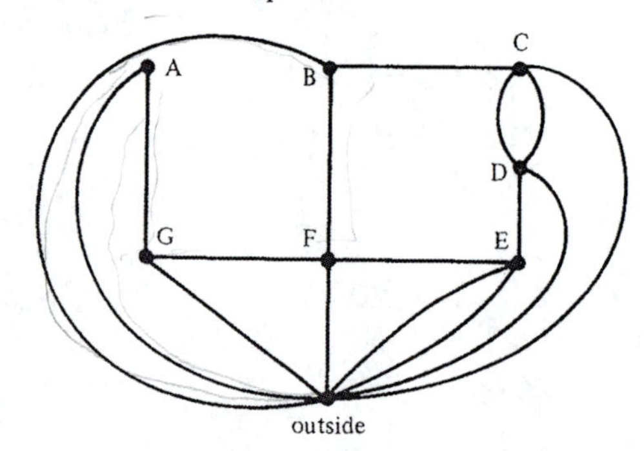

All vertices except B and G have even degree, and so such a tour is possible from B to G, or from G to B, but not from or to any other rooms.

25. $v_1 \to v_2 \to v_3 \to v_4 \to v_8 \to v_7 \to v_6 \to v_5 \to v_1$ is a Hamiltonian cycle.
27. If $n \geq 3$ and $v \in V(K_n)$ then $deg(v) = n - 1 \geq n/2$, and so the graph is Hamiltonian by Corollary 6.3.4. However, if $n < 3$ then K_n is not Hamiltonian.

Section 6.4 (page 348)

1. Not isomorphic. Invariant 1. (G_1 has 4 vertices but G_2 has 6 vertices.)
3. Isomorphic. $f(a) = z$, $f(b) = x$, $f(c) = y$, $f(d) = w$.
5. Isomorphic. $g(a) = p$, $g(d) = q$, $g(b) = x$, $g(e) = y$, $g(c) = r$, $g(f) = s$.
7. Not isomorphic. Invariant 3. (G_1 has a cycle of length 3 but G_2 does not.)
9. Not isomorphic. Invariant 5. (G_1 is connected but G_2 is not.)
11. Not isomorphic. Invariant 4. (G_1 has a vertex of degree 5 but G_2 does not.)
13. Let $deg(u) = n$, and suppose that u is connected to the distinct vertices $u_1, u_2, \ldots, u_n$, and no others. Then $f(u)$ is connected to the distinct vertices $f(u_1), f(u_2), \ldots, f(u_n)$, and no others. Hence, $f(u)$ also has degree n.

19.

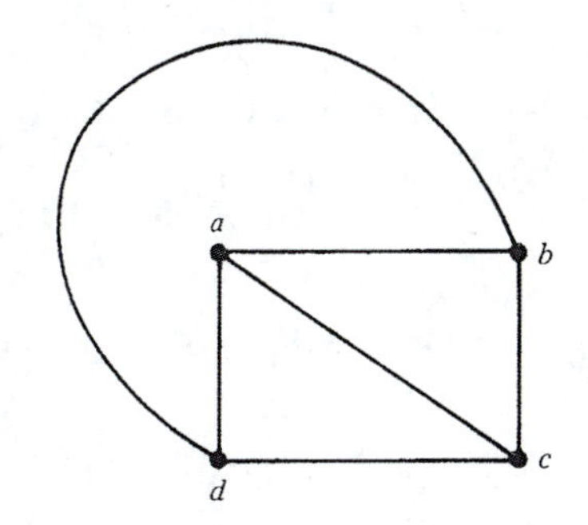

21.

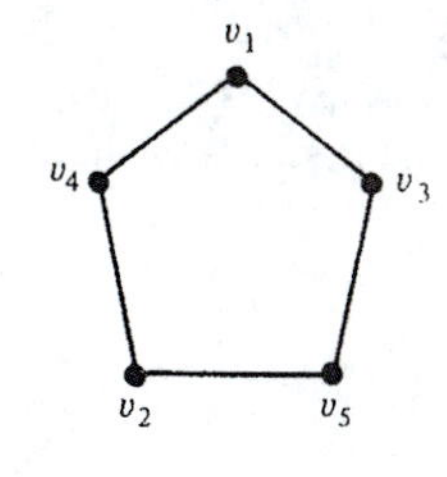

23.

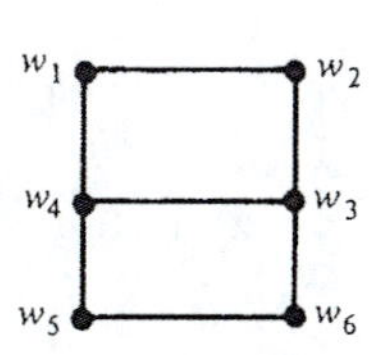

25.

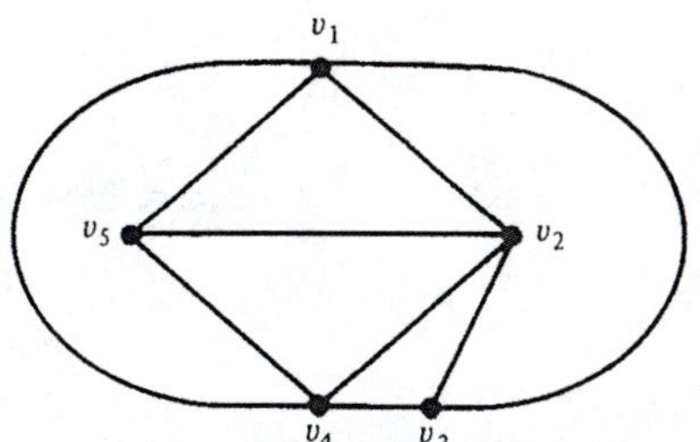

27. No, since $7 - 16 + 10 = 1 \neq 2$, and so Euler's formula does not hold.

Section 6.5 (page 361)

1.

3. The entire graph is a spanning tree.

5.

7.

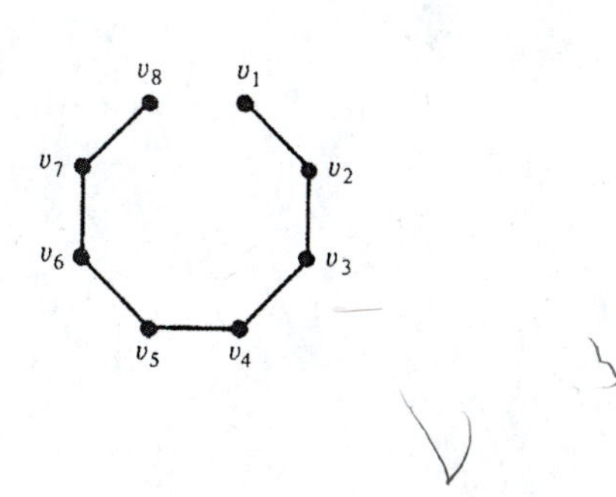

9. The edges in the following depth first search spanning tree were obtained in the order indicated, starting at vertex v_1:

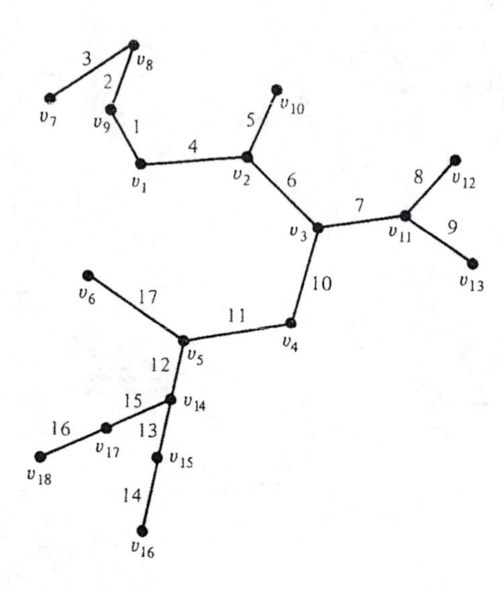

21. If $k > n - 1$, then the connected graph must have a cycle, and so it is possible to remove an edge and still have a connected graph. Repeating this process, we can remove a total of $k - (n - 1)$ edges, leaving a graph with $k - (k - (n - 1)) = n - 1$ edges. Such a graph is a tree, and no further edges can be removed.

Section 6.6 (page 372)

1. **3.**

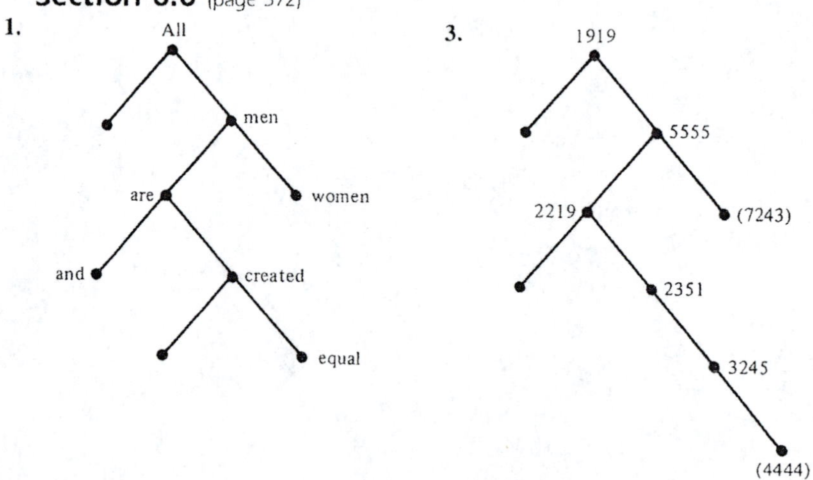

5.

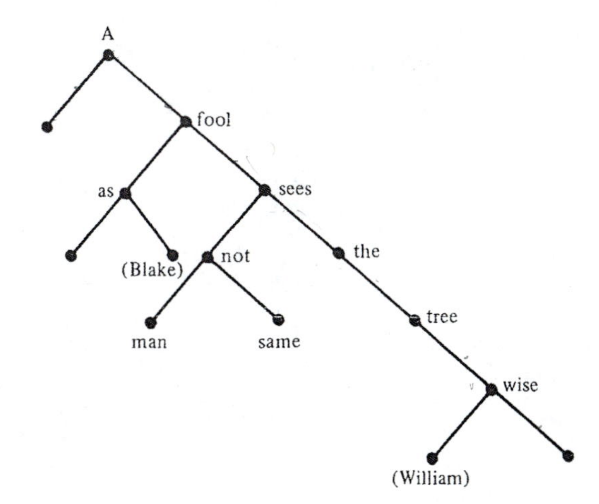

7.

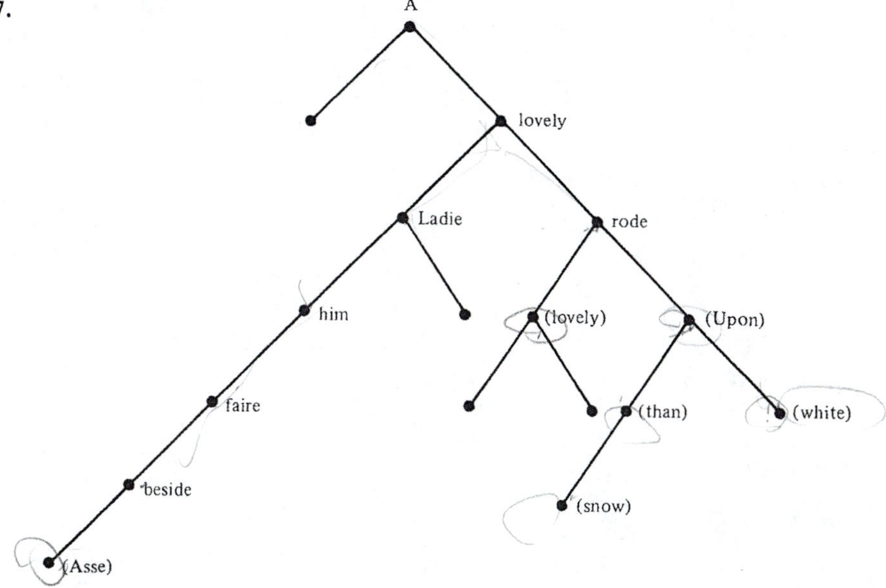

9. If there are only two characters in the character set, then the answer is yes. Otherwise, the answer is no.

13. a) h → 0
p → 11
e → 100
l → 101

b) 19

c) 010010011

d) help

The Huffman tree that led to this code is:

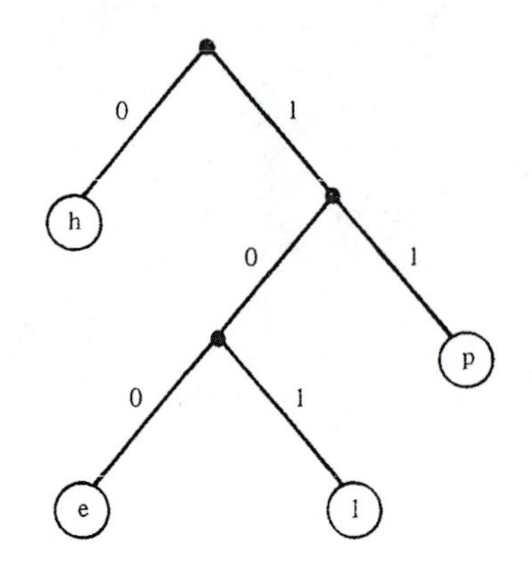

15. a) $) \to 00$
$+ \to 01$
$- \to 10$
$(\to 110$
$x \to 1110$
$y \to 1111$

b) 23

c) 110111001111100

17. a) $p \to 0$
$a \to 100$
$h \to 101$
$g \to 110$
$r \to 111$

b) 196

c) 1011001110

d) graph

19. a) $d \to 00$
$a \to 01$
$* \to 10$
$h \to 110$
$p \to 1110$
$y \to 1111$

b) 249

c) happy

Section 6.7 (page 385)

1. a)

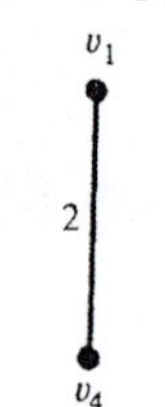

b)

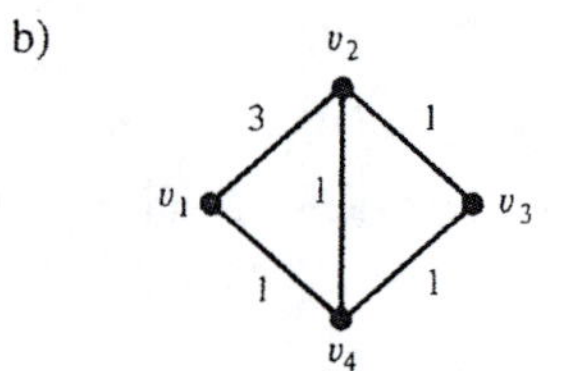

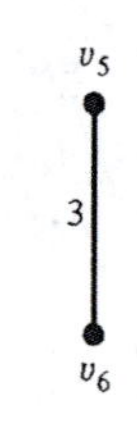

c)

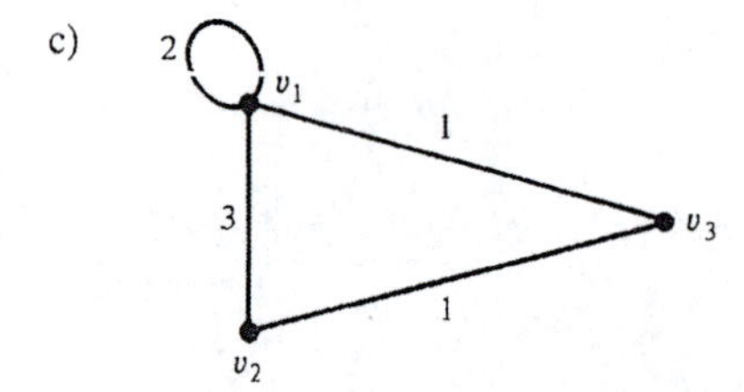

3.

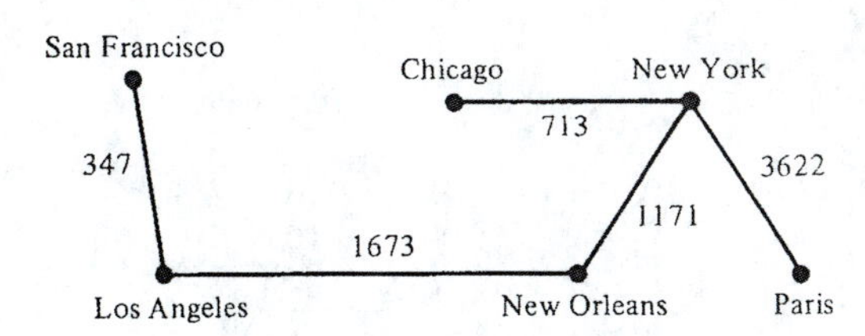

5.

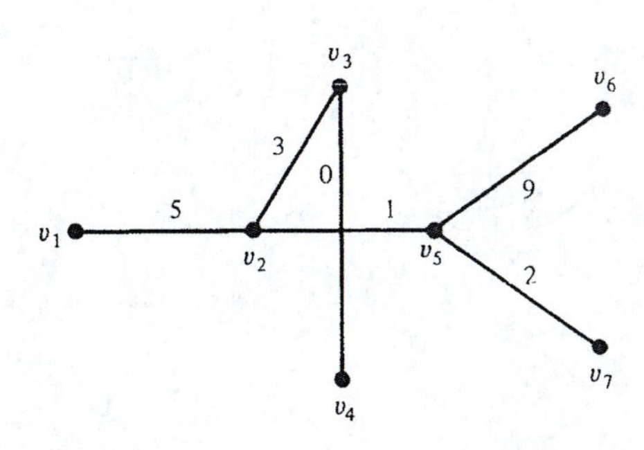

7.

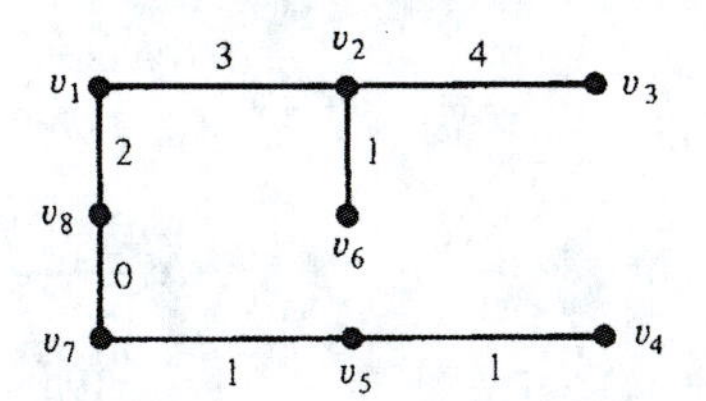

13. No. The minimum spanning subgraph must contain all edges with negative weights.

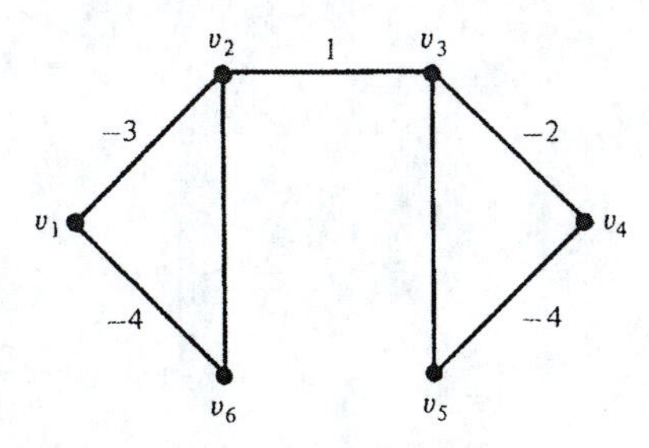

17.

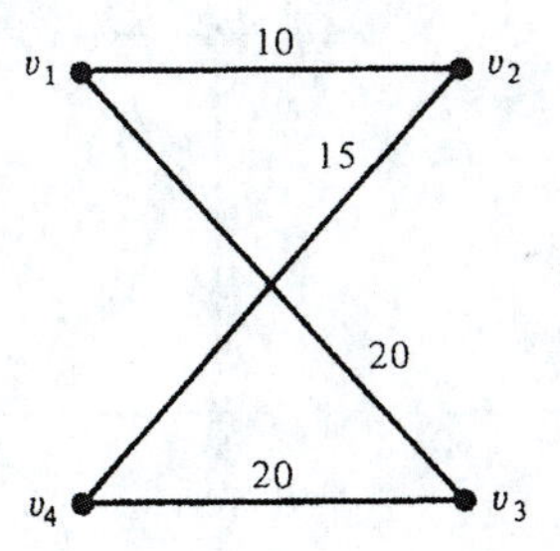

19.

Start at A:

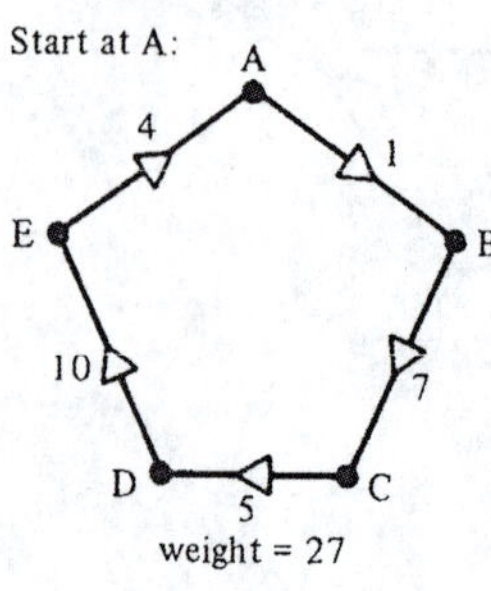

weight = 27

Start at B:

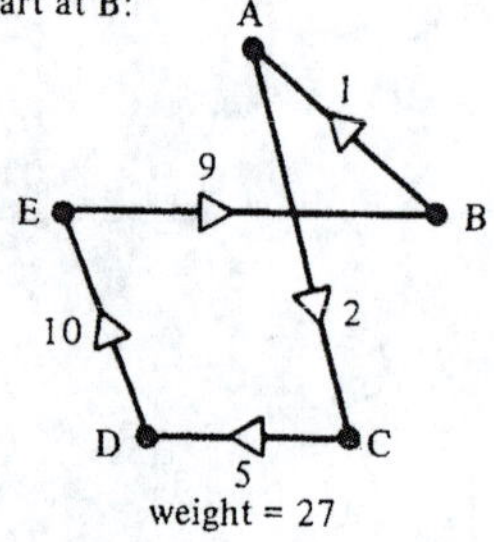

weight = 27

Start at C:

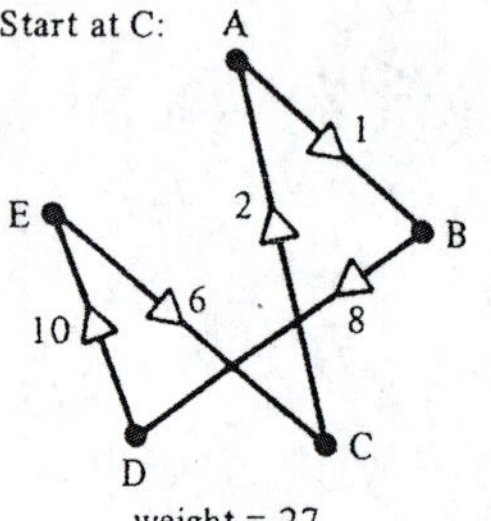

weight = 27

Start at D:

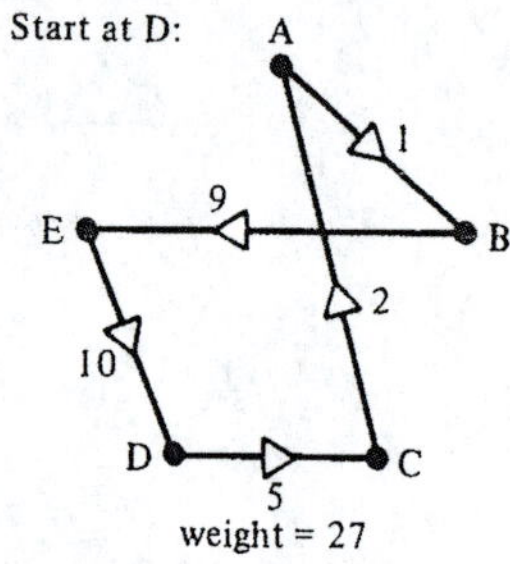

weight = 27

Start at E:

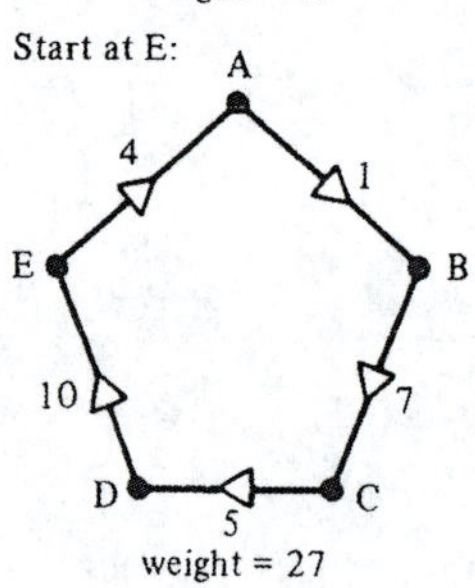

weight = 27

The solution:

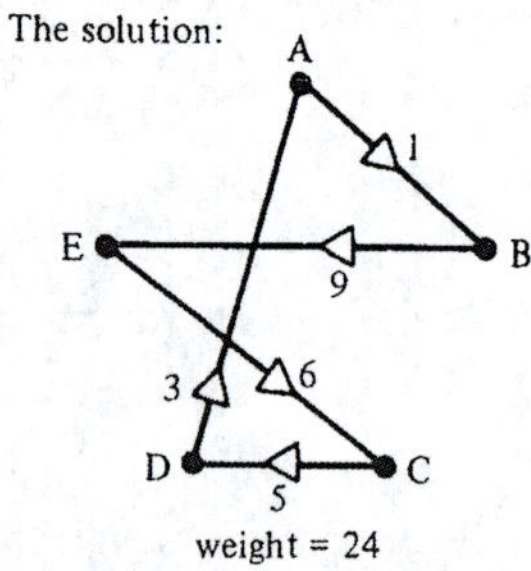

weight = 24

21.

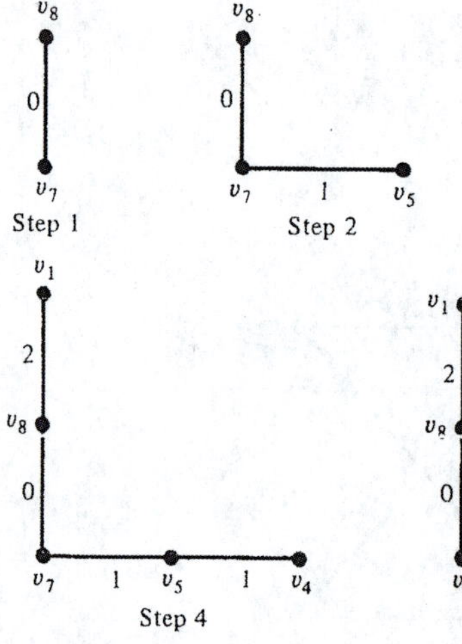

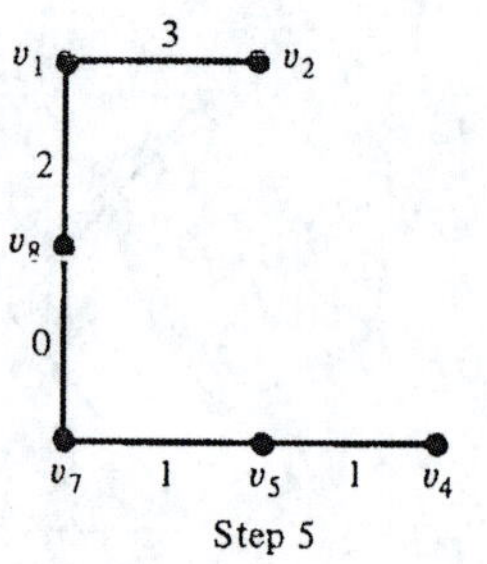

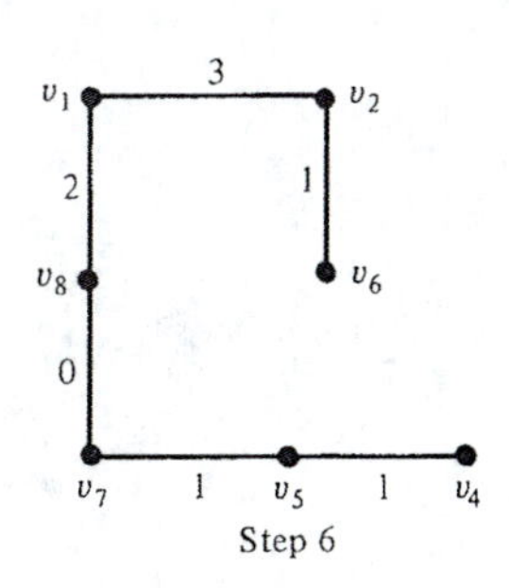

Step 6

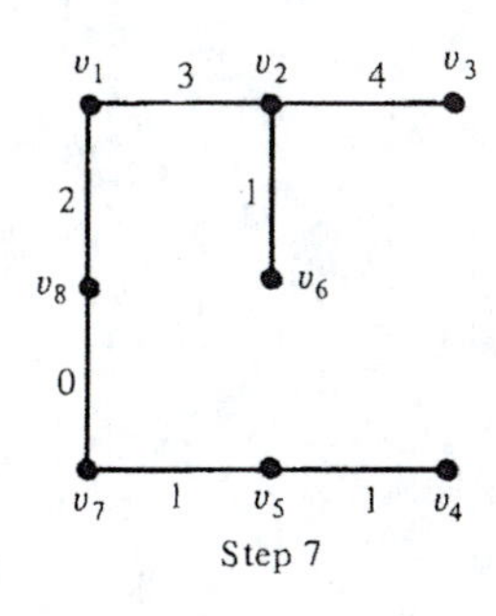

Step 7

Section 6.8 (page 397)

1. b) Yes, $v_1 \to v_{10} \to v_9 \to v_{11} \to v_8 \to v_6$.
   Yes, $v_6 \to v_{11} \to v_9 \to v_1$.
   c) No. Yes, $v_{10} \to v_9 \to v_{11}$.
3. Yes. A directed path from w to v followed by a directed path from v to u forms a directed path from w to u.
5. The graph is not strongly connected since vertex v_6 cannot be reached from any other vertex.
7. This graph is strongly connected.
11. No person can get a message to person v_6.
15. Strongly orientable.
17. Strongly orientable.
19. Not strongly orientable.

Section 6.9 (page 410)

1. $v_1 \to v_1$, $v_1 \to v_5 \to v_2$, $v_1 \to v_5 \to v_2 \to v_4$, $v_1 \to v_5$.

3.

	$S_1 = \{v_1\}$ $D_1(v)$	$P_1(v)$	$u = v_3$ $D_1(v_3) + w(v_3, v)$	$S_2 = \{v_1, v_3\}$ $D_2(v)$	$P_2(v)$	$u = v_4$ $D_2(v_4) + w(v_4, v)$	$S_3 = \{v_1, v_3, v_4\}$ $D_3(v)$	$P_3(v)$	$u = v_5$ $D_3(v_5) + w(v_5, v)$	$S_4 = \{v_1, v_3, v_4, v_5\}$ $D_4(v)$	$P_4(v)$	$D_5(v)$	$P_5(v)$
v_1	0	v_1		0	v_1		0	v_1		0	v_1	0	v_1
v_2	∞	v_1	$0 + \infty$	∞	v_1	$0 + 3$	3	v_4	$0 + \infty$	3	v_4	3	v_4
v_3	0	v_1		0	v_1		0	v_1		0	v_1	0	v_1
v_4	∞	v_1	$0 + 0$	0	v_3		0	v_3		0	v_3	0	v_3
v_5	1	v_1	$0 + \infty$	1	v_1	$0 + \infty$	1	v_1		1	v_1	1	v_1

Shortest paths:
1) $v_1 \to v_3 \to v_4 \to v_2$
2) $v_1 \to v_3$
3) $v_1 \to v_3 \to v_4$
4) $v_1 \to v_5$

5.

	$S_1 = \{v_1\}$ $D_1(v)$ $P_1(v)$		$u = v_5$ $D_1(v_5) +$ $w(v_5, v)$	$S_2 = \{v_1, v_5\}$ $D_2(v)$ $P_2(v)$		$u = v_3$ $D_2(v_3) +$ $w(v_3, v)$	$S_3 =$ $\{v_1, v_5, v_3\}$ $D_3(v)$ $P_3(v)$		$u = v_4$ $D_3(v_5) +$ $w(v_4, v)$	$S = \{v_1, v_5, v_3, v_4\}$ $D_4(v)$ $P_4(v)$	
$v = v_1$	0	v_1		0	v_1		0	v_1		0	v_1
$v = v_2$	∞	v_1	$1 + \infty$	∞	v_1	$3 + \infty$	∞	v_1	$1 + \infty$	∞	v_1
$v = v_3$	∞	v_1	$1 + 2$	3	v_5		3	v_5		3	v_5
$v = v_4$	∞	v_1	$1 + \infty$	∞	v_1	$3 + 1$	4	v_3		4	v_3
$v = v_5$	1	v_1		1	v_1		1	v_1		1	v_1
$v = v_6$	∞	v_1	$1 + \infty$	∞	v_1	$3 + \infty$	∞	v_1	$1 + \infty$	∞	v_1

Shortest paths:
1) v_2 not reachable from v_1
2) $v_1 \rightarrow v_5 \rightarrow v_3$
3) $v_1 \rightarrow v_5 \rightarrow v_3 \rightarrow v_4$
4) $v_1 \rightarrow v_5$
5) v_6 not reachable from v_1

Section 6.10 (page 419)

1. a) s_2, s_3

b) w_1 and w_3 are accepted, w_2 is rejected

c)

	0	1
s_1	s_3	s_2
s_2	s_2	s_1
s_3	s_2	s_5
s_4	s_5	s_4
s_5	s_2	s_2

3. a) s_3 b) s_3 c) rejected

d)

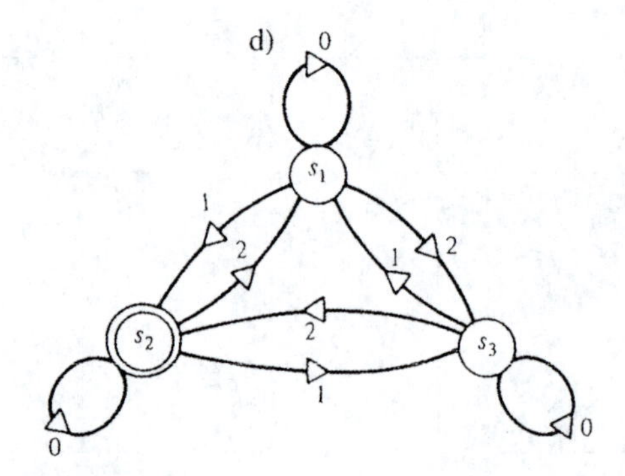

5.

	0	1
s_1	s_1	s_2
s_2	s_3	s_2
s_3	s_1	s_4
s_4	s_3	s_2

7. The language consists of all binary words with the property that the number of 1's in the word is congruent to either 0 or 1 modulo 3.

11.

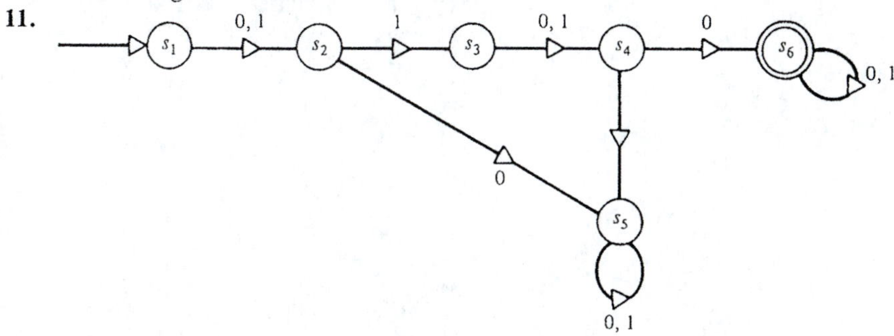

13.

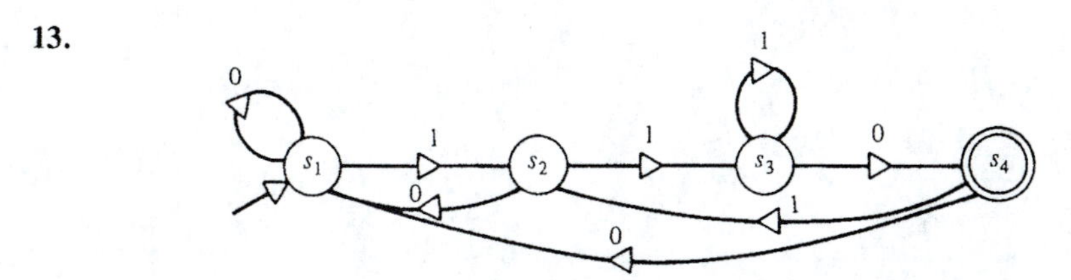

Index

If you don't find it in the index,
look very carefully through the entire
catalogue.
—Sears, Roebuck and Co., Consumer's Guide,
1897